D0322592

Annals of Mathematics Studies

Number 115

ON KNOTS

BY

LOUIS H. KAUFFMAN

PRINCETON UNIVERSITY PRESS

PRINCETON, NEW JERSEY

1987

Clothbound editions of Princeton University Press books are printed on acid-free paper, and binding materials are chosen for strength and durability. Paperbacks, while satisfactory for personal collections, are not usually suitable for library rebinding.

ISBN 0-691-08434-3 (cloth)
ISBN 0-691-08435-1 (paper)

Printed in the United States of America
by Princeton University Press, 41 William Street
Princeton, New Jersey

☆

Library of Congress Cataloging in Publication data will
be found on the last printed page of this book

To the Memory of

Andrés Reyes

CONTENTS

PREFACE

These notes on the theory of knots comprise an expanded version of a seminar held in the Departmento de Geometria y Topologia at the Universidad de Zaragoza, Zaragoza, Spain during the winter of 1984. Due to the supernatural enthusiasm and persistence of the members of the seminar, this author was given the energy to record a (we believe!) careful set of notes, and to relish the process.

The notes begin with the most elementary concepts of diagram moves, and linking numbers. Then they move quickly (perhaps steeply), using minimal technical apparatus, to problems of knot cobordism and the Arf invariant (Chapters 1 through 5).

Chapter 6 is a miscellany, compiled throughout the course. It contains ideas, sidetrips, and special topics. The last sections of this chapter contain an exposition of the author's geometric musings on the first generalized polynomial ([HOMFLY], [JO1], [JO2], [JO3]).

This polynomial is a powerful invariant that generalizes the classical Alexander and Conway polynomials. We show how the generalized polynomial arises as an ambient isotopy invariant of knotted, twisted bands, and how the new polynomial can be used to distinguish many knots from their mirror images.

Chapters 7 through 18 then develop more technical geometric knot theory—with covering spaces and branched covering spaces. By Chapter 6 the reader has already been introduced (combinatorially) to the Conway (Alexander) polynomial and to skein theory. The ascent into geometric knot theory begins in Chapter 7 with the introduction of spanning surfaces and the Seifert pairing. We discuss S-equivalence and prove that ambient isotopic knots have S-equivalent Seifert pairings. Then the potential function $\Omega_K = \mathrm{Det}(t\theta - t^{-1}\theta')$ is introduced, producing a model for the Conway polynomial in terms of the Seifert pairing. The signature σ_K of a knot K is also introduced in Chapter 7. It is easy to prove that the signature changes sign when the knot is replaced by its mirror image. Chapter 7 ends with a key exercise, asking the reader to prove the identity (∇_K is the Conway polynomial) $\nabla_K(2\sqrt{-1})/|\nabla_K(2\sqrt{-1})| = (\sqrt{-1})^{\sigma_K}$. This is then used to show that the knot 9_{42} is not amphicheiral. The exercise goes on to show that 9_{42} cannot be distinguished from its mirror image by the generalized polynomial. This exercise points to the open question of settling amphicheirality and to the possibilities inherent in new invariants such as the generalized polynomial.

Chapter 8 returns to knot cobordism and discusses the relationship with the Seifert pairing and with surgery curves on the spanning surfaces. Chapter 9 relates the Alexander polynomial to branched covering spaces and to the infinite cyclic covering of the knot complement. We discuss Seifert's original method for computing homology of cyclic (branched) coverings, and its relation to the Seifert pairing. Chapter 10 discusses the Arf invariant in terms of the Seifert pairing, and proves Levine's Theorem relating the Arf invariant to the value of the Alexander polynomial at minus one (taken modulo eight). Chapter 11 is a brief introduction to the free differential calculus of Ralph Fox. It ends with an exercise about affine representations of knot groups. This is another key exercise, providing an entry point into a classical paper by DeRahm, and to the work of Riley on hyperbolic structures on knot complements (via representations of knot groups). We only indicate a direction here. Nevertheless, this exercise is the link between these notes and Riley's work and the spectacular results of Thurston and his school.

Chapter 12 introduces cyclic branched coverings again, and shows how to compute their homology via the intersection pairing on associated manifolds. This leads to another (more modern) explanation of the relationship of Seifert pairing and branched covering. The chapter ends with a computation of signatures for torus knots. This is a key example both for signature computations, and for an entrance into the study of the topology of algebraic singularities (which we discuss in Chapter 19). Within Chapter 12 we introduce the concept of the signature of a manifold and show how the signatures and ω-signatures of knots and links are in fact signatures of branched cyclic covering spaces (or of appropriate eigenspaces in the case of ω-signatures).

In Chapter 13 we prove the Novikov Addition Theorem, giving the signature of a union of two manifolds in terms of the pieces. We also prove that signatures of bounding manifolds vanish, and use these general results to show that signatures of knots and links are concordance invariants. We prove the product theorem for signatures, and go on to discuss g-signatures in terms of signatures of eigenspaces in the case of cyclic actions. We compute g-signatures (case of skew forms) for cyclic actions on surfaces, and then assemble these results to obtain the basic signature defects for four-manifolds.

In Chapter 14 we build on the results of Chapter 13 and an argument of Cameron Gordon to give a proof-sketch of the G-signature theorem for 4-manifolds. This exposition gives a complete picture of the G-signature theorem in this dimension.

Chapters 15 through 18 are an exposition of the work of Casson and Gordon on slice knots. This depends upon all the previous work, particularly upon the cyclic branched covering spaces and G-signature theorem.

In Chapter 19 we give an introduction to the topology of algebraic singularities and its relation with the theory of knots and links. We discuss Brieskorn varieties and their relationship with branched coverings. This discussion proceeds by detailed examination of examples. We give specific discussions of $\Sigma(2,2,2)$ as projective three-space, and $\Sigma(2,3,5)$ as dodecahedral space. We then discuss the Milnor fibration, the empty knots, the cyclic suspension and product constructions. (The latter are geometric constructions in knot theory derived from algebraic ideas.) With the cyclic suspension and the product construction many results about cyclic branched covers find their proper setting. In particular, it is possible to see how the constructions of exotic spheres via Brieskorn manifolds are intimately related to the classical branched covering methods of Seifert. We conclude this chapter with an example of the eight-fold periodicity in the list $\Sigma(k,2,2,\cdots,2)$ (large odd number of twos).

The appendix contains a sketch of further developments about generalized polynomials. In particular, a description is given of the author's two-variable polynomial and its relation to his states model for the Jones polynomial. We discuss applications to alternating knots that settle century-old conjectures (work due to the author, to Kunio Murasugi, and to Morwen Thistlethwaite—all using this states model for the Jones polynomial).

Knot theory comprises a deep underpinning of geometric topology. It is a beautiful subject in its own right, with ramifications that spread throughout all of geometry. May these pages reflect this spirit!

With great pleasure I thank José Montesinos, Maité Lozano, Carmen Safont, Pilar del Valle, Alvaro Rodes, Esteban Indurian and Elena Martin for sharing this journey through tangled terrain. Special thanks to Dale Rolfsen for kind permission to reproduce his drawing of an infinite cyclic covering space, and to Ken Millett, Ray Lickorish, Vaughan Jones, Larry Siebenmann, Jon Simon, Dennis Roseman, Keith Wolcott, Joan Birman, Hugh Morton, Massimo Ferri, Mario Rasetti, Sostenes Lins,and Corrado Agnes for helpful conversations. Special thanks also to my students, Ivan Handler, Randall Weiss, and Steve Winker for listening and contributing, to Frederick Joseph Staley for introducing me to new realms of nature and mathematics, and to all those friends that love a tangled tale. I am greatly indebted to Cameron Gordon for his generosity in allowing me to weave the texts of his work on the g-signature theorem and slice knots into the expositions of Chapters 14 to 18. To Ms. Shirley Roper, Head Math Typist at the University of Illinois at Chicago, my thanks for an excellent typing job of the first draft, and to Ada Burns, mathematics typist at the University of Iowa for extraordinary collaboration in producing the final text. The last stages of this project were partially supported by ONR Grant No. N0014-84-K-0099 and the Stereochemical Topology Project at the University of Iowa, Iowa City, Iowa.

Chicago, February 1986
and
Iowa City, December 1986

On Knots

I

INTRODUCTION

These notes constitute a leisurely introduction to knot theory that is (we hope!) in the spirit of Fox's Quick Trip [F1]. We shall also feel free to digress occasionally, sometimes in the direction of applications, sometimes with an analogous structure, sometimes with philosophy or general ideas.

What is knot theory about? One answer is that knot theory studies the placement problem: Given spaces X and Y, classify how X may be placed within Y. Here the _how_ is usually an embedding, and classify often means up to some form of movement of X in Y (isotopy, for example). If X is the circle $S^1 = \{(x,y) | x^2+y^2 = 1,\ x \text{ and } y \text{ real}\}$ and Y is Euclidean three-space

$$\mathbb{R}^3 = \{(x,y,z) | x,y,z \text{ real numbers}\},$$

then we have _classical knot theory_. Classical knot theory studies the embeddings of S^1 in $\mathbb{R}^3$ or in $S^3 = \{(x,y,z,w) | x^2+y^2+z^2+w^2 = 1;\ x,y,z,w \text{ real}\}$. Note that the three-sphere S^3 is the one-point compactification of $\mathbb{R}^3$.

Of course, a "knot on a rope" may follow a pattern like the one below. The corresponding classical knot is obtained by splicing the ends of the rope together (without introducing new tangling):

T (trefoil)

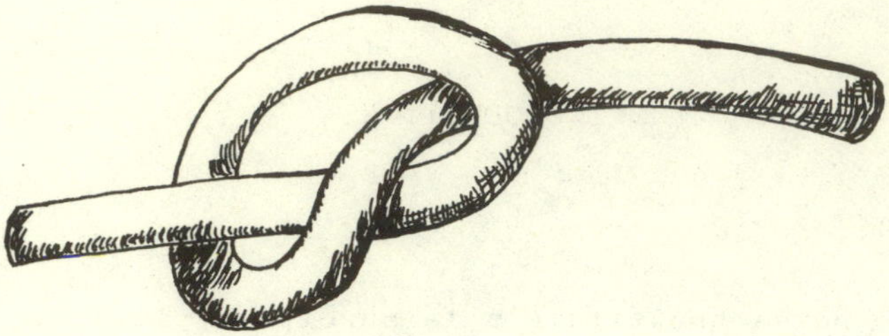

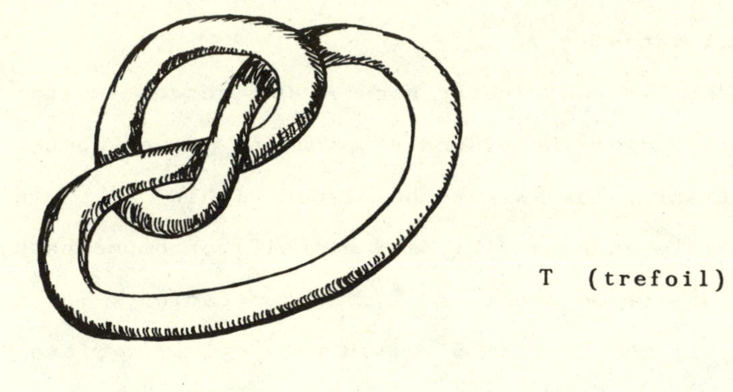

Once the ends are spliced together it is possible to define <u>knottedness</u>. The <u>unknot</u> U is represented by

U

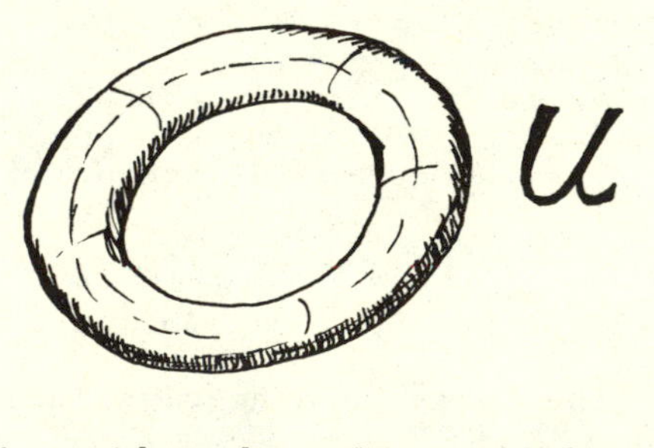

and a knot is said to <u>be</u> <u>unknotted</u> if it can be deformed,

without tearing the rope, until it turns into the unknot.

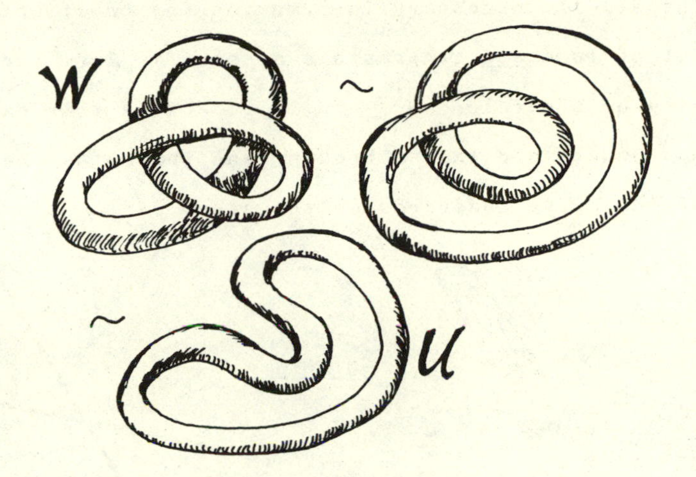

Thus the form W above is unknotted.

The trefoil T (two sketches above) is actually knotted. This requires proof! And it is the question of proving knottedness, and of classifying types of knottedness that gives rise to the need for a theory of knots.

There are different ways to approach such a theory, and we shall use more than one approach. The first approach will be <u>combinatorial</u> and <u>pictorial</u> knot theory. (I called part of it Formal Knot Theory in [K1].) The idea here is to closely abstract the rope drawings that represent knots.

becomes

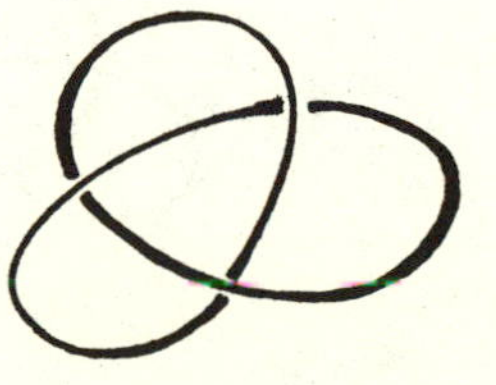

We call the picture on the right a knot diagram. It contains all the necessary information for constructing the knot out of rope and it presents a specific form for an embedding of a circle, S^1, in $\mathbb{R}^3$. To see this embedding you must understand that a broken line indicates where one part of the curve undercrosses the other part.

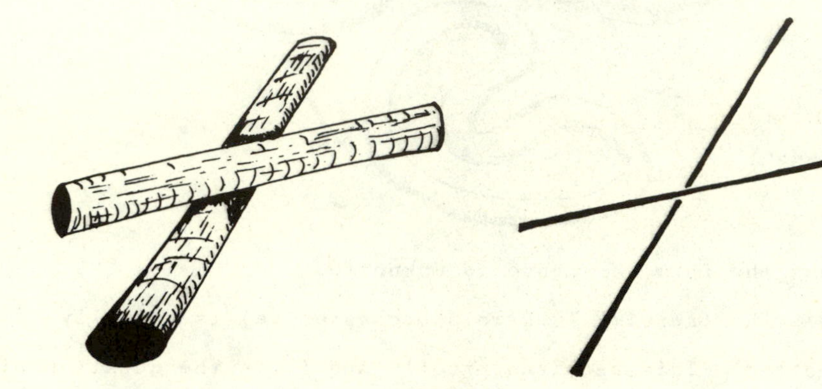

Thus we start with a planar graph

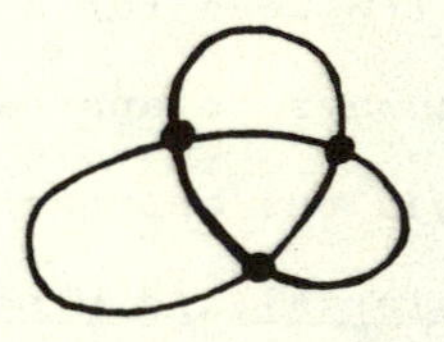

and lift at the crossings to form a curve which dips above or below the plane at these crossings:

<u>There are two choices at each crossing</u>:

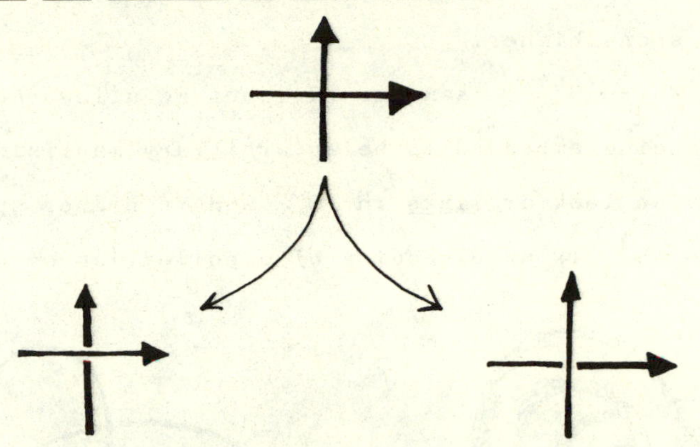

and hence 2^n potential knots for each planar graph with n crossings.

The theory commences once we explain how to deform these diagrams. We begin in this way in the next section.

Another approach to the theory is to take the abstract notion of the knot as embedding $\alpha:S^1 \longrightarrow S^3$ and to study the topology of the complementary space $S^3-\alpha(S^1) = Z$. We can then apply the apparatus of algebraic topology to Z. And we can also construct spaces (such as branched covering spaces) associated with Z, and work with the algebraic topology of these spaces. This approach also has a long history, and will be used later in these notes.

Many generalizations of the initial placement problem $S^1 \longrightarrow S^3$ are possible. Two are particularly worth mentioning.

a) $S^n \longrightarrow S^{n+2}$, the study of the embeddings of an n-dimensional sphere into a sphere of two dimensions higher.

b) $W^n \longrightarrow S^{n+2}$, same as (a), but we allow the space being embedded to be an arbitrary manifold.

One can also look at links in S^3 and in higher spheres. A link in S^3 is an embedding of a collection of circles. Thus

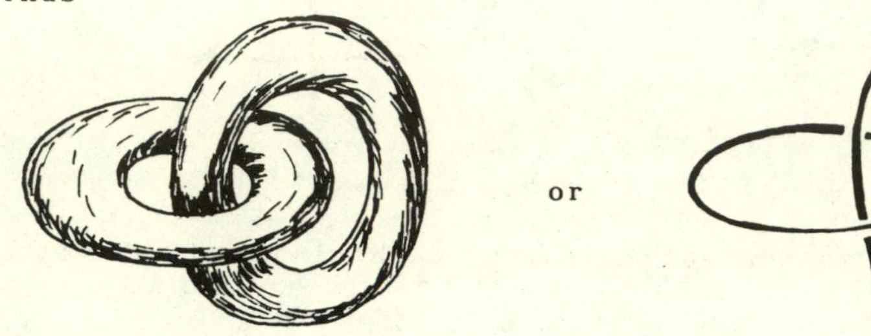

is the simplest example of a linked link. Linking is a fascinating phenomenon of three-dimensional space. We conclude this introduction with a picture of the Borromean Rings, a link that exhibits a triadic relation: any two of its components are unlinked. It is itself linked.

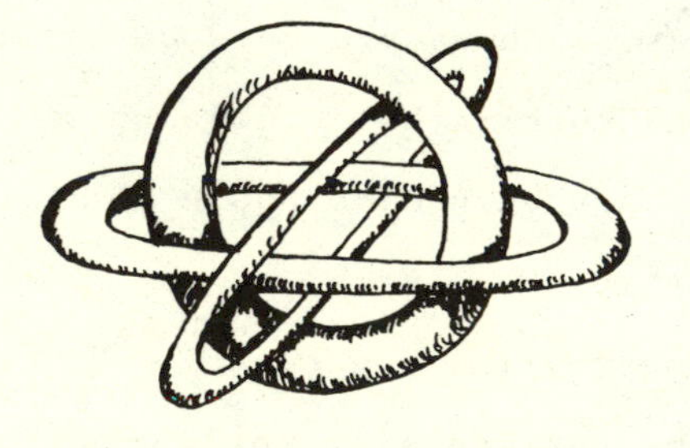

or

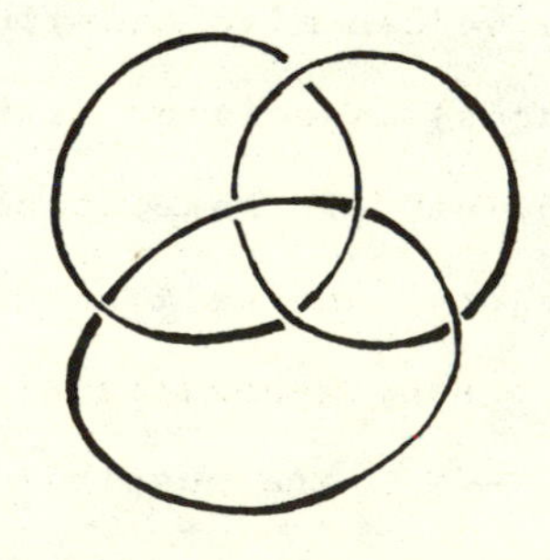

II

LINKING NUMBERS AND REIDEMEISTER MOVES

Our first model for the theory of knots and links in $\mathbb{R}^3$ is combinatorially based. We say that two knot or link diagrams K and K′ are <u>equivalent</u> if there exists a sequence of <u>Reidemeister moves</u> (see [R1]) changing K into K′. Equivalence is denoted by the symbol ~, as in K ~ K′. Equivalent diagrams are <u>ambient isotopic</u>, meaning that there is a continuous deformation through embeddings from one to the other. Reidemeister proved the converse—making equivalence and ambient isotopy identical. There are three basic types of Reidemeister moves:

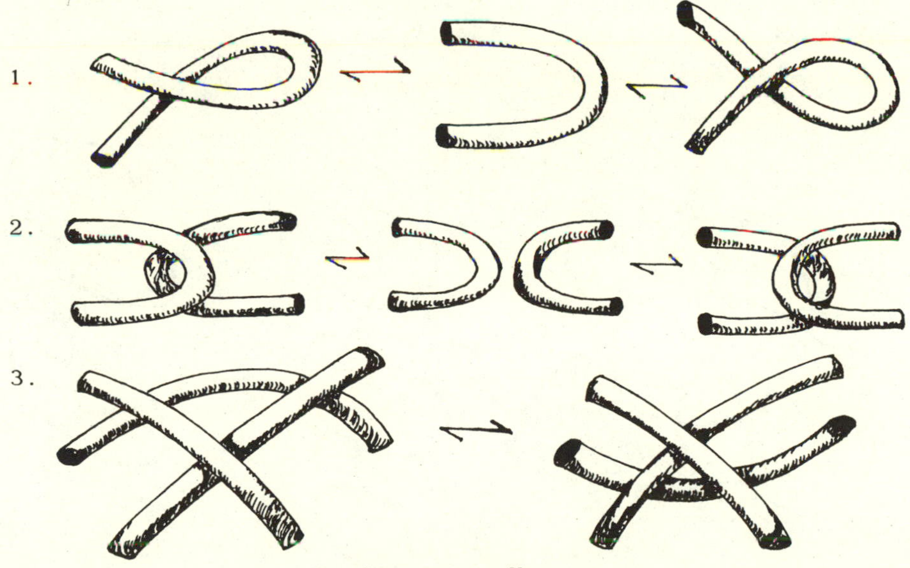

<u>Reidemeister Moves</u>

It is understood that these moves are to be performed locally on the knot diagram. It is also understood that no other strands are present locally other than those depicted in the moves. Thus

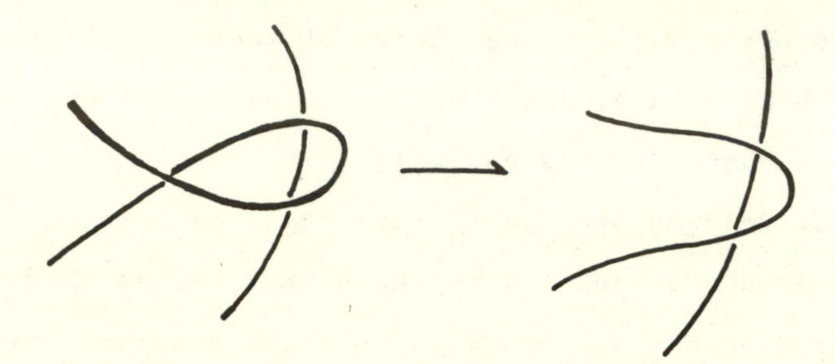

is not a legitimate move of type 1. The reader may feel that this example is a "higher order" move of type 1. I agree! And I invite him/her to formulate a theory of higher order moves. We insist here on pure moves not out of pedantry, but to simplify the theory (read on!).

Example:

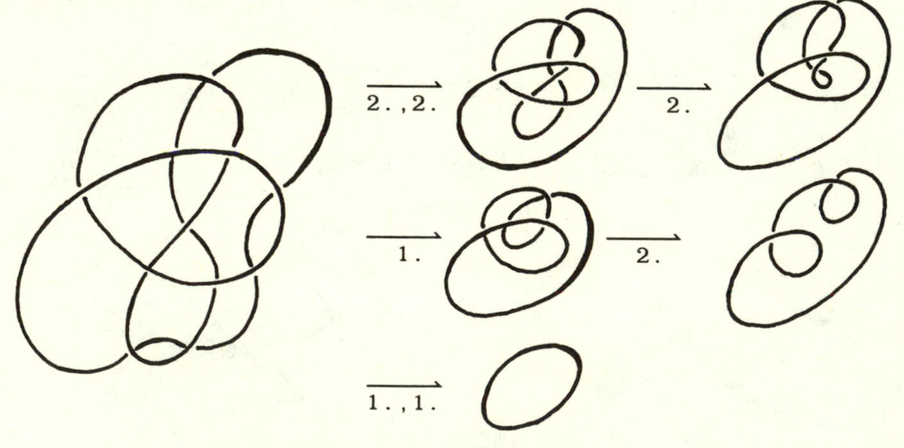

It is also understood that two diagrams that are "topologically equivalent" are equivalent. Thus

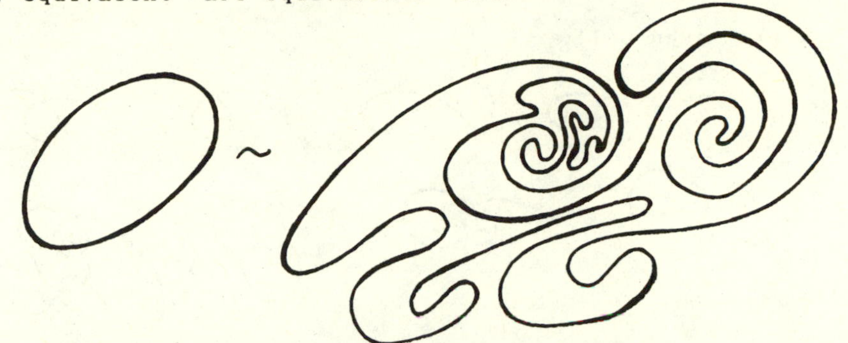

and

This part of the equivalence goes beyond the Reidemeister moves. We really mean that there exists a homeomorphism of $\mathbb{R}^2$ that throws one underlying graph to the other.

<u>Exercise 2.1</u>. Give an example of a knot diagram K such that $K \sim 0$ (the unknot) and K requires a type 3. move to get the equivalence started.

<u>Exercise 2.2</u>. Prove that the following process will always

produce an unknotted diagram: Start drawing. Whenever you encounter a previously drawn line, undercross it. Return to start, eventually,

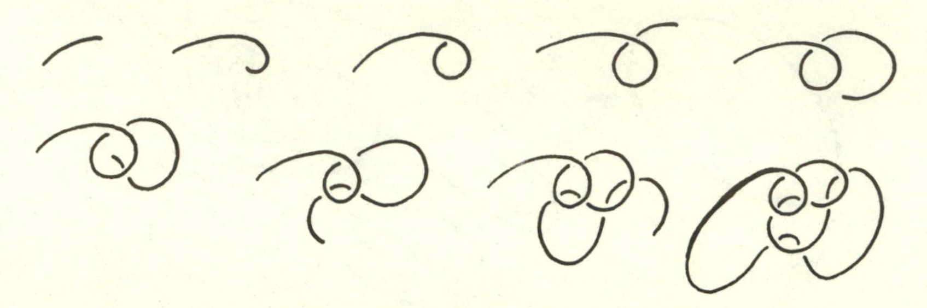

Exercise 2.3. Call a planar graph with 4-valent vertices ( ) a universe. Thus is a trefoil universe. Such a graph may have more than one component as in Here by component I mean a curve obtained by walking along the graph and always crossing at a crossing. The components of a universe are the potential components of a knot or link of the projection.

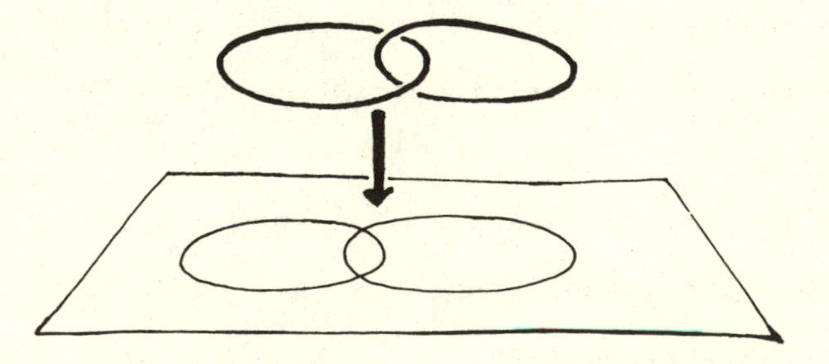

Also, we say that a knot or link is alternating if its crossings alternate under -- over -- under -- over - ••• as

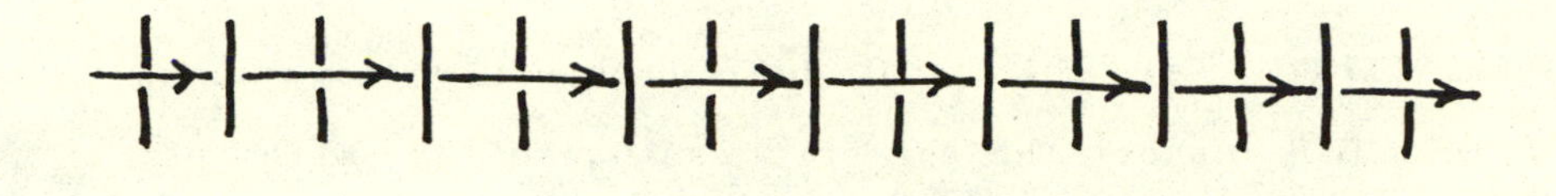

you traverse any component. Thus  is alternating.

Prove: Any universe is the projection of an alternating knot or link.

We now consider 2-component links. If α and β are the components, and we orient them, then we wish to define a linking number $lk(\alpha,\beta) = lk(L)$ (where $L = \alpha \cup \beta \subset S^3$) so that

lk[ ] = +1

This will conform with the usual right-hand-rule. In order to do this we associate a sign ϵ to each crossing.

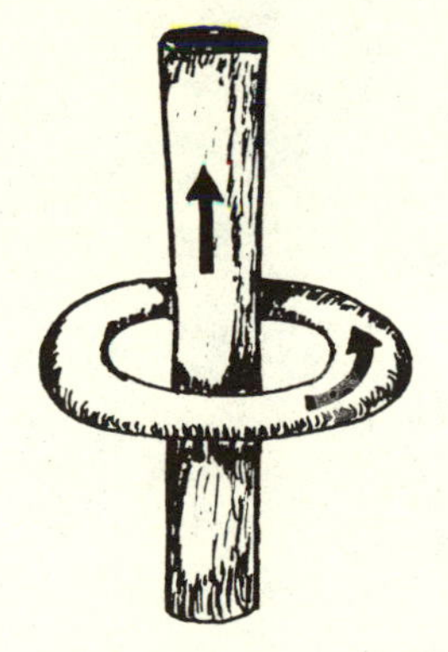

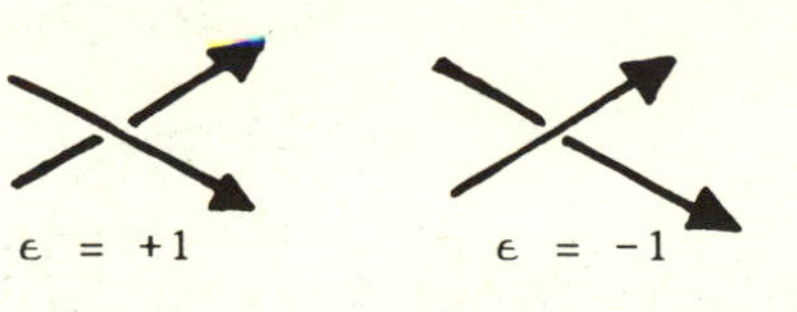

$\epsilon = +1$ $\epsilon = -1$

DEFINITION. *Let* $L = \alpha \cup \beta$ *be a link of two components. Let* $\alpha \sqcap \beta$ *denote the set of crossings of* α *with* β.

Then

$$\boxed{\mathrm{lk}(\alpha,\beta) = \frac{1}{2} \sum_{p \in \alpha \sqcap \beta} \epsilon(p)} \, .$$

This formula defines the linking number for a given diagram.

Example:

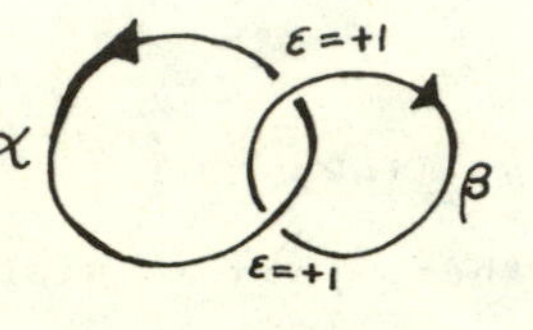

$\mathrm{lk}(\alpha,\beta) = \frac{1}{2}(1+1) = 1.$

Example:

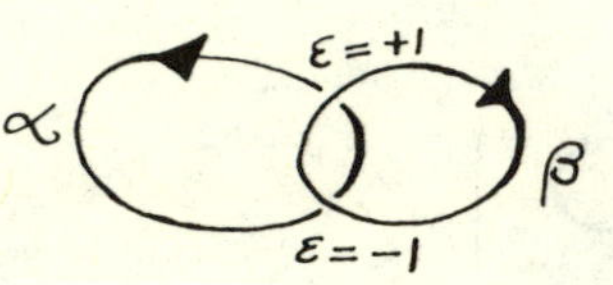

$\mathrm{lk}(\alpha,\beta) = \frac{1}{2}(1-1) = 0.$

Example:

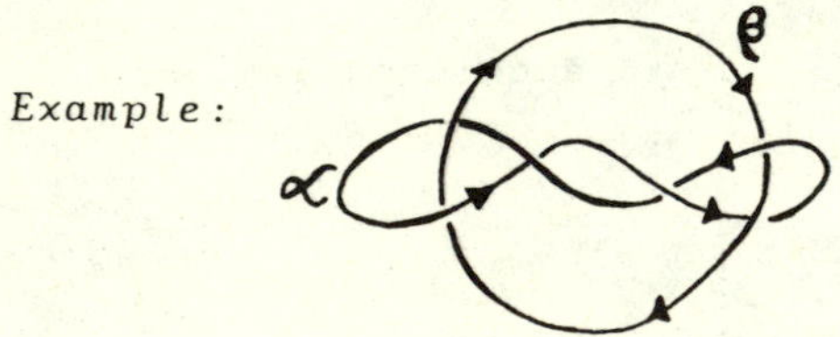

$\mathrm{lk}(\alpha,\beta) = +2.$

Example:

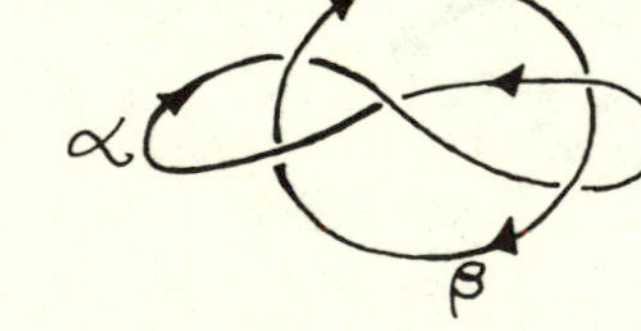

$\mathrm{lk}(\alpha,\beta) = 0.$

This last example is J.H.C. Whitehead's link. Links can be linked even when their linking number is zero.

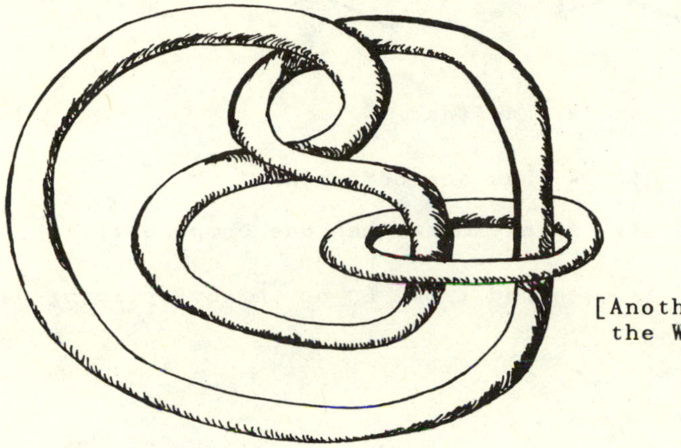

[Another version of the Whitehead link]

Exercise 2.4. Prove the

THEOREM. *If* $L \sim L'$, L *and* L' *oriented 2-component links, then* $lk(L) = lk(L')$. *Linking number is an invariant of ambient istopy.*

This last exercise shows that lk is a topological invariant of links and gives us our first proof that some links are really linked! But the Whitehead link and the Borromean rings are not yet captured.

Exercise 2.5. Let L and $\overline{L}$ be two two-component links that differ at the site of one crossing, as shown below.

Given a knot or link W, define

$$C(W) = \begin{cases} 1 & \text{if } W \text{ has one component} \\ 0 & \text{if } W \text{ has more than one component.} \end{cases}$$

Suppose that W is obtained from L or $\overline{L}$ by splicing out the crossing:

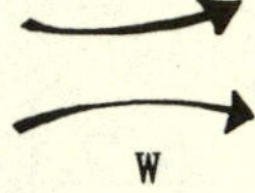

Show:

$$\boxed{\mathrm{lk}(L) - \mathrm{lk}(\overline{L}) = C(W)}$$

(This is a trivial exercise, but the pattern will generalize to give new invariants!)

<u>Exercise 2.6</u>. Prove that there does not exist an orientation preserving homeomorphism $h: \mathbb{R}^3 \longrightarrow \mathbb{R}^3$ such that $h(M_R) = M_L$ where M_R and M_L are right and left handed Mobius bands, respectively.

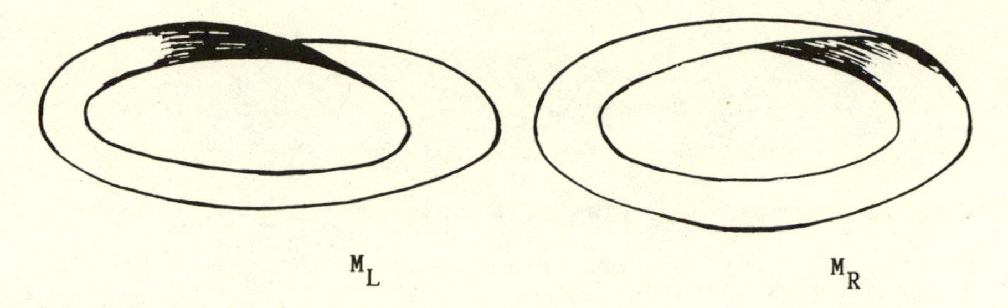

(*Hint*: Consider a linking number of the core of the band with its edge.)

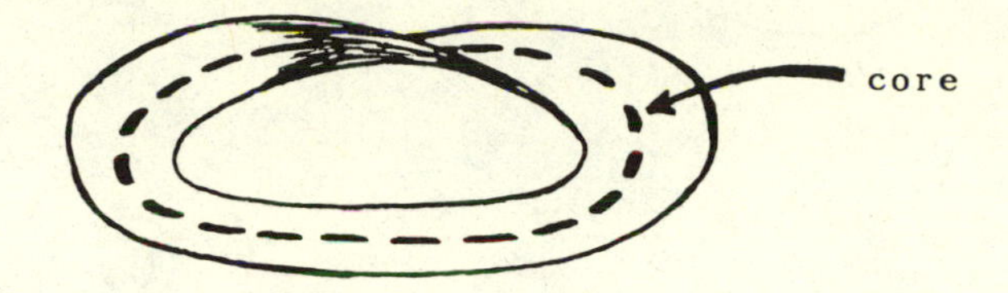

Exercise 2.7. Link, Twist, Writhe.

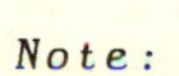

Note: ~ 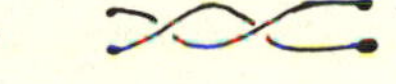

and 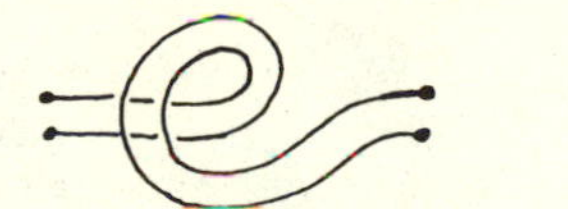 ~

(Isotopies relative to the end-points)

Hence we can have situations with a double-stranded link where twisting

is exchanged for writhing

For an appropriate class of 2-component links L define $T(L)$ = twist of L and $W(L)$ = writhe of L so that you can prove $lk(L) = T(L)+W(L)$. You should have

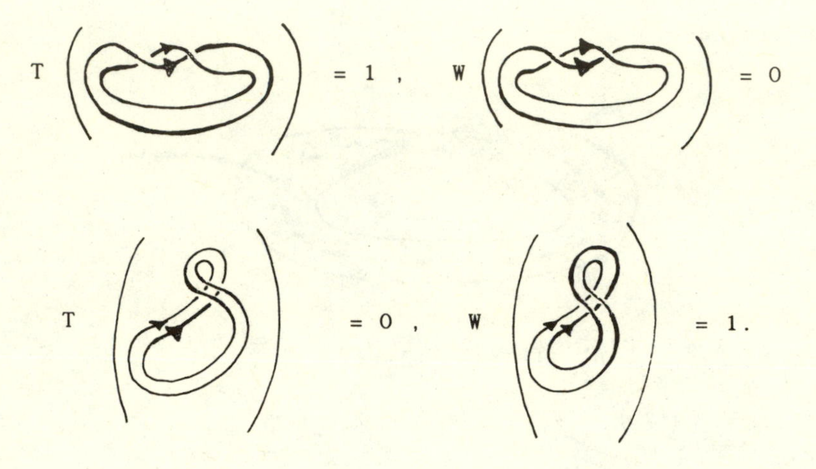

This formula, $lk(L) = T(L)+W(L)$, has been used by biologists studying closed, double-stranded DNA. (See [WH], [FB] and [BCW].) You can see the exchange phenomenon between T and W by playing with a rubber band or a telephone cord.

Exercise 2.8. Classify the ways a rubber band can be wrapped around a cylinder. [This is called the rubber band, battery problem by Brayton Gray. (He wraps his band around a small transistor battery.)]

III

THE CONWAY POLYNOMIAL

We now introduce a more powerful invariant of oriented knots and links. It is the Conway polynomial, a refined variant of the classical Alexander polynomial (see [A1] and [C1], [K1], [K2]). This polynomial invariant is described by three axioms:

AXIOM 1. *To each oriented knot or link* K *there is associated a polynomial* $\nabla_K(z) \in Z[z]$. *Equivalent knots and links receive identical polynomials:* $K \sim K' \Rightarrow \nabla_K = \nabla_{K'}$.

AXIOM 2. *If* $K \sim 0$ *(the unknot) then* $\nabla_K = 1$.

AXIOM 3. *Suppose that three knots or links differ at the site of one crossing as shown below:*

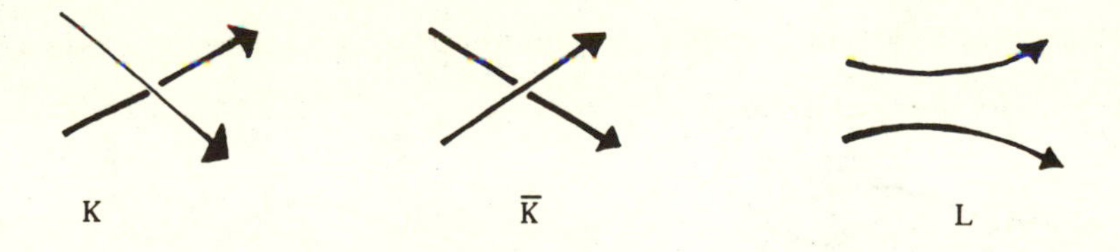

then $\nabla_K - \nabla_{\overline{K}} = z\nabla_L$. (We call this the exchange identity.)

Remarks. The ring $Z[z]$ is the ring of polynomials in z

with integer coefficients. Thus these axioms assert that $\nabla_K(z) = a_0(K)+a_1(K)z+a_2(K)z^2+\cdots$ where, for each $n = 0,1,\cdots,$ $a_n(K) \in Z$ is an invariant of K. Since $\nabla_K(z)$ is a polynomial, these integer invariants are all zero on a given knot for n sufficiently large.

As we shall see, except for a_0, a_1 and possibly a_2, these invariants are mysterious! What do they mean geometrically?

The axioms do assert that each invariant $a_{n+1}(K)$ is related to the invariant $a_n(K)$ by a corresponding exchange relation: $a_{n+1}(K)-a_{n+1}(\bar{K}) = a_n(L)$ (translate Axiom 3). We shall use this property to interpret a_0 and a_1, and to prove that the axioms are consistent.

For now, we assume consistency and set up some calculations.

LEMMA 3.1. *If* L *is a split link then* $\nabla_L = 0$.

[Recall that a link is split if it is equivalent to a link with diagram containing two nonempty parts that live in disjoint neighborhoods. Thus

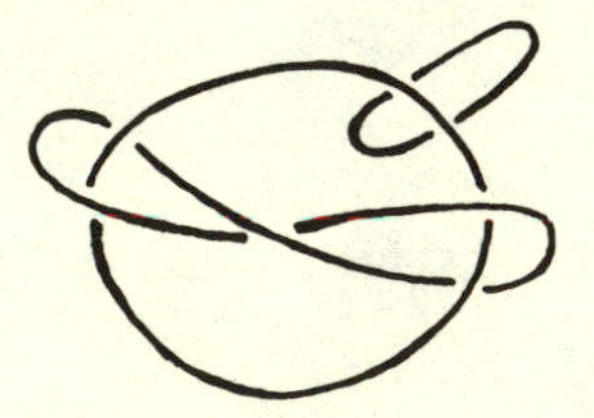

~

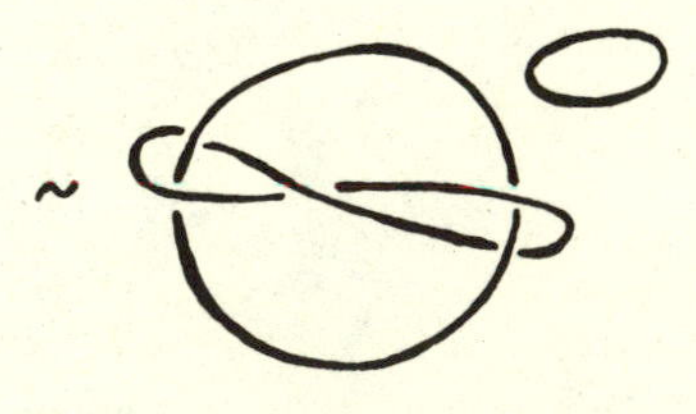

is a split link.]

Proof. If L is split then we may assume its diagram has two parts with strands related as shown below on the right. We may then form associated links K and $\overline{K}$ as shown below. K and $\overline{K}$ are equivalent via a 2Π-twist. Hence $\nabla_K = \nabla_{\overline{K}}$ by Axiom 1. Therefore $0 = \nabla_K - \nabla_{\overline{K}} = z\nabla_L$, hence $\nabla_L = 0$. ■

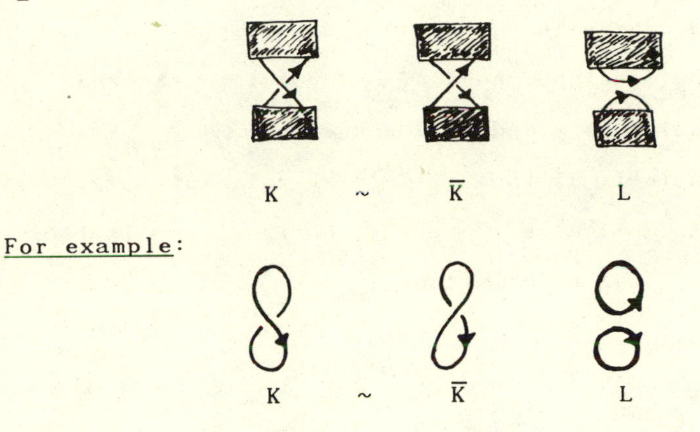

For example:

Remark. You may enjoy proving the lemma that the 2Π-twist

(or Π twist)

can be accomplished via Reidemeister moves.

Hint. Prove a generalization of type 2 move:

We use the fact that $\sim$ is equivalent to ambient isotopy. For a proof, see [R1]. Two knots or links K, K' are ambient isotopic if there is a family of embeddings K_t varying continuously in t such that $K_0 = K$ and $K_1 = K'$. Here we may assume that the embeddings in $\mathbb{R}$ or S^3 are differentiable and that the family varies smoothly. Knots arising from diagrams in our context are <u>tame</u> (in a sufficently small neighborhood of any point on the knot or link the embedding is standardly unknotted).

The lemma tells us that all of the unlinks: ○○, ○○○ ○○○○ ,··· receive the value 0 from ∇. It is now easy to do recursive calculations:

<u>Example 3.2</u>.

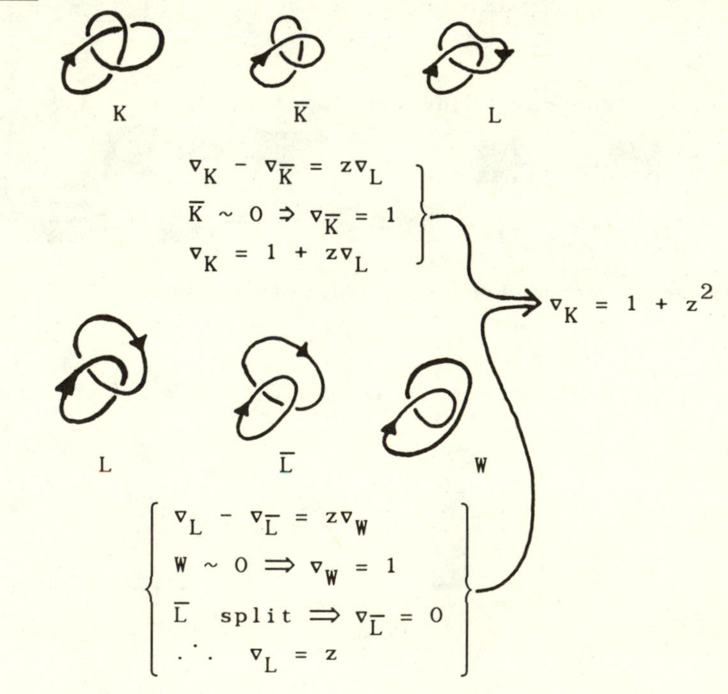

Example 3.3:

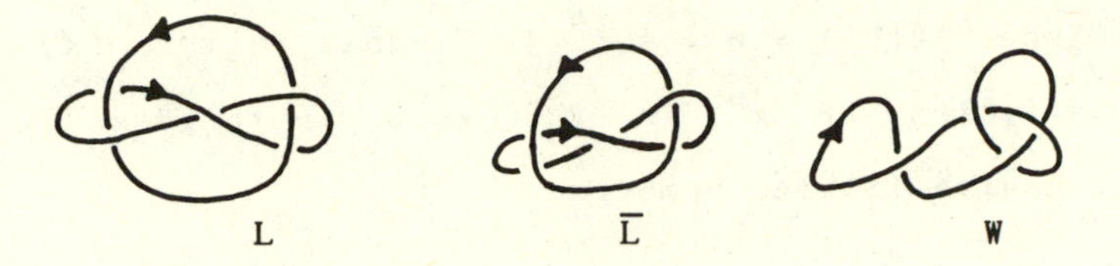

Here L is the Whitehead link from Chapter II. Thus $lk(L) = 0$. We see that $\bar{L} \sim$ and thus (by calculation similar to Example 3.1) we have $\nabla_{\bar{L}} = -z$. $W \sim$ and hence $\nabla_W = 1+z^2$. Putting this information together, we get $\nabla_L-(-z) = z(1+z^2)$.

$$\therefore \quad \nabla_L = z^3.$$

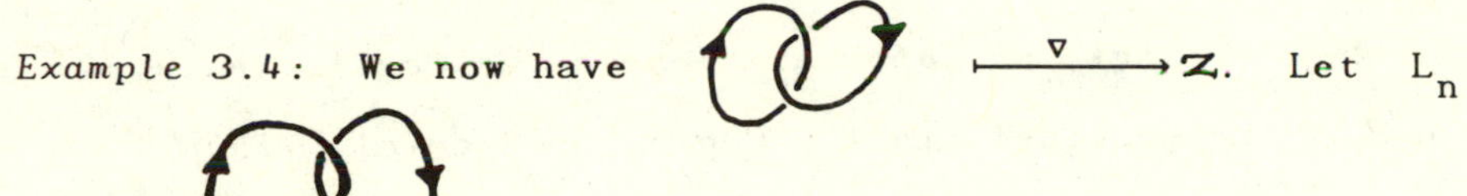

Example 3.4: We now have $\overset{\nabla}{\longmapsto} z$. Let L_n

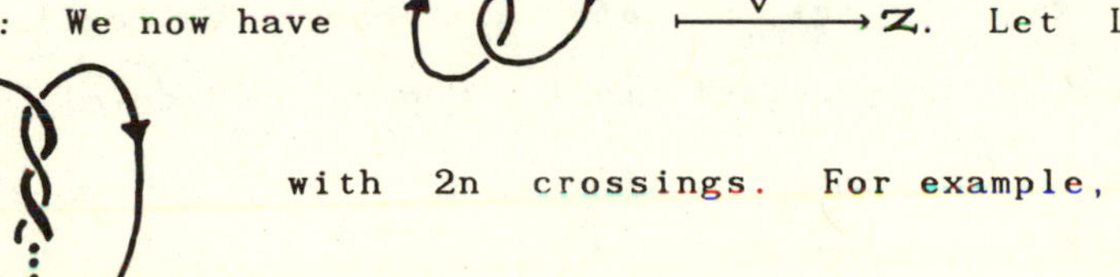

denote with $2n$ crossings. For example,

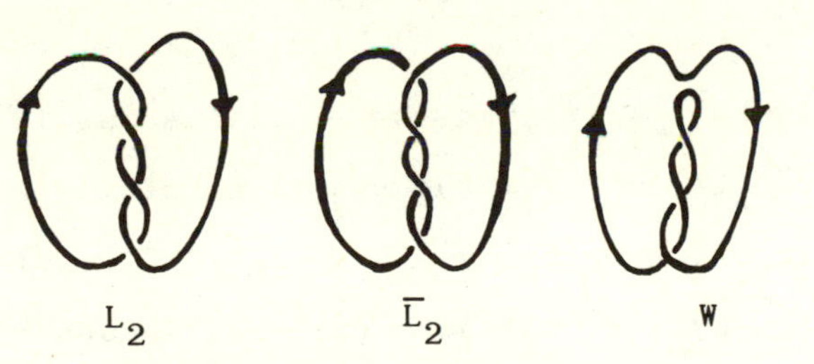

Thus $\bar{L}_2 \sim L_1$ and $W \sim U$. Therefore

$$\nabla_{L_2} - z = z \cdot 1$$

$$\nabla_{L_2} = 2z.$$

The same reasoning (by induction) shows that $\nabla_{L_n} = nz$.

Since $lk(L_n) = n = a_1(L_n)$, [recall that $a_n(K)$ is the coefficient of z^n in ∇_K] we begin to guess a relation with linking numbers.

DEFINITION 3.5. *Let* L *be any knot or link. Define* C(L) *by the formula*

$$C(L) = \begin{cases} 1 & \text{if } L \text{ has one component,} \\ 0 & \text{if } L \text{ has more than one component.} \end{cases}$$

Thus C(L) *is an invariant of* L *and it distinguishes knots and links.* (Repeated from Exercise 2.5.)

LEMMA 3.6. *Let* a_0 *and* a_1 *denote the coefficients of* 1 *and* z *respectively in the Conway polynomial. Then*

(i) $a_0(K) = C(K)$ *for all knots and links* K,

(ii) $$a_1(K) = \begin{cases} lk(K) & \text{when } K \text{ has two components,} \\ 0 & \text{otherwise.} \end{cases}$$

Proof. (i) Let K, $\overline{K}$ and L be related as in Axiom 3. Then $a_0(K)-a_0(\overline{K}) = 0$ is the statement of the axiom for this coefficient. This says that a_0 is invariant under strand switching. Since there exists a sequence of switches that transmute K to an unknot or unlink (For example, use Exercise 2.2.), we see that $a_0(K) = 1$ or 0 according to whether it has one or more components. Hence $a_0(K) = C(K)$. (ii) Again let K, $\overline{K}$, L be as in Axiom 3.

Then $a_1(K)-a_1(\overline{K}) = C(L)$ when K has two components. We leave the rest of the proof as an exercise [Exercise 3.6].

This section will end with a discussion of the validity of our axioms via an inductive definition of the coefficients. We work first with more examples.

Example 3.7: Watch the calculation of $a_2(K)$ for the trefoil. We have

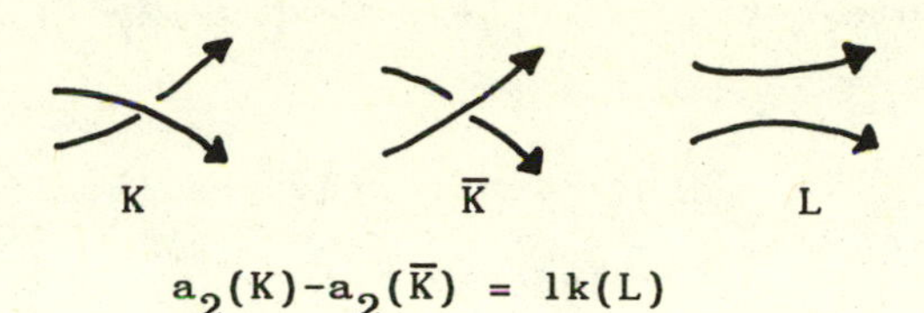

$$a_2(K)-a_2(\overline{K}) = \text{lk}(L)$$

[*Notation*: Let $\text{lk}(L) \equiv a_1(L)$ for any knot or link. Thus $\text{lk}(L) = 0$ if L does not have two components.]

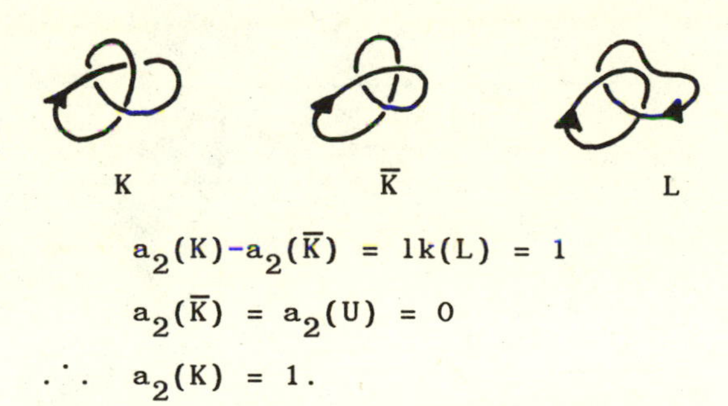

$$a_2(K)-a_2(\overline{K}) = \text{lk}(L) = 1$$

$$a_2(\overline{K}) = a_2(U) = 0$$

$$\therefore \ a_2(K) = 1.$$

In this sense, $a_2(K)$ computes a kind of "self-linking" number that is obtained from links made by splicing crossings on K.

Example 3.8: Problem: Calculate $a_2(K)$ for

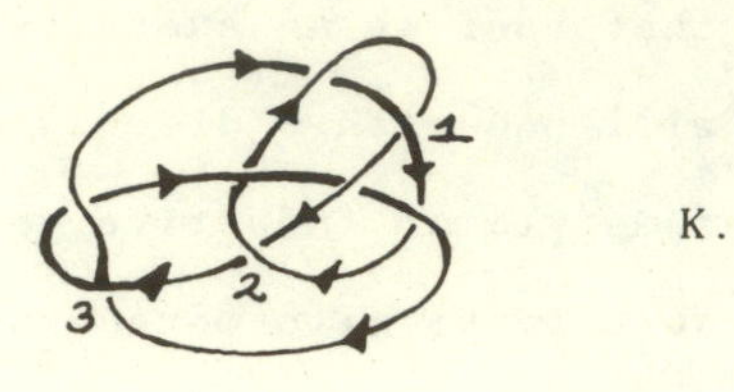

Note that K becomes unknotted if we switch crossings numbered 1, 2, and 3. Obviously at this point there is the need for some notation to help keep track of the calculation.

(i) Let $S_i(K)$ denote the result of switching the ith crossing of K.

S[ ] =

(ii) Let $E_i(K)$ denote the result of eliminating the ith crossing of K, by a splice.

E[ ] = = E[ ]

(iii) Let $\epsilon_i(K)$ denote the sign of the ith crossing of K.

ϵ[ ] = +1 , ϵ[ ] = -1

Then for any knot or link K with indexed crossings, Axiom 3 becomes

$$a_{n+1}(K)-a_{n+1}(S_iK) = \epsilon_i(K)a_n(E_iK).$$

Using this notation we have :

$$(\epsilon_i = \epsilon_i(K))$$

$$a_2(K)-a_2(S_1K) = \epsilon_1 lk(E_1K)$$

$$a_2(S_1K)-a_2(S_2S_1K) = \epsilon_2 lk(E_2S_1K)$$

$$a_2(S_2S_1K)-a_2(S_3S_2S_1K) = \epsilon_3 lk(E_3S_2S_1K).$$

Since $S_3S_2S_1K \sim 0$ we conclude that

$$a_2(K) = \epsilon_1 lk(E_1K)+\epsilon_2 lk(E_2S_1K)+\epsilon_3 lk(E_3S_2S_1K).$$

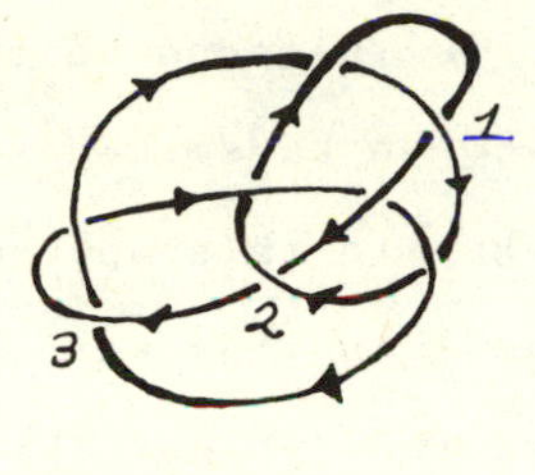

K

$$\begin{cases} \epsilon_1 = -1 \\ \epsilon_2 = +1 \\ \epsilon_3 = -1 \end{cases}$$

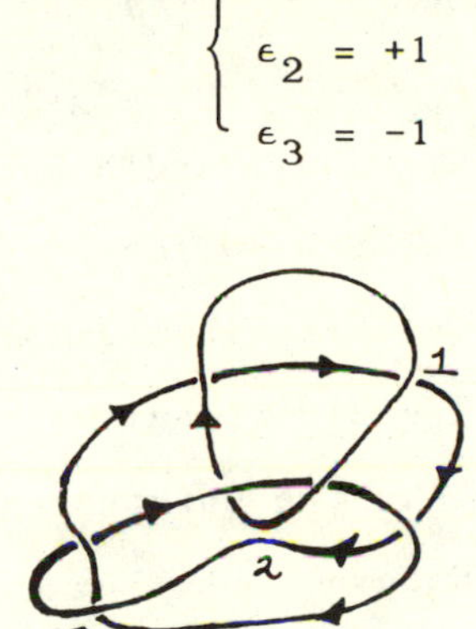

$X_1 = E_1K$ $X_2 = E_2S_1K$

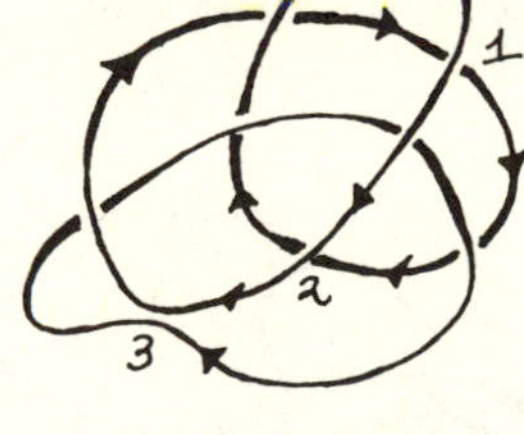

$$\begin{cases} lk(X_1) = 0 \\ lk(X_2) = +1 \\ lk(X_3) = 0 \end{cases}$$

$X_3 = E_3S_2S_1K$ $\therefore$ $\underline{a_2(K) = 1}$

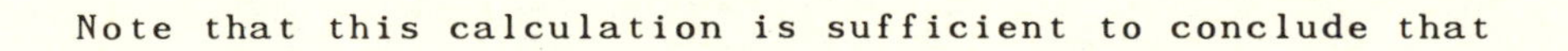

Note that this calculation is sufficient to conclude that

K is knotted.

In general, $a_2(K)$ can be expressed as a sum of signed linking numbers as in the formula above.

CREATING $a_2(K)$

Since we understand how $lk(K)$ is an invariant via the Reidemeister moves, it is natural to try to define $a_2(K)$ as a certain sum of linking numbers and then to prove that it is an invariant. To this end we shall define a "candidate" $\alpha(K)$ to be proved to have the properties and invariance of $a_2(K)$. The key to this approach (compare with Ball and Mehta in [BaM]) is that we shall restrict the switching sequences that unknot the knot to exactly those that arise from a standard unknot as in Exercise 2.2. That is, we do the following:

(a) Take the universe corresponding to the knot K. Choose a base-point on it that is not a crossing. (Denote the base-point as ——•——.) Walk along the universe in the direction of the knot's orientation and create an over-crossing each time you cross a crossing for the first time.

Example:

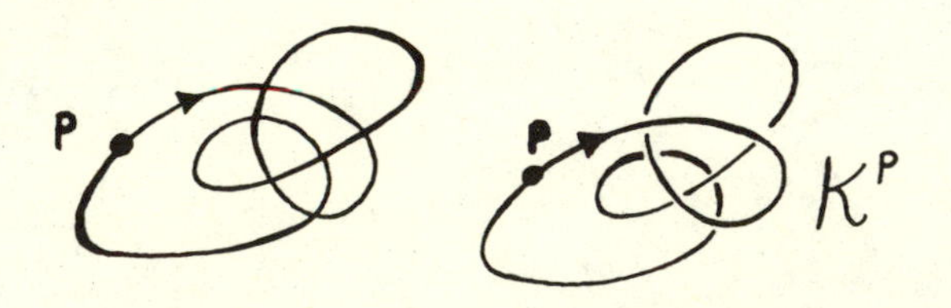

Let K^P denote the unknot that is produced by this operation. The specific unknotted diagram depends upon the choice of base-point and orientation.

(b) Compare K and K^P. We will label by $1,2,\cdots,n$ those crossings that are different. To form this labelling traverse the knot in its direction of orientation <u>from the base-point</u>. Label the first crossing from the set

D = {crossings in K that differ from crossings in K^P}

by n, the second by $n-1$, and generally label by $n-i+1$ the ith new crossing from D that is met in this traverse.

This gives a switching sequence $S_1,S_2,\cdots,S_n$ such that $S_nS_{n-1}\cdots S_1K = K^P$ is unknotted. Call this a <u>standard sequence</u> for the oriented knot K. The standard sequence depends upon K and the choice of base-point.

Example:

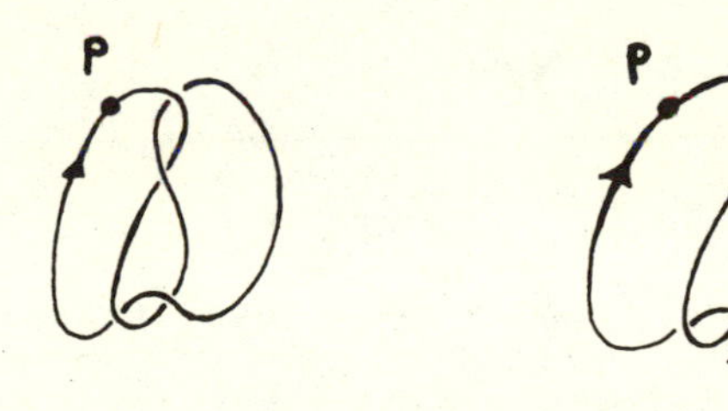

$$S_2S_1K = K^P.$$

Each crossing is labelled (in the order $n,n-1,\cdots,1$) where it is first traversed.

DEFINITION 3.9. *Let* K *be an oriented knot. Let* $S_1, \cdots, S_n$ *be a standard unknotting sequence for* K. *Let* $\epsilon_i = \epsilon_i(K)$ *and* $X_i = E_i S_{i-1} S_{i-2} \cdots S_1 K$ $(i = 1, 2, \cdots, n)$. *Define* $\alpha(K)$ *by the formula* $\alpha(K) = \Sigma_{i=1}^{n} \epsilon_i \operatorname{lk}(X_i)$.

We must show that $\alpha(K)$ is independent of the choice of base-point, and that it is a topological invariant. In order to do this a preliminary discussion of the unknot K^p is required.

UNKNOT DISCUSSION

Let's think about how K and K^p differ. As we traverse from p, K may look like:

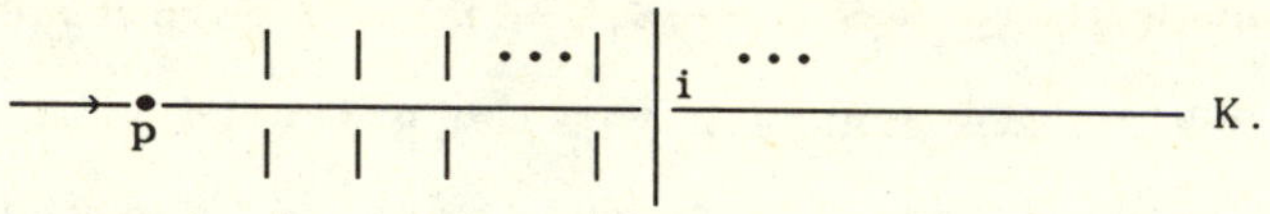

The crossing labelled i is the first undercrossing encountered when travelling from p. As a result, K^p will look like this:

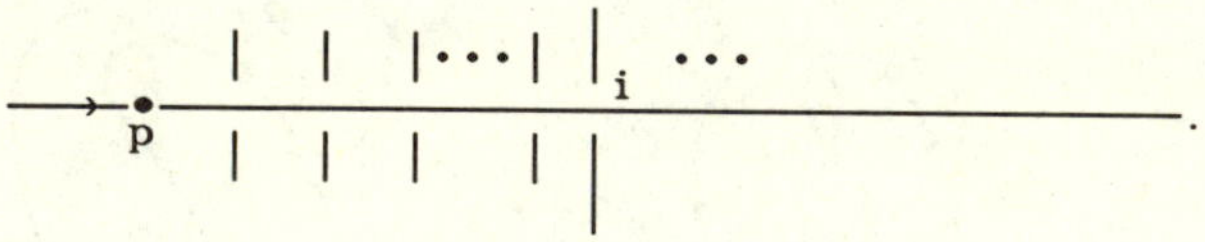

That is, i will be changed to an over-crossing in K^p, and i labels the first crossing change between K and K^p.

Let's consider the following situation:

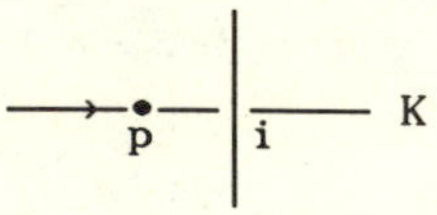

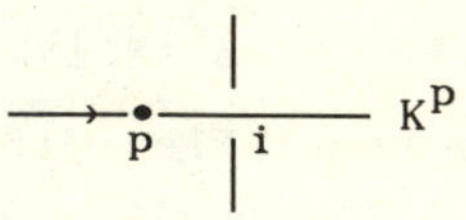

Here the first crossing that occurs after p is also the site of the first crossing change. Under these circumstances E_iK^p is a split unlink.

Example:

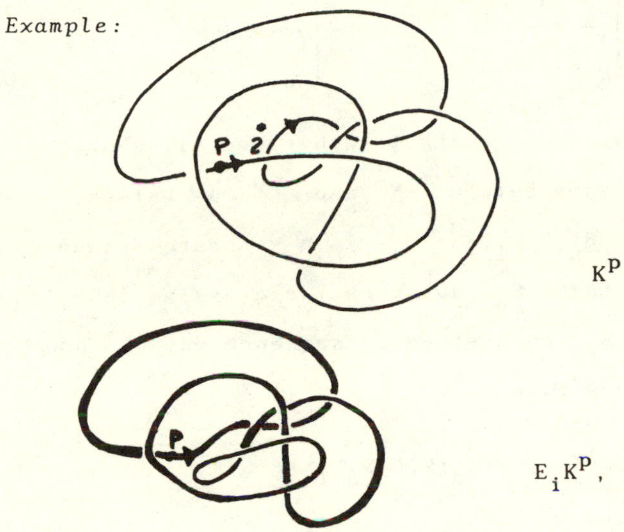

E_iK^p, a split unlink.

We leave the proof of this assertion as an exercise.

PROPOSITION 3.10. *Let* K *be an oriented knot diagram. Then* $\alpha(K)$, *as defined above, is independent of the choice of base-point on* K.

Proof. It will suffice to show that we can slide the base-point through a crossing in K without changing the value of α. To emphasize the possible dependence on base-point we will write $\alpha(K,p)$. Now the base-point may slide over or it may slide under. This leads to two cases:

Case 1.

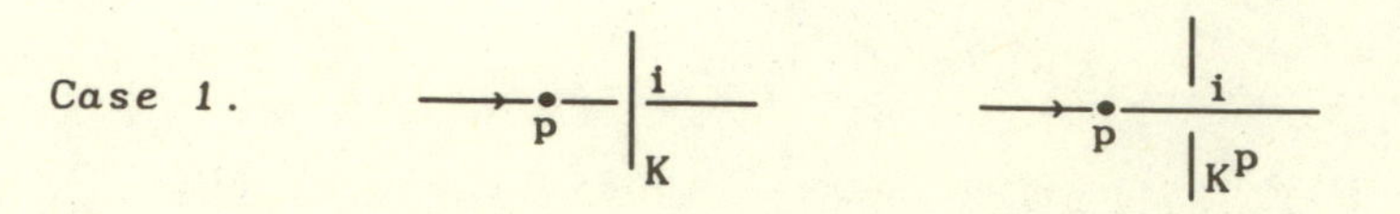

Here K and K^p differ at i. If we slide the base-point under the crossing to q then

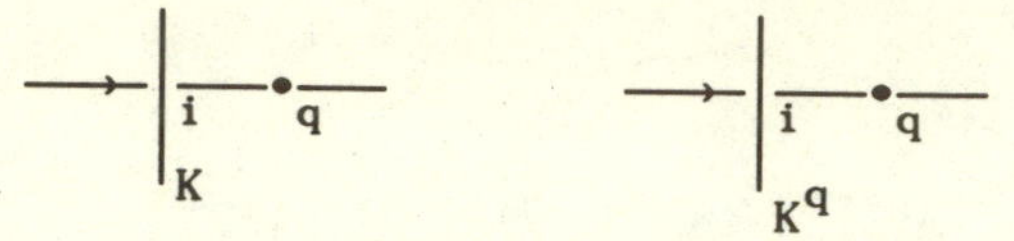

the crossing i does not change. Otherwise all changing crossings are the same between K and K^p as between K and K^q. Thus if $S_n, S_{n-1}, \cdots, S_1$ is a standard sequence for K and K^p, then S_n switches the crossing labelled i. And $S_{n-1}, \cdots, S_1$ is a standard sequence for K and K^q. Consequently we have

$$\alpha(K,p)-\alpha(K,q) = \epsilon_n lk(E_n S_{n-1} \cdots S_1 K).$$

But $S_n S_{n-1} \cdots S_1 K = K^p$ and $E_n S_{n-1} \cdots S_1 K = E_n K^p$. Since $E_n K^p = E_i K^p$, we know from the unknot discussion that $E_n K^p$ is a split unlink. Therefore $lk(E_n K^p) = 0$. Hence $\alpha(K,p) = \alpha(K,q)$, proving Case 1.

Case 2.

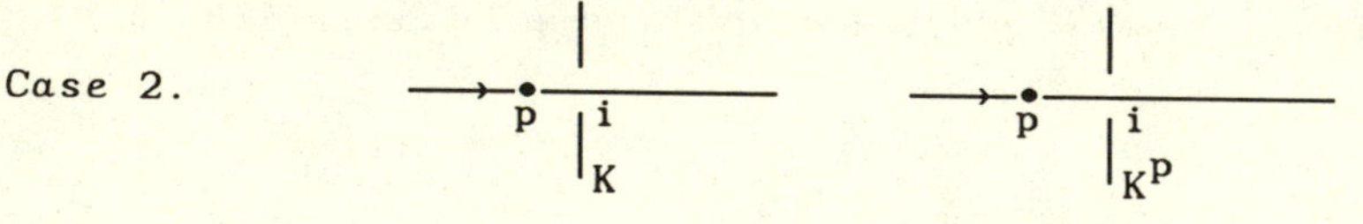

Here p lies on the over-crossing line just prior to the crossing. The crossing i does not change from K to K^p. If we slide p across then

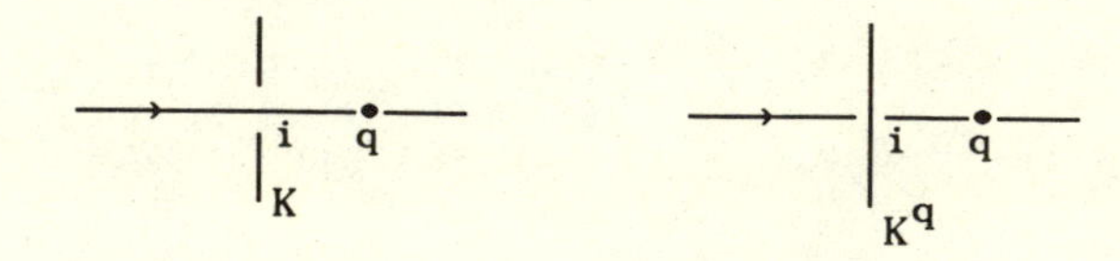

the crossing does change between K and K^q. Thus we have added one crossing to the set of changing crossings of K^p. If the switching sequence for K, K^q is

$$S_n, S_{n-1}, \cdots, S_{i+1}, S_i, S_{i-1}, \cdots, S_1$$

with S_i switching the crossing labelled i, then $S_n, S_{n-1}, \cdots, S_{i+1}, S_{i-1}, \cdots, S_1$ is a standard sequence for K, K^p.

Thus the first $i-1$ terms in the sums for $\alpha(K,q)$ and $\alpha(K,p)$ are identical. Then $\alpha(K,q)$ has the term $\epsilon_i \mathrm{lk}(E_i S_{i-1} \cdots S_1 K)$. This term is missing from the sum for $\alpha(K,p)$ and all the remaining terms differ by a switch at the ith crossing. Note that such a switch will affect the linking numbers

$$\epsilon_{k+i} \mathrm{lk}(E_{k+i} S_{i+k-1} \cdots S_{i+1} S_{i-1} \cdots S_1 K) \longleftrightarrow K, K^p$$

$$\epsilon_{k+i} \mathrm{lk}(E_{k+i} S_{i+k-1} \cdots S_{i+1} S_i S_{i-1} \cdots S_1 K) \longleftrightarrow K, K^q$$

only if the crossing labelled i is a crossing of two components of the resulting link. And this is true if and only if $(k+i)$ labels a crossing of two components of $E_i K$.

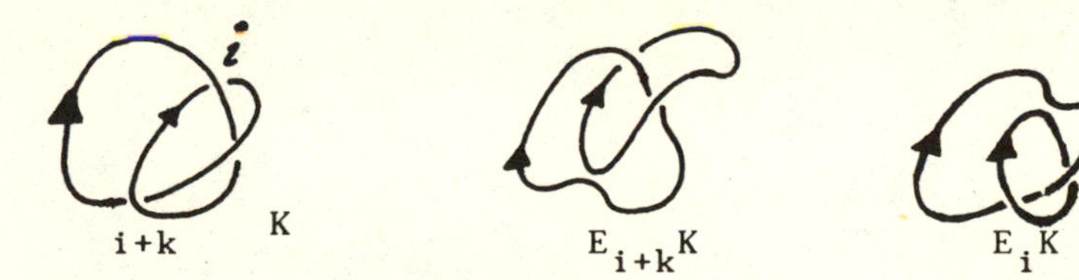

Note that $\mathrm{lk}(X) - \mathrm{lk}(S_i X) = \epsilon_i C(E_i X)$. Therefore we conclude that the sum of the differences yields:

$$\alpha(K,p) - \alpha(K,q) = -\epsilon_i \mathrm{lk}(E_i S_{i-1} \cdots S_1 K) + \epsilon_i \Delta$$

where Δ = the sum of ϵ_{k+i}, $k = 1,2,\cdots$ such that $k+i$ is a crossing of two components of E_iK.

We <u>also</u> know that $E_iS_nS_{n-1}\cdots S_{i+1}S_{i-1}\cdots S_1K$ is a split unlink for the same reasons as before (i is the first return just <u>prior</u> to the base-point). Since the linking number of a link is the sum of any set of crossing indices whose switchings unlink it, this implies that $\Delta = lk(E_iS_{i-1}\cdots S_1K)$. Hence $\alpha(K,p) = \alpha(K,q)$. This completes Case 2 and the proof of the proposition.

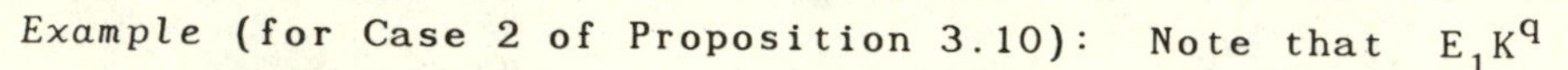

Example (for Case 2 of Proposition 3.10): Note that E_1K^q

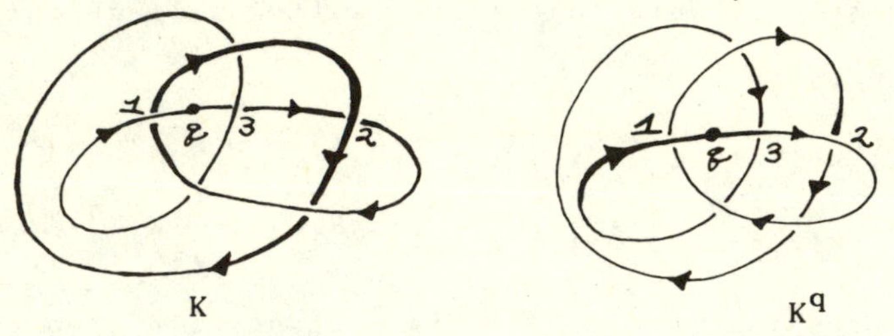

is a split unlink. Here $S_3S_2S_1K \sim K^q$. If we slide q backwards through 1 to p (as below) then we will lose S_1 from the switching sequence.

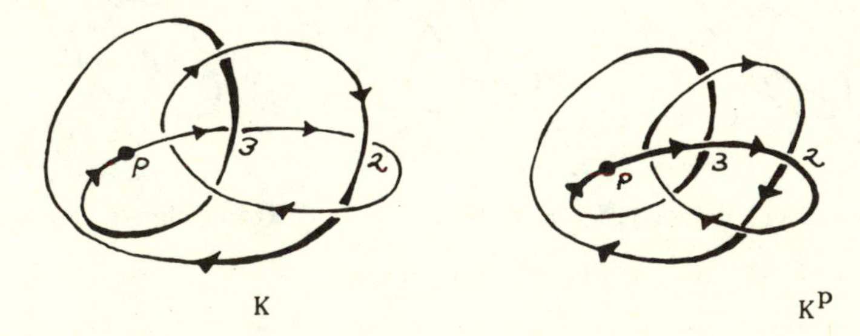

Here $S_3S_2K \sim K^p$.

Now compare the computations for q and p.

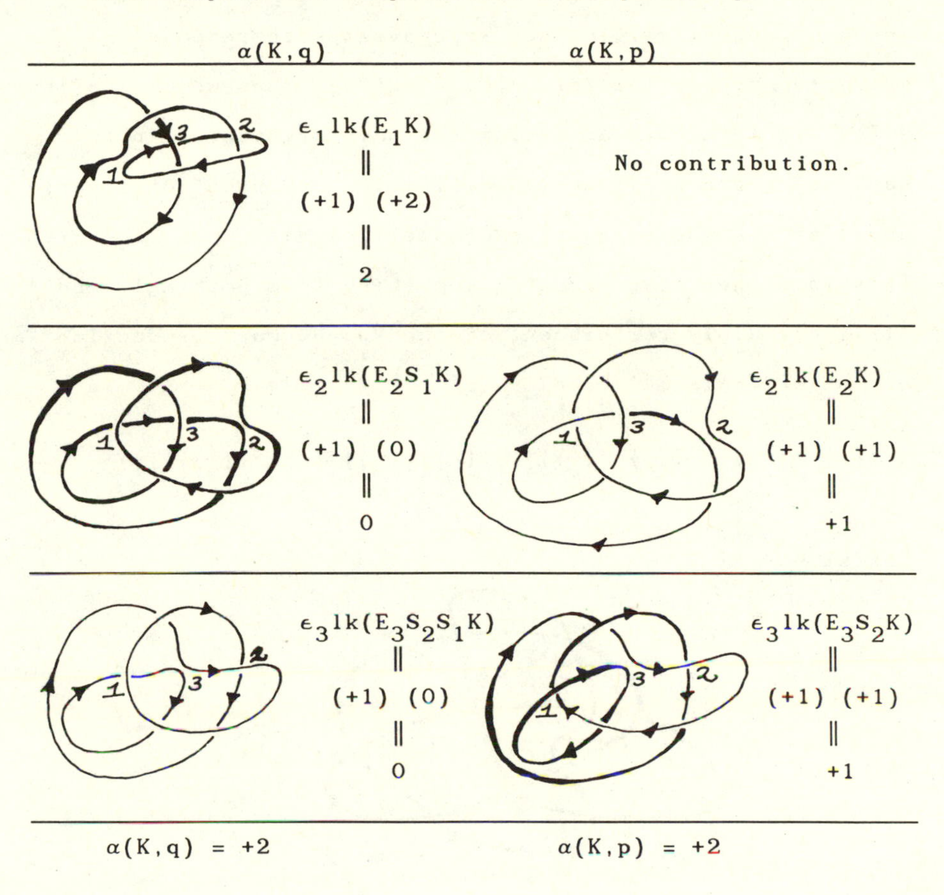

We see clearly how the linking number $lk(E_1K)$ is distributed across the other terms in the calculation of $\alpha(K,p)$.

Remark. If you examine the proof of Proposition 3.10 carefully, you will see that we actually used exactly the axiomatic properties of the linking number. That is, that

$lk(X)-lk(S_iX) = \epsilon_i(X)C(E_iX)$, $lk(\mathbf{OO}) = 0$, and topological invariance. As a result, this proof can be generalized and used inductively! Once we prove the corresponding axiomatic properties for $\alpha(K) = a_2(K)$, then we can define $a_3(K)$ in terms of a_2 using the same argument. In this way, we create an inductive definition and proof of invariance for the whole sequence of coefficients $a_0, a_1, a_2, \cdots$. This is rather like creating something from nothing! And it is certainly reminiscent of the Von Neumann production of the ordinals

{ }, {{ }}, {{ } {{ }}}, {{ } {{ }} {{ } {{ }}}}$\cdots$.

<u>Exercise 3.11</u>.

(a) Work out $\alpha(K,q)$ from this diagram.

(b) Work out $\alpha(K,p)$ and compare your calculation with (a).

(c) Simplify the diagram by Reidemeister moves to a diagram K' with fewer crossings. Find $\alpha(K')$. Find $\nabla_{K'}$.

<u>Exercise 3.11a</u>. This exercise is designed to help you reexamine both linking numbers and the proof of Proposition 3.10.

(i) Let $L = \alpha \cup \beta$ be a link of two components α, β. Let $1, 2, \cdots, n$ be a set of crossings of α with β such that $S_n S_{n-1} \cdots S_1 L$ is a split link. <u>Show that</u> $lk(L) = \epsilon_1(L) + \epsilon_2(L) + \cdots + \epsilon_n(L)$.

Example:

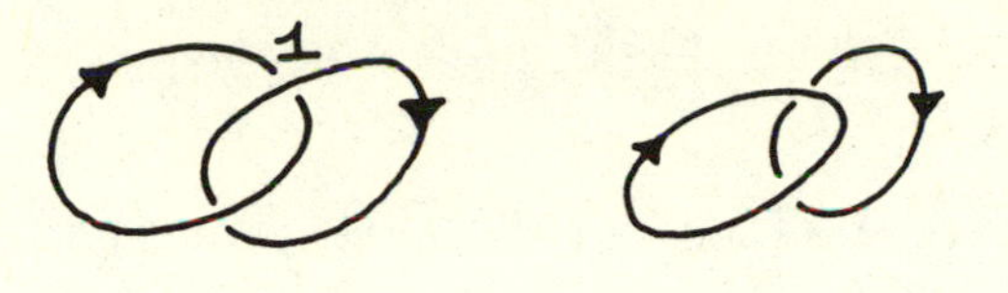

L $\qquad\qquad$ $S_1 L$ <u>split</u>

$$lk(L) = -1 = \epsilon_1(L).$$

(ii) Let $L = \alpha \cup \beta$ be a link of two components α, β. Let $1, 2, \cdots, n$ be <u>any</u> set of crossings in L such that $S_n S_{n-1} \cdots S_1 L$ is a split link. <u>Show that</u> $lk(L) = \Sigma_{i=1}^{n} \epsilon_i(L)\ C\ (E_i L)$ where C as defined in 3.5, is 1 on knots and zero on links.

(iii) Give some examples of unknot diagrams that are not standard unknots.

(iv) If you had started with the formula in (ii) (or (i)) as the definition of linking number, how would you show that it was well-defined and a topological invariant?

Example:

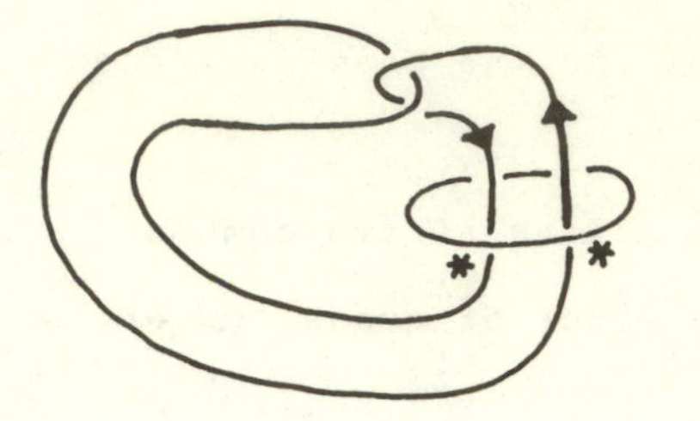

L

Switching the two starred crossings unlinks L. They have opposite sign. Therefore $lk(L) = 0$.

Now onward to the proof of invariance.

Remark. We have defined $\alpha(K)$ only when K is a knot (one component). However, everything we have said works just as easily for links. (Define a switching sequence by untangling one component. Show this is independent of basepoint. Now define α' by summing over all components and taking the average value of α. Then α' is the extension of α to links; it will be denoted by α as before. Note that this procedure is exactly what we do when we define linking number as <u>one-half</u> of the sum of the crossing signs.)

Note. $\alpha_2(L)$ can be nonzero on a 3-component link:

is an example.

From now on, we assume $\alpha(K)$ is well-defined on knot <u>and</u> link diagrams.

PROPOSITION 3.12. *Let* K *be an oriented knot or link diagram. Then*

(i) $\alpha(K)$ *is invariant under the Reidemeister moves.*

(ii) *If links* K, $\overline{K}$ *and* L *are related as*

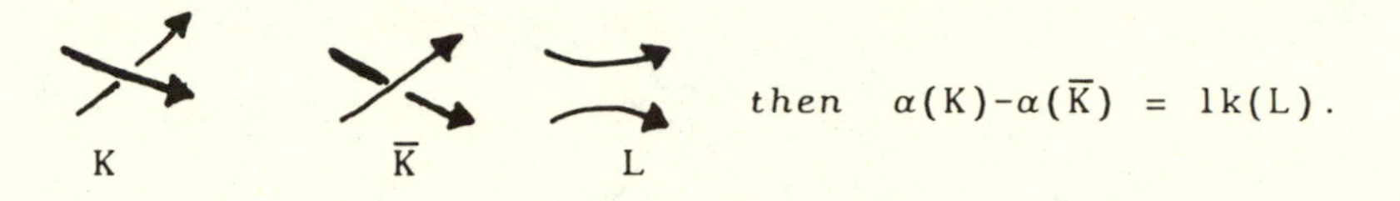

then $\alpha(K)-\alpha(\overline{K}) = lk(L)$.

Thus $\alpha(K)$ *satisfies the axioms for* $a_2(K)$. *Hence these axioms are consistent and* $\alpha(K) = a_2(K)$.

Proof. <u>The idea</u>: Position the base-point so that "none" of the crossings involved in a given Reidemeister move are part of the switching sequence.

This idea works for moves of type 1 and type 2 as shown below:

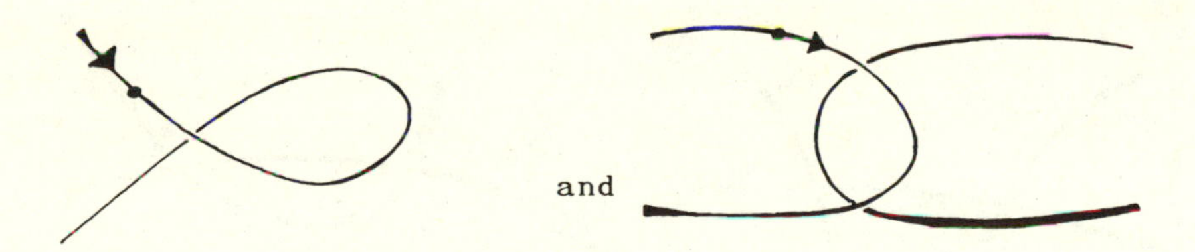

Since α is defined by a sum of invariants, we can perform the R-move for each form in the sum without changing the values.

This idea almost works for the type 3 move:

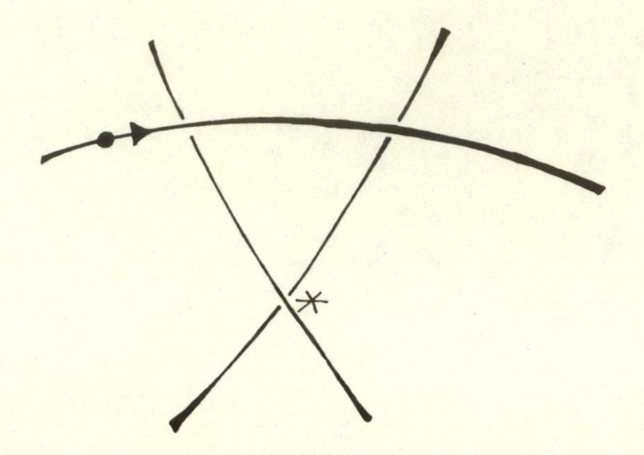

The starred (*) crossing may still be involved in the switching sequence. (Hence we have "none" in quotes above.) However, this presents no problem! In a given term $\epsilon_i \mathrm{lk}(E_i S_{i-1} \cdots S_1 K)$, the starred crossing may be switched or it may be spliced. If switched, we can still do the move. And if spliced, we are looking at a simple equivalence, or two type two moves as in

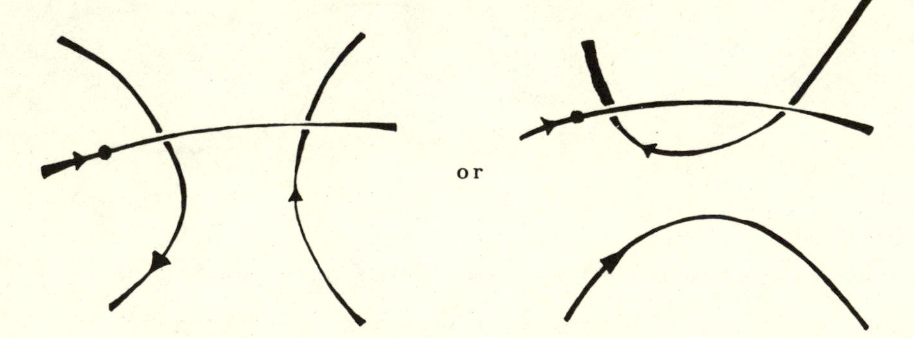

Thus in any case, any move we want to do can be inherited by every term $\epsilon_i \mathrm{lk}(E_i S_{i-1} \cdots S_1 K)$ and hence will not change $\alpha(K)$. This proves invariance.

Part (ii) will be left as Exercise 3.13.

Hint: Position the base-point correctly! This completes the proof. ■

Exercise 3.13 (A small project). Rewrite the theory (Proposition 3.10-3.13) so it inductively defines all of the polynomial coefficients.

Exercise 3.14. Investigate the consequences of replacing the − sign in Axiom 3 for the Conway Polynomial ($\nabla_K - \nabla_{\overline{K}} = z\nabla_L$) by a + sign! Prove that the resulting polynomial is a topological invariant of knots and links. (This is a special case of the first generalized polynomial. See [JO1], [JO2], [JO3] and [HOMFLY].)

IV

EXAMPLES AND SKEIN THEORY

Here we continue calculating via the axioms of Chapter 3. Note that the end of Chapter 3 has provided a proof of the consistency of the axioms. First some recursions, and then some skein theory (see Conway [C1]).

Exercise 4.1.

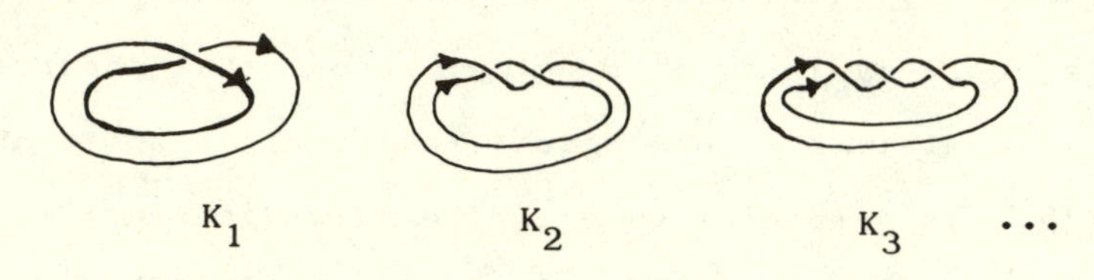

These are alternately knots and links.

If $\nabla_n = \nabla_{K_n}$ then $\nabla_n - \nabla_{n-2} = z\nabla_{n-1}$. Thus we have

$$\boxed{\begin{aligned} \nabla_1 &= 1 \\ \nabla_2 &= z \\ \nabla_n &= z\nabla_{n-1} + \nabla_{n-2} \end{aligned}}$$

For $z = 1$, this yields the Fibonacci Series $1,1,2,3,5,\cdots$. Here Fibonacci shows that none of these are equivalent (along with component count for the first two).

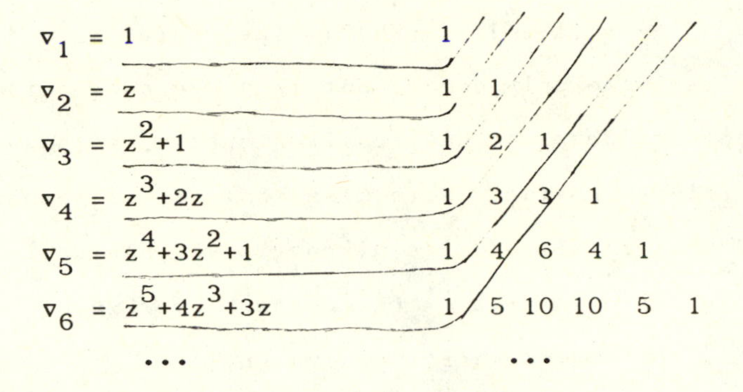

$\nabla_1 = 1$ 1

$\nabla_2 = z$ 1 1

$\nabla_3 = z^2+1$ 1 2 1

$\nabla_4 = z^3+2z$ 1 3 3 1

$\nabla_5 = z^4+3z^2+1$ 1 4 6 4 1

$\nabla_6 = z^5+4z^3+3z$ 1 5 10 10 5 1

...   ...

Example 4.2.

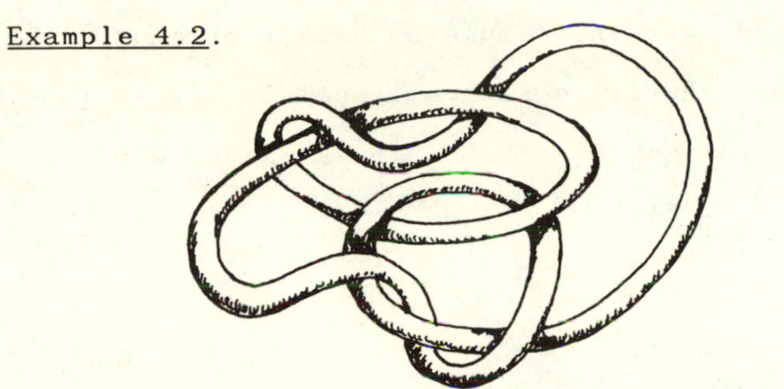

K

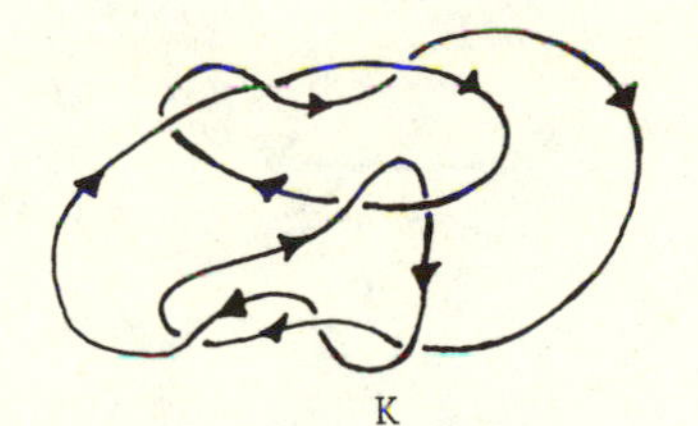

K

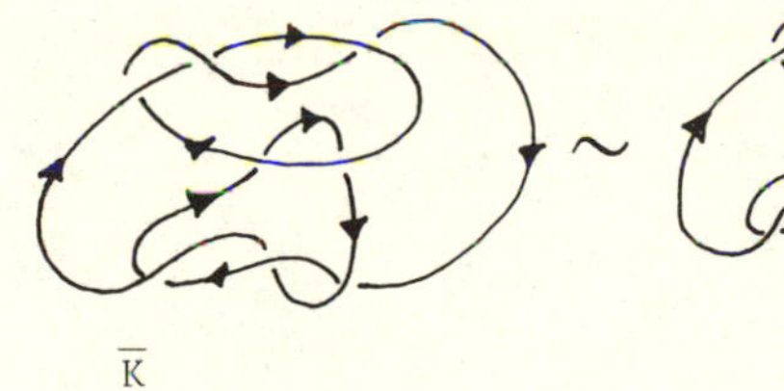

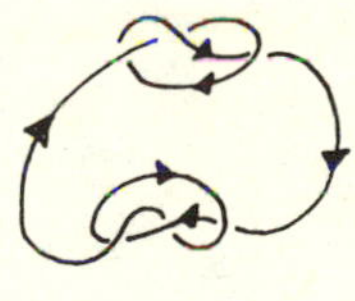

$\overline{K}$

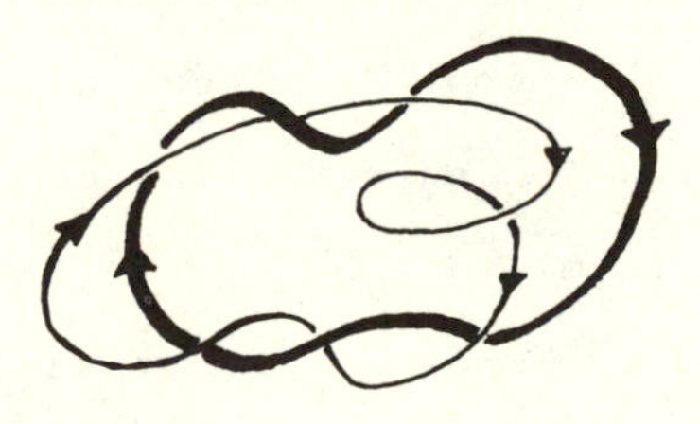

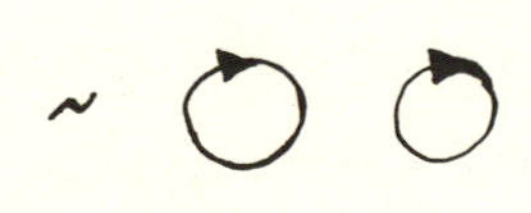

L

Thus $\nabla_K = \nabla_{\overline{K}} = (1+z^2)^2$. (Why?) The knots K and $\overline{K}$ receive the same polynomial, but they are not equivalent. The proof of inequivalence requires subtler methods than we have developed so far. Note also that $\overline{K}$ is composed of a copy of the trefoil and its mirror image. We shall return to this example later on. (Exercise: Use Exercise 3.15 to show that K and $\overline{K}$ are inequivalent!)

Exercise 4.2. Find another example of a pair of apparently distinct knots that share the same polynomial. [Research Problem: Given a polynomial $f(z)$, investigate $\mathbb{K}(f)$ where $\mathbb{K}(f) = \{K | \nabla_K = f\}$.]

SKEIN NOTATION

When links A, B and C are diagrammatically related in the form

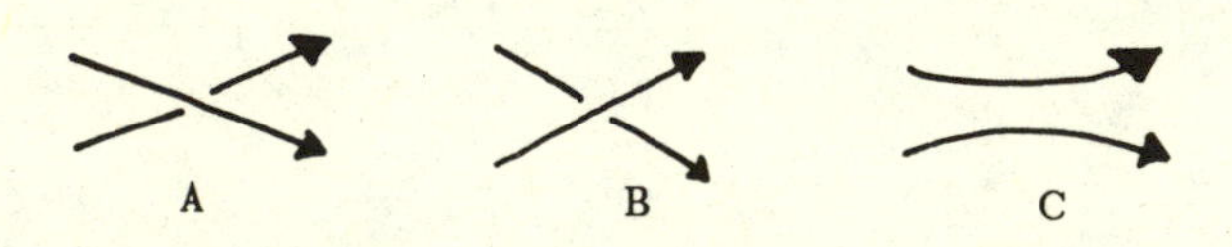

we shall write $A = B \oplus C$ and $B = A \ominus C$. The operations $\oplus$ and $\ominus$ are nonassociative, and they provide a convenient notation for these relationships. [You may think of C as a switching operator in this form.]

By Axiom 3 we have $\nabla_A - \nabla_B = z\nabla_C$. Hence $\nabla_{B\oplus C} = \nabla_B + z\nabla_C$ and $\nabla_{A\ominus C} = \nabla_A - z\nabla_C$. Thus we could also define $\oplus$ and $\ominus$

in $Z[z]$ by $f \oplus g = f+zg$ and $f \ominus g = f-zg$. This notation, and the related ideas are due to John Conway. Then $\nabla_{A \oplus B} = \nabla_A \oplus \nabla_B$.

Example:

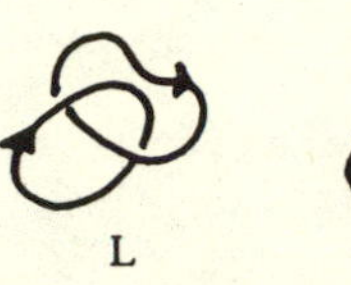
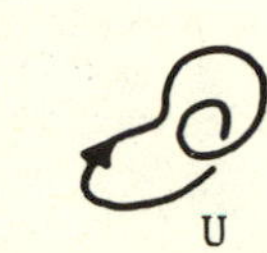

K U L $\overline{L}$ U

$$K = U \oplus L, \qquad L = \overline{L} \oplus U$$

$$K = U \oplus (\overline{L} \oplus U)$$

$$\nabla_K = \nabla_U \oplus (\nabla_{\overline{L}} \oplus \nabla_U)$$

$$\nabla_K = 1+z(0+z) = 1+z^2.$$

We shall use $\oplus$ and $\ominus$ primarily for notational convenience. However, Conway creates what he calls <u>Skein Theory</u> as follows: Define $A = B \oplus C$ whenever $A \sim A'$, $B \sim B'$ and $C \sim C'$ where A', B' and C' are diagrams that fit together as above. By including equivalence in this way we allow the possibility that we may do this in more than one way! Thus there may be $B'' \sim B$, $C'' \sim C$ with B'' and C'' also fitting together. The resulting compositions $B'' \oplus C''$ and $B' \oplus C'$ may not be topologically equivalent. Thus we say that $B'' \oplus C''$ is <u>skein equivalent</u> to $B' \oplus C'$ if these compositions are defined and $B' \sim B''$, $C' \sim C''$. We write $B'' \oplus C'' \underset{sk}{\sim} B' \oplus C'$ (and let skein equivalence be the equivalence relation so generated).

More generally, two algebraic compositions involving $\oplus$ and $\ominus$ are skein equivalent if their individual terms are topologically equivalent. Clearly, if $L \underset{sk}{\sim} L'$ then $\nabla_L = \nabla_{L'}$. Thus if $\mathcal{K}$ denotes the set of equivalence (topological equivalence) classes of knots and links, then ∇ is a well-defined homomorphism from the skein $\mathcal{K}/sk$ to the skein $Z[z]$ (operations $\oplus$, $\ominus$). $\nabla : \mathcal{K}/sk \longrightarrow Z[z]$.

<u>Exercise</u>: Find an example of two skein-equivalent knots or links that are not topologically equivalent.

<u>Open</u> <u>Problem</u>: Is there a nontrivial knot K in $\mathcal{K}$ that is skein-equivalent to the unknot?

In the example above we had $K = U \oplus (\bar{L} \oplus U)$ where $U =$ and $\bar{L} =$  . This is a <u>skein-decomposition</u> of K into the <u>generators</u> of the skein $\mathcal{S} = \mathcal{K}/sk$. And this is just a fancy way of saying that, by recursion, all of our calculations bottom out on one of the list:

$U =$, , , ···. These are the generators of the skein $\mathcal{S}$ of knots and links. Let $U_n =$ ··· n components. Any knot or link can be written as a skein-decomposition involving the U_n's.

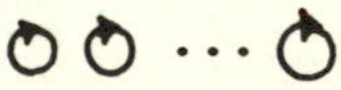

Example:

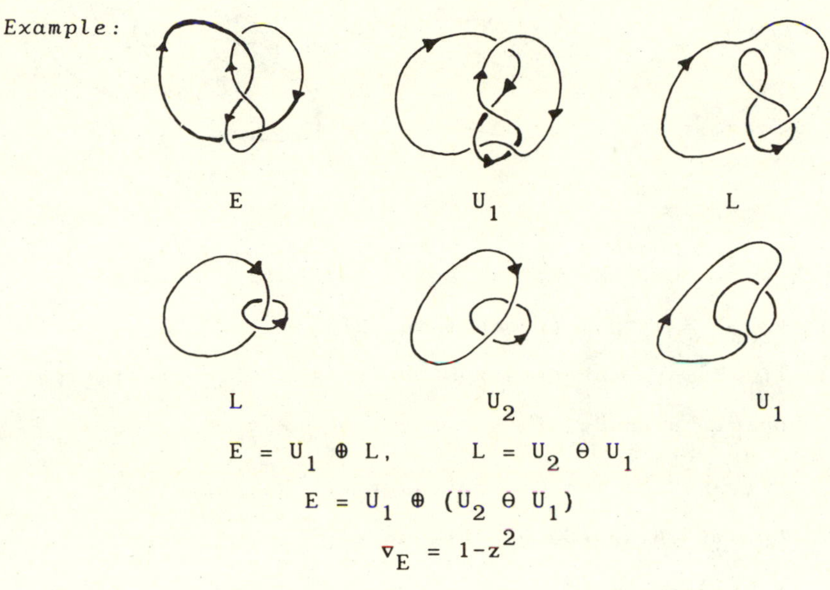

$E = U_1 \oplus L, \qquad L = U_2 \ominus U_1$

$$E = U_1 \oplus (U_2 \ominus U_1)$$

$$\nabla_E = 1-z^2$$

E is the figure-eight knot.

Note that we have distinguished it from the trefoil.

We will now state a theorem whose proof is made very simple by the existence of the skein decompositions. First recall that the <u>connected</u> <u>sum</u> of links K and K′, denoted K # K′, is defined by splicing two strands as shown below. This may depend upon the choice of strands.

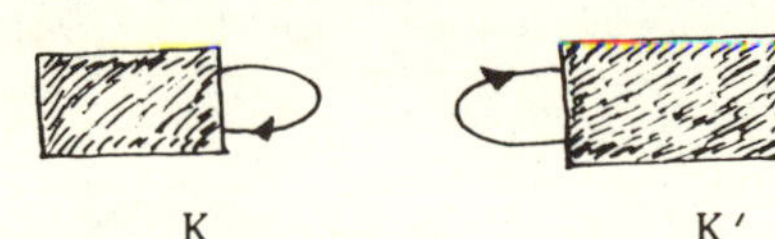

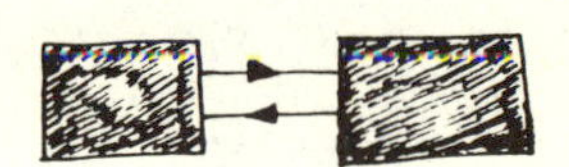

I prefer to think of it this way:

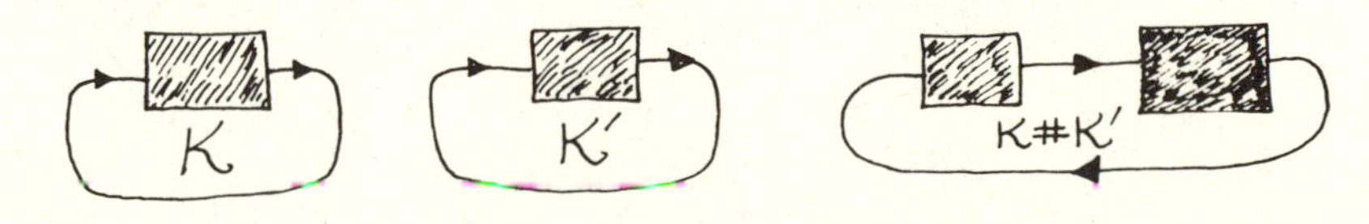

Thus

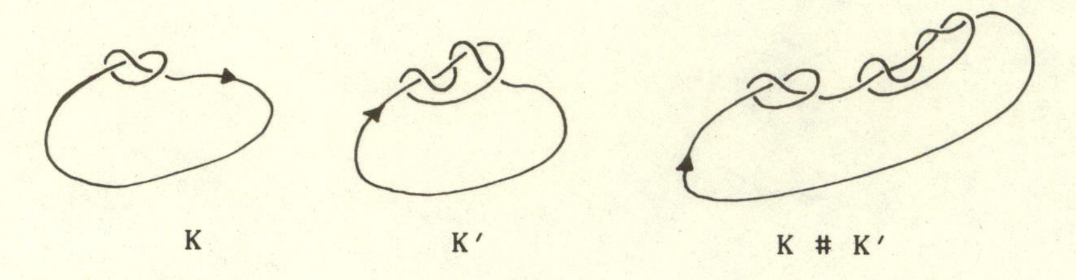

THEOREM 4.3. *Let* K, K′ *be oriented knots or links.*

(i) $\nabla_{K\#K'} = \nabla_K \nabla_{K'}$ *(product in* Z[z]).

(ii) *If* K^* *is the result of reversing all orientations on all strands of* K, *then* $\nabla_{K^*} = \nabla_K$.

(iii) *Let* $K^!$ *be the mirror image of* K *(obtained by switching every crossing of* K). *Then* $\nabla_{K^!}(z) = \nabla_K(-z)$,

(iv) *If* L *is a link with* λ *components then* $\nabla_L(-z) = (-1)^{\lambda+1}\nabla_L(z)$. *Hence* $\nabla_{L^!} = (-1)^{\lambda+1}\nabla_L$.

Proof: <u>Exercise 4.3</u>.

COROLLARY 4.4. *If* L *is a link with an even number of components, and* $\nabla_L \neq 0$, *then* L *and its mirror image* $L^!$ *are not equivalent.*

Example:

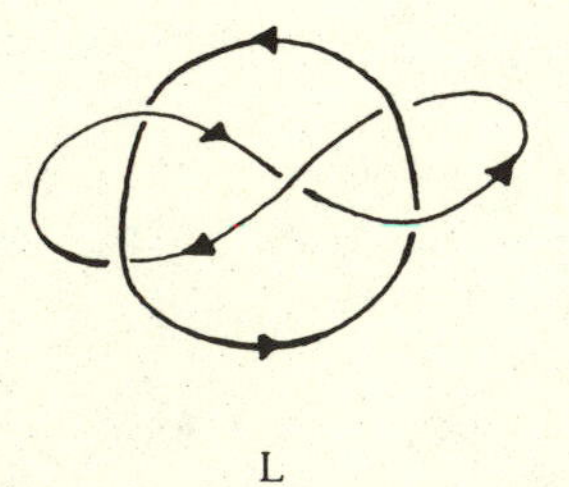

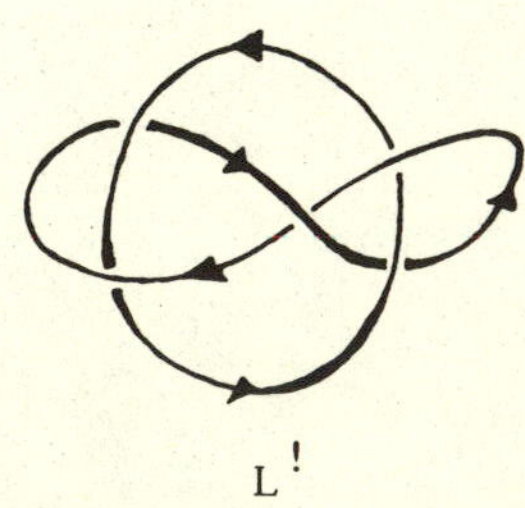

These links are not equivalent.

Note: We have already shown that

$\overset{\nabla}{\longmapsto} z^3$

(one strand has orientation reversed from L)

Now

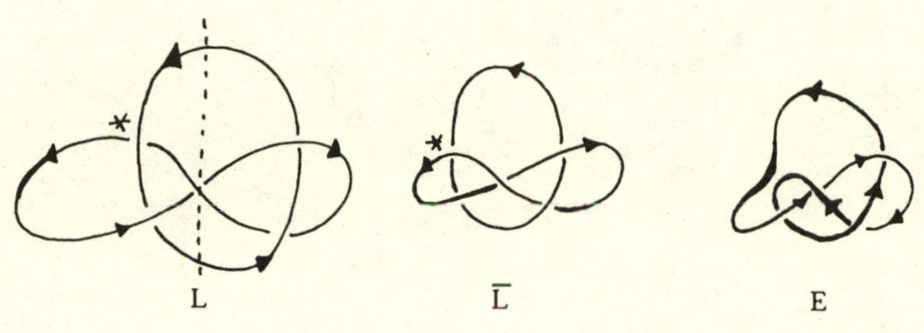

(figure eight knot)

$\therefore \quad -\nabla_L + z = z(1-z^2).$

$\therefore \quad \nabla_L = z^3$ also! [This follows by symmetry around the vertical dotted axis.] We therefore have the following table:

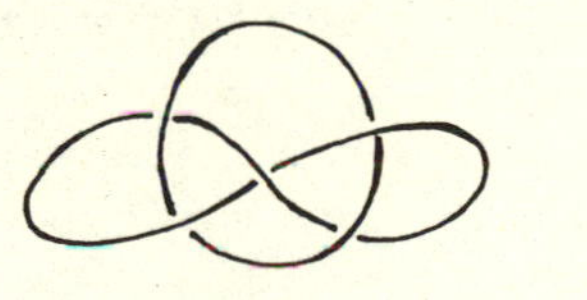

W

with <u>any</u> component orientation has $\nabla = +z^3$.

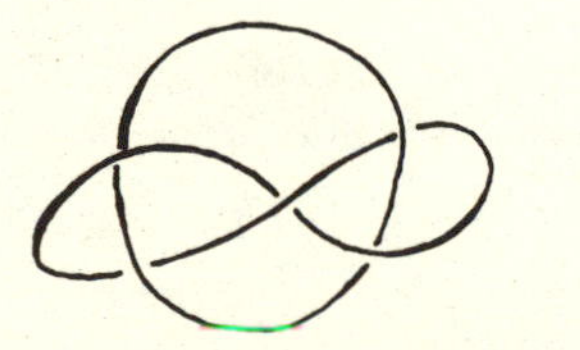

$W^!$

with <u>any</u> component orientation has $\nabla = -z^3$. <u>Therefore the</u>

unoriented Whitehead link W and her mirror image $W^!$ are inequivalent.

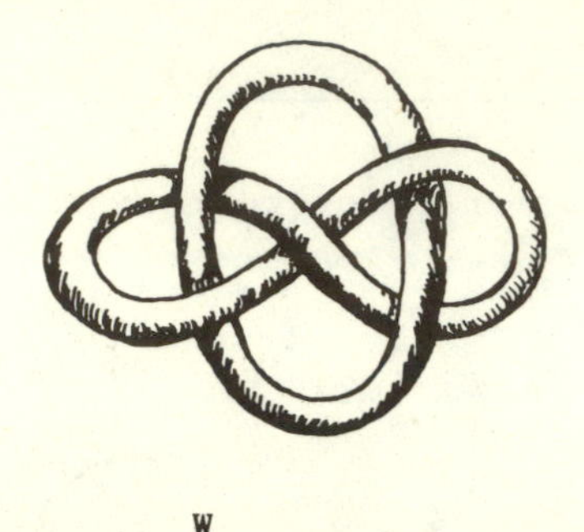
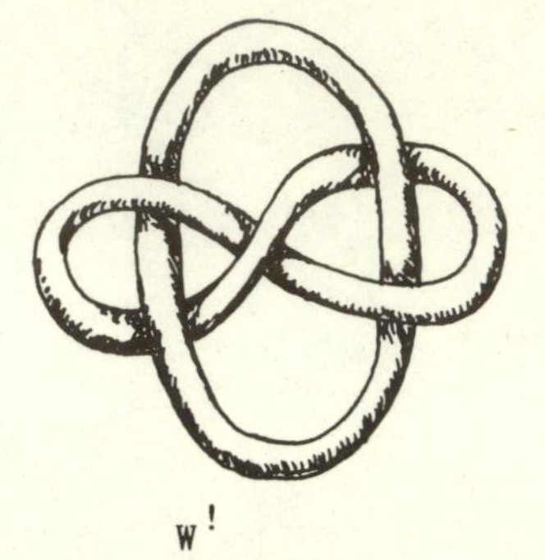

W $\qquad$ $W^!$

Example 4.5. Weaving

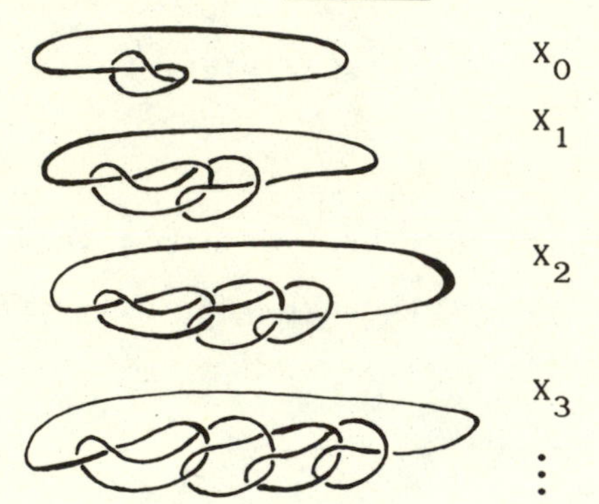

X_0

$X_1 \qquad X_1 = RX_0$

$X_2 \qquad X_2 = RX_1$

$X_3 \qquad X_3 = RX_2$

$\vdots \qquad \vdots$

This is a simple chain stitch. The basic recursion that generates the weave is of the form

$\wedge \qquad R\wedge$

Thus $X_{n+1} = RX_n$. From this we can calculate $\nabla_n = \nabla_{X_n}$:

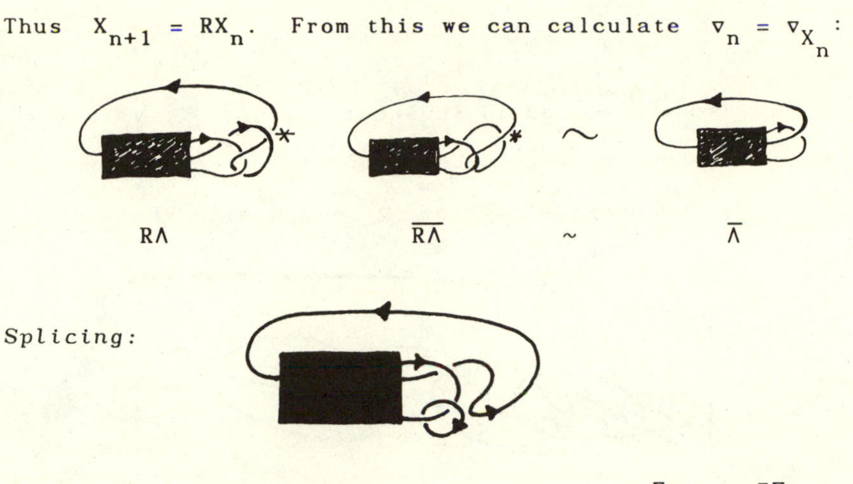

$$\nabla_{L\Lambda} = -z\nabla_\Lambda$$

LΛ

Thus $\nabla_{R\Lambda} = \nabla_{\overline{\Lambda}} - z^2\nabla_\Lambda$. In our case, $X_n = R^nX_0$. Hence $\overline{X}_n = \overline{R^nX_0} = \overline{R(R^{n-1}X_0)} \sim \overline{R^{n-1}X_0} \sim\cdots\sim \overline{RX_0} \sim \overline{X_0} \sim 0$. That is, if you switch the right-most crossing of X_n, then it becomes unknotted. $\therefore$ $\nabla_{X_n} = 1-z^2\nabla_{X_{n-1}}$.

$$\nabla_0 = 1+z^2$$
$$\nabla_1 = 1-z^2-z^4$$
$$\nabla_2 = 1-z^2+z^4+z^6 \qquad\qquad \nabla_n = 1-z^2+z^4-\cdots\pm z^{2n+2}.$$
$$\nabla_3 = 1-z^2+z^4-z^6-z^8$$

You may be tempted by this formalism to go to infinity! Then there would be an "infinite knot" X_∞. And $RX_\infty = X_\infty$. In place of $\nabla_{X_n} = 1-z^2\nabla_{X_{n-1}}$, we would have $\nabla_{X_\infty} = 1-z^2\nabla_{X_\infty}$. Whence $(1+z^2)\nabla_{X_\infty} = 1$ and

$$\nabla_{X_\infty} = \frac{1}{1+z^2} = 1-z^2+z^4-\cdots .$$

This formal power series is indeed the limit of $\nabla_n = 1-z^2 +\cdots\pm z^{2n+2}$. But is there such an entity as this infinite knot X_∞?? One possibility for X_∞ is a wild knot in the form

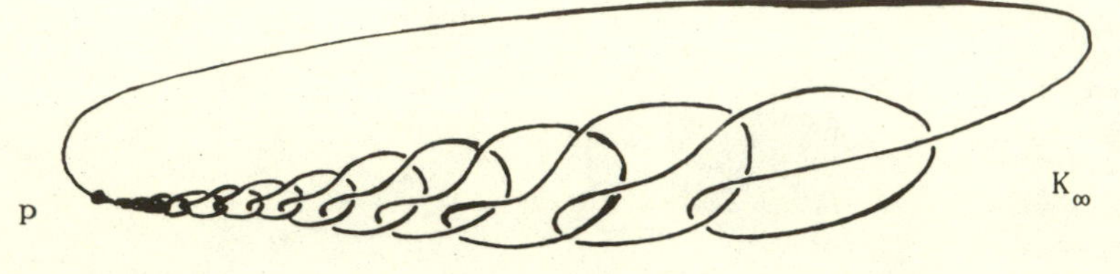

The "wild point" p.

Here K_∞ represents a homeomorphic image of S^1 in $\mathbb{R}^3$ such that the image has a wild point p. Every neighborhood of p contains infinitely many crossings from K and the pair (Neighborhood, Neighborhood $\cap K_\infty$) does not straighten out (under a homeomorphism) to the standard (ball, arc) pair. Nevertheless, K_∞ does have the weaving stability: $RK_\infty = K_\infty$.

However, there is a sense in which switching the starred crossing in K_∞ results in an unknot! If you call this switched version $\overline{K_\infty}$, then $\overline{K_\infty}$ is a curious animal. You can do any finite amount of unravelling of him without making him disappear. In this case there is an infinite composition of elementary moves corresponding to a homeomorphism $h : \mathbb{R}^3 \longrightarrow \mathbb{R}^3$ that unknots $\overline{K_\infty}$. In general, the topological considerations for wild knots are more complicated. Instead, we may construe the infinite knot differently as in X_∞ below:

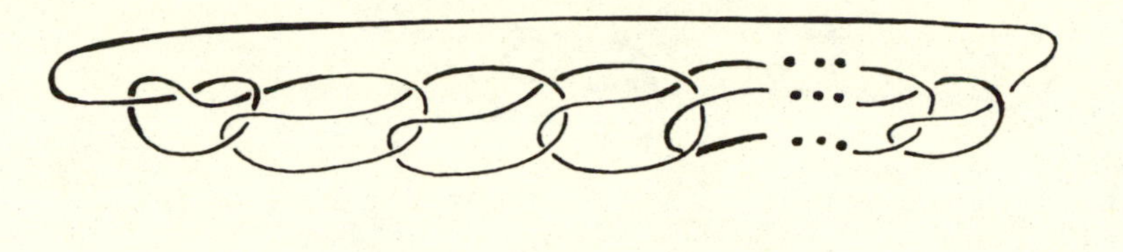

X_∞

with infinitely many tieings in the region indicated by the three dots. But this beast doesn't live in Euclidean space!

I suggest the following way out. Think of X_∞ as represented by a picture similar to that above; define X_∞ as $X_\infty = [X_0, X_1, X_2, X_3, \cdots]$ in a category of infinite sequences of ordinary knots where $[a_0, a_1, \cdots] = [a]$ denotes a stable equivalence class of infinite n-tuples. That is, $[a] = [b]$ if $a_{n+k} = b_{n+\ell}$ for all $n = 0, 1, 2, \cdots$ and some k, ℓ. Then if we set $R[a_0, a_1, \cdots] = [a_0, Ra_0, Ra_1, \cdots]$ then $RX_\infty = X_\infty$.

We can use the same stable classes for the polynomials. Thus

$$\nabla_{X_\infty} = [\nabla_{X_0}, \nabla_{X_1}, \nabla_{X_2}, \cdots]$$

$$= [1+z^2, 1-z^2-z^4, \cdots]$$

and this can be taken as a representative of the formal

power series $1+z^2-z^4+\cdots$.

Other infinite knots have ∇'s that only exist as stable sequences. Thus if

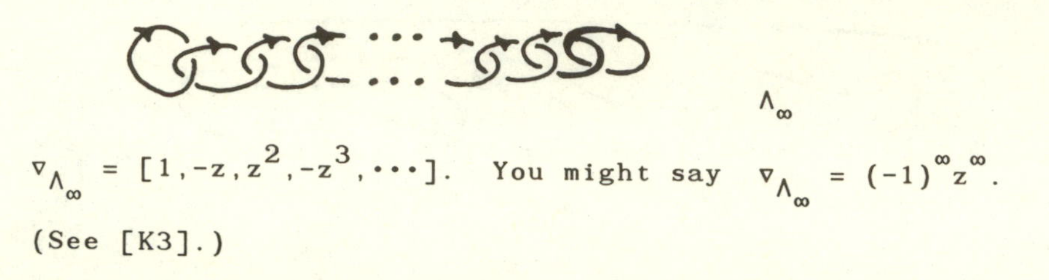

then

$\nabla_{\Lambda_\infty} = [1,-z,z^2,-z^3,\cdots]$. You might say $\nabla_{\Lambda_\infty} = (-1)^\infty z^\infty$. (See [K3].)

Exercise 4.5. Choose your own recursion and analyze the weaves and knots so obtained.

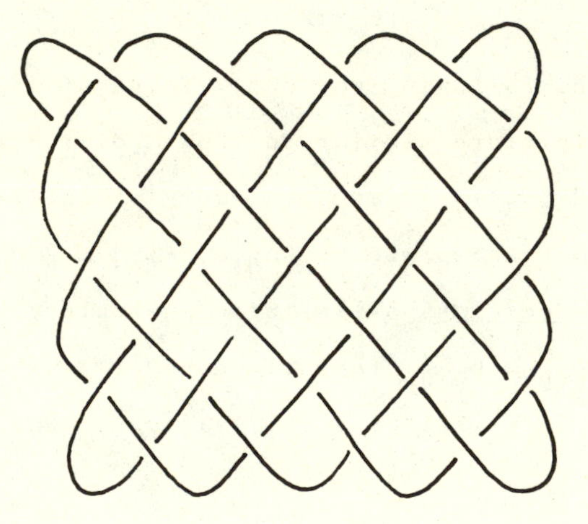

Remark. We have defined R on sequences by $R(a_0,a_1,\cdots) = (a_0,Ra_0,Ra_1,Ra_2,\cdots)$. This has the nice property that the limit of $R^n(a)$ makes sense as the sequence $\lim_{n\to\infty} R^n(a) = (a_0,Ra_0,R^2a_0,R^3a_0,\cdots)$ and this sequence is invariant under R! Then taking stable equivalence classes [a] has the effect that

$(a_0, Ra_0, R^2a_0, \cdots) \sim (Ra_0, R^2a_0, \cdots) \sim (R^2a_0, R^3a_0, \cdots) \sim \cdots$

$\sim (R^k a_0, R^{k+1}a_0, \cdots), \cdots$. In other words $[a_0, Ra_0, R^2a_0, \cdots]$ is a formal way to say "$R^k a_0$ for k indefinitely large." Thus $R^\infty a_0$ is a good <u>name</u> for the sequence

$$[a_0, Ra_0, R^2a_0, \cdots].$$

Remark. We could have stayed with the wild knot K_∞. Let $\overline{K}_\infty$ denote K_∞ with crossing labelled * switched. The question is, what is $\nabla_{\overline{K}\infty}$?

Research Problem. Generalize the Conway polynomial into an invariant for wild knots (say using equivalences that use only finitely many Reidemeister moves, or infinitely many moves with some control).

Digression 4.6. While we're talking wild knots it is worth recording a proof that <u>you can't cancel knots under connected sum</u>. That is, we wish to prove

THEOREM 4.6. *Let* K *and* K′ *be two knots. If* K # K′ *is unknotted, then* K *and* K′ *are each unknotted. That is,* $K \# K' \sim 0 \Rightarrow K \sim 0$ *and* $K' \sim 0$.

The proof will take the form of a detour:

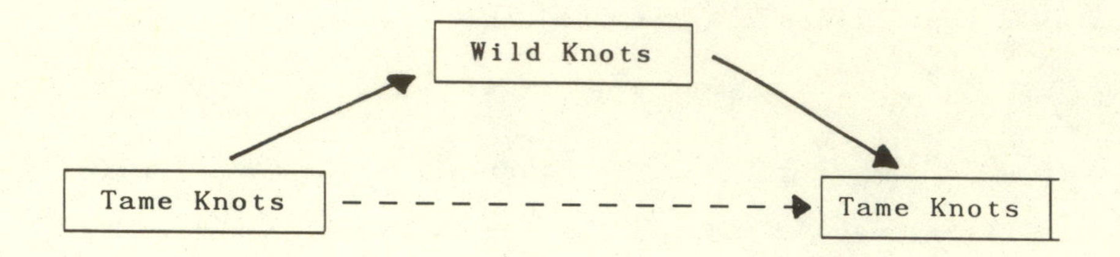

So we have to say something about the relationship of these categories.

A (possibly wild) knot is represented by any homeomorphism $\alpha : S^1 \longrightarrow S^3$ of the circle into S^3. Given two knots $\alpha, \beta : S^1 \longrightarrow S^3$ we say that $\alpha \cong \beta$ if there is an orientation preserving homeomorphism $h : S^3 \longrightarrow S^3$ such that $\beta = h \circ \alpha$.

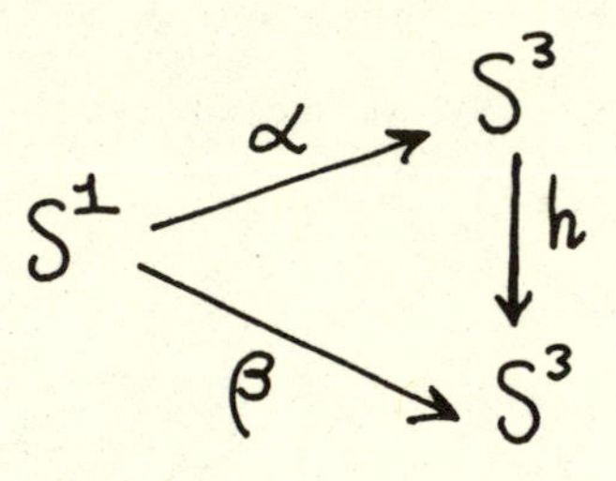

In other words, $h(\alpha(S^1)) = \beta(S^1)$. The homeomorphism h throws one knot to the other and transforms the surrounding space as well.

One says that a knot is <u>tame</u> if it is $\cong$ to a knot without any wild points. Thus it must be equivalent to a knot that, up to homeomorphism, looks locally like the standard (ball, arc) pair.

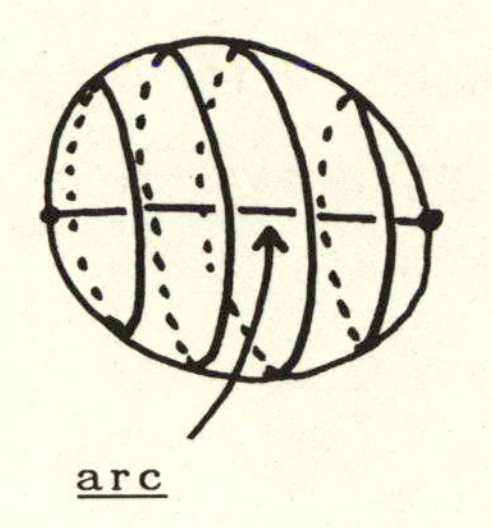

ball

Standard Pair

Fact: Let $K, K' \subset S^3$ be a tame knot represented by diagrams K and K'. Then $K \cong K'$ if and only if $K \sim K'$. That is, equivalence via Reidemeister moves is the same as the topological equivalence $\cong$ defined above. See [R1] for a detailed discussion of this point.

This fact allows us to venture out into the wild territory and return safely!

Proof of the Theorem. Suppose that $K \# K' \sim 0$. Form the wild knot $\mathcal{K}$ indicated below:

Then we may write $\mathcal{K} = K \# K' \# K \# K' \# \cdots$. Hence

$$\mathcal{K} = (K\#K') \# (K\#K') \# \cdots$$

$\therefore \quad \mathcal{K} \cong$ ○ # ○ # ○ # $\cdots$

$\therefore \quad \mathcal{K} \cong$ ○ .

On the other hand, $\mathcal{K} = K \# (K'\#K) \# (K'\#K) \# \cdots$

$$\therefore \quad \mathcal{K} \cong K \# \bigcirc \# \bigcirc \# \cdots \cong \mathcal{K}$$

$$\therefore \quad \mathcal{K} \cong K.$$

Hence, by <u>Fact</u>, $\bigcirc \sim K$. This completes the proof. ∎

Remark. This same idea can be used to show that if M^n and N^n are compact closed n-dimensional manifolds, then $M^n \# N^n \cong S^n \Rightarrow M^n \cong S^n$ and $N^n \cong S^n$ where $\cong$ denotes homeomorphism, and $\#$ is connected sum for manifolds (remove n-balls from each and paste together along the resulting boundaries).

Furthermore, the idea of going off into some place where "illegitimate" limits live works in some other realms as well. Let R be a <u>noncommutative</u> ring with unit 1. Let a,b be elements of R such that $1-ba$ is invertible.

Problem: Find a formula for $(1-ab)^{-1}$ thereby showing that it too is invertible.

Solution:

$$\begin{aligned} (1-ab)^{-1} &= \frac{1}{1-ab} \\ &= 1 + ab + abab + ababab + \cdots \\ &= 1 + a(1+ba+baba+\cdots)b \\ \therefore \quad (1-ab)^{-1} &= 1 + a(1-ba)^{-1}b. \end{aligned}$$

<u>Exercise</u>. Show that this formula really works.

Example 4.7.

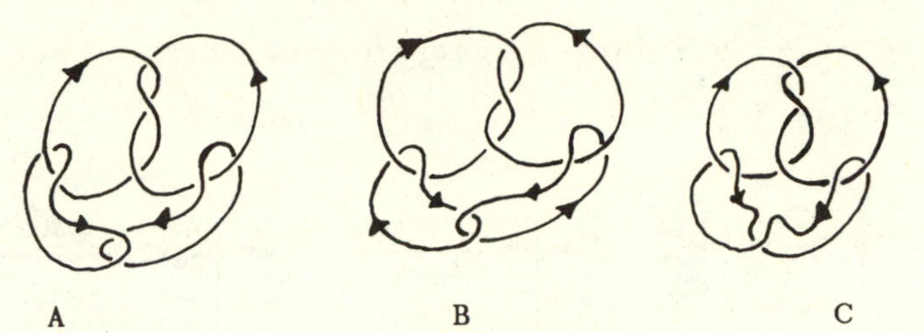

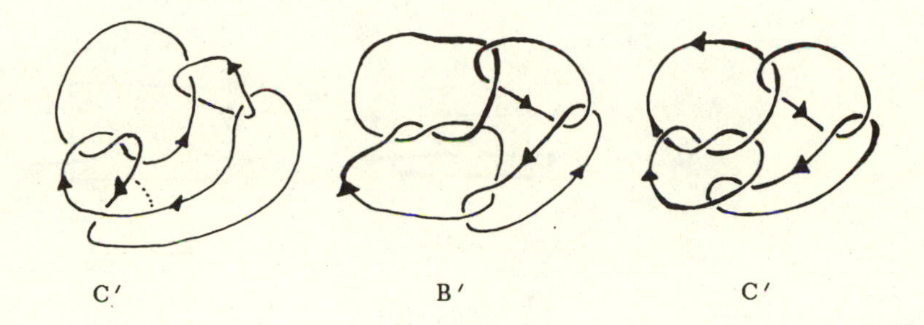

Thus A and A′ are skein-equivalent.

Example 4.8: Tangle Theory. There are many more tricks available for polynomial calculation. We refer the reader to [G1], [C1], [K1], [K2] and [LM] for more information.

It is worthwhile to explain here the elements of tangle theory. A tangle is a knot diagram with "inputs" and "outputs" in this form:

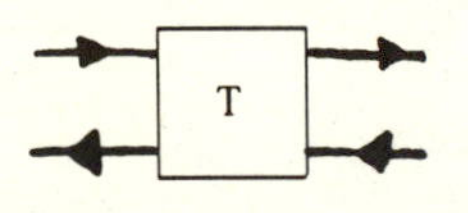

Thus

is a simple tangle. Assuming that the tangle is actually connected in with a larger knot diagram, you can decompose it skein-wise.

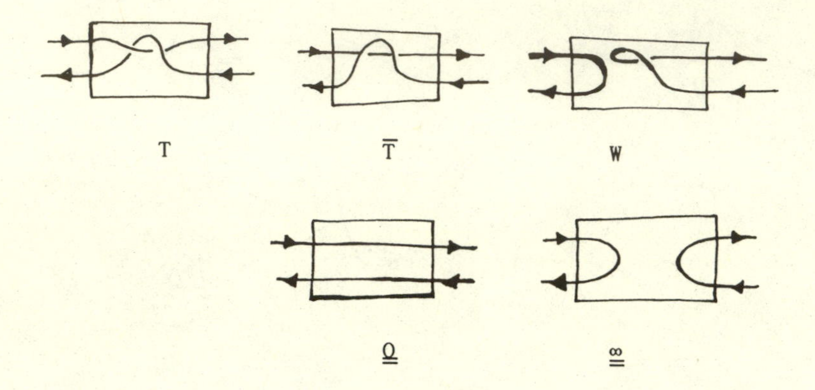

Here we have that $T = \underline{\underline{0}} \oplus \underline{\underline{\infty}}$ where we have called the two tangles above $\underline{\underline{0}}$ and $\underline{\underline{\infty}}$ for reasons about to be revealed.

Clearly, the skein of tangles is generated by $\underline{\underline{0}}, \underline{\underline{0}}_1, \underline{\underline{0}}_2, \cdots$ and by $\underline{\underline{\infty}}, \underline{\underline{\infty}}_1, \underline{\underline{\infty}}_2, \cdots$ where the subscript indicates the presence of many (n for 0_n) unknotted, unlinked circles in the tangle box.

Now we define addition of tangles in the obvious way:

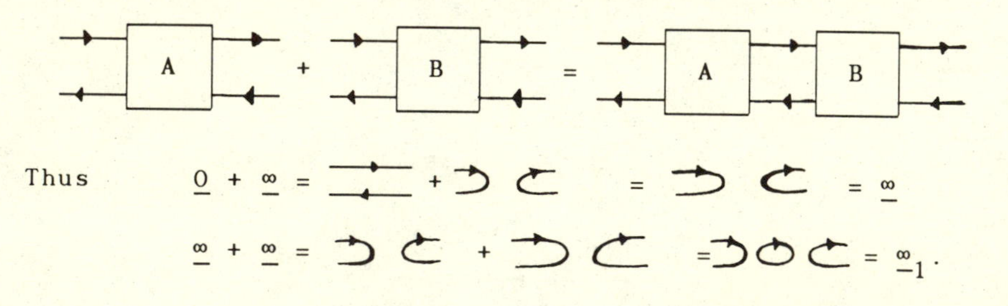

Thus $\quad \underline{0} + \underline{\infty} =$ + $=$ $= \underline{\infty}$

$\underline{\infty} + \underline{\infty} =$ + $=$ $= \underline{\infty}_1 .$

Finally, we define two operations that associate knots and links to tangles:

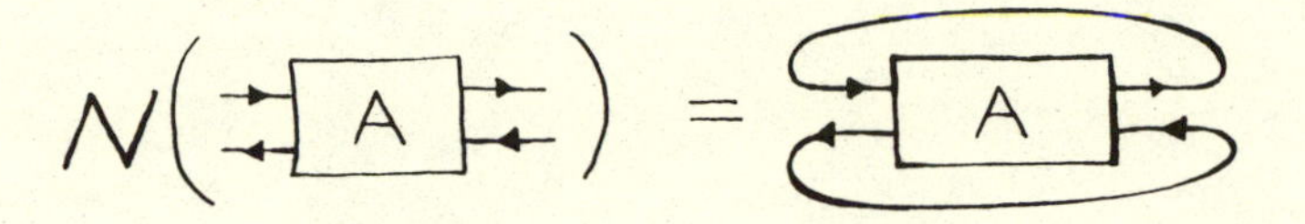

This is denoted $N(A)$, and called the numerator of A.

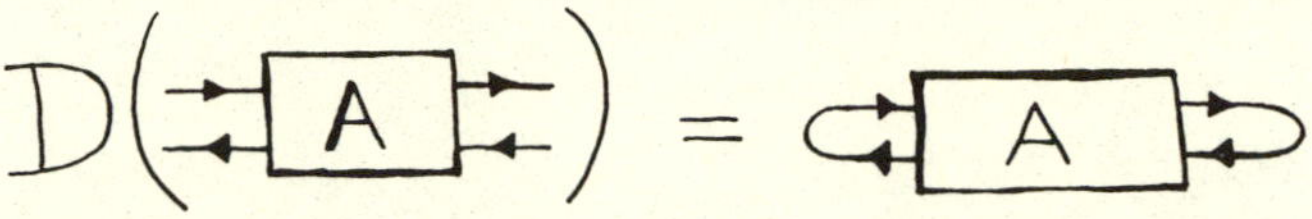

This is denoted $D(A)$, and called the denominator of A.

We also define a quotient of polynomials the fraction of A, $F(A)$, by the

$$F(A) = \nabla_{NA} / \nabla_{DA}.$$

Thus, the numerator and denominator of the fraction of the tangle are the Conway polynomials of the numerator and denominator of A.

THEOREM 4.9 (Conway). *The fraction of a sum is the (formal) sum of the fractions: Explicitly,*

$$\nabla_{N(A+B)} = \nabla_{NA}\nabla_{DB} + \nabla_{DA}\nabla_{NB}$$

$$\nabla_{D(A+B)} = \nabla_{DA}\nabla_{DB}.$$

Note: $\frac{x}{y} + \frac{z}{w} = \frac{xw+yz}{yw}$ is the formal sum of fractions. (If you like, the fraction $\frac{x}{y}$ is an ordered pair $[x,y]$.)

Example: $\underline{0} + \underline{\infty} = \underline{\infty}$

$$F(\underline{0}) = \frac{\nabla}{\nabla} = \frac{0}{1}$$

$$F(\underline{\infty}) = \frac{\nabla}{\nabla} = \frac{1}{0}$$

$$\frac{0}{1} + \frac{1}{0} = \frac{0 \cdot 0 + 1 \cdot 1}{1 \cdot 0} = \frac{1}{0}$$

Example: $\underline{\infty} + \underline{\infty} = \underline{\infty}_{-1}$

$$F(\underline{\infty}) = \frac{1}{0}, \quad F(\underline{\infty}_{-1}) = \frac{0}{0}$$

$$\frac{1}{0} + \frac{1}{0} = \frac{1 \cdot 0 + 0 \cdot 1}{0 \cdot 0} = \frac{0}{0}.$$

A little thought shows that it is sufficient to check Conway's theorem for the skein generators as we have done! Thus, its proof is easy. But it is a powerful tool for calculation.

Note: T = should be called [-3] since $\nabla_{NT} = -3z$, $\nabla_{DT} = 1$.

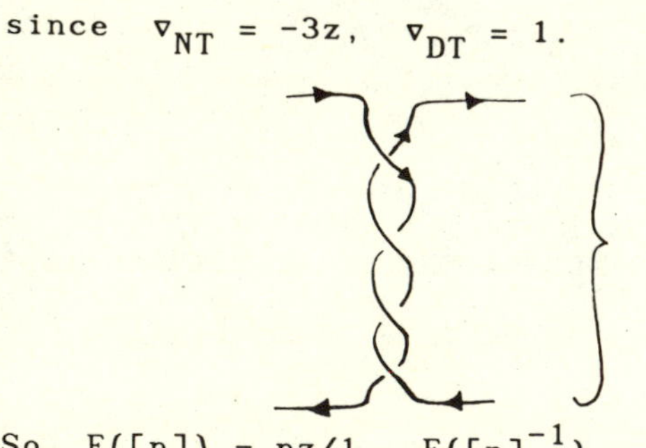

Call this $[2]^{-1}$.

So $F([n]) = nz/1$, $F([n]^{-1}) = 1/nz$.

But now

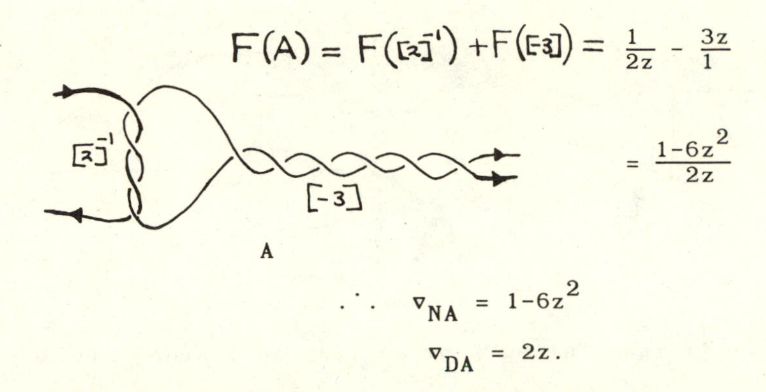

$$F(A) = F([2]^{-1}) + F([-3]) = \frac{1}{2z} - \frac{3z}{1} = \frac{1-6z^2}{2z}$$

$$\therefore \quad \nabla_{NA} = 1-6z^2$$
$$\nabla_{DA} = 2z.$$

Thus we conclude that if $K = NA$

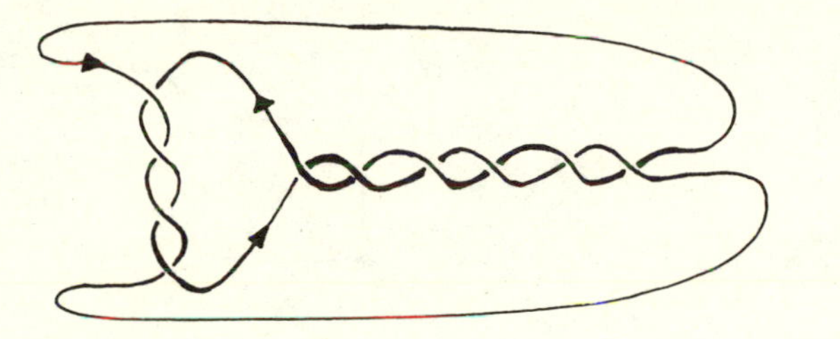

then $\nabla_K = 1-6z^2$.

In general, the fraction theorem (4.9) allows us to quickly calculate polynomials of the numerators and denominators of <u>rational tangles</u> where a rational tangle is a tangle obtained by addition and <u>inversion</u> from the integral tangles $[n]$ $(n \in Z)$. We have already seen one example of inversion, namely, $[n]^{-1}$.

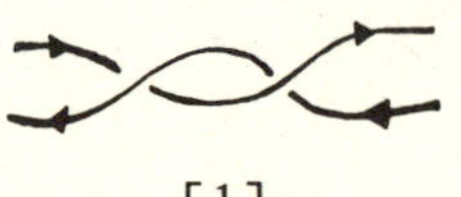
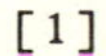

$[1]$

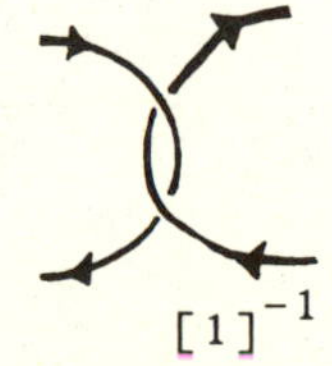

$[1]^{-1}$

In general,

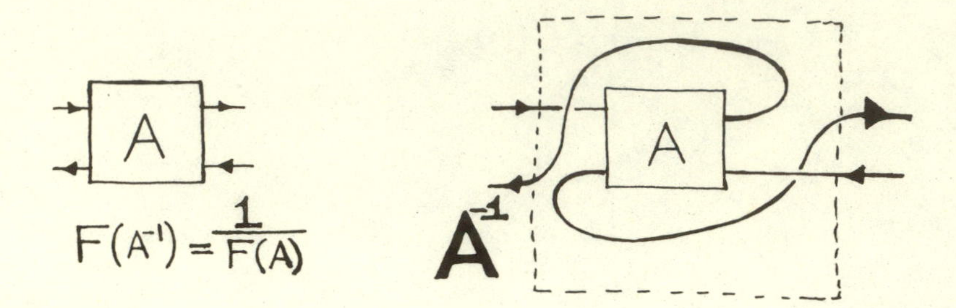

Note that A^{-1} is obtained from A (up to isotopy fixing the inputs and outputs) by rotating A by 180° around the upper-left/lower-right diagonal.

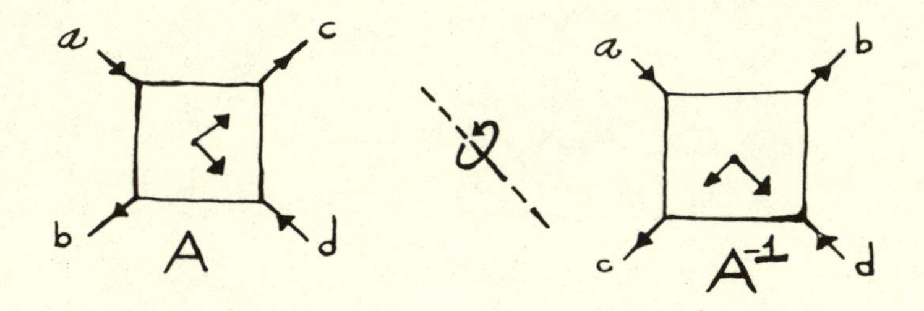

Now we have that $N(A^{-1}) = D(A)$ and $D(A^{-1}) = N(A)$. Hence $F(A^{-1}) = 1/F(A)$.

For example we can form continued fractions.

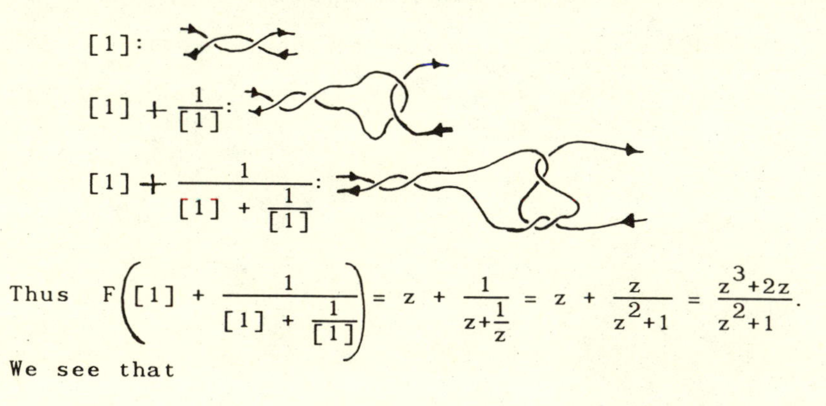

Thus $F\left([1] + \dfrac{1}{[1] + \dfrac{1}{[1]}}\right) = z + \dfrac{1}{z+\frac{1}{z}} = z + \dfrac{z}{z^2+1} = \dfrac{z^3+2z}{z^2+1}.$

We see that

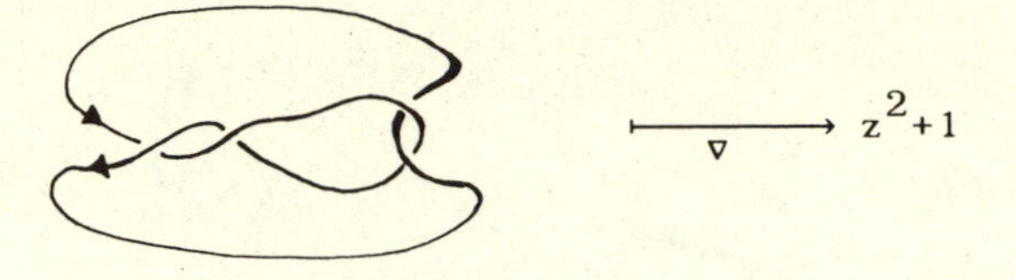

since this is the numerator of $F([1]+1/[1])$. And

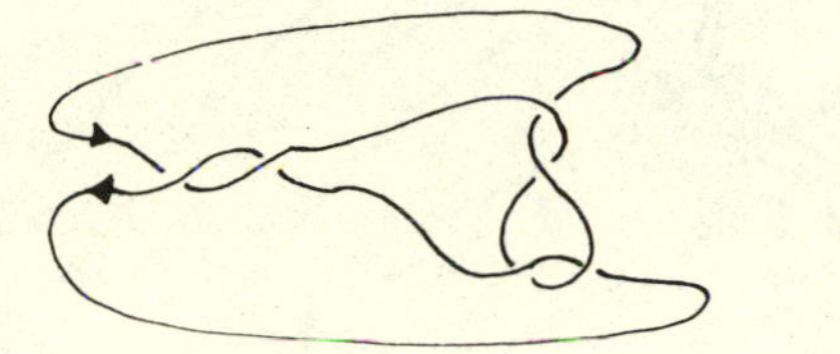
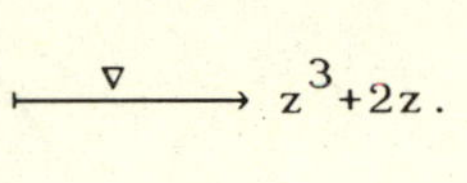

<u>Exercise</u>. Let $C_n = N\left([1] + \cfrac{1}{[1] + \cfrac{1}{\cdots + \cfrac{1}{[1]}}}\right)$ with n appearances of [1]. Show that $C_n \sim K_{n+1}$ where the knots and links K_n are defined in Example 4.1.

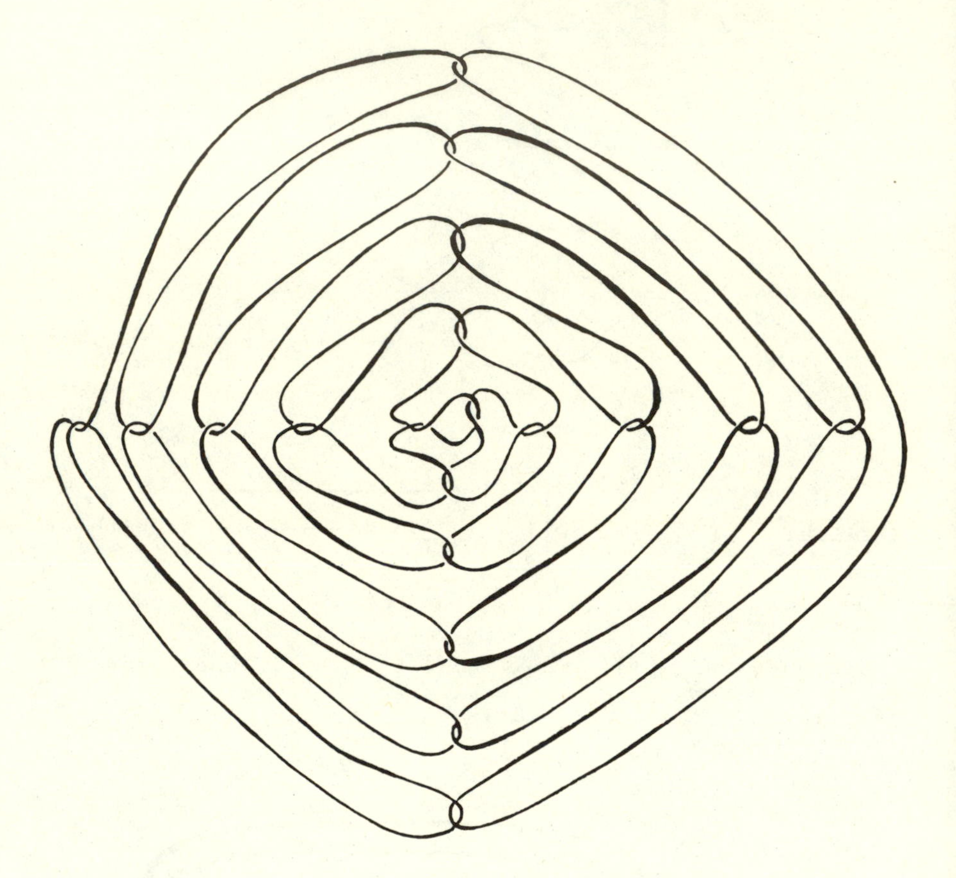

C_{19}

Just as the snail builds his shell, so do these continued fraction knots spiral outwards,

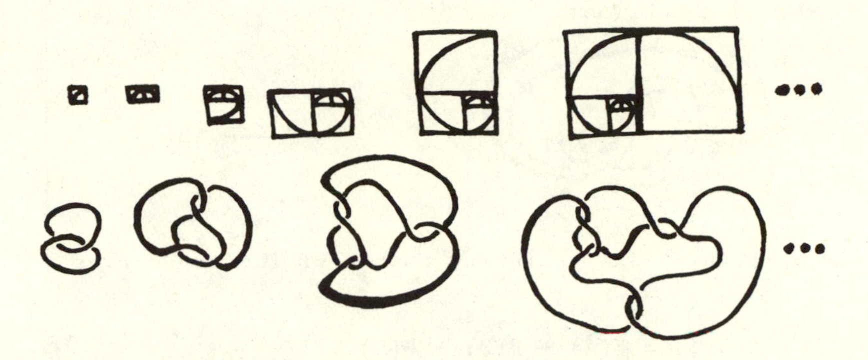

Rational knots and tangles are important in studying lens spaces ([ST], [S1]) as branched covering spaces.

One more example. A knot K whose polynomial is equal to 1. K is actually knotted, but we can't prove it yet (unless you've done Exercise 3.15!)

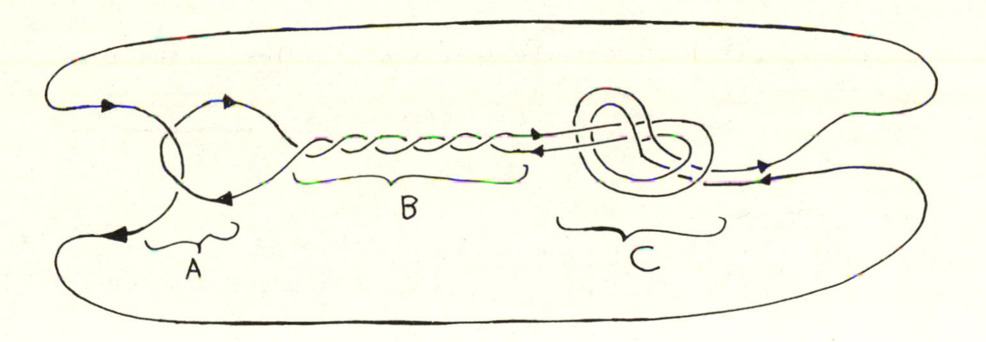

K

$$K = N(A+B+C)$$

$$F(A) = 1/z, \quad F(B) = 3z/1, \quad F(C) = -3z/1$$

You should check that

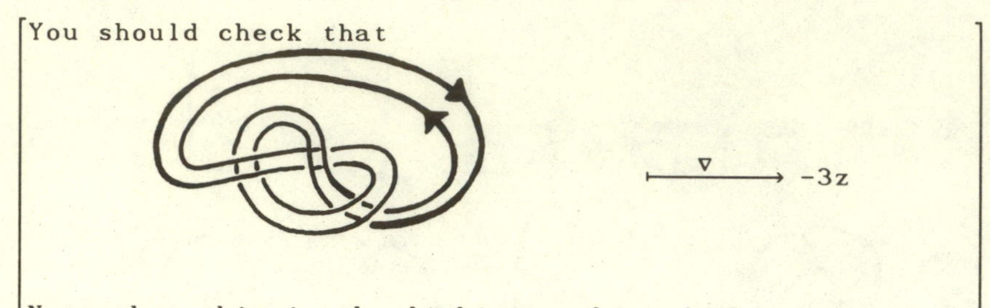

Note that this is the linking number of the two curves.

Hence $F(A+B+C) = F(A) + F(B) + F(C) = 1/z$,

$\therefore \quad \nabla_{N(A+B+C)} = 1$,

$\therefore \quad \underline{\nabla_K = 1}$.

Remark: The numerator of C is called a <u>cable</u> of the trefoil knot. K itself is an <u>untwisted</u> <u>double</u> of the trefoil.

<u>Exercise</u>. (a) Work out the theory of tangles of the form

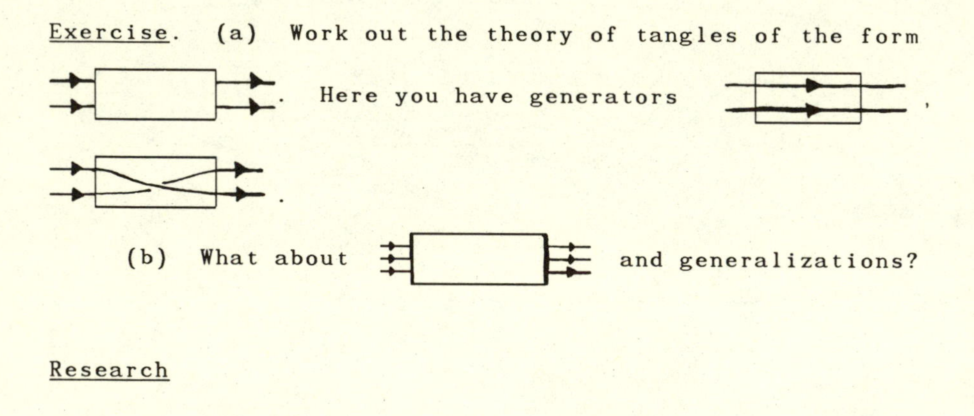

. Here you have generators , .

(b) What about and generalizations?

<u>Research</u>

<u>Exercise</u>. Take a knot

. Form the cable with reverse orientation.

K

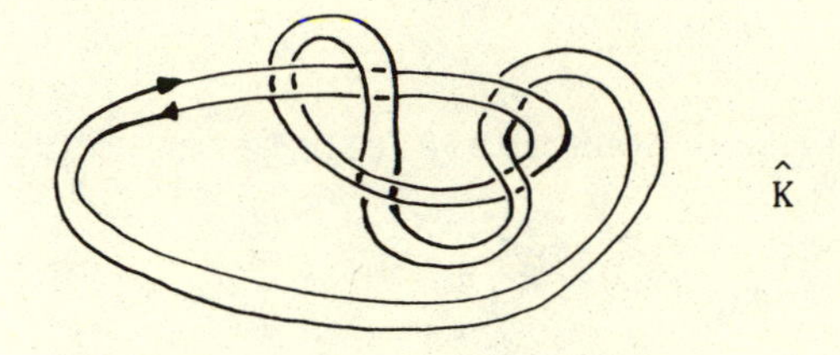

Prove that $\nabla_{\hat{K}} = \mathrm{lk}(\hat{K})z$. [In this problem, we shall see an easy proof later. There <u>should</u> be a good skein-theoretic proof.]

V

DETECTING SLICES AND RIBBONS—A FIRST PASS

Here is a ribbon knot:

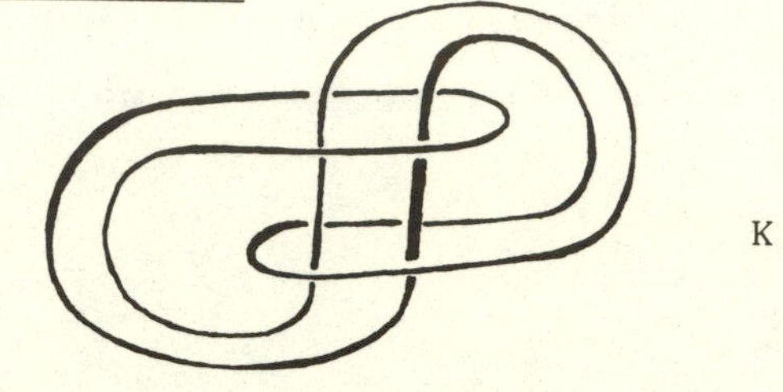

It is called a ribbon knot because it forms the boundary of a "ribbon" that is immersed into three-dimensional space with ribbon singularities:

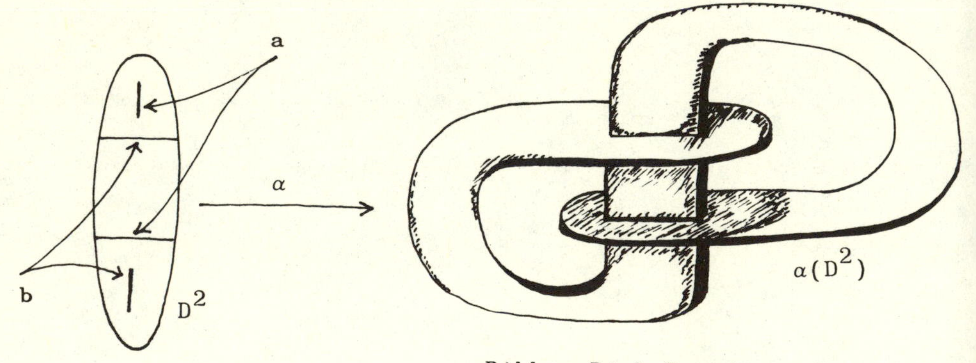

Ribbon Disk Bounding K

The ribbon, or ribbon disk, is the image $\alpha(D^2)$ of a mapping $\alpha : D^2 \longrightarrow \mathbb{R}^3 \subset S^3$ whose only singularities are of the form illustrated above. Thus each component of the image singular set consists in a pair of closed intervals in D^2: one with end points on the boundary of D^2, one entirely interior to D^2.

Exercise. Show that every knot $K \subset S^3$ bounds a disk with *clasp* singularities.

A clasp singularity.

But not every knot is a ribbon knot, as we shall see. The first nonribbon is the trefoil . But how do we know that no projection of the trefoil will have ribbon form? Read on.

Ribbon knots are examples of *slice knots*. A knot is said to be (smoothly) slice if it *bounds a differentiable disk in the 4-ball* D^4. More specifically this means that it is possible to arrange this disk $D^2 \subset D^4$ so that concentric 3-spheres move through it (intersect it) to produce either

a) an ordinary nonsingular knot or link, or

b) a knot or link with singularities corresponding to one of

   i) simple maximum or minimum,

   ii) saddle point.

See the figure below for illustrations of these singularities.

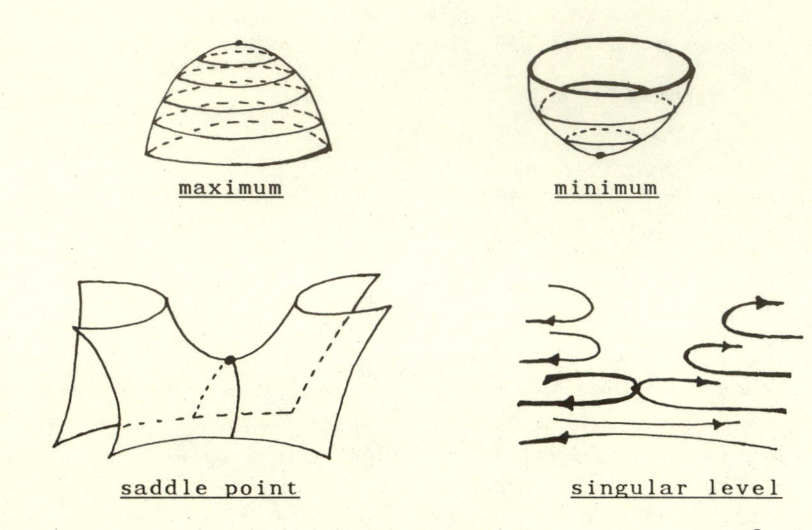

Thus the sequence of intersections of concentric S^3's with $D^2 \subset D^4$ will provide a "movie" of how the slice knot K bounds the embedded disk D^2. Here is an example.

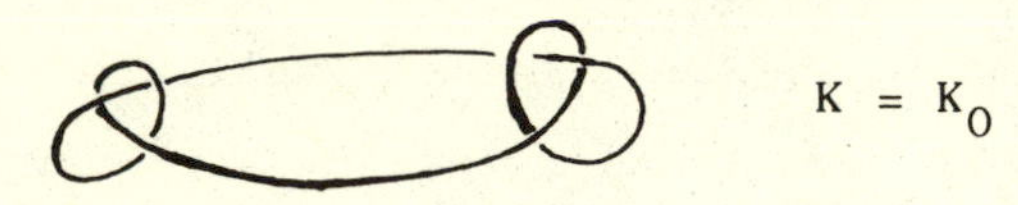

$K = K_0$

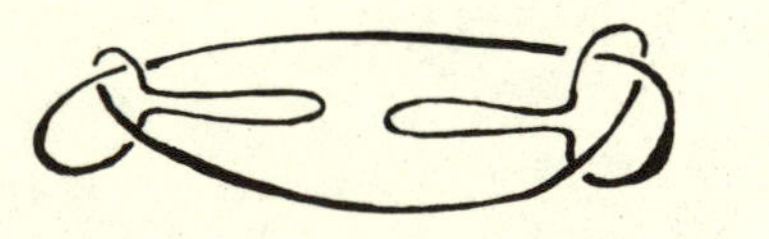

K_1

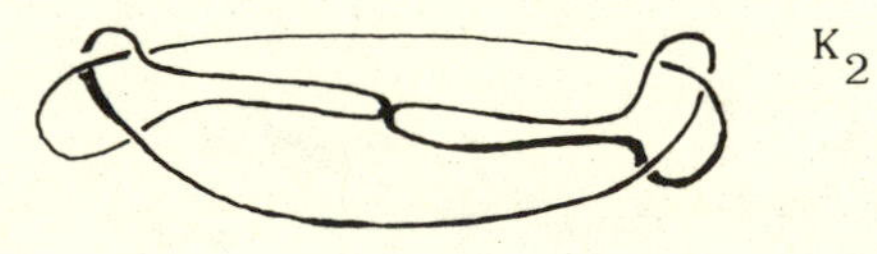

K_2

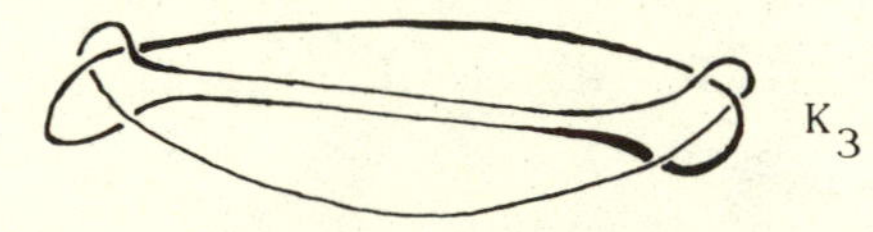

K_3

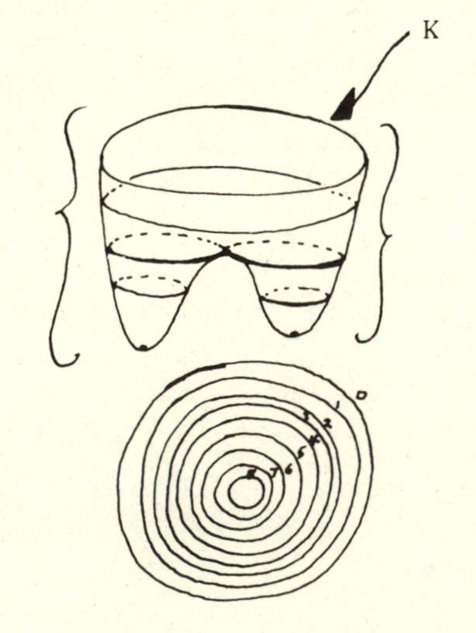

D^2

D^4

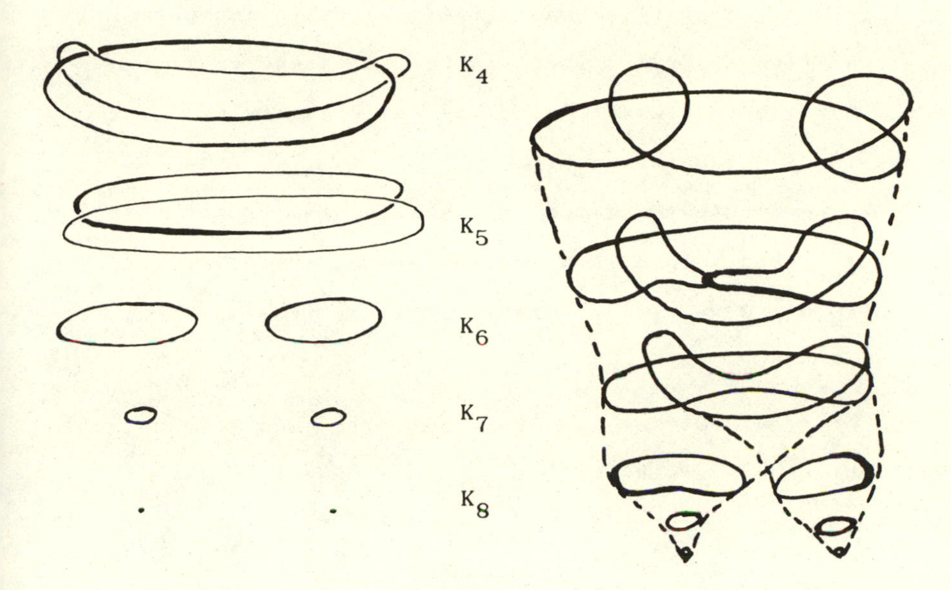

K_r indicates the intersection of the rth concentric S^3 with $D^2 \cup D^4$. We see that K undergoes one saddle-point singularity and two minima to produce the slice disk.

Fact. The existence of such a movie for a slice knot follows from Morse Theory (see [F1]). Ribbons are slice because you can push the singularities (ribbon singularities) into the 4-ball, thereby obtaining an embedded $D^2 \subset D^4$ spanning the knot. Can you find a max-min-saddle movie (henceforth called just movie) for our example of a ribbon?

It has long been conjectured that ribbon and differentiably slice are equivalent. (See [F1].)

The notion of slice knot first occurs in a paper of Fox and Milnor ([F2]). There is now an extensive literature, but the problem of detecting and classifying slice knots in S^3 remains unsolved.

We now present a method for showing that some knots are not ribbon. This is a version of the Arf invariant (see [K5]) and we will return to it later from another point of view.

The idea: Devise a "move" that will undo ribbons, but will not undo every knot.

First guess: Add 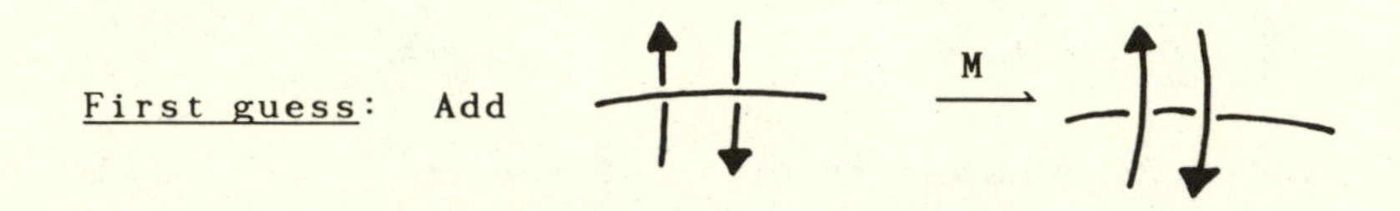

to the list of elementary moves.

Exercise: Sorry! The first guess doesn't work. You can untie any knot if you use M and the Reidemeister moves.

Solution to the exercise:

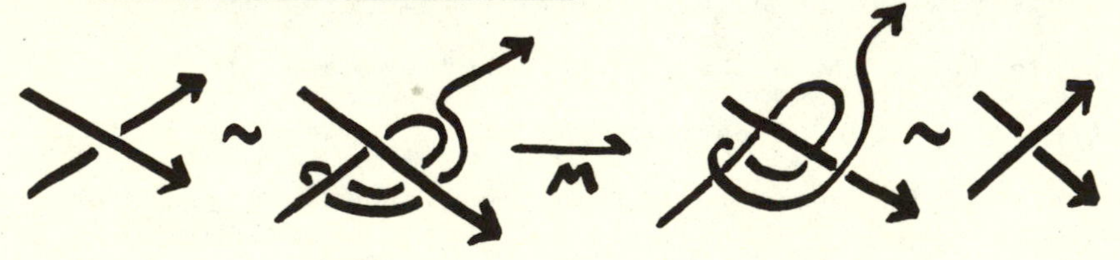

(The first isotopy is called a "finger move.") The way that the first guess goes wrong then suggests that we modify M to move Γ:

DEFINITION 5.1. *Two knots are Γ-equivalent if there exists a sequence of Reidemeister moves, combined with Γ-moves taking one to the other. If* K *and* K′ *are Γ-equivalent, we will write* $K \underset{\Gamma}{\sim} K'$.

PROPOSITION 5.2. *If* K *is ribbon, then* K *is Γ-equivalent to the unknot.*

Proof. Remove ribbon singularities with Γ-moves:

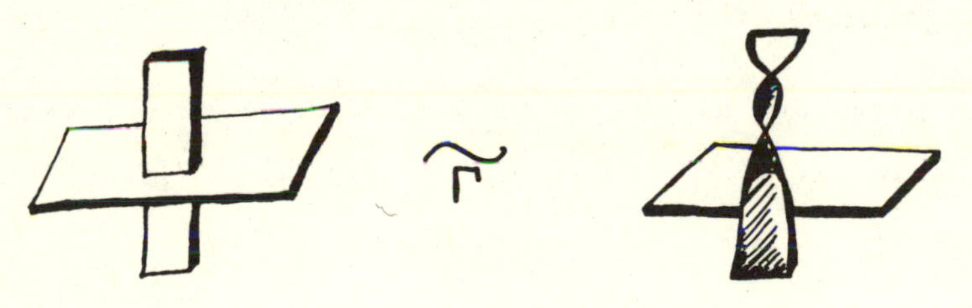

Eventually, you arrive at an embedded disk. ■

THEOREM 5.3. *Any knot* K *is Γ-equivalent to either the unknot or the trefoil. The trefoil and the unknot are not Γ-equivalent.*

This is our fundamental result. We prove it through a reformulation:

DEFINITION 5.4. *Two knots are pass-equivalent if one can be obtained from the other by a combination of Reidemeister moves and* <u>*pass-moves*</u>. *The forms for a pass-move are as shown below:*

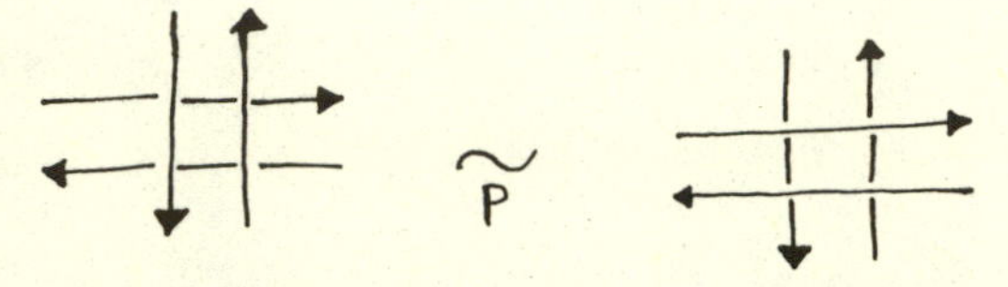

<u>pass-moves</u>

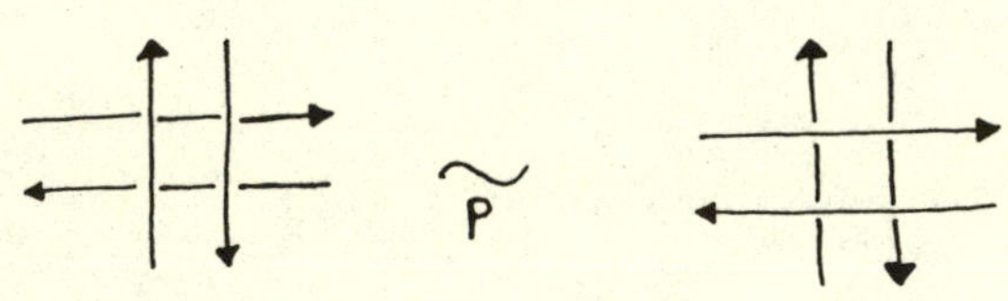

Thus a pass-move allows us to switch oppositely oriented strands in pairs. If K and K' are pass-equivalent, we shall write $K \underset{p}{\sim} K'$.

LEMMA 5.5. *Two knots* K *and* K' *are pass-equivalent if and only if they are* Γ*-equivalent.*

Proof. We first show that a Γ-move can be accomplished by pass-equivalence:

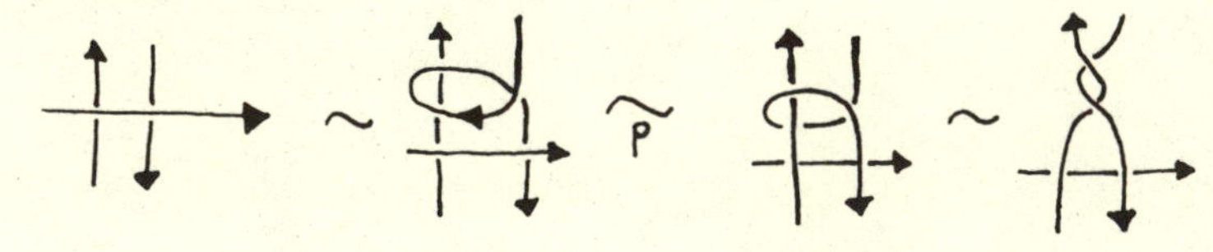

Then we show that a pass-move can be accomplished by Γ-equivalence:

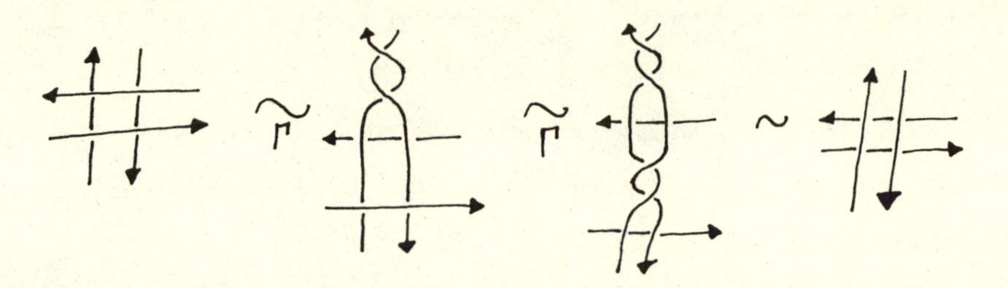

This completes the proof of the lemma.

Now let $\alpha(K)$ denote $a_2(K)$, the second coefficient of the Conway polynomial.

PROPOSITION 5.6. *Let* K *and* K′ *be two pass-equivalent knots. Then* $\alpha(K) \equiv \alpha(K')$ (modulo 2).

Proof. This is an interesting exercise in the techniques of Section 3. We leave it as an exercise and refer the reader to [K1] or [K2] for details.

COROLLARY 5.7. *The trefoil is not ribbon.*

Proof. Let T denote the trefoil. Then $\alpha(T) = 1$ while $\alpha(0) = 0$. Hence, by 5.6, T is not pass-equivalent to 0 and therefore, by 5.5, T is not Γ-equivalent to 0. Since T ribbon $\Rightarrow T \underset{\Gamma}{\sim} 0$ (5.2), we are done. ∎

It remains to show that any K is Γ-equivalent to the trefoil or to the unknot. In order to do this, we need to bring in more geometry. In particular,

PROPOSITION 5.8. *Any oriented knot or link* K *bounds a connected oriented surface* $F \subset \mathbb{R}^3$ *such that* F *has oriented boundary* K.

Proof. We produce the surface by an algorithm due to H. Seifert [S]. Accordingly, the surface will be called the Seifert surface (when it is produced by this algorithm).

SEIFERT'S ALGORITHM (illustrated).

1. Draw the planar diagram for your knot.

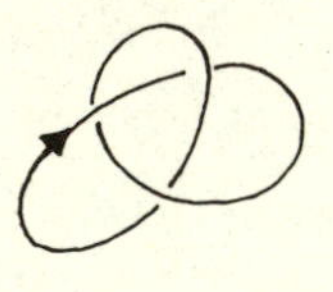

K

2. Draw the corresponding universe k.

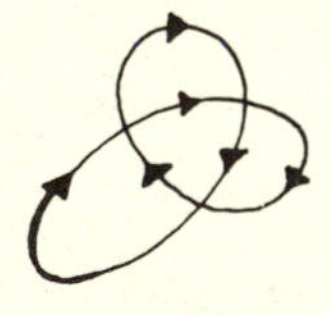

k

3. Split every crossing of k in oriented fashion:

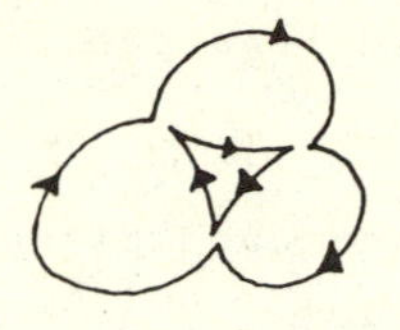

The resulting set of disjoint closed curves is called the Seifert circles for K.

4. Attach a disjoint collection of disks above the plane to the Seifert circles. One disk per circle.

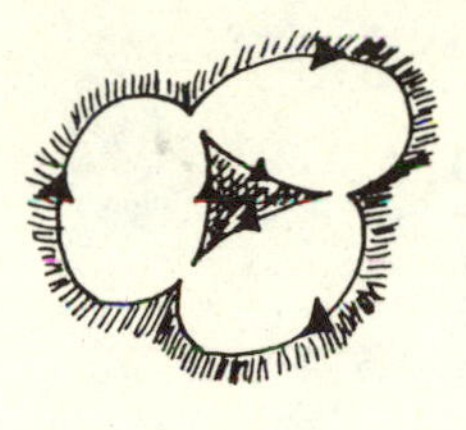

5. Between split-points  add twisted bands  according to the crossing in K. This is the surface F.

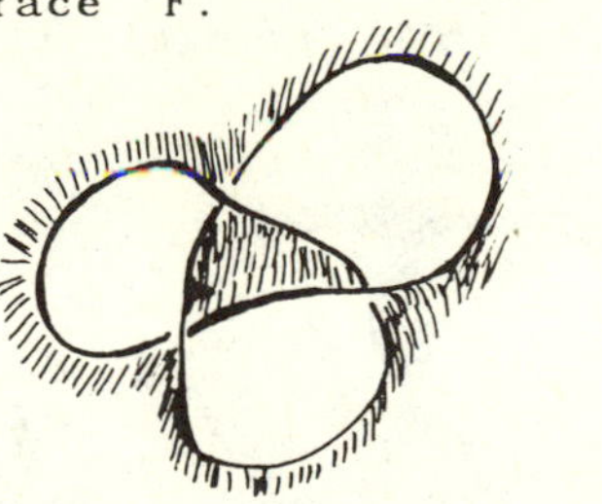

F

Orientability follows from the Jordan curve theorem. That is, if you start in a given domain in the plane and pass through a number of crossings in the Seifert surface only to return to that domain, you must pass through an even number of crossings. This geometry passes over to the surface itself.

The surface can be sketched directly on the knot diagram by traversing Seifert circles: jump at each crossing to avoid crossing it!

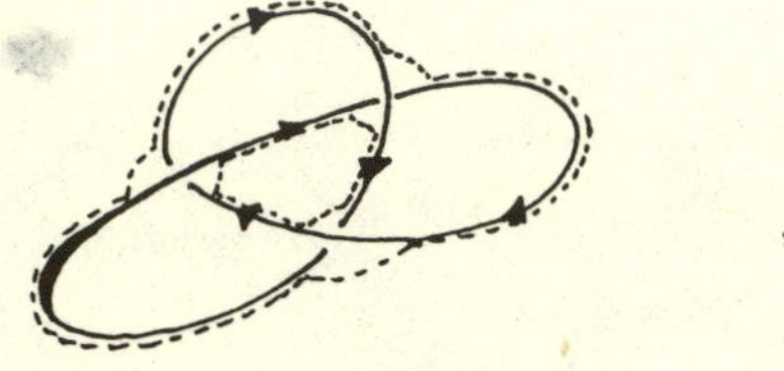

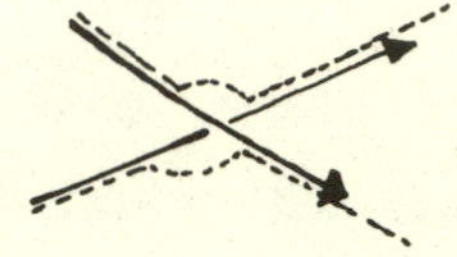

Seifert Circles

The disks may be nested as in

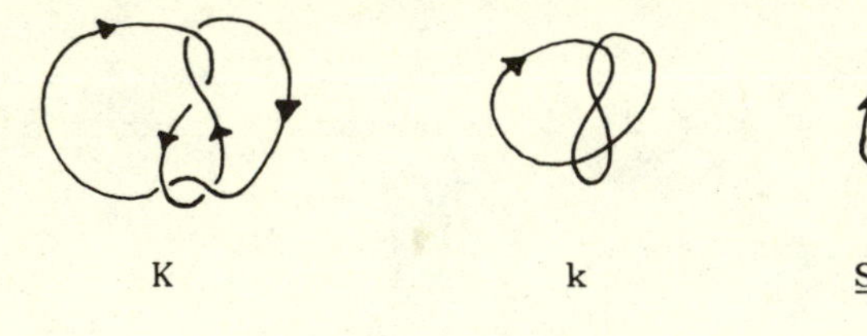

K k Seifert Circles

The disk corresponding to this "fringe" is not so easy to draw.

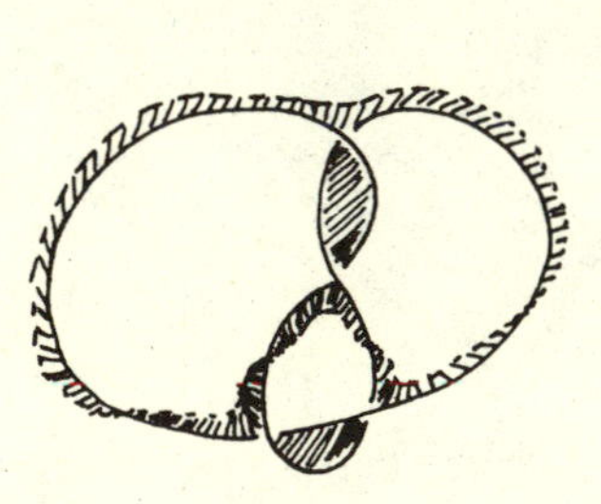

Exercise. Draw a good picture of the Seifert surface for the figure-eight knot (above).

Now we also need to recall the canonical representatives for orientable surfaces (abstract representatives):

$g = 1$, F_1

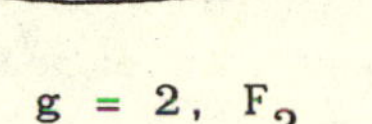

$g = 2$, F_2

For a single boundary component, $\{F_0, F_1, F_2, \cdots\}$, $F_0 = D^2$, is a complete list of homeomorphism types of orientable surfaces. (See [MA].) Each of these representatives takes the form of a disk with attached bands.

Up to ambient isotopy, an embedding of F_g looks like a standardly embedded disk with twisted, knotted and linked bands. (That is, given any embedding $F \subset \mathbb{R}^3$ then F fits into a time-parameter family of continuously changing embeddings $F(t)$ such that $F \subset \mathbb{R}^3$ is $F(0) \subset \mathbb{R}^3$ and $F(1) \subset \mathbb{R}$ meets the description above.) For example:

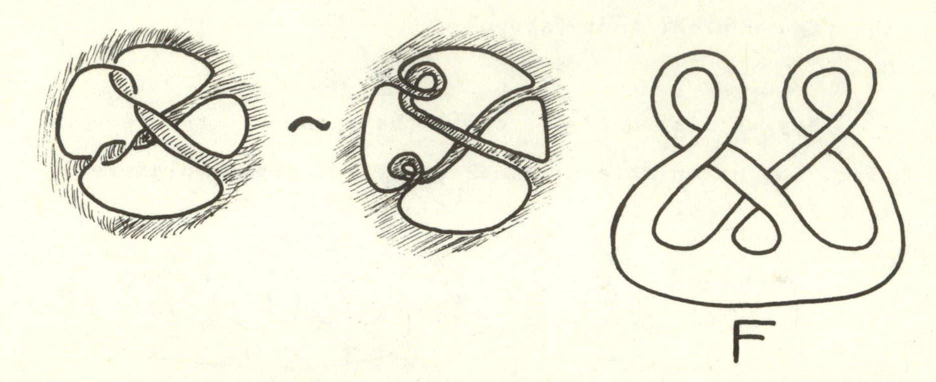

Here we have an embedding as disk-with-bands whose boundary is a trefoil knot.

Exercise. Show that the Seifert surface for the trefoil is ambient isotopic to F above.

Now let's call a disk with twisted, knotted, linked bands a standard embedding. Since there is an even number of twists per band (orientability), these can be arranged as curls.

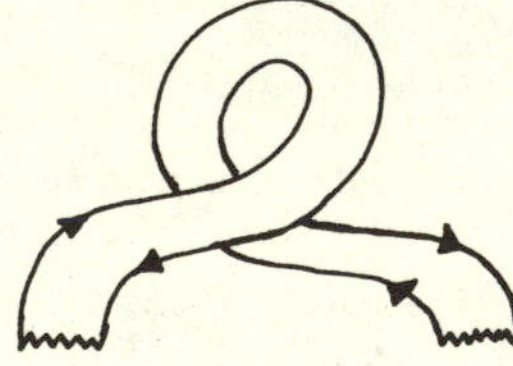

(Compare Exercise 2.7.)

On the boundary, each curl affords the opportunity for a pass-move:

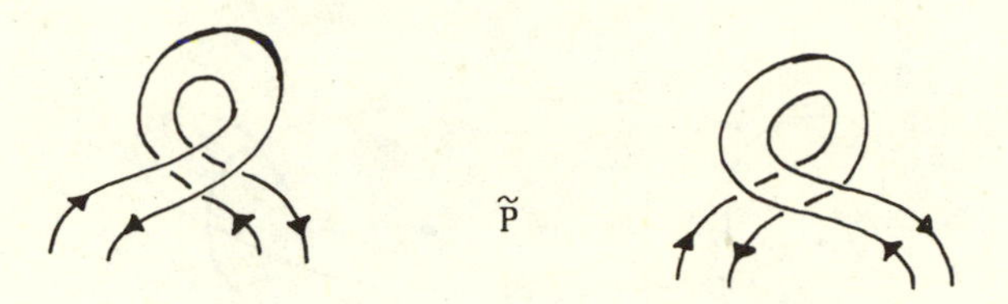

This means that we can reduce the number of curls per band modulo two by band-passing. [$\sim$]

Since it is obvious that we can get rid of knotting and linking of bands by passing, we can assume that our knot bounds a surface that is a boundary connected sum of the following culprits:

trefoil

unknot

unknot

Hence we have shown:

LEMMA 5.9. *Any knot is pass-equivalent to a sum of trefoils. Any knot is pass-equivalent to its mirror image* $(K \underset{p}{\sim} K^!)$.

Proof: To see the last part, pass all the bands! ■

The *coup de grace* is:

PROPOSITION 5.10. *For any knot* K, $K \# K^!$ *is a ribbon.*

Proof:

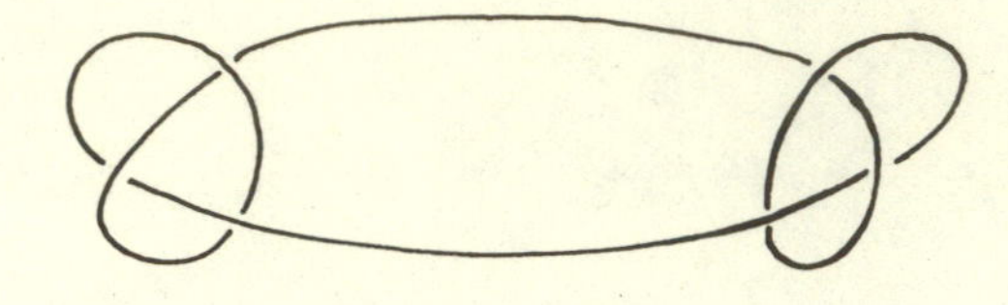

Connect corresponding points across the mirror, and the ribbon surface appears.

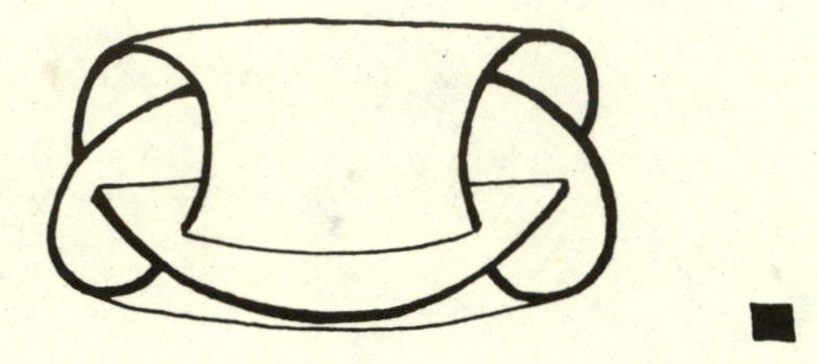

∎

Proof of Theorem 5.3: By the discussion, it suffices to show that $K \# K \underset{P}{\sim} 0$ when K is a trefoil. In fact, this is true in general since $K \underset{P}{\sim} K^!$ implies

$$K \# K \underset{P}{\sim} K \# K^! \underset{P}{\sim} 0$$

since ribbon implies $\underset{\Gamma}{\sim} 0$ implies $\underset{P}{\sim} 0$. ∎

COROLLARY 5.11. *Two knots,* K, K', *are pass-equivalent* (*Γ-equivalent*) *if and only if* $\alpha(K) \equiv \alpha(K')$ (modulo 2) *where* $\alpha(K) = a_2(K)$.

Remark: We shall later see that $\alpha(K)$ taken modulo two can be identified with what is called the Arf invariant, ARF(K), of the knot K. $\alpha(K)$ is easily calculated using the techniques of Chapter III. Thus in Example 3.8, we calculated $\alpha(K) = 1$ for K.

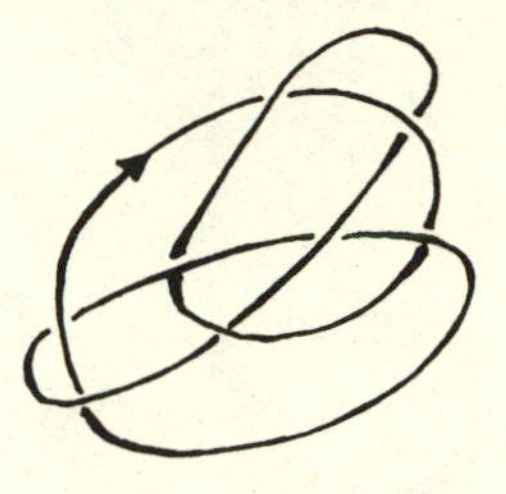

As a result, we know that this knot is not ribbon.

<u>Exercise</u>. Show that the following knot is not ribbon (this is a 3-4 torus knot).

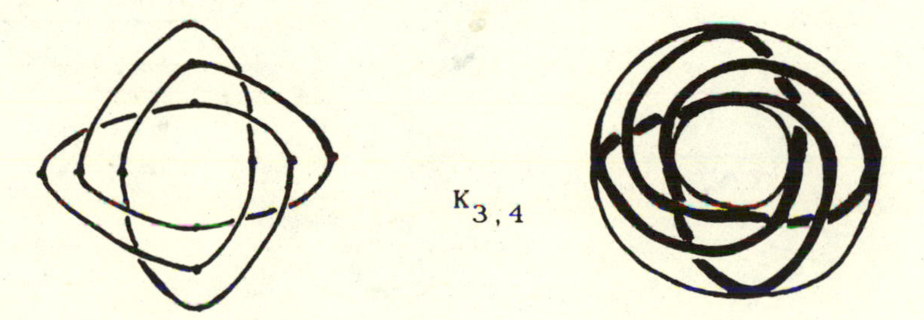

It is called a torus knot because it lies on the surface of a torus.

Remarks on Slice Knots. We have seen that $K \# K^!$ is slice (in fact, ribbon) for any knot K. There are many slice knots that are not of this form. For example, let K be the <u>stevedore's knot</u>:

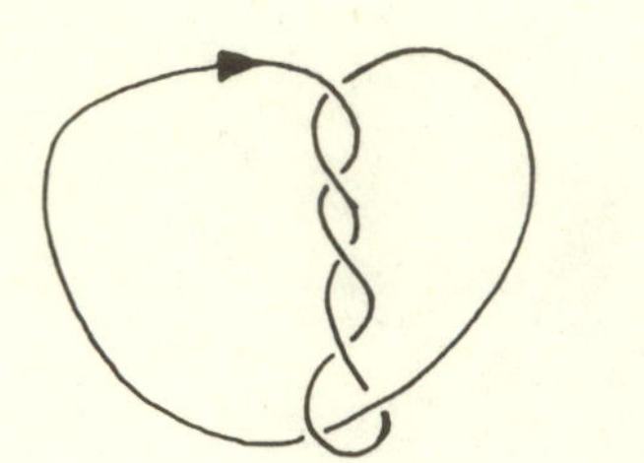

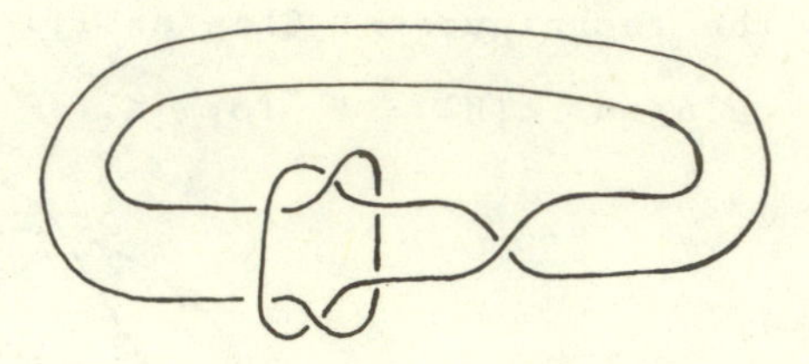

One can see a move for the slice disk as follows:

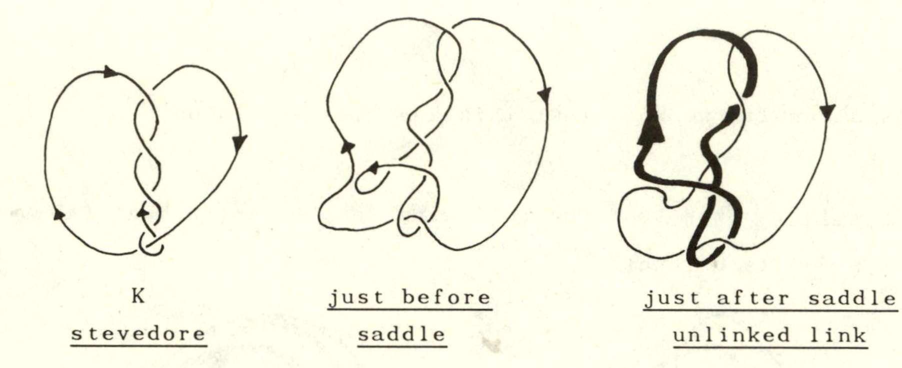

Hence K bounds a disk with singularities in the form

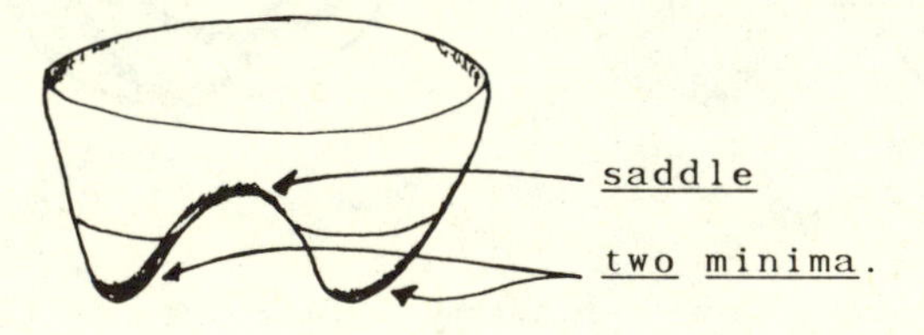

This shows you one way to create volleys of slice knots: Take two unlinked circles. Wrap them around each other a bit. Go through a saddle to form a knot. If knotted it be, then 'tis a slice knot for sure.

Example:

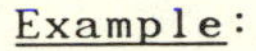

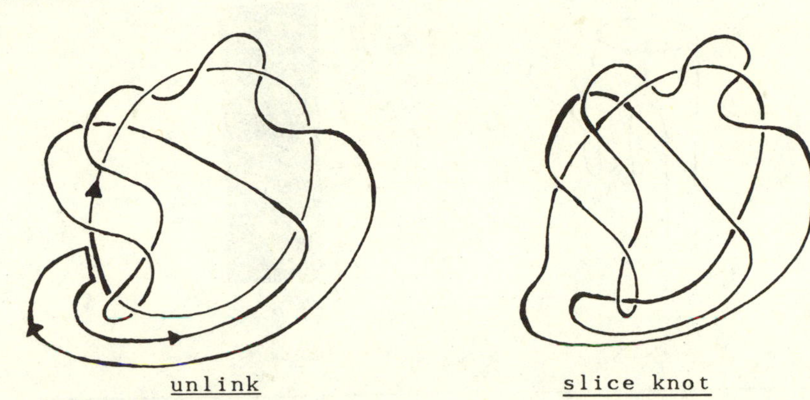

unlink slice knot

Exercise 5.12. Prove that a knot is ribbon if and only if it has a movie with saddle points and minima, but no maxima.

Exercise 5.13. Prove that the Stevedore's Knot is prime (not a connected sum of two nontrivial knots).

THE KNOT CONCORDANCE GROUP

Two knots $K, K' \subset S^3$ are said to be (smoothly) concordant ($K \underset{c}{\sim} K'$) if there exists a differentiable embedding $F : S^1 \times I \longrightarrow S^3 \times I$ ($I = \{t \in \mathbb{R} \mid 0 \leq t \leq 1\}$, the unit interval) such that $F|S^1 \times 0 : S^1 \longrightarrow S^3$ represents K, and $F|S^1 \times 1$ represents K'.

For example, any slice knot is concordant to the unknot. Just run the movie until a single unknotted component is left!

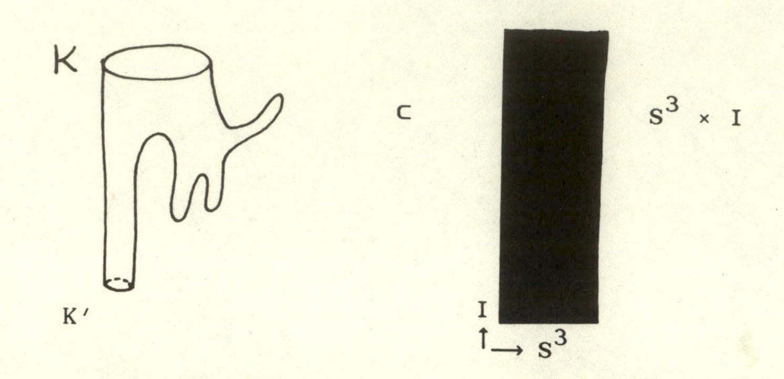

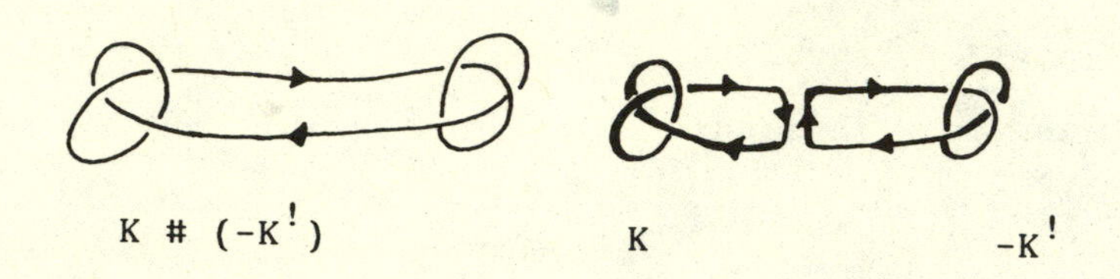

Now let's be careful about orientations. We have actually shown that $K \# (-K^!)$ is ribbon, hence slice. Here $-K^!$ denotes $K^!$ with the reversed orientation. Let $\overline{K} = -K^!$. Then we have $K \# \overline{K} \underset{c}{\sim} 0$.

The # operation is invertible up to concordance. As a result, the collection of classical knots, modulo concordance, forms a group under connected sum. This group, <u>the knot concordance group</u>, is denoted by C_3.

If we define $A(K) \equiv$ the mod-2 residue class of $a_2(K)$, so that $A(K) \in Z_2 = \{0,1\}$, then we have shown that $A : C_3 \longrightarrow Z_2$ is a well-defined homomorphism of groups, and that it is nontrivial. <u>A</u> will later be identified with the Arf invariant. For classical knots, the full structure of the knot concordance group remains a mystery.

We shall have more to say about C_3 and its generalizations.

Remark 5.14. I want to record a few comments here about how to prove Proposition 5.6: We want to determine the difference between $\alpha(K)$ and $\alpha(K')$ when K and K' locally differ as shown below:

If we choose the switching sequence S_1, S_2, S_3, S_4 corresponding to the crossings labelled 1,2,3,4 above, then $K' = S_4S_3S_2S_1K$. Using the technique of Section 3, we then have

$$\alpha(K)-\alpha(K') = \mathrm{lk}(X_1)-\mathrm{lk}(X_2)+\mathrm{lk}(X_3)-\mathrm{lk}(X_4)$$

where the links X_i, $i = 1,\cdots,4$ are as below:

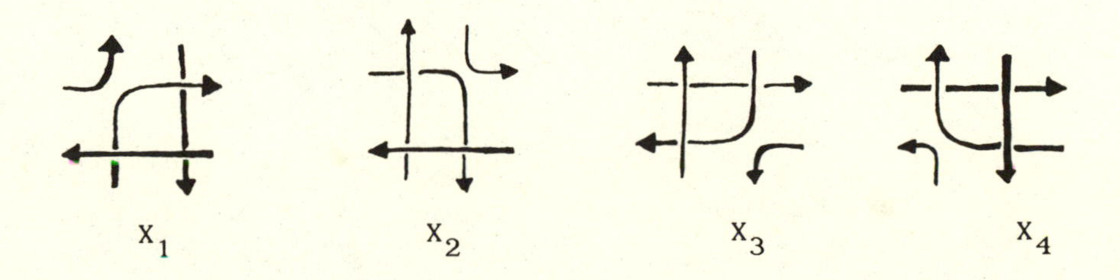

We want to see that this sum of linking numbers is zero modulo two.

In order to do this, first remember that K is a knot. Hence the strands that we have indicated must be connected in a pattern like

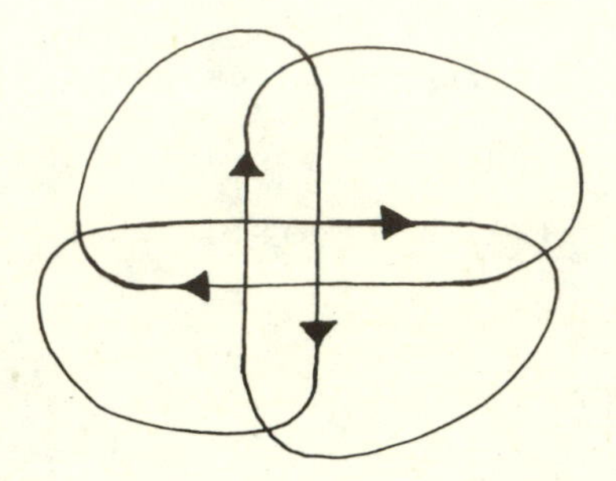

There are other patterns of connection (list them!), and the argument we show will work for them as well. Now contemplate linking number contributions that arise using this pattern:

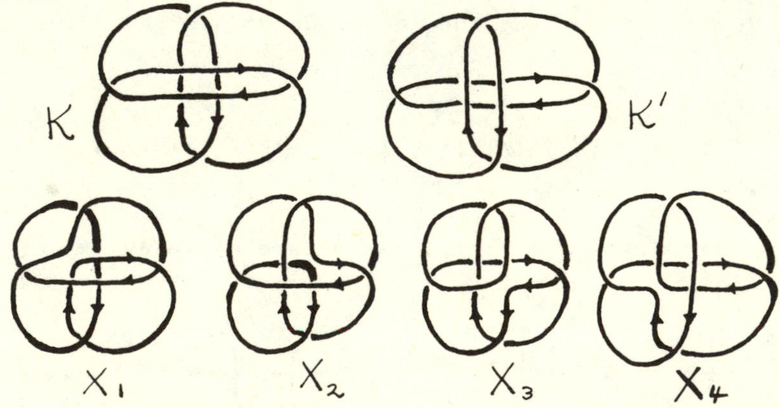

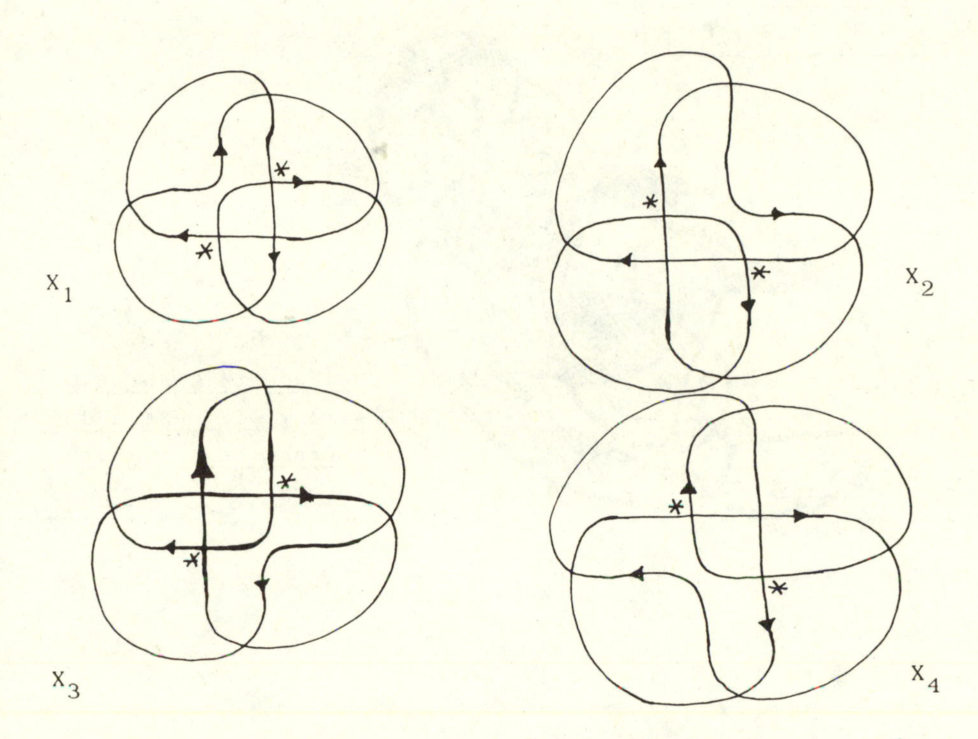

For example, I have starred (*) all the crossings that contribute to the sum of linking numbers from the configuration. Note that each crossing appears twice, hence cancelling in the mod-2 summation. The same double-contribution will hold throughout, and this completes the sketch of the proof.

It appears to be much more difficult to obtain <u>integral</u> (rather than mod-2) information about the difference $\alpha(K)-\alpha(K')$. In fact, the invariant $\alpha(K)$ remains as mysterious as ever!

6 = 1 + 2 + 3

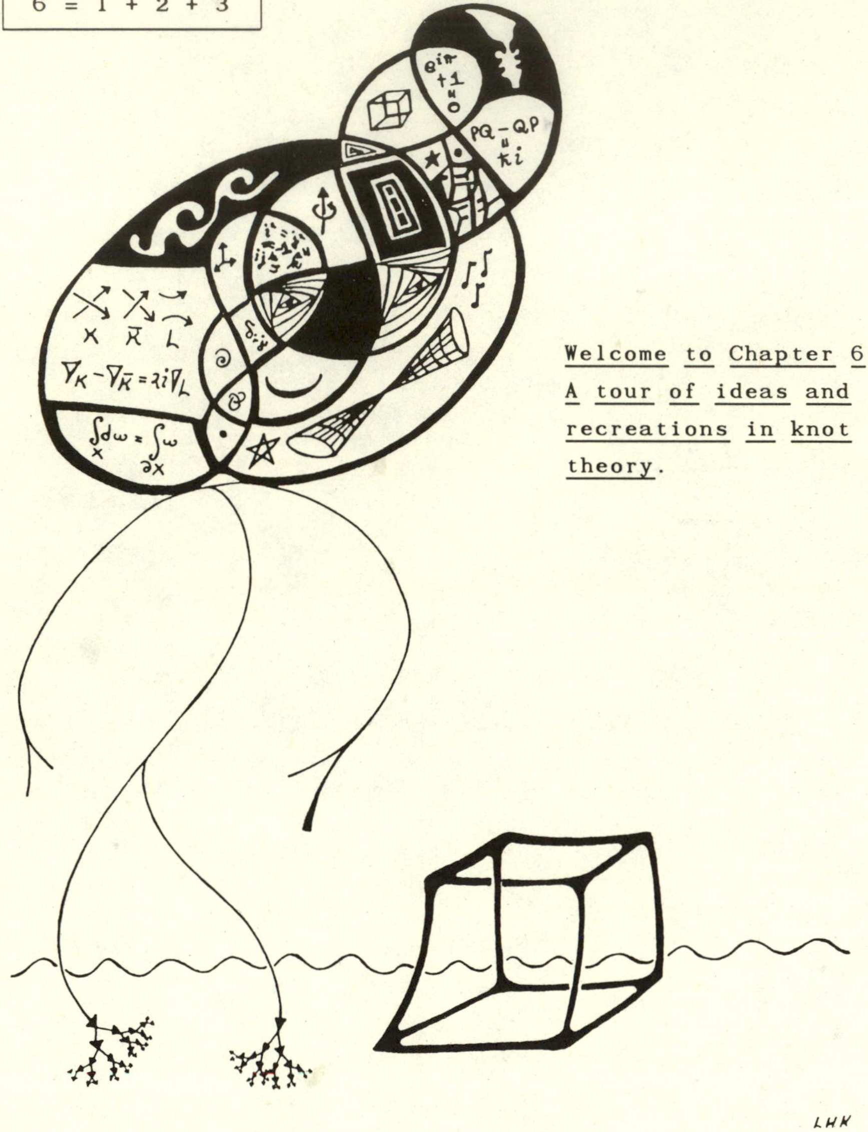

<u>Welcome</u> <u>to</u> <u>Chapter</u> <u>6</u>: <u>A</u> <u>tour</u> <u>of</u> <u>ideas</u> <u>and</u> <u>recreations</u> <u>in</u> <u>knot</u> <u>theory</u>.

VI

MISCELLANY

This chapter is a grab bag of bits of mathematical knottery, pictures, tricks, observations.

§1. *QUATERNIONS AND THE BELT TRICK*

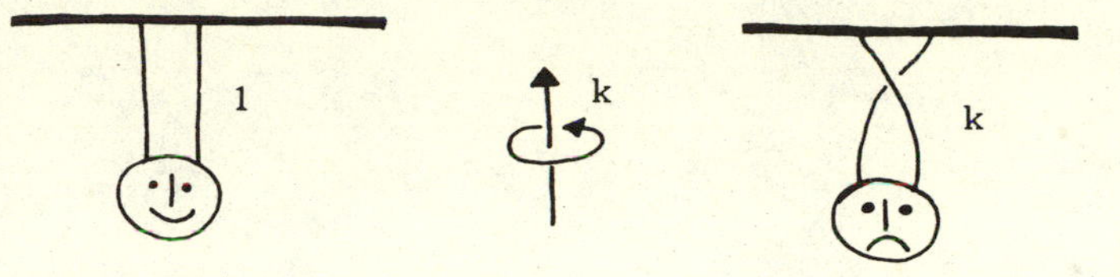

We take this disk and hang him free, one side smile and one side frown. And yet we'll turn him around and around.

k = turn 180° about the vertical

Four turns by k return a smile, but quite a tangle up above.

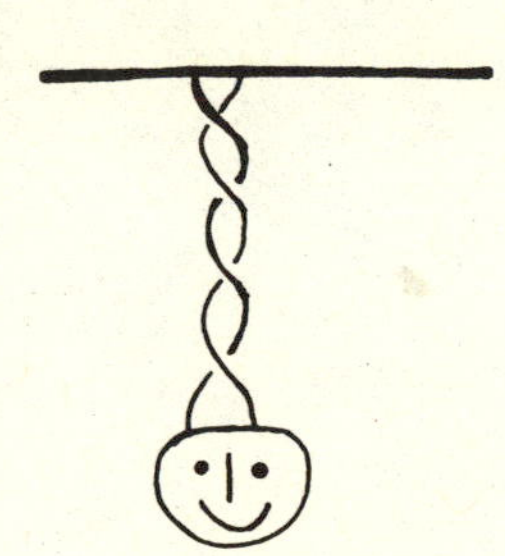

k^4

[Label tangle by the rotation producing it from ]

Recall the curl that two twists make:

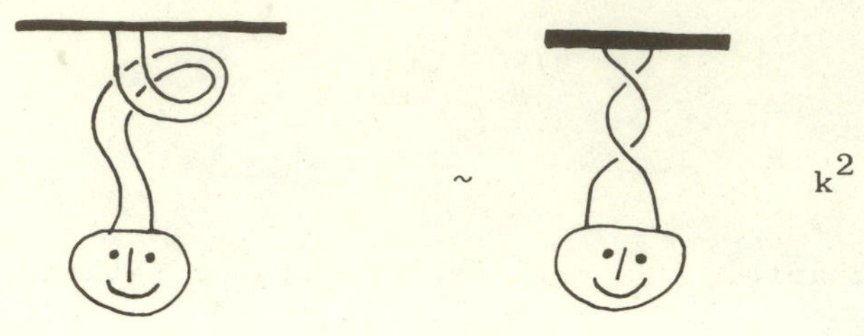

Then k^4 is two curls, and thence this isotopy:

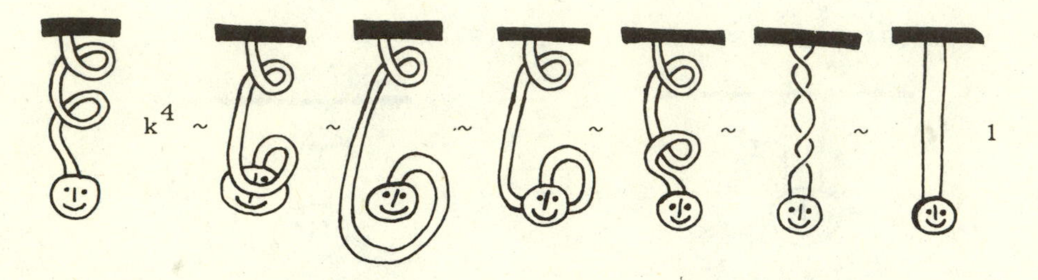

So k^4 is equivalent to 1, and not a motion of his face!

And now to i and j as well. Each has his own small tail (tale) to tell.

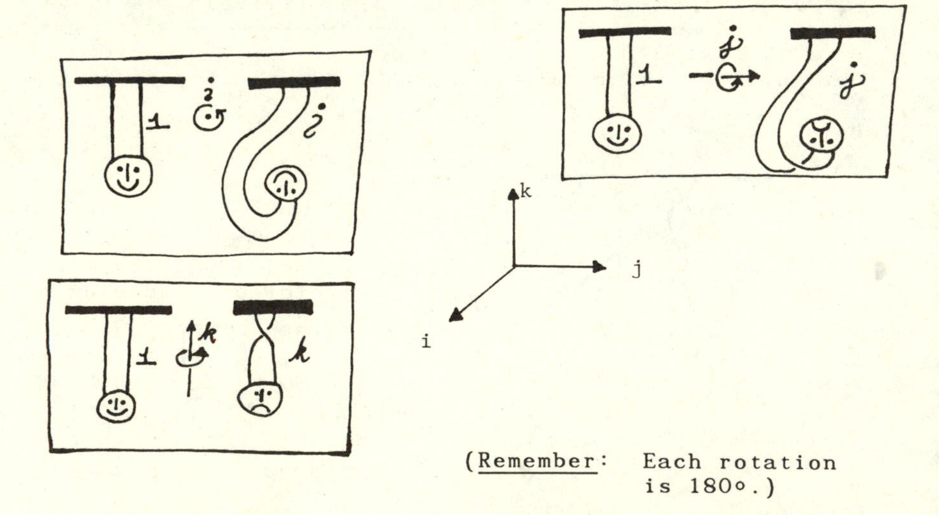

(Remember: Each rotation is 180°.)

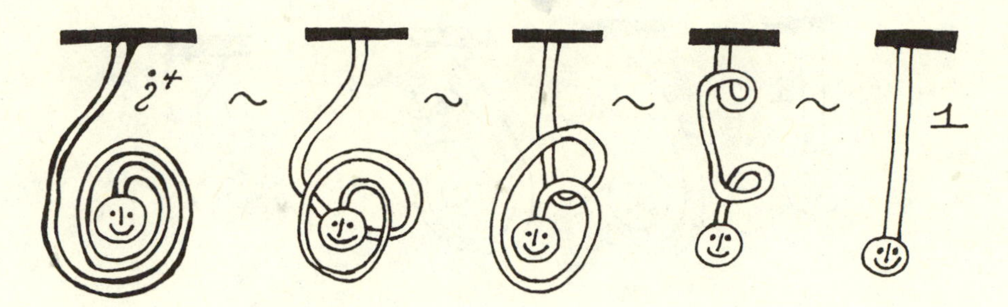

Big curl over. And second one goes under, To make this i^4 isotopy. And j^4 we'll leave to thee.

But there are products here. For combinations also dear:

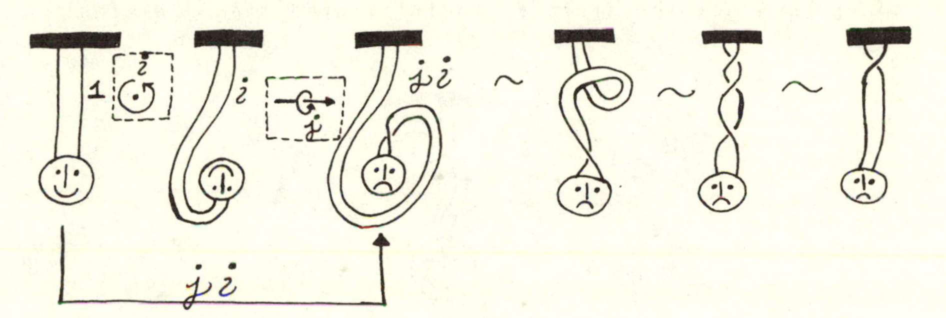

What should we call this fellow on the right? k is not quite right. For k has crossing t'other way. And therefore $\bar{k}$ I say:

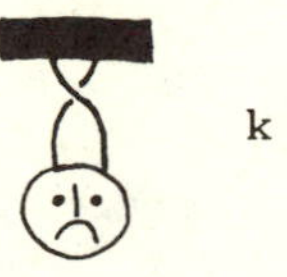

k

$\bar{k} = ji$.

And what about ij?

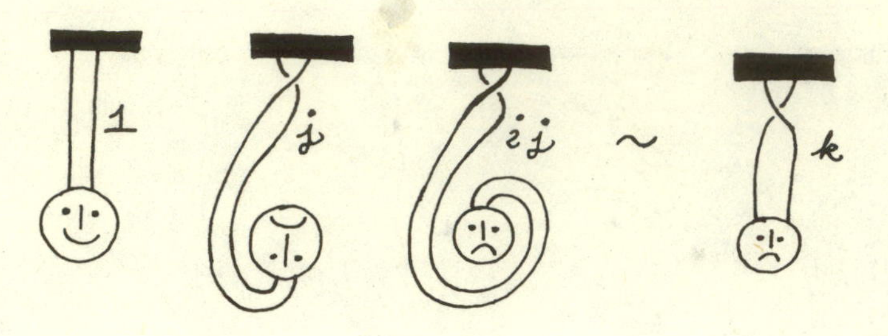

Thus we have: $i^4 \sim j^4 \sim k^4 \sim 1$

$ij \sim k, \quad ji \sim \overline{k}.$

What of i^2, j^2, k^2? In each case the disk returns to place, but not the strings; their twisted trace remains:

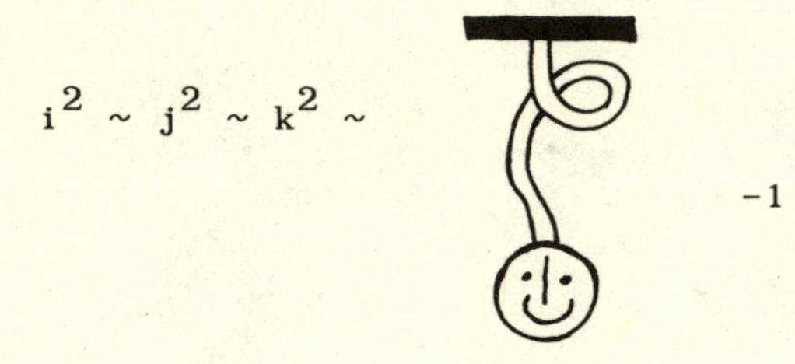

We'll call this one minus-one (−1).

Note: $(-1)i \sim \overline{i}$

$(-1)j \sim \overline{j}$

$(-1)k \sim \overline{k}.$

So finally, the <u>quaternion group</u> is revealed and from this rope trick takes a bow

And gives his thanks to that from which he/she sprung: a crossing and the square roots of minus one:

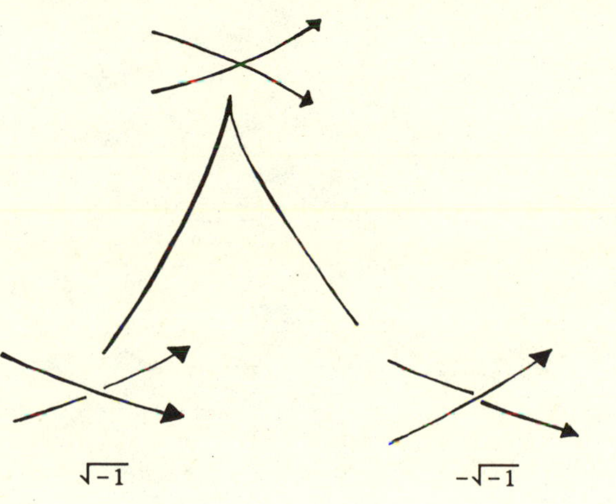

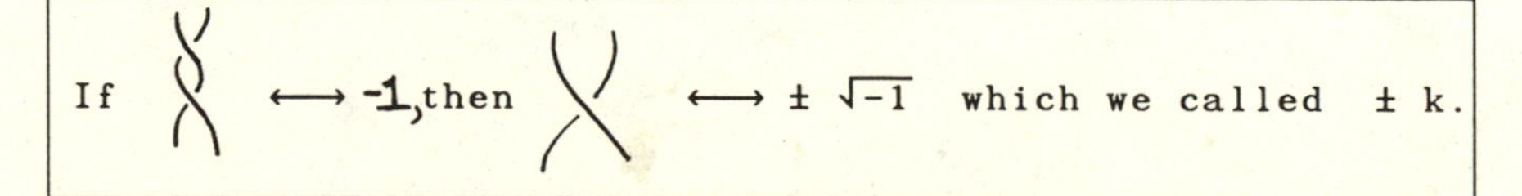

Remark: There is more to say about the belt trick ($k^4 \sim 1$) and various aspects of topology. For one thing, it is a visualization of the fact that $\pi_1(SO(3)) \cong Z_2$ where SO(3) is the group of orientation-preserving, orthogonal linear transformations of $\mathbb{R}^3$. It is also no accident that the unit quaternions is S^3 and forms the double cover of SO(3)! Here we have exhibited part of this correspondence directly. It is really amazing to contemplate the fact that if you are, puppet-like, connected to all-the-world by a thousand strings, then all those strings can be untangled if you but turn twice around!

Turn and turn again.

Yea, once more.

And

Return

To

One.

§2. *THE ROPE TRICK*

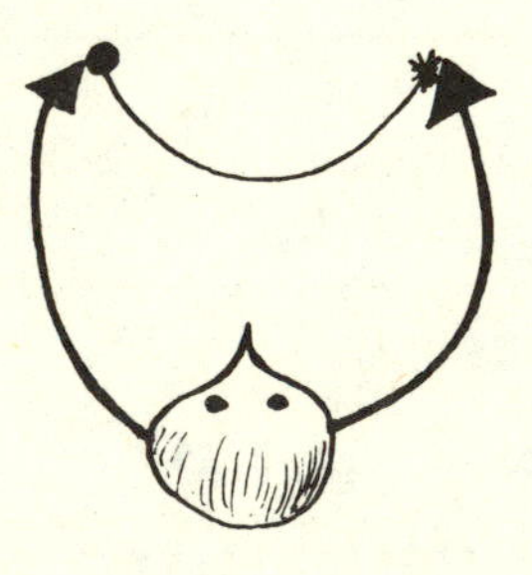

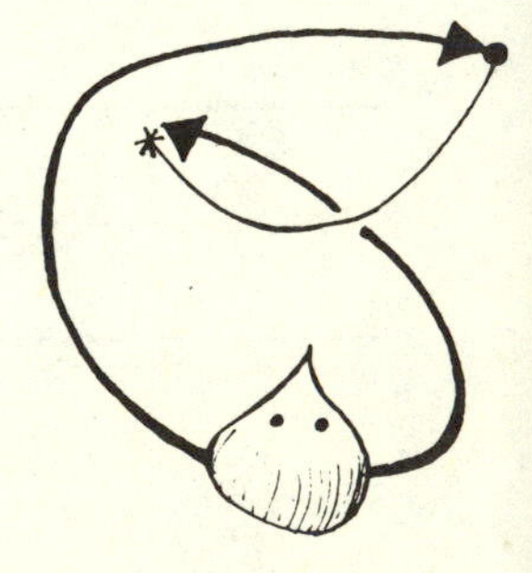

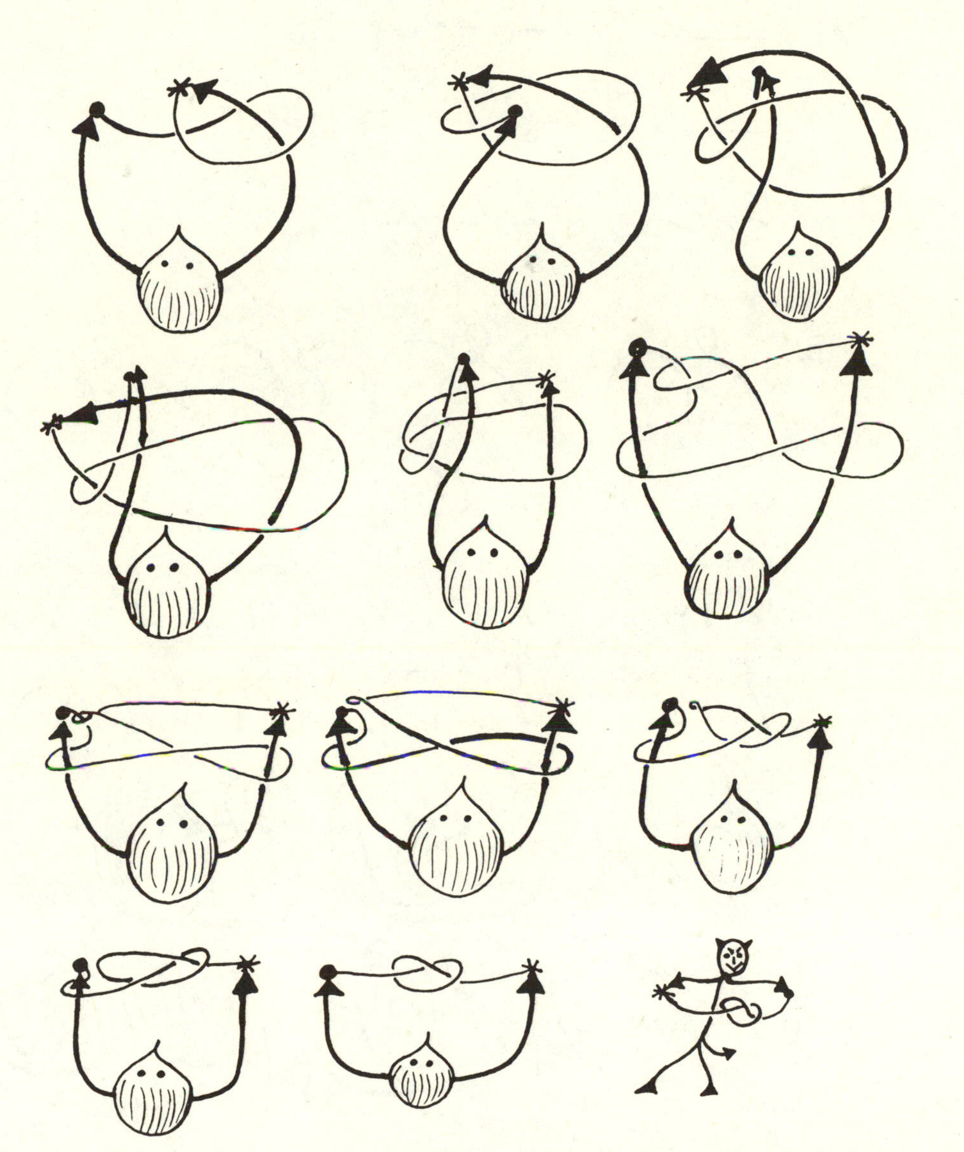

§3. TOPOLOGICAL SCRIPT

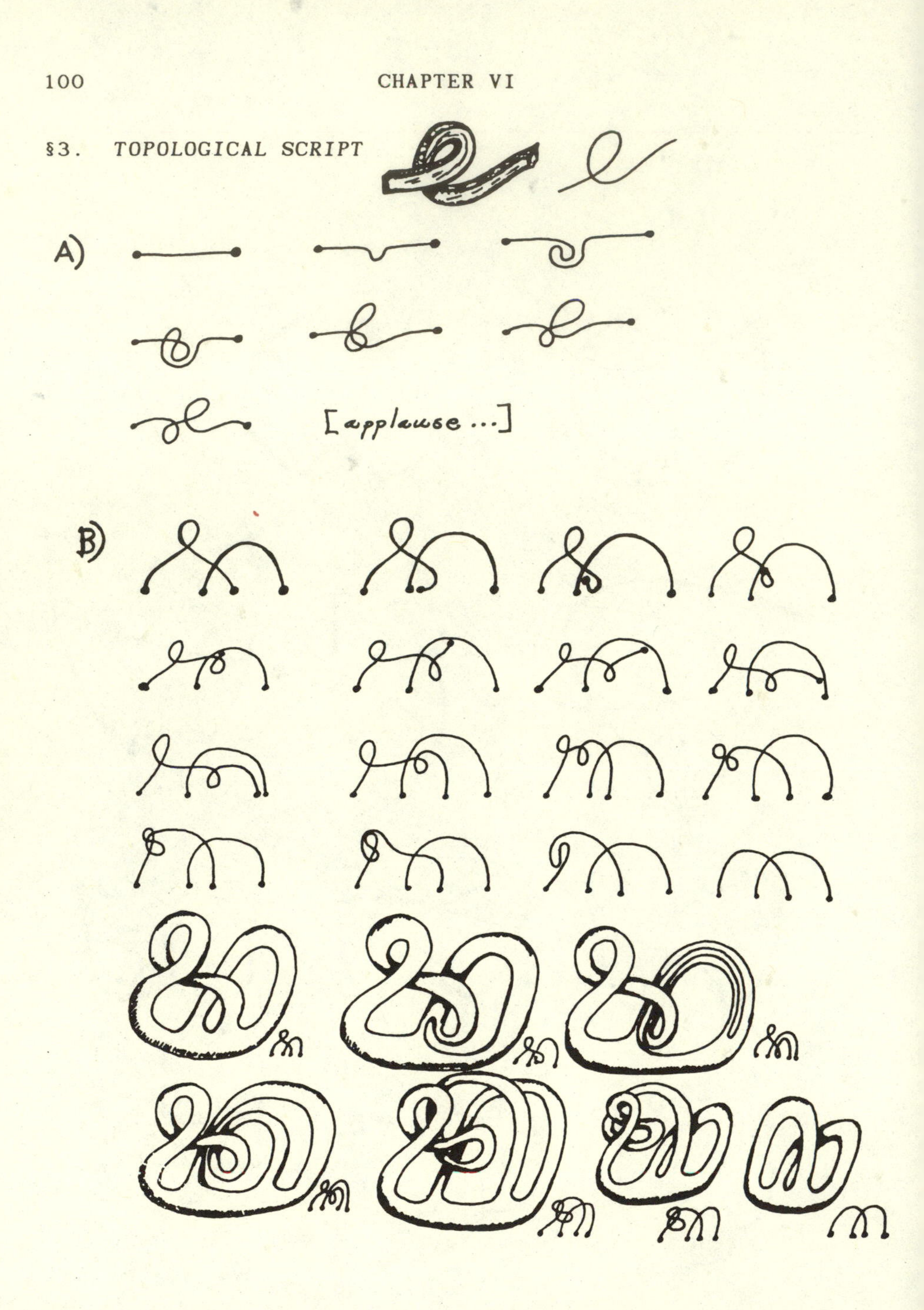

(B)
(A)

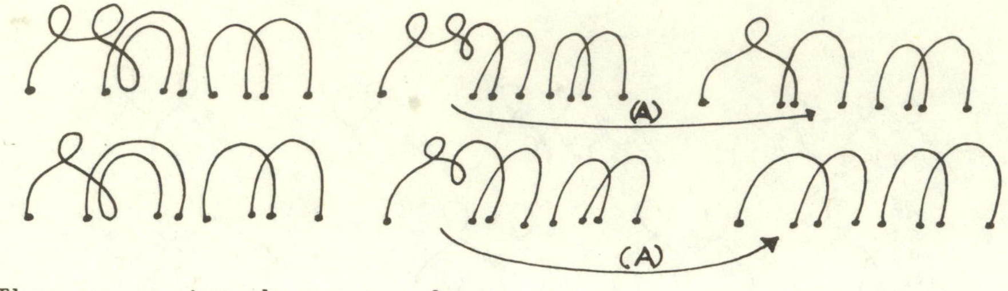

Thus we scripted our way from

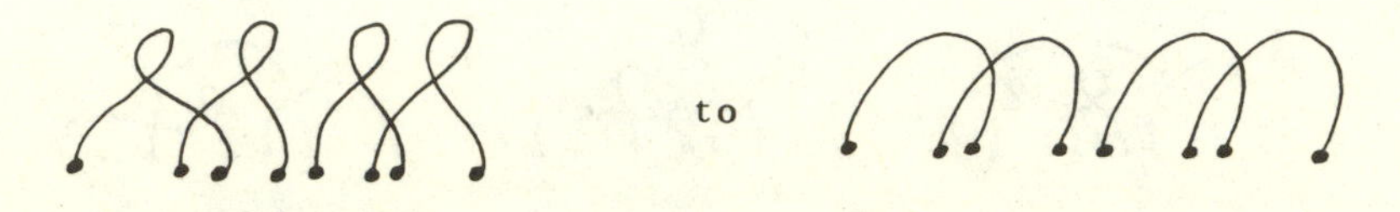

Of what use might this strange topologist's script be? It was conceived as a short-hand for isotopies and regular homotopies of surfaces wherein these moving intervals live.

Thus

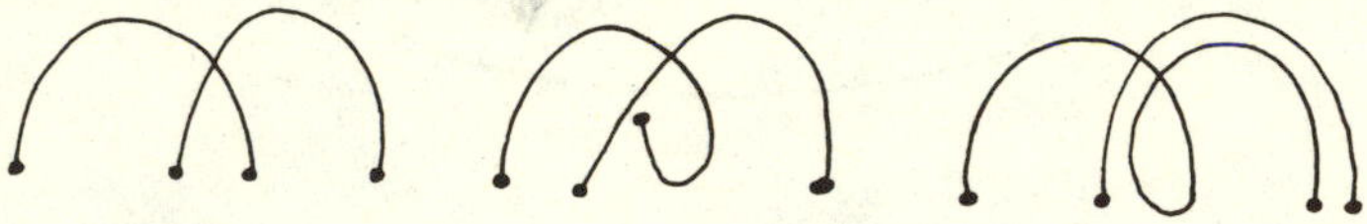

is short-hand for

or possibly the projection of this isotopy seen as the trace of a regular homotopy.

But the key ingredient in this script is the Whitney Trick. (See [W1].)

I happen to think that this is superior to the bending of spoons.

§4. *THE SPENCER-BROWN CALCULUS*

We have introduced various calculi for knots, starting with the Reidemeister moves

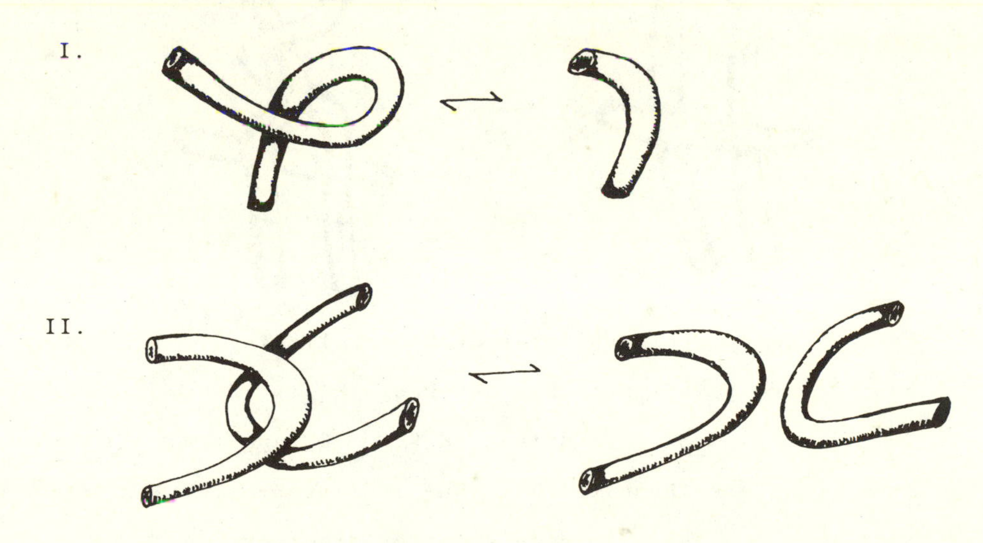

III.

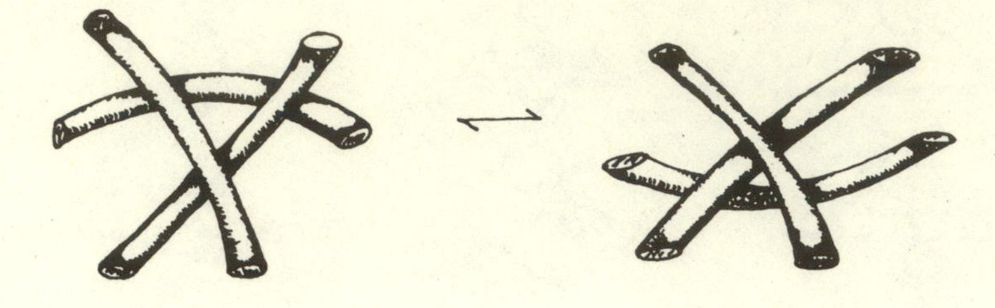

These produce the system of calculation we call $\mathcal{K}$ (knots up to ambient isotopy), the subject of all our deliberations. Then we constructed many calculi related to this, such as the switching calculus of the Conway polynomial $[\nabla(\quad) - \nabla(\quad) = z\nabla(\quad)]$, the Skein and Tangle Theories. But these are calculi <u>about</u> $\mathcal{K}$. We also introduced a "quotient calculus" of $\mathcal{K}$ by adding the Γ-move:

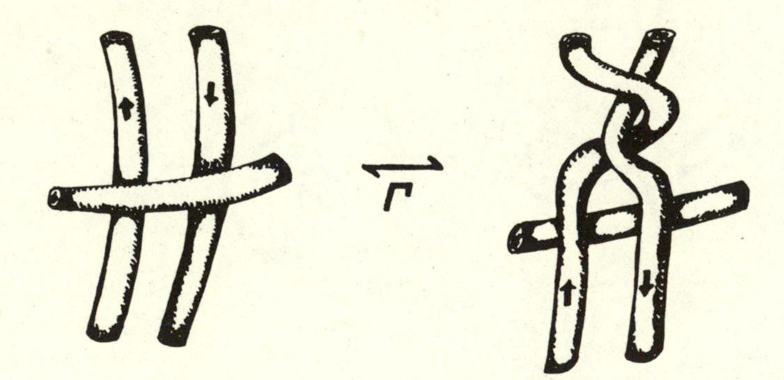

Let $\tilde{\Gamma}$ denote this calculus of knots allowing I, II, III, and Γ. We have shown that any knot K is Γ-equivalent to or , and that $K \# K \sim 0$ <u>and</u> that and are inequivalent. But the important thing is that $\tilde{\Gamma}$ is a two-valued calculus of diagrams. Any expression (knot) reduces in $\tilde{\Gamma}$ to one of two possible inequivalent forms. Nevertheless, there are more than two elements to consider. We often want to know how a knot reduces, but the knot is a form in its own right.

In his book, *Laws of Form* [SB], G. Spencer-Brown constructs what must be the simplest possible two-valued diagrammatic calculus. It is not knot theory, but the structural resemblance to $\mathcal{K}$ and $\tilde{\Gamma}$ is striking! The expressions in Spencer-Brown's calculus are (changing his notation from $\urcorner$ to $\square$) disjoint collections (finite at first) of rectangles in the plane. Thus

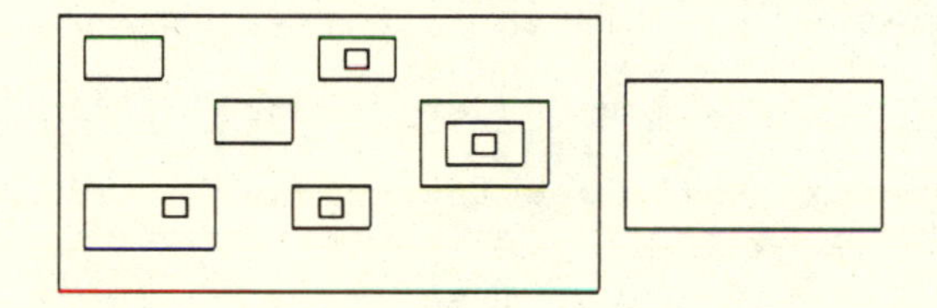

is an expression in the <u>calculus of indications</u>. The moves in the calculus are

I. $\square\,\square \longleftrightarrow \square$ and

II. $\boxed{\square} \longleftrightarrow$

In this calculus the "unknot" is the empty expression. Thus II replaces two "concentric" rectangles (with the inner rectangle empty) by the empty expression. Number I amalgamates two empty rectangles into one. Call $\square$ the <u>marked state</u> and $\boxed{\square}$ (pardon me!) the <u>unmarked state</u>. Then you can easily see that any expression in $\mathcal{C}$ (calculus of indications) reduces to either $\square$ or $\boxed{\square}$ ($\boxed{\square}$ is a convenient expression for the unmarked state).

<u>For example</u>:

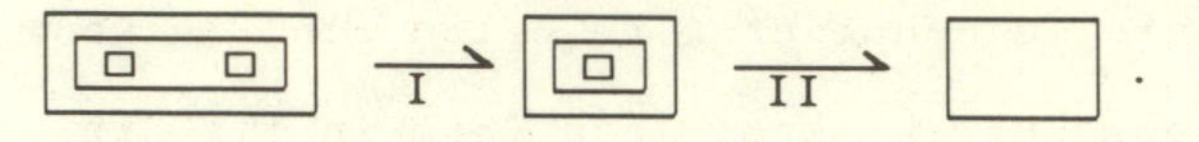

For $\mathscr{C}$, let $\equiv$ denote the equivalence relation generated by I and II (plus the usual topological equivalence of diagrams).

<u>Exercise</u>. Prove that in $\mathscr{C}$, $\square \not\equiv \boxed{\square}$. *Hint:* Devise a way to figure out the "value" ($\square$ or $\boxed{\square}$) of an expression from the form of the expression. Let e be any expression. You need to be able to calculate $V(e) = \square$ or $\boxed{\square}$ such that V is invariant under the use of the moves I and II. Of course V satisfies the axioms

$$V(\boxed{e}) \equiv \boxed{V(e)}$$

$$V(\square e) \equiv \square$$

$$V(\quad) \equiv \boxed{\square}\,.$$

§5. *OFF TO INFINITY*

In Chapter 11 of his book, Spencer-Brown goes off to infinity. Here his calculus begins to look a little bit more like knot theory. Infinite expressions don't reduce in a finite number of steps. So they have a singular character. For example:

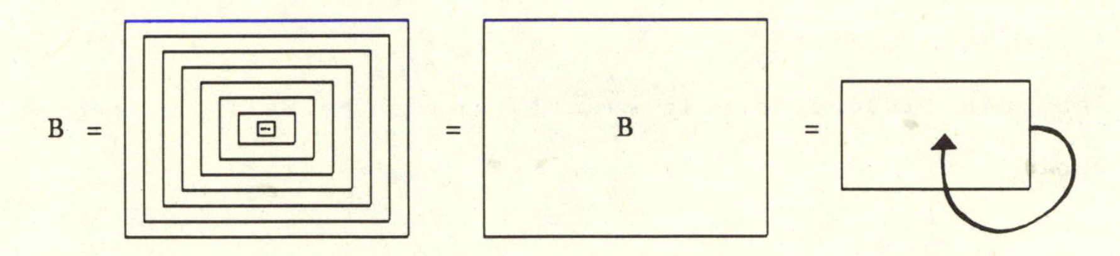

Here B is an infinite box of boxes and consequently contains a topological image of itself. Hence we write B = [B] (strictly speaking, up to similarity). And we write B = to indicate how and where the form B reenters its own indicational space. This is rather like a knot!

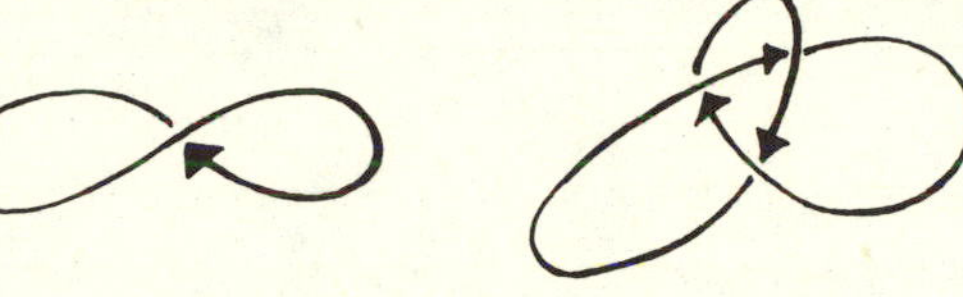

But knots don't re-enter themselves, you say. Well, but maybe they do!

R. D. Laing wrote a book called Knots [L], which is about human knots. Things like

A wins over B

B wins over C

C wins over A

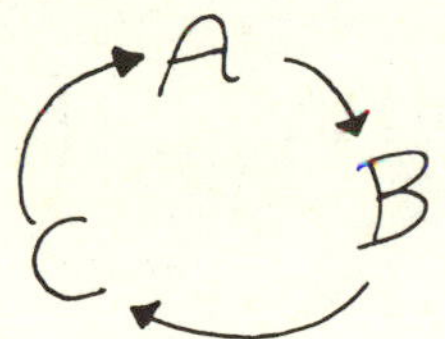

and the re-entry is this coming back to where you started. More seriously, three entities can all be mutually defined:

A balances B's dominance of C.

B balances C's dominance of A.

C balances A's dominance of B.

A family pattern!

And this last pattern is exactly the pattern of the trefoil knot:

$$\begin{array}{c} \uparrow Z \\ \longrightarrow Y \\ \uparrow X \end{array} \qquad \boxed{Z = X*Y} \qquad (\underline{\text{Notation}})$$

Interpret the crossing diagram as Z balances Y's dominance of X. Then our last triplet becomes:

$$\begin{array}{c} \uparrow A \\ \longrightarrow B \\ \uparrow C \end{array} \qquad \begin{array}{c} \uparrow B \\ \longrightarrow C \\ \uparrow A \end{array} \qquad \begin{array}{c} \uparrow C \\ \longrightarrow A \\ \uparrow B \end{array}$$

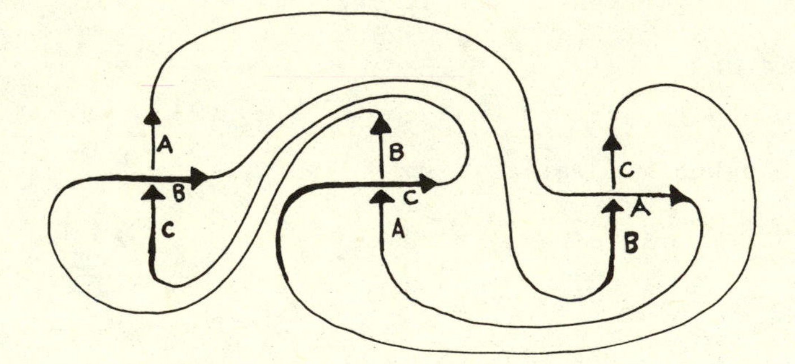

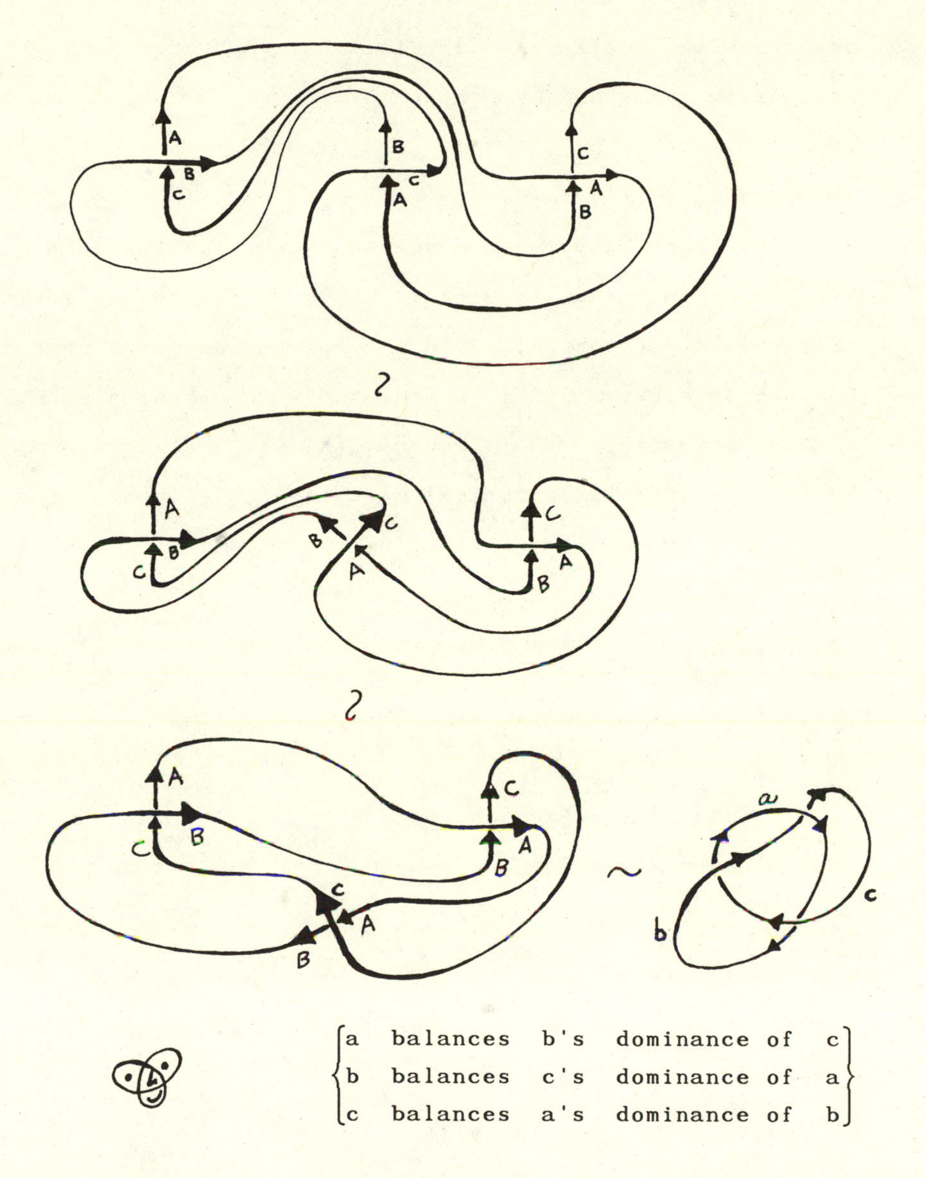

Trefoil as eternal triangle.

The matter of mutual definitions has its mathematical

counterparts in many ways, of course. For example, in an algebra with generators and relations, each element is defined in terms of the others.

§6. *QUANDLES*

Just a quick comment about an algebraic approach to knot invariants that is very close in spirit to the ideas of the last section (5) of this chapter. David Joyce [J] defines an algebra $Q(K)$ associated with a knot diagram in the following way: The algebra $Q(K)$ is a <u>quandle</u>. Quandles are (nonassociative) algebras Q with two binary operations $*$ and $\bar{*}$ satisfying the axioms:

$$\text{I.}\quad \begin{Bmatrix} a * a = a \\ a \mathbin{\bar{*}} a = a \end{Bmatrix} \quad \forall\, a \in Q.$$

$$\text{II.}\quad \begin{Bmatrix} (a*b) \mathbin{\bar{*}} b = a \\ (a \mathbin{\bar{*}} b) * b = a \end{Bmatrix} \quad \forall\, a,b \in Q.$$

$$\text{III.}\quad \begin{Bmatrix} (a*b) * c = (a*c) * (b*c) \\ (a \mathbin{\bar{*}} b) \mathbin{\bar{*}} c = (a \mathbin{\bar{*}} c) \mathbin{\bar{*}} (b \mathbin{\bar{*}} c) \end{Bmatrix} \quad \forall\, a,b,c \in Q.$$

For example: Let G be a group with group operation $a,b \longmapsto ab$. Define on the set G the operations $a * b = b^{-1}ab$ and $a \mathbin{\bar{*}} b = bab^{-1}$. This makes $(G,*,\bar{*})$ a quandle.

Here's how this is related to knots: Label the strands in the diagram $a,b,c,\cdots$. Thus

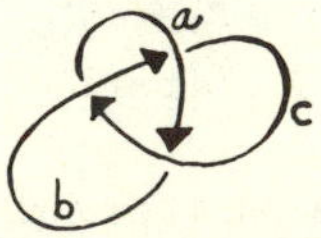

for the trefoil.

<u>Code each crossing by an equation.</u>

$b = a*c.$ $\qquad\qquad$ $z = x\bar{*}y.$

Regard the set of equations as relations for a quandle whose generators are the strand labels.

<u>Two comments</u>:

(i) The set of equations has exactly the form of our set of circular descriptions:

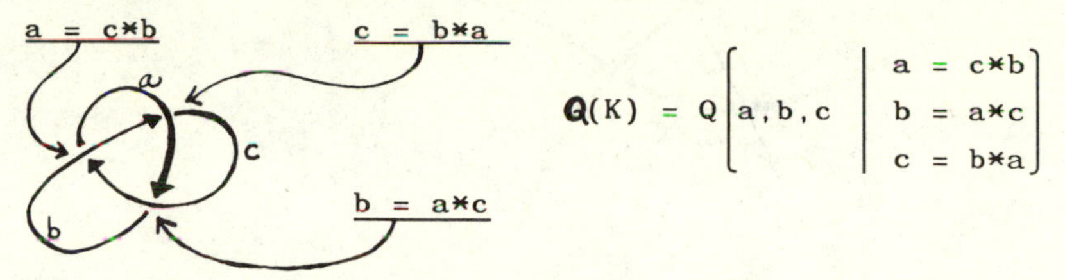

(ii) <u>The quandle axioms are designed so that Q(K) can't "see" the Reidemeister moves.</u>

I.

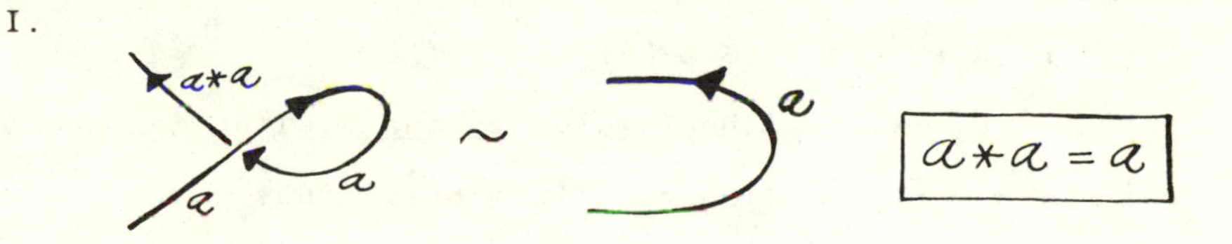

II.

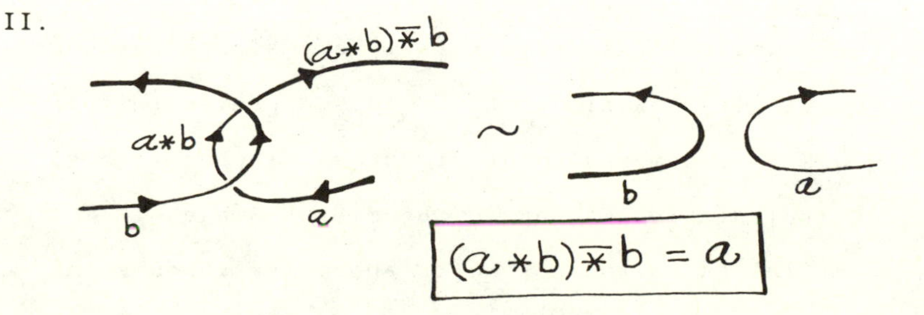

III.

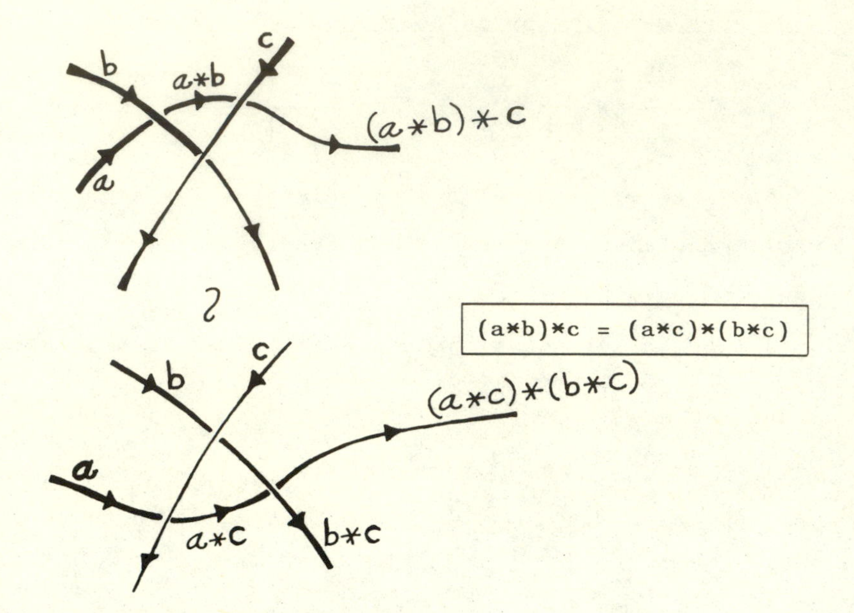

This proves that the quandle is an invariant of K.

Exercise. 1. Compare Q(K) and the Wirtinger presentation for $\Pi_1(S^3-K)$. (See [F1].)

2. Let W(K) (the involutory quandle) be defined by setting $* = \bar{*}$ in all our definitions. Thus $(a*b)*b = a$ in W(K). Show that W( ) is a finite algebraic system, and compute its multiplication table.

The quandle Q(K) itself is a very strong (hence not very computable) invariant of K. Joyce proves that if $Q(K) \cong Q(K')$ (isomorphism of algebras), then K is either ambient isotopic to K′ or to the mirror image of K′. His proof relies on known results about the fundamental group and its subgroups. It would be very useful to have a direct proof of his theorem. For more information about quandles, see Steve Winker's thesis [W]. Winker has diagrammatic techniques to spot infinite involutory quandles.

§7. *THE TOPOLOGY OF DNA*

Actually, this is the solution to Exercise 2.7. Suppose you have a link L of two components that is represented by a planar diagram that looks locally like

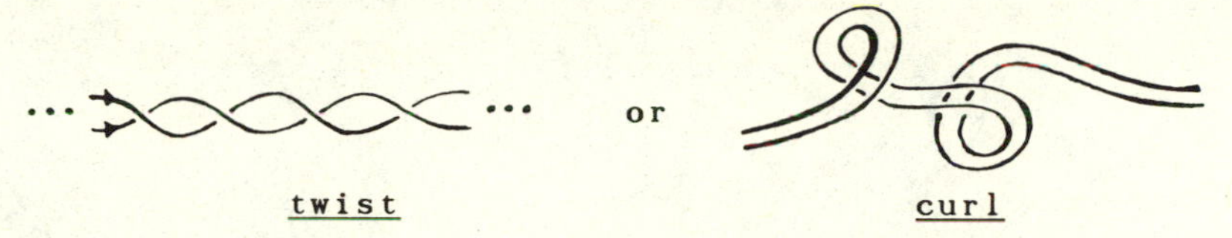

Then we can work out the individual twisting and curling contributions to the linking number, lk(L).

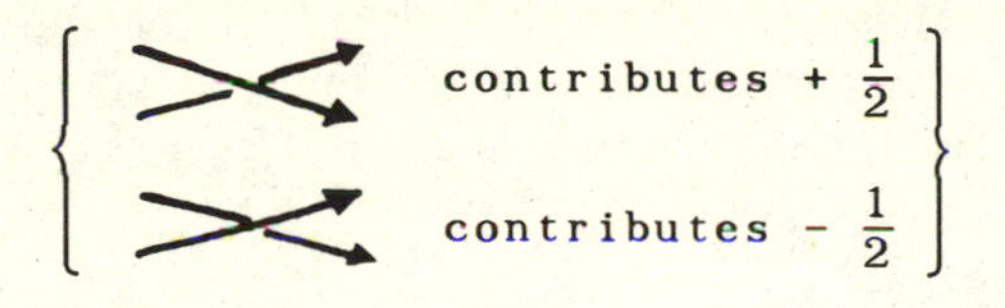

contributes -1

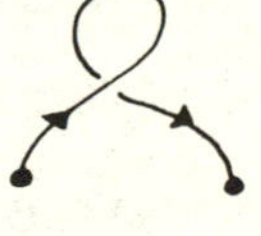

contributes + 1

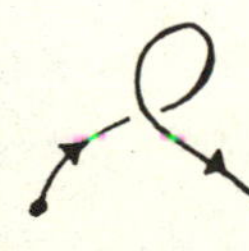

Let $T(L)$ = sum of twist contributions,

$W(L)$ = sum of curl contributions.

Then $\boxed{lk(L) = T(L) + W(L)}$

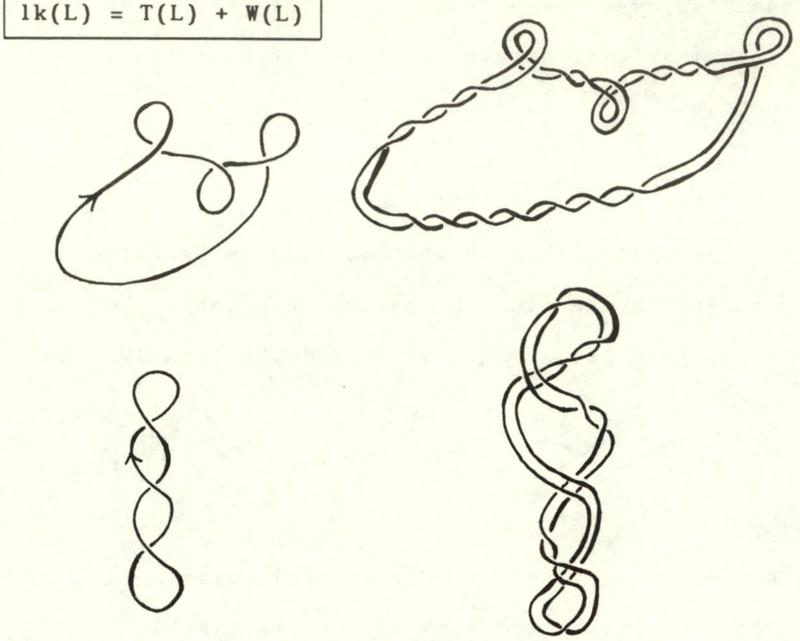

patterns links

This formula for the linking number in terms of twist T and writhe W has been used by biologists to argue that certain forms of DNA are in closed, linked, double stranded form, somewhat like the links pictured above. It is possible to see these patterns and estimate linking and writhing from electron microscopy and other methods. The formulae themselves have been generalized, using differential geometry, to integrals involving curves in three-dimensional space. (See [WH], [FB], [SU].)

§8. *KNOTS ARE DECORATED FIBONACCI TREES*

Here is the Fibonacci Tree:

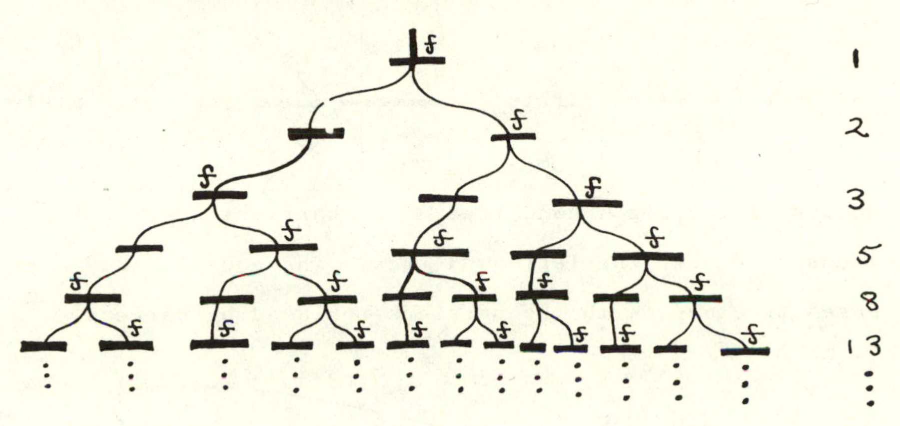

At each (descending) level it has a Fibonacci number $(1,1,2,3,5,8,13,\cdots : f_{n+1} = f_n + f_{n-1},\ f_0 = 1,\ f_1 = 1)$ of branches. This tree is infinite and it satisfies the equation $f = \overline{\overline{f\,}\rceil\, f\,}\rceil$!

Let me explain what I mean. Look at the tree and notice that you see

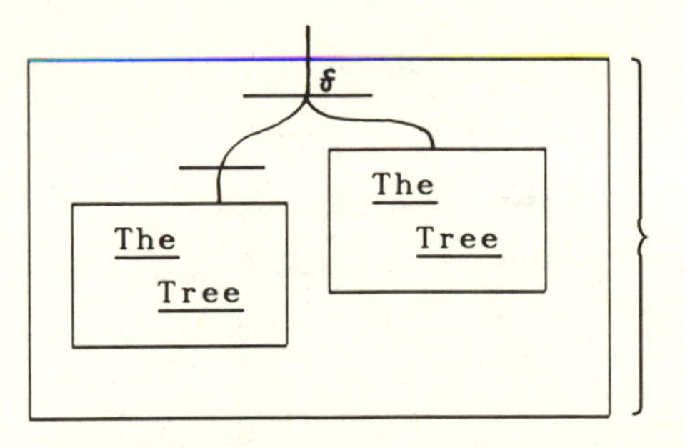

The Tree

Thus if you look at the whole tree, then you see that it divides itself conveniently into two copies of itself! As you pass the first barrier you immediately encounter the right-hand tree. Two barriers have to be crossed to meet the left-hand tree. The equation has the same structure, with the barriers replaced by markers $\rceil$. Thus

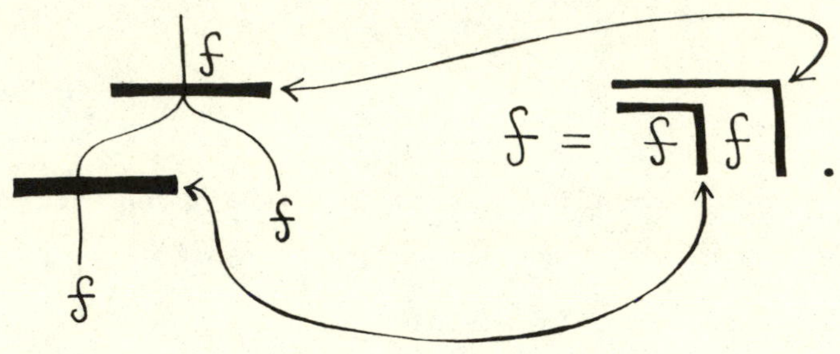

Given an equation in the notation at the right you can draw a corresponding tree. Thus $g = \overline{gg}\rceil$ corresponds to

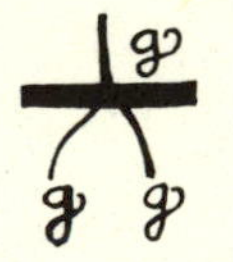

which yields the <u>doubling tree</u>

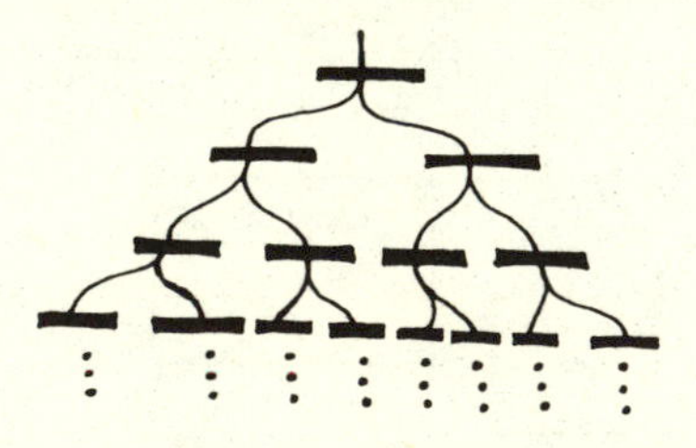

Now let's look at the trefoil (or any knot) and code its crossings via

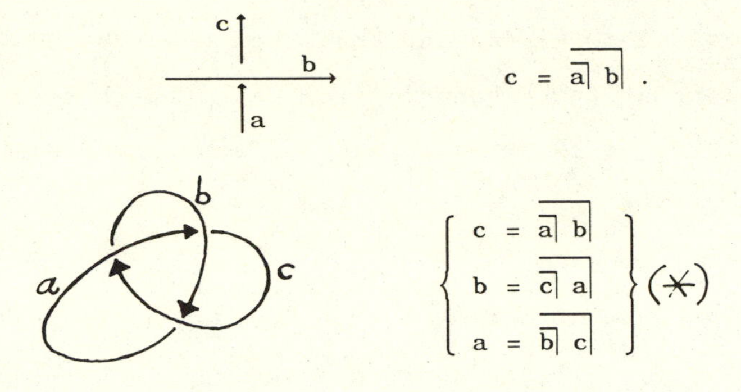

These equations will create a tree:

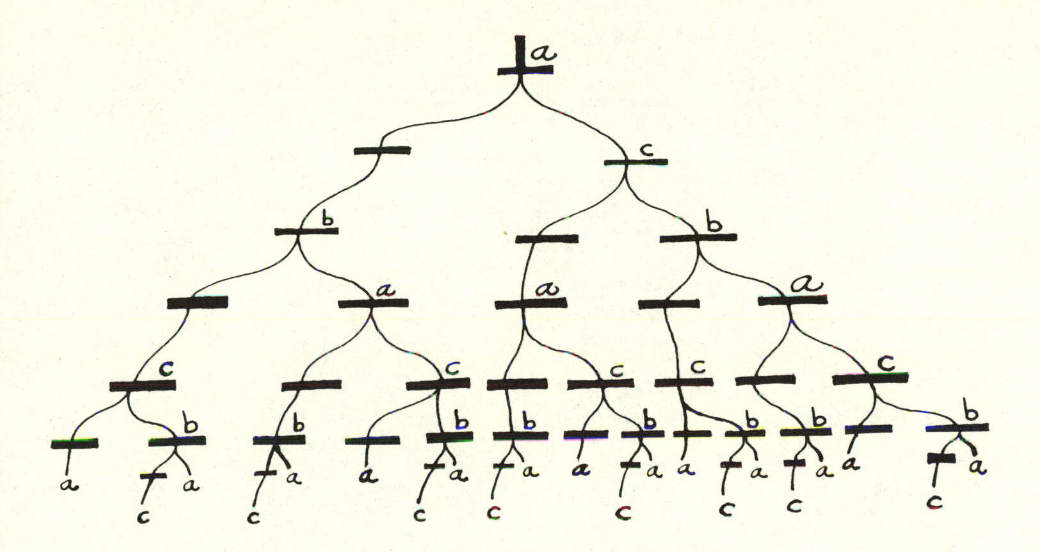

<u>Research</u> <u>Question</u>: If <u>Knots</u> are reformulated as labelling patterns on the Fibonacci tree, what do the invariants we have constructed so far (Conway polynomial, hints of the generalized polynomial, the quandles) look like in this context? Invent new invariants!

The equations (*) for the knot then describe an infinite labelling of the Fibonacci tree into (in this case) subtrees a, b, c each of which can be seen to contain all the others. Each knot contrives its own way of labelling the tree.

Pathways down through the tree correspond to journeys on the knot where <u>at each confrontation with a crossing, you may cross or walk the crossline</u>.

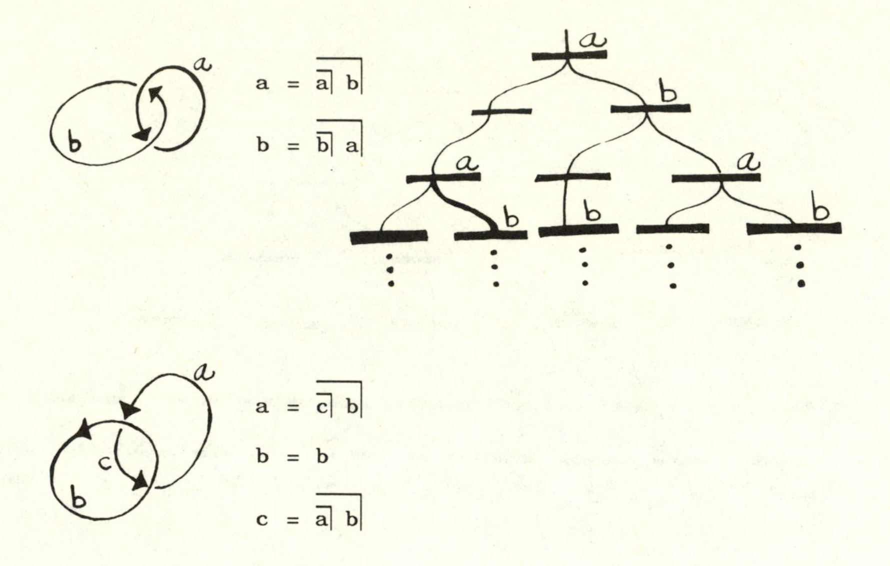

<u>Exercise</u>. Construct a tree for this un-link. *Hint:* Some pruning may be needed!

Of course, the original tree is the unknot.

$f = \overline{\overline{f}\rceil\; f}\rceil$. Infinitree.

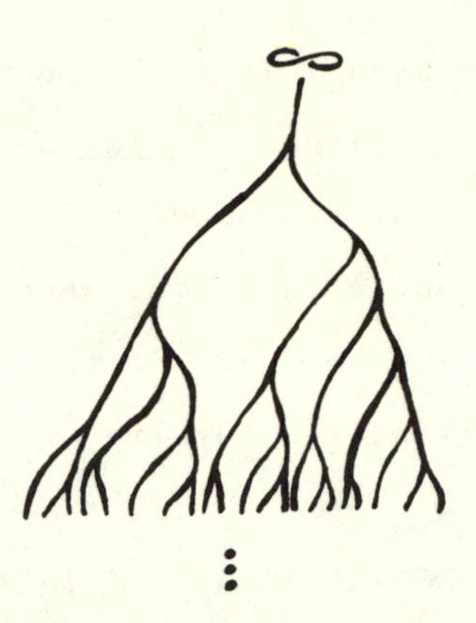

Exercise. A variant of the notation of this section is useful for working with the quandle (see section 6 of this chapter). Let $a*b = a\overline{b}|$ and $a\bar{*}b = a|\overline{b}$ and regard $\overline{b}|$ ($|\overline{b}$) as operating on everything to its left. Thus $(a*b)*c = a\overline{b}|\,\overline{c}|$ while $a*(b*c) = a\,\overline{b\,\overline{c}|}|$. Show that, in this noncommutative, associative formalism, the axioms for the quandle (e.g., $a*a = a$, $(a*b)\bar{*}b = a$, $(a*b)*c = (a*c)*(b*c)$) may be replaced by:

ab⌉ b a

1. $a\,\overline{a}| = a\,|\overline{a} = a$
2. $\overline{b}|\,|\overline{b} = |\overline{b}\,\overline{b}| =$ (empty word)
3. $a\,\overline{b\,\overline{c}|}| = |\overline{b}\,a\,\overline{b}|$

(and obvious variants).

For example, using 1., 2. and 3. above we have:

$$(a*b)*c = a\overline{b}|\,\overline{c}|$$

$$(a*c)*(b*c) = a\overline{c}|\,\overline{b\overline{c}|}| = a\overline{c}|\,|\overline{c}\,\overline{b}|\,\overline{c}| = a\overline{b}|\,\overline{c}| = (a*b)*c.$$

Call a formal system with axioms 1., 2., 3. a crystal. The

crystal of a knot has associative, but noncommutative multiplication. Let the <u>light</u> <u>crystal</u> be the corresponding structure for the involutory quandle (Exercise 2, Section 6 of this chapter). Compute the light crystal of the figure eight knot. (We leave it as part of the exercise to provide a complete definition for the crystal.)

§9. *A PATTERN SEEN DECORATING A WALL IN THE ALHAMBRA*

(and also on an inner door in gold trim in the church La Seo de Zaragoza, Zaragoza, Spain).

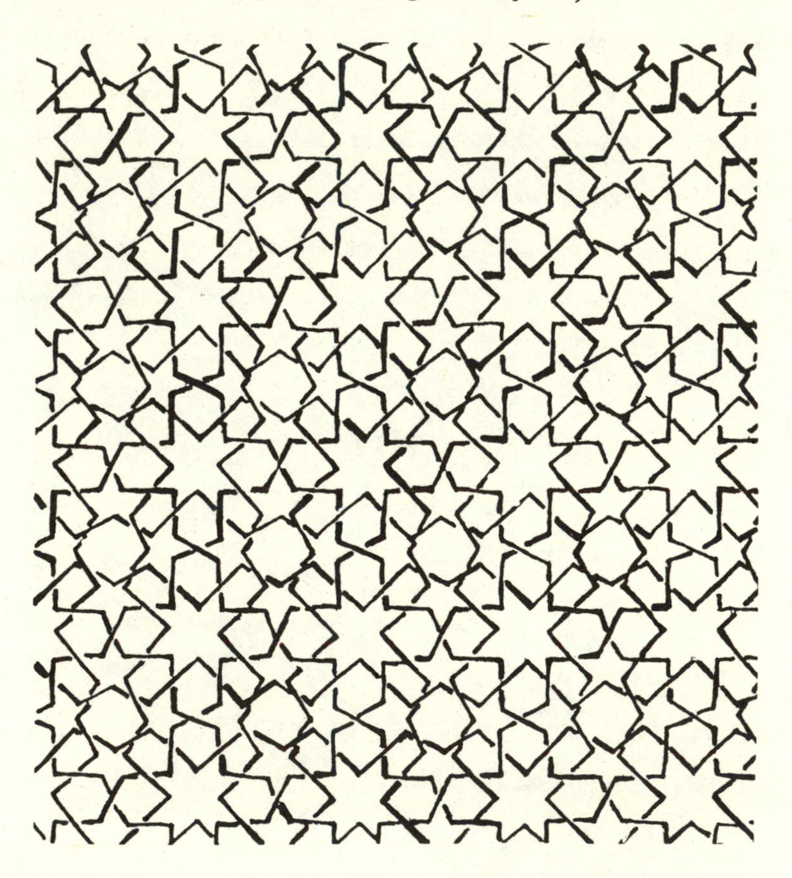

§10. *FROM THE ASHLEY BOOK OF KNOTS* [A2]

"This is one of the queerest knots. Some days I can tie it, some days I can't."

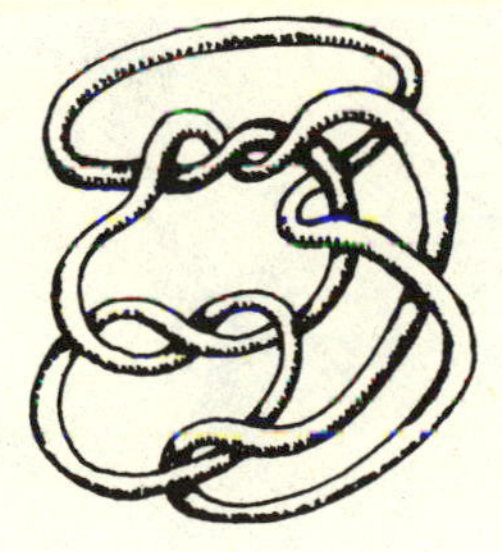

$$\nabla(\;) = \nabla(\;) + z\nabla(\;)$$
$$\nabla(\;) = \nabla(\;) - z\nabla(\;)$$

§11. *PILAR'S FAMILY TREE* [* = +crossing, $\bar{*}$ = -crossing]

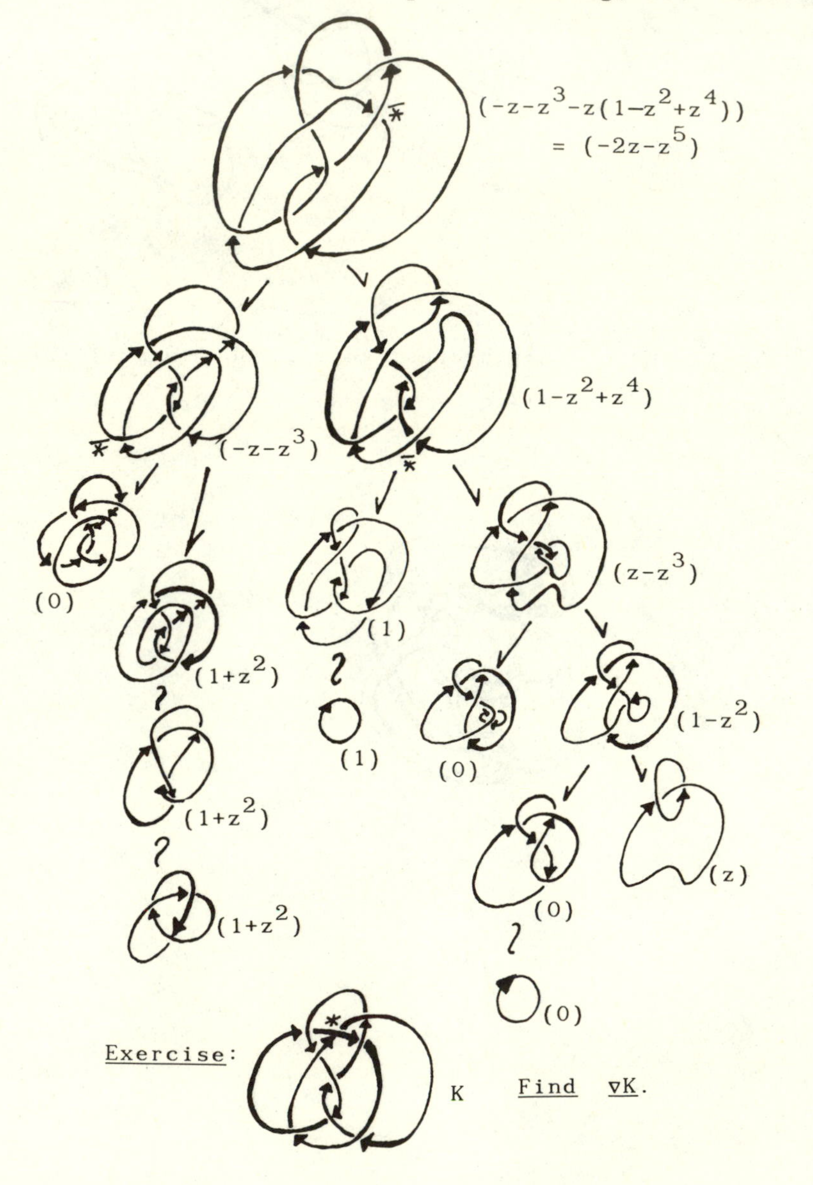

Exercise: K Find ∇K.

§12. *THE UNTWISTED DOUBLE OF THE DOUBLE OF THE FIGURE EIGHT KNOT.*

This knot is known to be topologically slice as a consequence of Michael Freedman's work [F]. No direct construction for a topological slicing disk is yet known. It is also unknown whether the knot is differentiably slice (or ribbon).

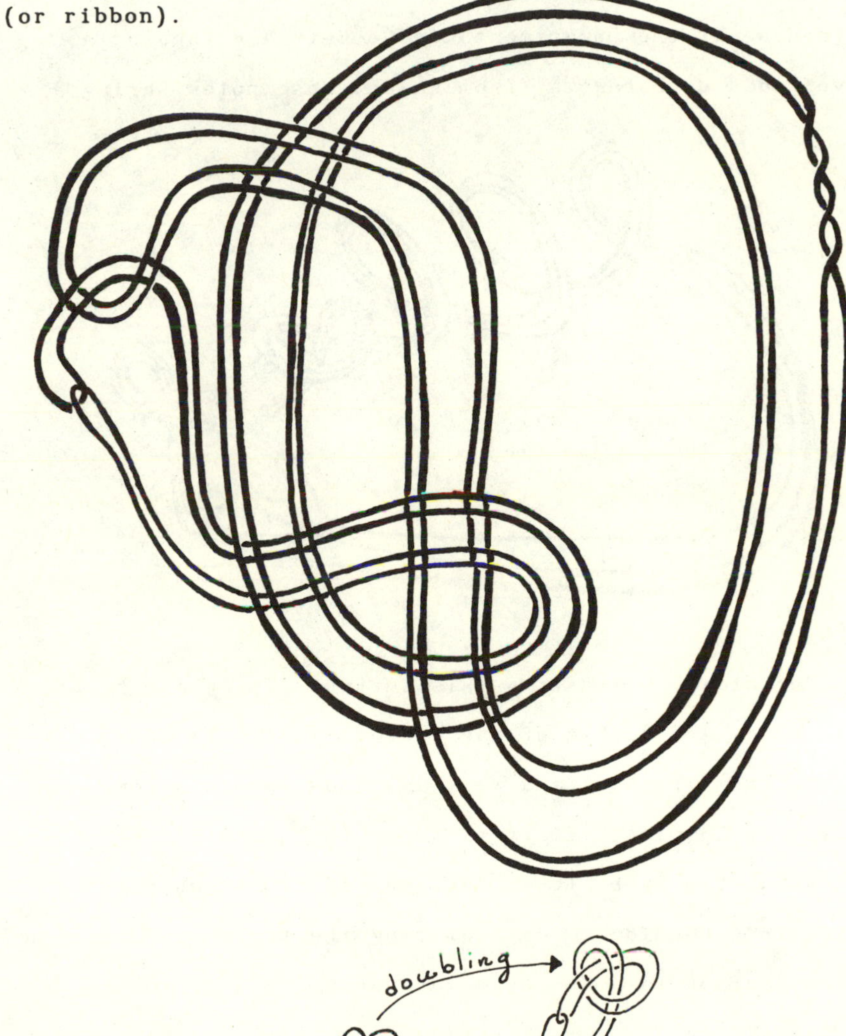

§13. *APPLIED SCRIPT—A RIBBON SURFACE*

Here we exhibit an application of topological script to the discovery of a ribbon surface for a twisted double of the trefoil knot (see Chapter VIII for theoretical details). These pictures are the joint work of Carmen Safont and Pilar Del Valle. (Both active participants in the course at the Departmento de Geometria y Topologia, Universidad de Zaragoza, from which these notes spring.)

Our story begins with this surface F, whose boundary K is a twisted double of the trefoil knot. It turns out that K is a ribbon, and we shall show how this comes about. In Chapter VIII it is shown that for this surface, the curve $\alpha = 3a+b$ (when represented as an embedded curve) is a candidate for modifying the surface F into a ribbon surface for K. As we are about to see, α is a connected sum of a trefoil and its mirror image (as $\alpha \subset S^3$). Hence α itself bounds a ribbon surface. Furthermore, α lies next to a parallel copy of itself on F with no twisting. Hence, we can remove an annulus from F and replace with two copies of the ribbon surface for α. By keeping track of how the remainder of F intersects these parallel surfaces, a ribbon surface for K is produced.

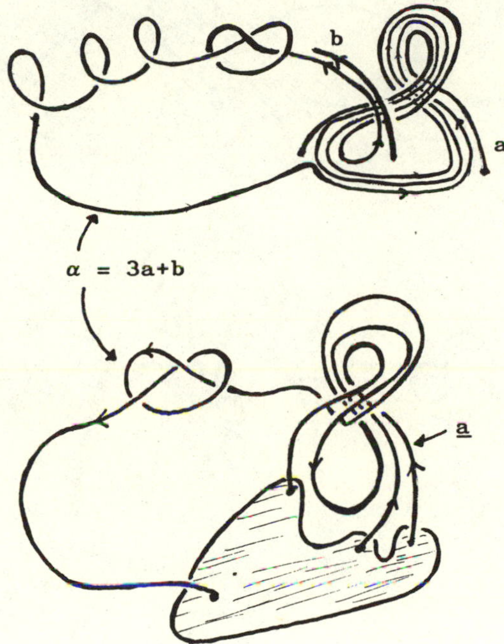

Here we show α = 3a+b, first in relation to script for a and b, then as script itself in relation to a. And we begin to isotope the linking between a and α.

Note that a actually represents the handle on which

a lives. Also, since the total self-linking of α is zero (i.e., if you take two parallel copies of α on F, then they have zero linking number), we can ignore curls on the script for α (and tally up a zero total twist plus writhe later on).

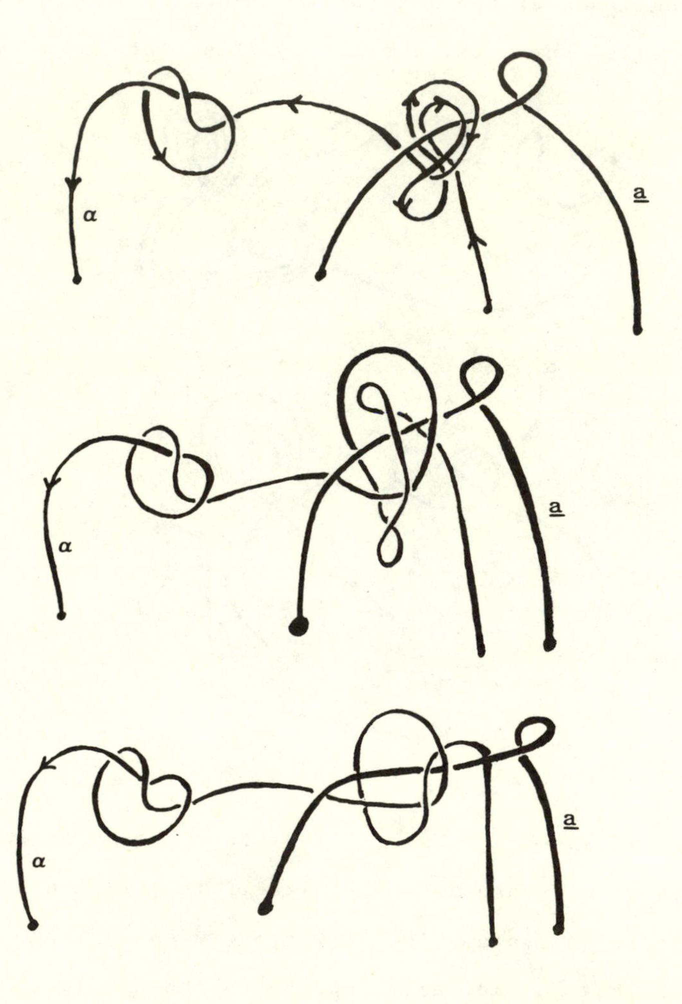

This scripting shows how α becomes a connected sum of trefoil and mirror trefoil. Just as

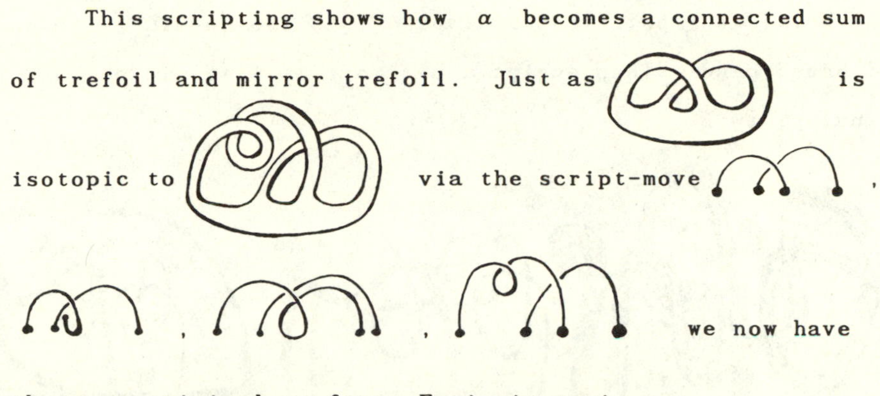

is isotopic to via the script-move , , , we now have that our original surface F is isotopic to:

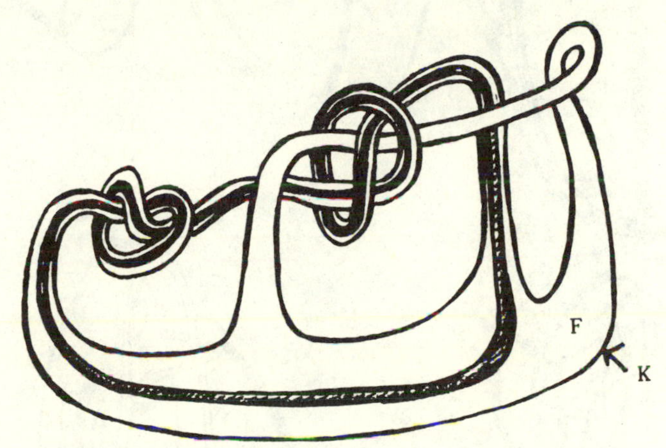

The way forward is now clear. The shaded annulus must be excised from F, and two parallel ribbon surfaces for $T \# T^!$ placed on the resulting parallel boundaries.

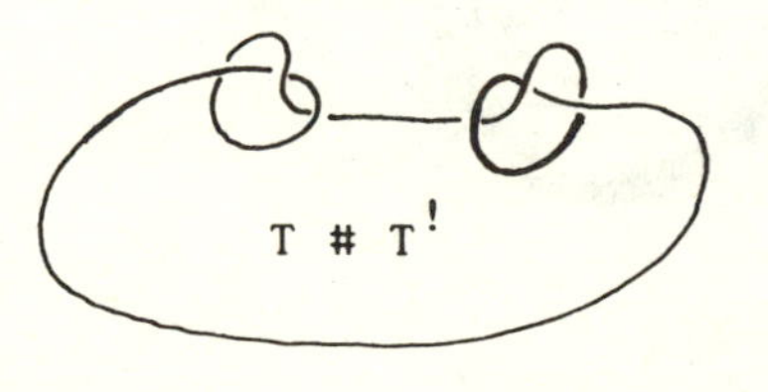

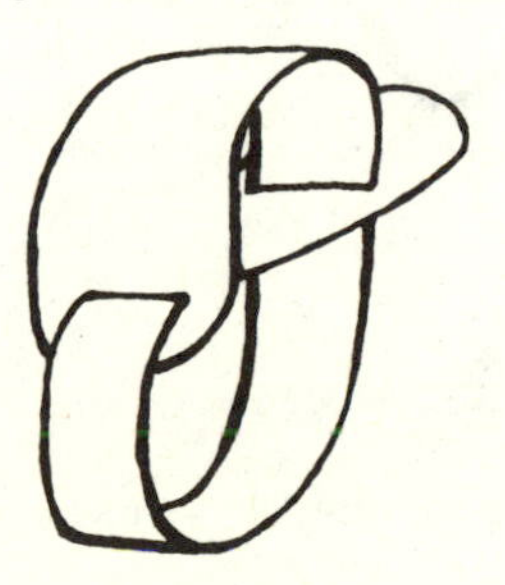

Ribbon for $T \# T^!$

But wait! Why not first isotope F to the annulus with an attached band! Then script a bit more to slide into the final picture.

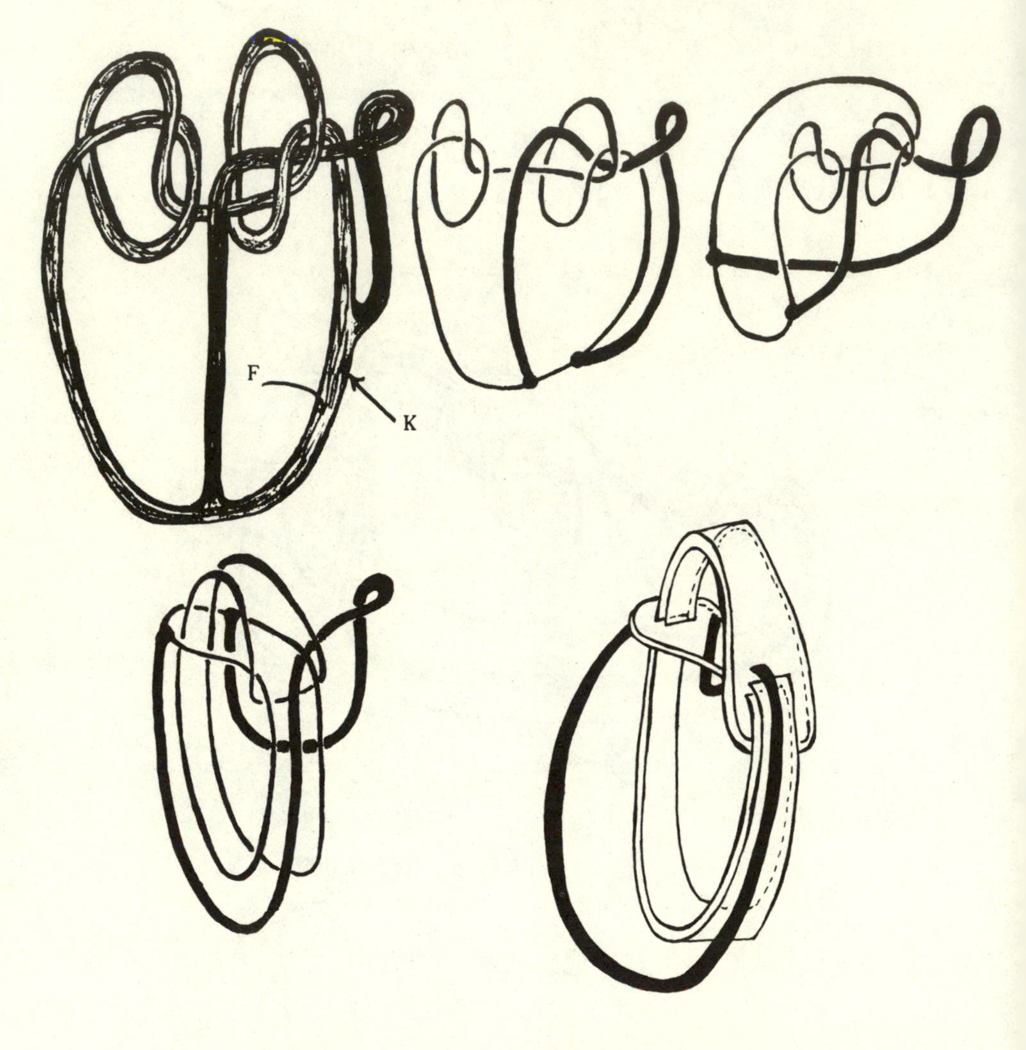

And here is a first version of the ribbon surface for K.

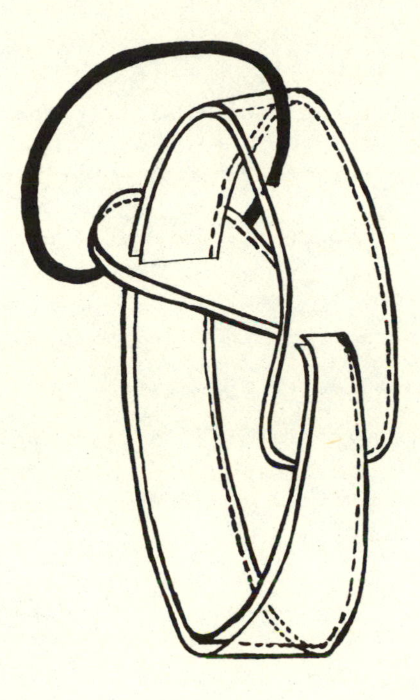

Final Version. A ribbon surface for the 6-twisted double of the trefoil knot. [© 1984. Carmen Safont y Pilar del Valle.]

§14. *KIRKOFF'S MATRIX TREE THEOREM*

Let G be a graph. We shall define the tree-polynomial $\nabla(G)$. It lists all the maximal trees in G, and it is a combinatorial relative of the Conway polynomial. First, label each edge of G with an algebraic variable (one variable per edge). Suppose $\{x_1, \cdots, x_n\}$ is the set of variables. Then $\nabla(G) \in Z[x_1, \cdots, x_n]$. It is defined as follows: For each maximal tree $T \subset G$ let $|T| = x_{i_1}x_{i_2}\cdots x_{i_r}$ where $\{x_{i_1},\cdots,x_{i_r}\}$ is the set of distinct edges in T. Then $\nabla(G) = \sum_{T\in\mathcal{T}} |T|$ where $\mathcal{T}$ denotes the set of maximal trees in G. For example:

$$G \Longrightarrow \nabla(G) = ab + ac + bc.$$

THEOREM 1. *Let* a *be an edge in a graph* G. *Let* $(G-a)$ *be the graph obtained by deleting* $\underline{a}$ *from* G. *Let* (G/a) *be the graph obtained by collapsing* $\underline{a}$ *to a point. Then*

$$\nabla(G) = \nabla(G-a) + a\nabla(G/a).$$

Example:

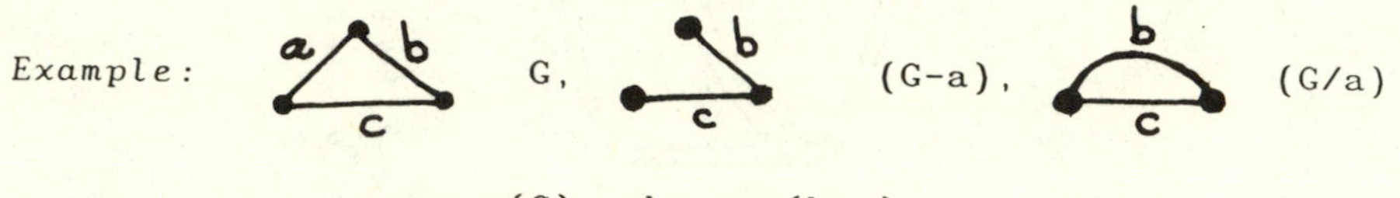

$$\nabla(G) = bc + a(b+c).$$

Proof of Theorem 1. Exercise.

DEFINITION. *Let* G *be a graph with vertices* $1,2,\cdots,m$ *and edges labelled from* $\{x_1,\cdots,x_n\} = X$. *Define the matrix of* G, $A(G) = A$ *by the formulas:*

(i) $A_{ii} = \sum_{x\in X_i} x$ *where* X_i = *those edges in* X *having* i *as a vertex.*

(ii) $A_{ij} = -\sum_{x\in X_{ij}} x$ *where* X_{ij} = *those edges whose vertices are* i *and* j *(when* $i \neq j$*).*

(An empty sum equals zero.)

Example:

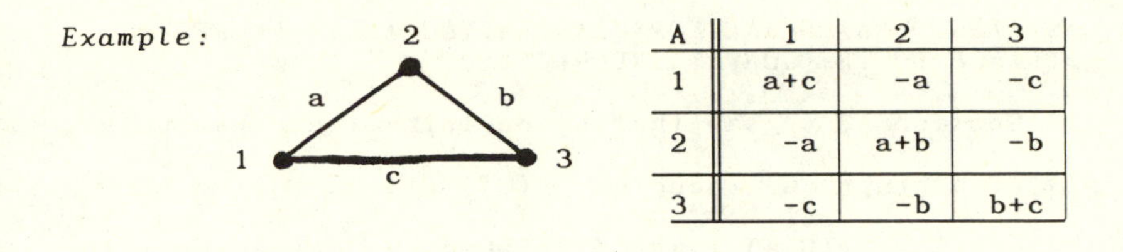

A	1	2	3
1	a+c	-a	-c
2	-a	a+b	-b
3	-c	-b	b+c

THEOREM 2 (Kirkhoff). *Let* $\hat{A}(G)$ *be any matrix obtained from* A *by striking our the* <u>*same*</u> *row and column (i.e., the* i^{th} *row and the* i^{th} *column). Then*

$$\nabla(G) = \mathrm{Det}(\hat{A}(G))$$

where Det *denotes determinant.*

Proof: <u>Exercise</u>. *Hint:* Expand by minors and use Theorem 1.

Remark: The idea for this proof of the Matrix Tree Theorem is due to George Minty, University of Indiana, Bloomington, Indiana.

Example: In the above example, $\hat{A} = \begin{bmatrix} a+b & -b \\ -b & b+c \end{bmatrix}$ and $|\hat{A}| = (a+b)(b+c) - b^2 = ab + ac + bc.$

Example: <u>Exercise</u>. Check out the results of this section on

G

$\mathcal{T}$

<u>Exercise</u>. Read about the <u>Goeritz</u> <u>form</u> in Fox's Quick Trip [F1] and in [GL2]. This is a knot-theoretic occurrence of the Kirkhoff patterns.

§15. *STATES, TRAILS AND FORMULAS HAVING A FAMILY RESEMBLANCE TO THE CONWAY INDENTITY*

In Section 13 we saw that by defining the "tree-polynomial" of a graph G (denoted $\nabla(G)$) it satisfied the identity $\nabla(G) = \nabla(G-a) + a\nabla(G/a)$ when a was an edge for G. This is very similar in form to the identity for the Conway polynomial: $\nabla_K - \nabla_{\overline{K}} = z\nabla_L$. In fact, the similarity runs deeper. There is a model for the Conway polynomial that comes from enumeration related to the knot diagram. Rather than enumerate trees directly, we enumerate states of a diagram (to be explained in a moment), and these states are in 1-1 correspondence with what I call Jordan trails. A Jordan trail is a traverse of the diagram that goes through every edge once, and does not cross any crossing:

For example,

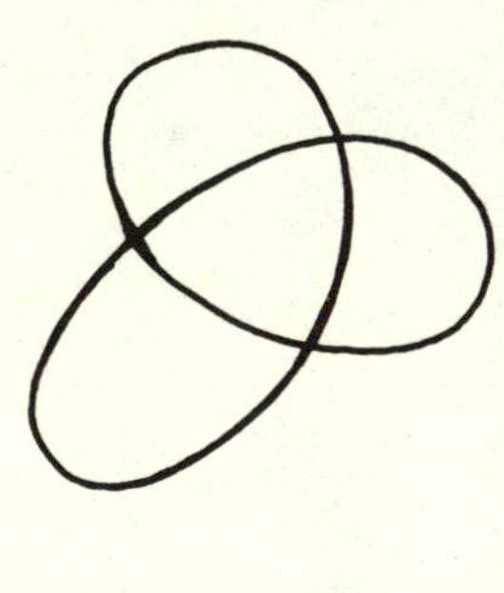

diagram

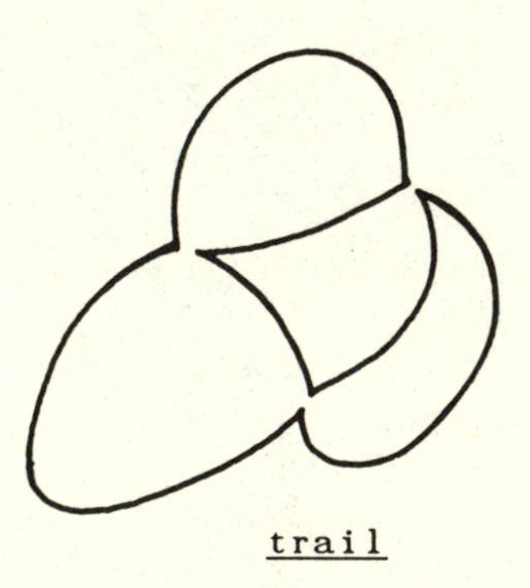

trail

It turns out that the collection of trails on a given diagram (note that here we do not have crossing choices so the diagram is actually a planar graph) can be enumerated by a polynomial that is the determinant of a certain matrix; and this matrix is actually a version of the classical Alexander matrix used by J.W. Alexander for the Alexander polynomial [A1].

The movement to a polynomial involves reformulating trails into <u>states</u>. A state is an extra structure on the knot diagram that takes the form of the addition of one <u>marker</u> at one of the four corners of each vertex.

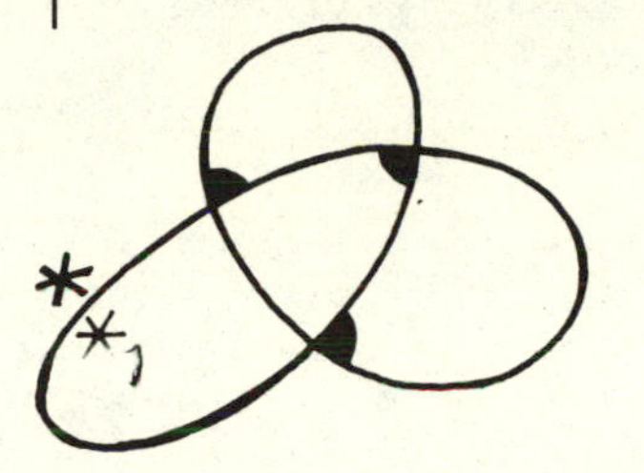

S

is a state for the trefoil diagram.

We further stipulate that

(1) No region contains more than one marker;

(2) The two regions missing a marker are adjacent (this last can be weakened). These regions are <u>starred</u> (*).

Then we have

PROPOSITION. *Let* $\mathcal{S}$ *be the set of states for a planar diagram* $\mathcal{D}$ *(with a given choice of fixed stars). Let* $\mathcal{T}$ *be the set of Jordan trails for* $\mathcal{D}$. *Then* $\mathcal{T}$ *and* $\mathcal{D}$ *are in 1-1 correspondence.*

One direction of this correspondence is as follows: <u>Split</u> each crossing at the state marker according to the scheme

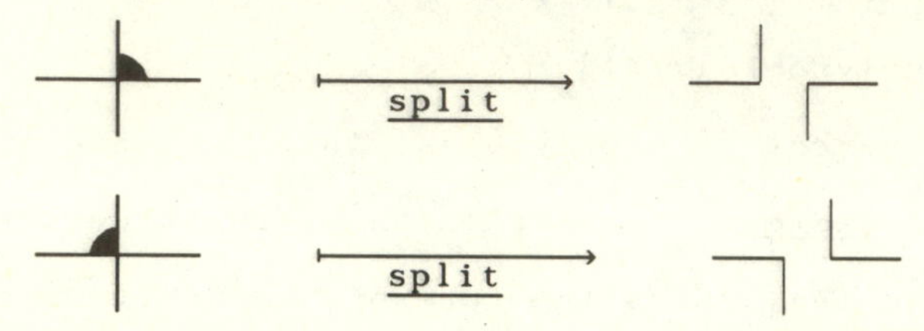

Then if S is a state, <u>Split</u> (S) is a trail!

Example:

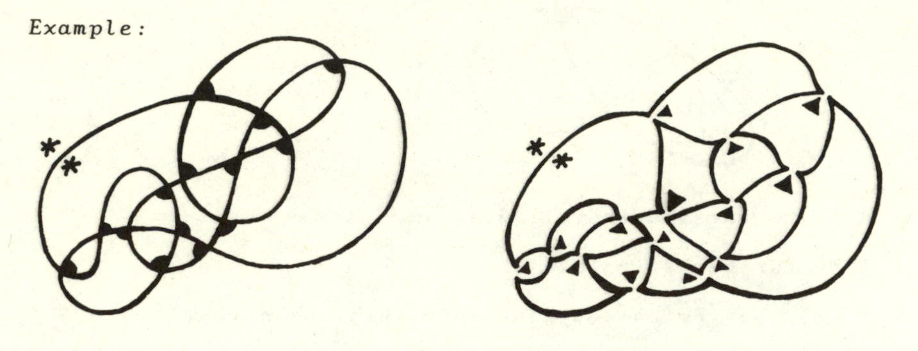

S <u>Split</u> (S)

States can be enumerated by a determinant because they correspond to certain ways that each vertex chooses one region. Thus we will use a vertex-region matrix. To form the entries of the matrix, label each vertex by the scheme:

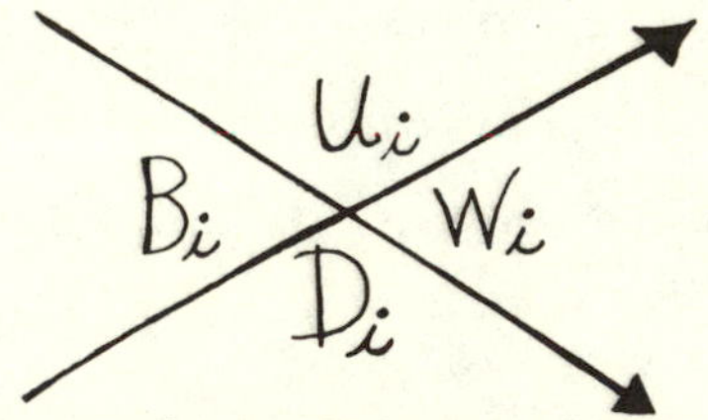

(i^{th} vertex)

(We orient the diagram.)

and place these labels under the columns corresponding to

their regions in the matrix. Place zeroes elsewhere in the row corresponding to the ith vertex.

Example:

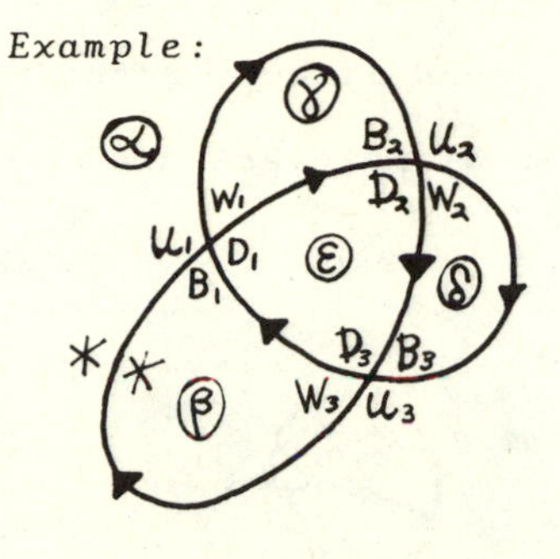

region / vertex	α	β	γ	δ	ϵ
1	U_1	B_1	W_1	0	D_1
2	U_2	0	B_2	W_2	D_2
3	U_3	W_3	0	B_3	D_3

If the stars are in α, β then we take the square submatrix whose columns are γ, δ, ϵ. Let M denote this submatrix.

$$\mathrm{Det}(M) = \begin{vmatrix} W_1 & 0 & D_1 \\ B_2 & W_2 & D_2 \\ 0 & B_3 & D_3 \end{vmatrix} = W_1 \begin{vmatrix} W_2 & D_2 \\ B_3 & D_3 \end{vmatrix} + D_1 \begin{vmatrix} B_2 & W_2 \\ 0 & B_3 \end{vmatrix}$$

$$= W_1 W_2 D_3 - W_1 D_2 B_3 + D_1 B_2 B_3.$$

These three terms are exactly the states of the diagram.

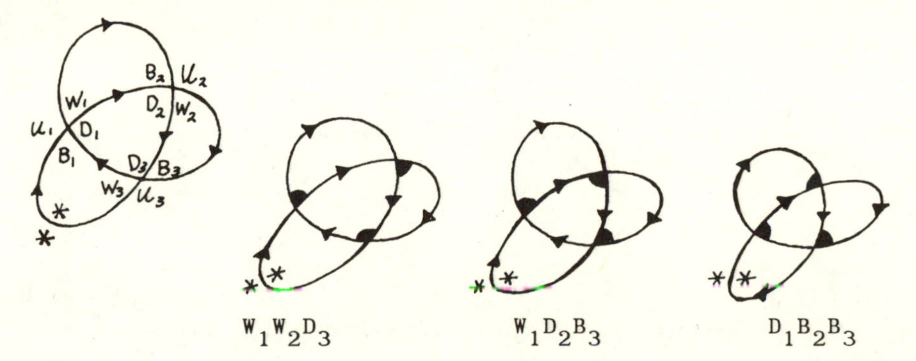

(Note that we produce the state corresponding to $W_1W_2D_3$ by placing markers at these labels. Thus states (hence trails) are enumerated by computing a determinant.

In applying this determinant, we found it convenient to orient the diagram. This results in strange orientations on the trails:

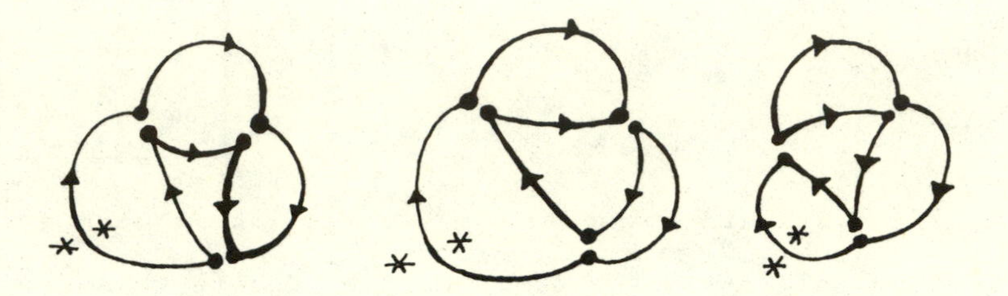

Trails

As you walk along the trail, you may be forced to change orientation when you come to a crossing: But it's all-right for a pedestrian to walk the wrong way on a one-way street!

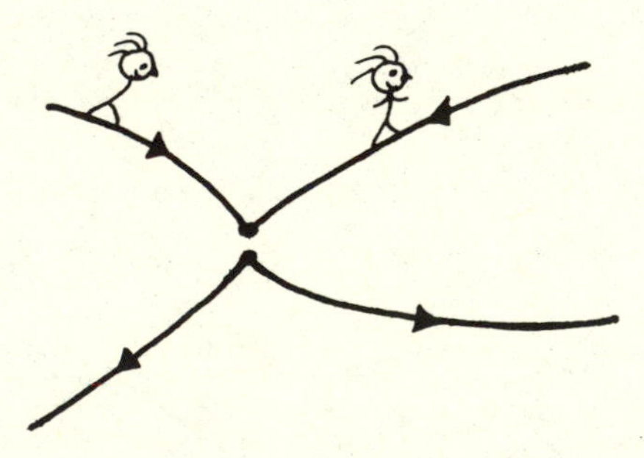

Let's concentrate on the oriented crossings:

with [W]

(<u>against</u>)(<u>back</u>) [B]

Even here, you may be walking with the orientation (as the little fellow on top) or <u>against</u> it (as the little fellow on bottom).

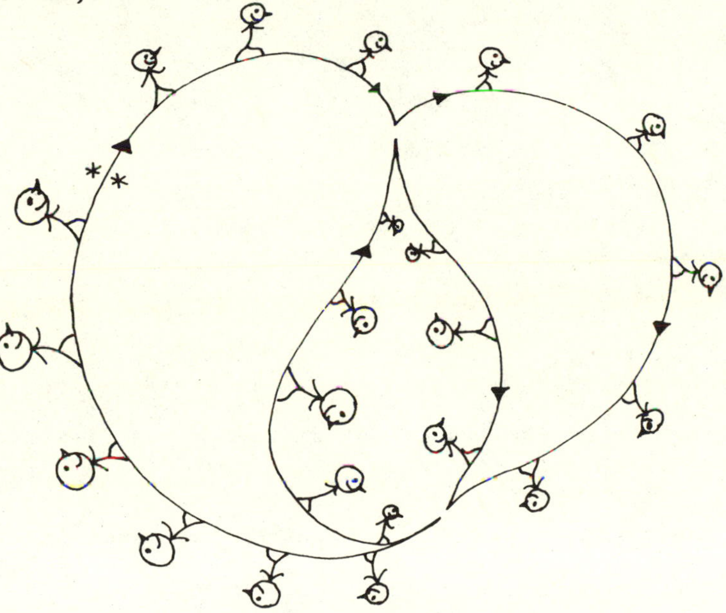

Letting this little fellow walk along the diagram we see that when he goes through an oriented crossing (in the trail) then on one trip he's with it, while on the second (return trip), he's against it. This can happen in either order, but if we start him from in-between the two stars, then we can classify crossings as W (<u>with</u> <u>it</u>, <u>white</u>) if the little fellow passes through in an oriented way the <u>first</u> <u>time</u>, and B (<u>back</u>, <u>against</u>, <u>black</u>) if he passes through against the orientation on the first time. <u>Thus</u>:

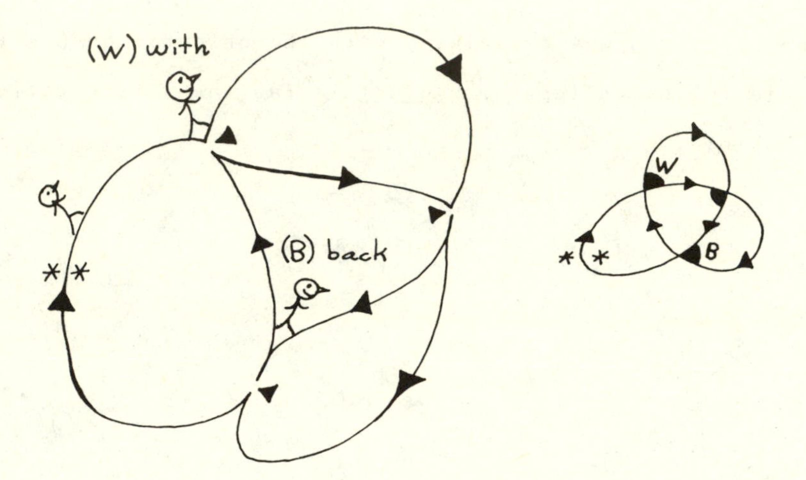

We have drawn the little fellow as he appears on his first traverse of the two oriented sites. His first traverse is <u>with</u> <u>it</u> (W) on one end and <u>backward</u> (B) on the other. Note that this is catalogued by the marker in the corresponding state:

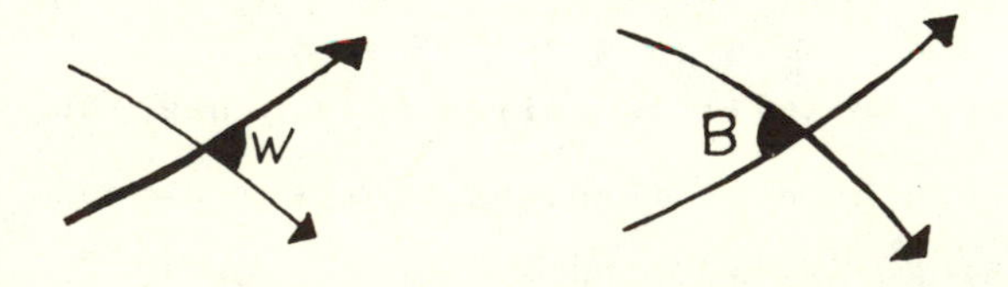

In the states form, I like to call these <u>white holes</u> (W) and <u>black holes</u> (B) because material is supposed to <u>come out</u> of a white hole

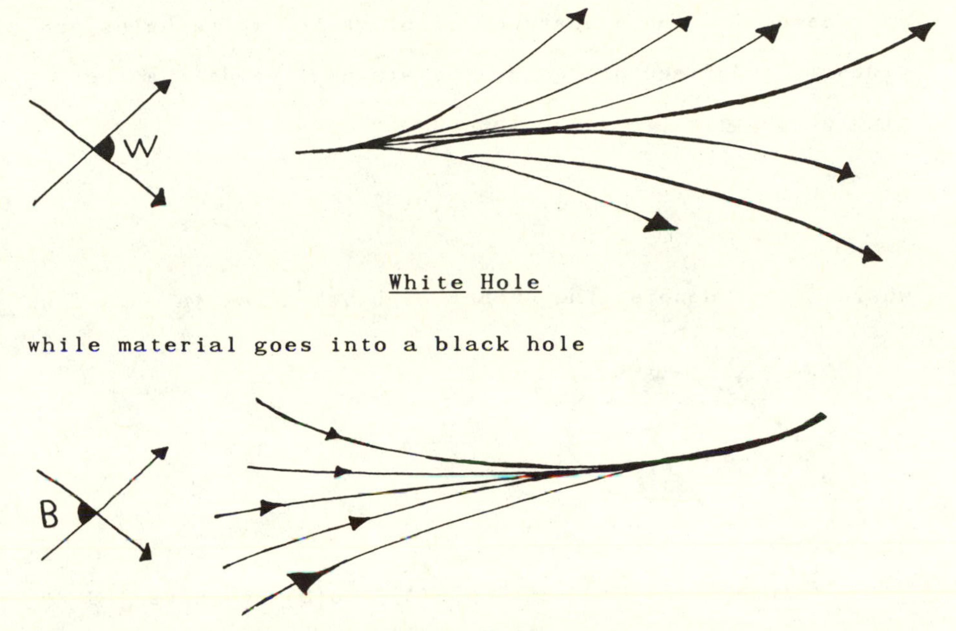

<u>White Hole</u>

while material goes into a black hole

<u>Black Hole</u>

Now some combinatorial miracles happen! First of all there is the <u>Duality Theorem</u> which we conjectured in [K1] and which was proved by P. Gilmer and R. Litherland [GL1]. Their proof involves some new and very beautiful identities for the state polynomial.

DUALITY THEOREM. *Let* $\mathcal{S}$ *be the set of states for an oriented diagram* $\mathfrak{D}$ *(with fixed adjacent stars). Let* $\mathcal{N}(r,s)$ *denote the number of states in* $\mathcal{S}$ *with* r *black holes and* s *white holes. Then* $\mathcal{N}(r,s) = \mathcal{N}(s,r)$ *for all* r *and* s.

Secondly, the occurrence of black and white holes provides well-defined <u>signs</u> for the states: We define the <u>sign of a state</u> $\mathcal{S}$ to be the number

$$\sigma(S) = (-1)^{b(S)}$$

where b(S) denotes the number of black holes in S. Thus

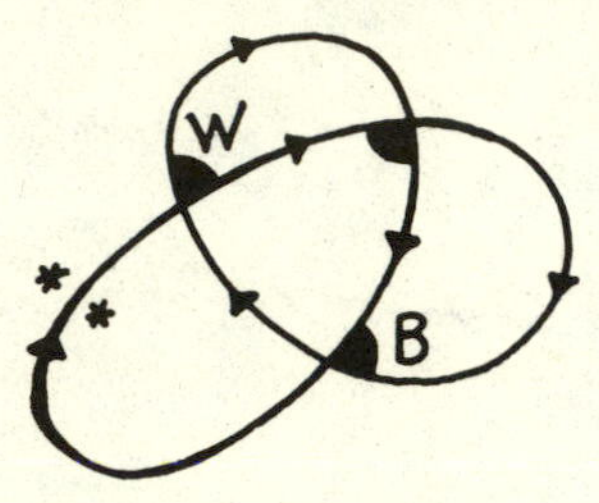

$\Longrightarrow \sigma(S) = -1.$

It then turns out that $\sigma(S') = -\sigma(S)$ whenever S′ is obtained from S by <u>transposition of markers</u>. By a transposition of markers, I mean a change of state in the form

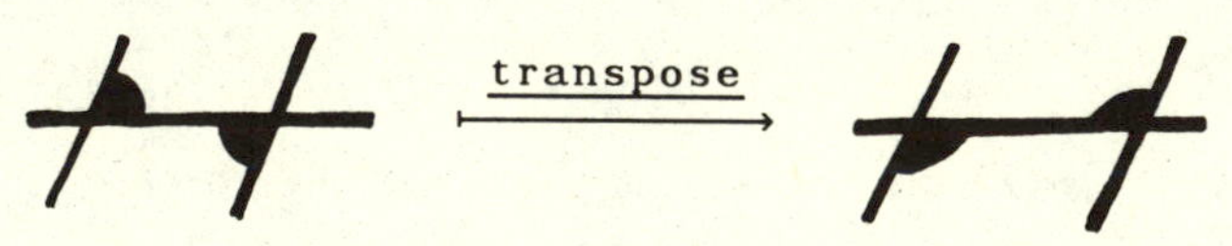

Even deeper is the fact that any two states can be obtained, one from another, by a sequence of transpositions of markers. This is the content of the <u>Clock Theorem</u> [K1].

Using these signs, we can define polynomials independently of the determinants: Given a labelled knot diagram

K (for example, labelled with B_i, U_i, W_i, D_i at the ith crossing) and a state S of this diagram, define

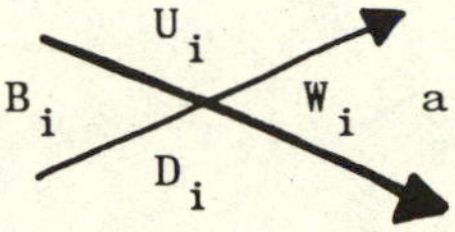

$\langle K|S\rangle$ = <u>the product of the labels that are touched by state markers</u>, where we think of the labels as elements of a commutative polynomial ring. Thus if K = [diagram with labels U_1, B_1, W_1, D_1, D_2, W_2, B_2, U_2]

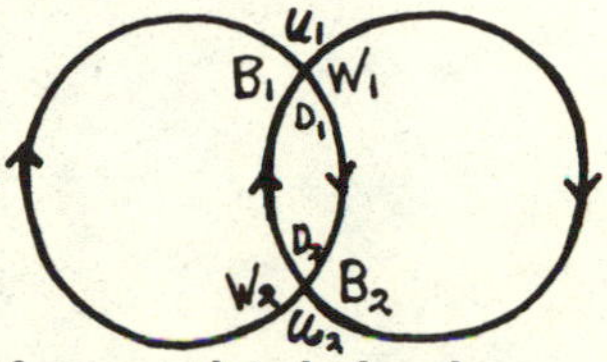

and S = [diagram], then $\langle K|S\rangle = D_1B_2$. If you think back to the example computation of Det(M), you will see that $\mathrm{Det}(M) = \sum_{S\in\mathcal{S}} \pm \langle K|S\rangle$ where the sign $\pm$ comes from the determinant expansion.

Accordingly, we define the <u>state polynomial</u>, ∇_K, by the formula

$$\nabla_K = \sum_{S\in\mathcal{S}} \sigma(S)\langle K|S\rangle .$$

This depends upon choosing a collection of states $\mathcal{S}$ (hence choosing fixed stars) and also upon the labellings for K.

<u>What labellings are appropriate for knot theory</u>? This is a long story. In Section 9.11 (Chapter IX of these notes) you can read an account of how Alexander invented a labelling in the form

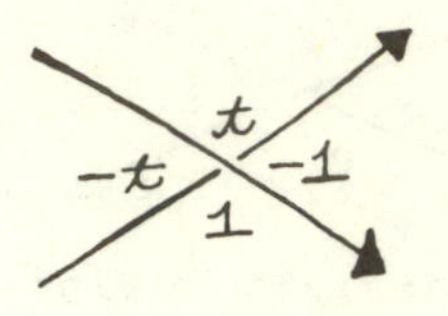

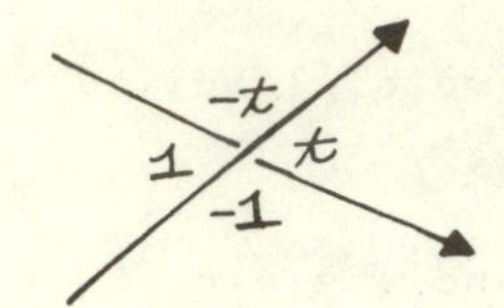

This gives rise to the Alexander polynomial $\Delta(t)$.

Thus

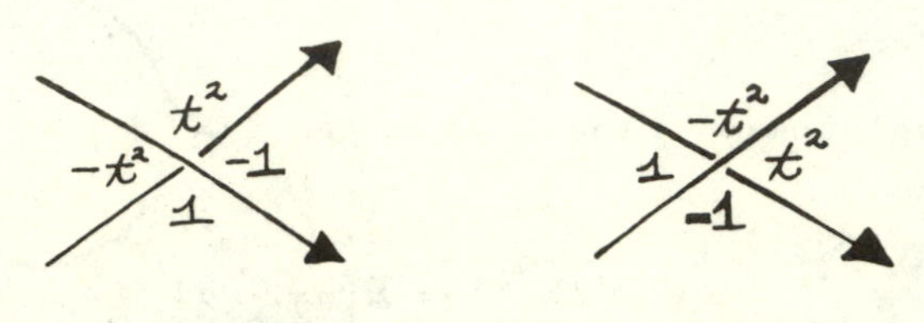

yields $\Delta(t^2)$. As it turns out, the Alexander polynomial is only well-defined up to sign and powers of t. Hence we can multiply every row of the matrix giving $\Delta(t^2)$ by t^{-1}. This changes the labelling to

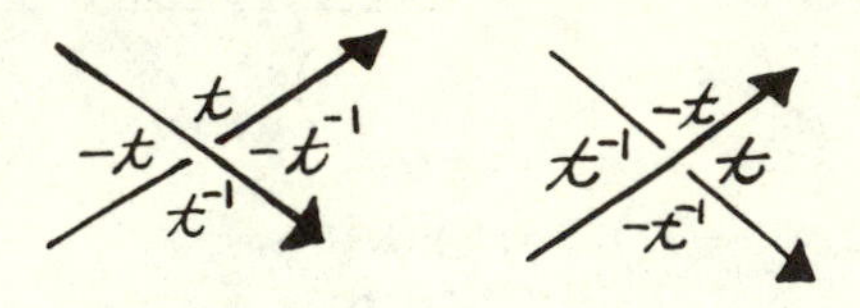

This still computes $\Delta(t^2)$.

Throwing away the signs is okay. So:

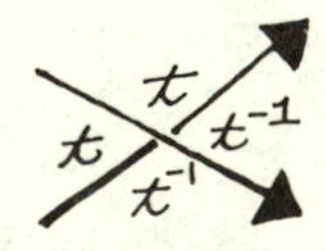

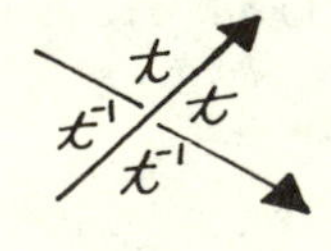

still computes $\Delta(t^2)$.

Then some thought shows that the vertical t's

actually contribute a factor $t^{d(k)}$ where

$d(k)$ is the so-called Whitney degree of the plane-curve [W1]. Thus

still computes $\Delta(t^2)$. But now everything is normalized, and diagrams using this labelling in fact compute, via $\nabla_K = \sum_{s \in \mathcal{S}} \sigma(s)\langle K|S\rangle$, the <u>Conway</u> <u>Polynomial</u> with $z = t^{-1} - t$.

Let's do an example: $(\bar{t} = t^{-1})$

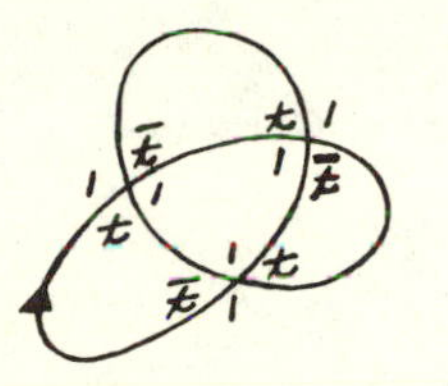

K

$\sigma = +1$

$\sigma = -1$

$\sigma = +1$

$$\nabla_K = +\bar{t}^2 - \bar{t}t + t^2 = (\bar{t}-t)^2 + \bar{t}t$$

$$\therefore \quad \nabla_K = z^2 + 1 \quad (z = \bar{t}-t).$$

This checks with what we know.

Thus we have a combinatorial model of the Conway polynomial that is obtained by taking the state polynomial

$$\nabla_K = \sum_{S \in \mathcal{S}} \sigma(S)\langle K|S\rangle$$

corresponding to the labelling

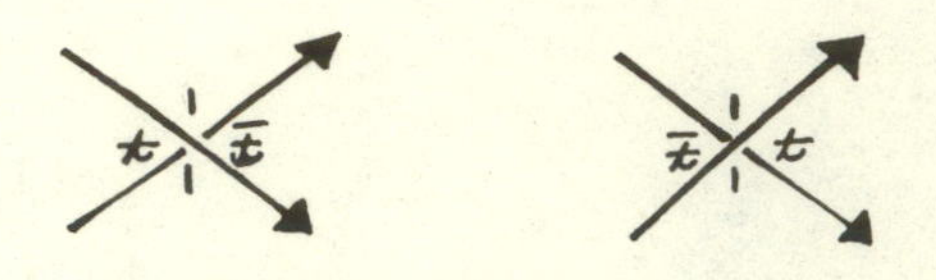

and $z = \bar{t} - t$, $\bar{t} = t^{-1}$. One can prove that ∇_K is a topological invariant and that it does not depend upon the choice of stars defining the state collection.

The most interesting aspect of this model for the Conway polynomial is how the proof of the exchange identity works. Not to spoil your fun, we leave this as an exercise.

<u>Exercise</u>. Use the state-polynomial model for the Conway polynomial to prove that $\nabla_K - \nabla_{\bar{K}} = (\bar{t}-t)\nabla_L$ when

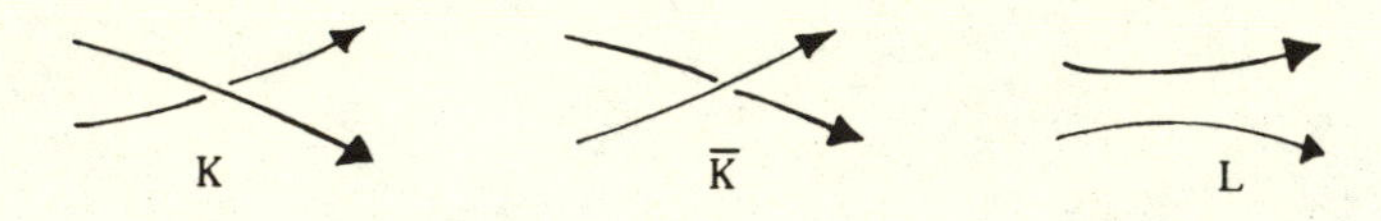

the three knots are links and are related as shown above.

Hint: Note how the states of are related to the states of . For example, look at and .

<u>Exercise</u>. In fact, the states are also in 1-1 correspondence with a collection of maximal trees. These are the trees in the <u>graph of the diagram</u> G(D). Given a knot

diagram D we can checkerboard color it with two colors as in

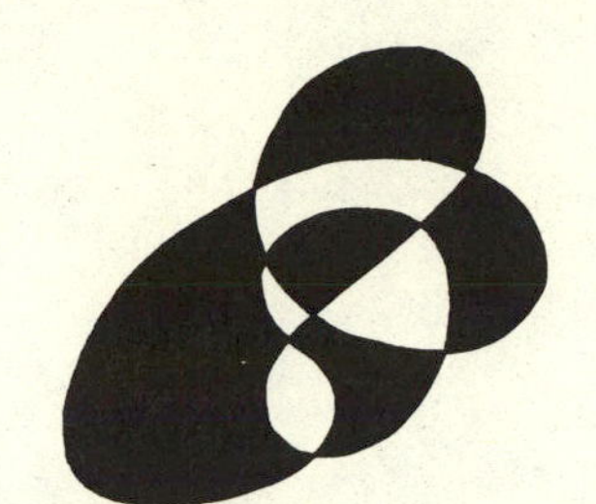

D

(Colored *black* and *white*.)

The graph G(D) is obtained by putting one vertex in the interior of each *white* region, and joining two vertices whenever there is a crossing in D that is incident to both their regions.

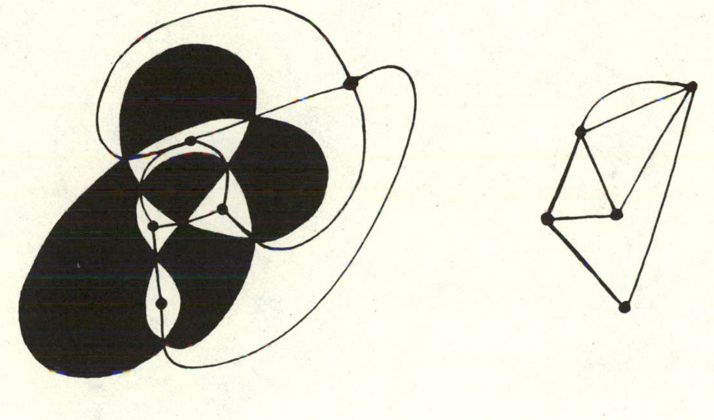

D G(D)

Maximal trees in G(D) *are in one-to-one correspondence with states* $\mathcal{S}$ *of* D (for a given choice of fixed stars).

The picture below showing how we construct a state from a tree should be a good hint for proving this statement.

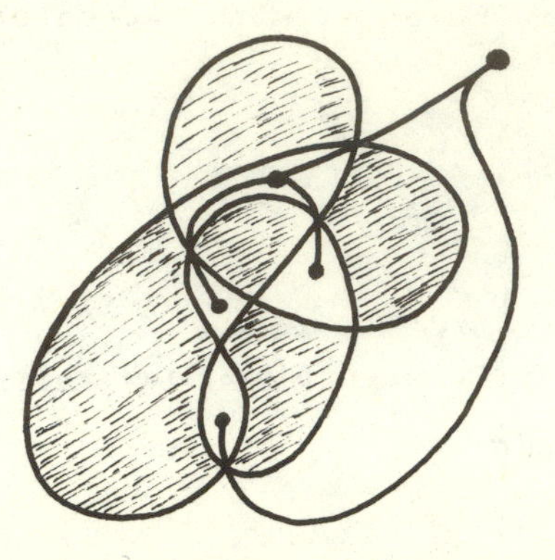

maximal tree
T in G(D).

<u>choice of stars</u>

Let $\hat{G}(D)$ = dual graph constructed from the black regions. Then T and choice of stars determines $\hat{T} \subset \hat{G}$.

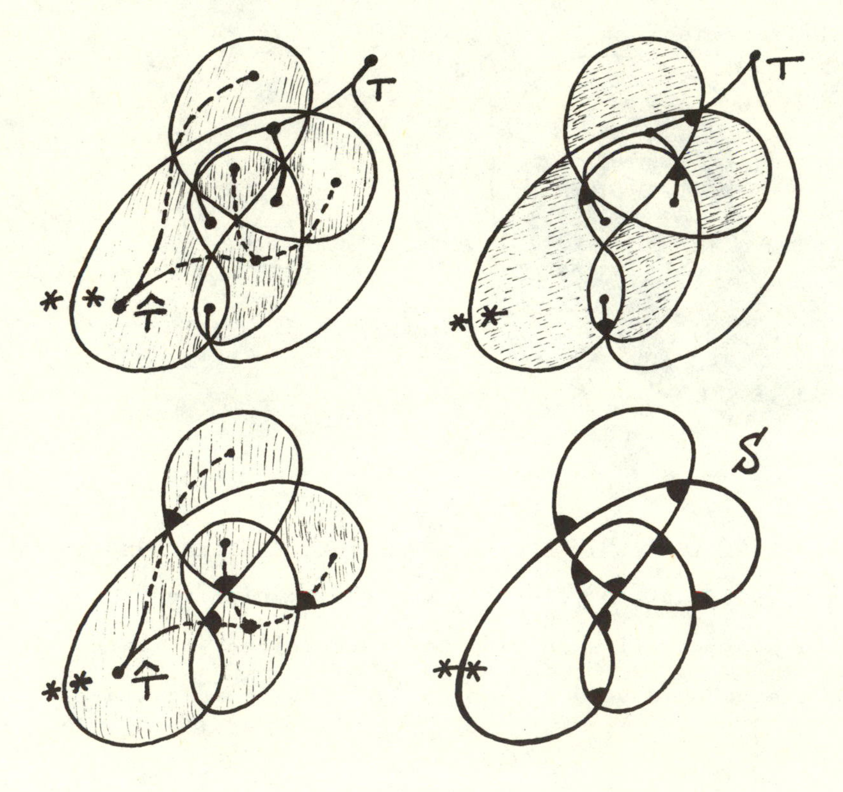

The two trees T and $\hat{T}$ cooperate to produce the state S.

§16. *THE MAP THEOREM*

Let G be a planar graph having no more than one edge between any two vertices. It is a well-known problem to show that there exists an assignment of four colors to the vertices of G so that no two vertices that are connected by an edge have the same color. Direct connection forces difference.

Let $F(G)$ denote the number of different ways to color G with four colors. For example, given the colors R (red), B (blue), P (purple), W (white) and $G = \bullet$, then $F(G) = 4$: $\underset{R}{\bullet}$, $\underset{B}{\bullet}$, $\underset{P}{\bullet}$, $\underset{W}{\bullet}$. If $G = \bullet\!\!-\!\!\bullet$, then $F(G) = 12$. Then it is easy to see that for any edge $\bullet\!\overset{a}{-}\!\bullet$ in an arbitrary graph G we have

$$(*) \qquad F(G-a) = F(G) + F(G/a)$$

since in $G-a$ the vertices joined by a may be <u>colored differently</u> (giving a coloring of G) or they may be <u>colored the same</u> (giving a coloring of G/a). We may call this the <u>chromatic formula</u> (*). (See [W2].)

If we rewrite the chromatic formula as

$$F(G) = F(G-a) - F(G/a),$$

then it can be used to inductively determine $F(G)$.

<u>For example</u>: $F(\bullet\!\!-\!\!\bullet) = F(\bullet\ \bullet) - F(\bullet)$

$$= 16 - 4$$

$$= 12.$$

Or $F(\ ) = F(\ ) - F(\ )$

$= F(\ ) - F(\ )$ (drop multiple edges)

$= F(\ ) - F(\ ) - F(\ )$

$= 4 \cdot 12 - 12 - 12$

$= 24.$

Certainly, *these calculations with the chromatic formula bear a family relationship with our calculations of knot polynomials via the Conway identity*. (See the appendix for a direct relationship with the Jones polynomial.)

What does the chromatic formula tell us about the possibility of $F(G) = 0$? If G was a non-four-colorable planar map, then we would have

$$0 = F(G-a) - F(G/a)$$

for some edge a in G. Let's suppose that G is a smallest noncolorable so that $F(G-a) \neq 0 \neq F(G/a)$. Then $F(G-a) = F(G/a)$ means that the two "exposed" vertices x and y of $G-a$ (formerly the ends of a) always receive the *same color* in every coloring of $G-a$. We will say that x and y are *forced*, and that $\Gamma = (G-a)$ is a *tyrant*.

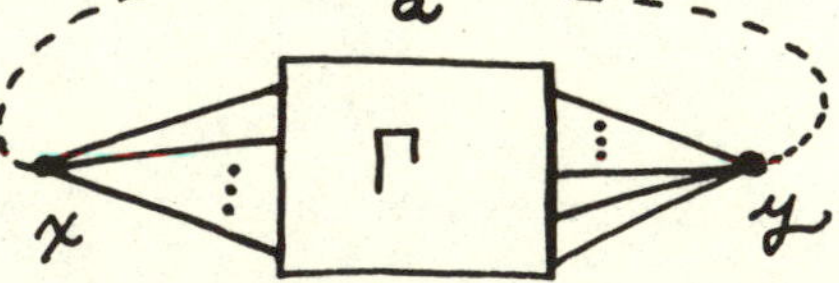

Thus the four-color problem is: *do tyrants exist*?

It is, in fact, very trying to try to create a tyrant. After all, if Γ is a tyrant, then Γ is also an extraordinarily good communicator between x and y. If we switch the color of x and start following down the consequences, they must eventually switch y as well, if Γ is to obtain a new coloring.

Let's try to make a tyrant:

R B R B R B R B R
x y

Here's a nontyrant, but if you switch x's red (R) to B and switch any B's that are forced to change, to R, then such switching eventually flips y to B as well. Thus, the above graph can be a tyrant in a very limited domain (of two colors).

The first thing we notice about a tyrant (for four colors) is that x <u>must</u> be connected directly to at least <u>3</u> other vertices. Otherwise x would be free to change color without affecting y. And these 3 (or more) vertices directly connected to x must always have three different colors among them (for the same reason). All this is true for y as well:

at least 3 at least 3

x ? y

We could try to do this as simply as possible. Suppose x is connected to exactly 3 other vertices:

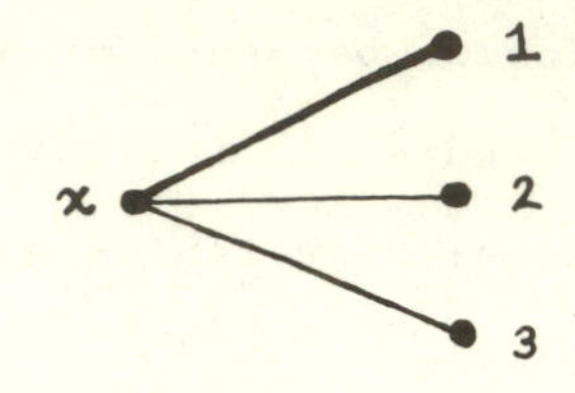

If we try to ensure that these (1,2,3) are all colored differently by directly connecting them we find:

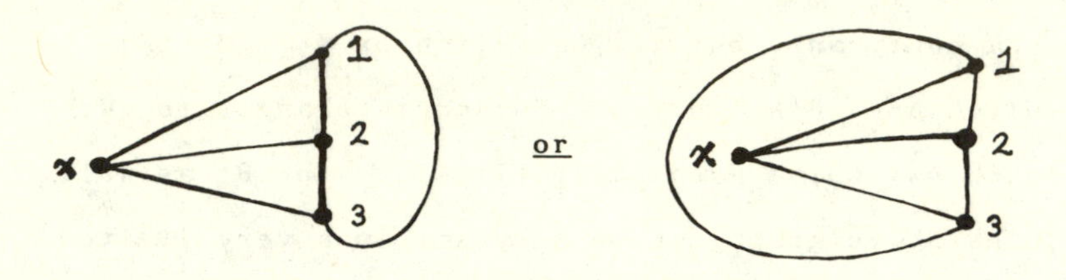

The second possibility is out since there must exist a path from x to y.

The first possibility doesn't look too promising. We've dropped 2 from further direct communication with the nether-reaches of the map. For example:

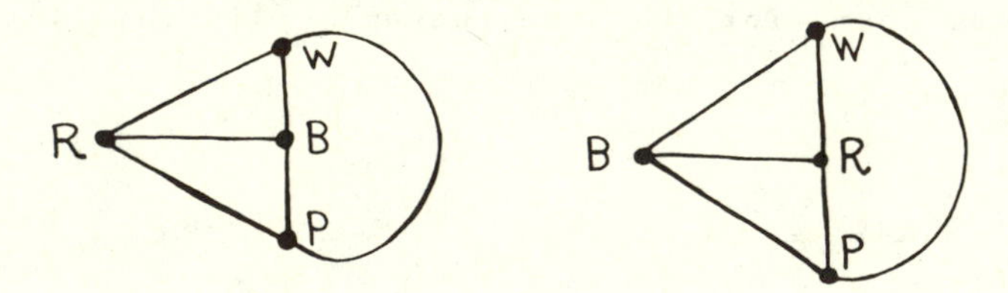

we change the color at x without disturbing 1 or 3. The result is disastrous for tyrants! Of course, we can try to bring 2 back into communication:

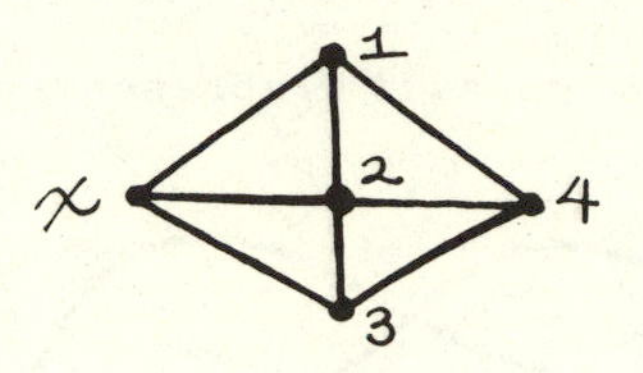

But this splits the communication between 1 and 3, allowing

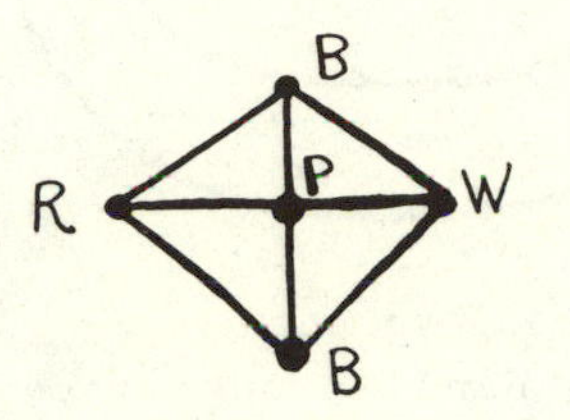

with x free to change colors.

Perhaps you begin to see the difficulty in constructing a tyrant. Proving that tyrants don't exist is, of course, a dangerous game. The best guarded proof at the moment is due to Haken and Appell, using much computer technique [AH].

Exercise. How many four colorings are there for the graph below? Say hello to the nonplanar tyrant:

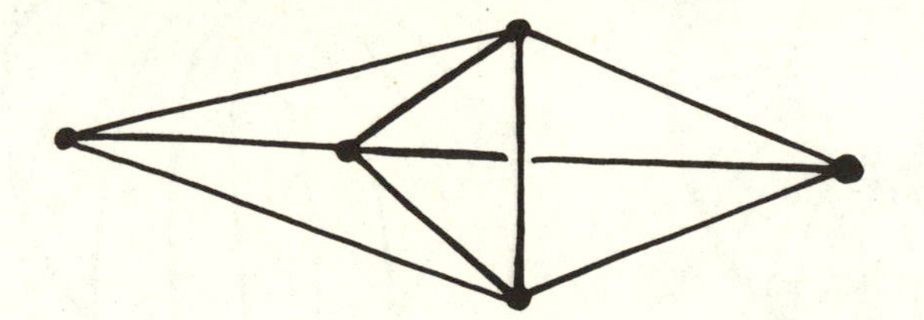

I sometimes dream of proving that any tyrant has $\mathcal{T}$ as a substructure. Ah so...

§17. *THE MOBIUS BAND*

The Mobius band is usually represented by a drawing such as:

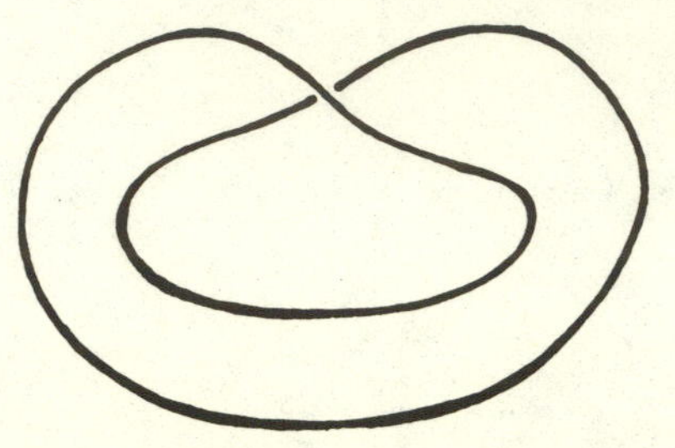

Since the boundary of the band is unknotted it makes a nice research problem in three-dimensional graphics to <u>find embeddings of the band with standardly unknotted boundary</u>. For a good history of this problem, see [S2].

Here follows an original solution by Carmen Safont (her drawings):

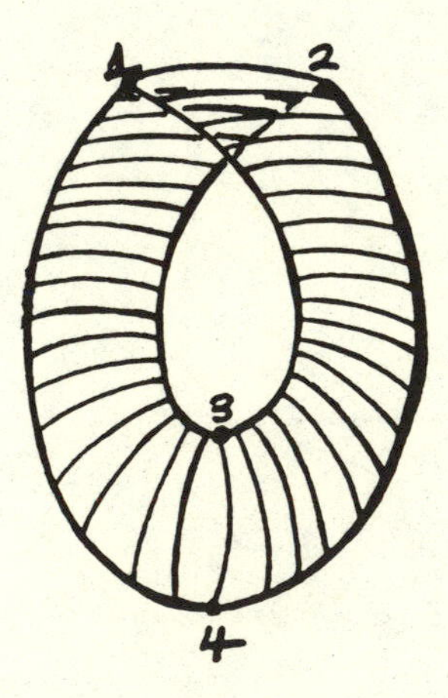

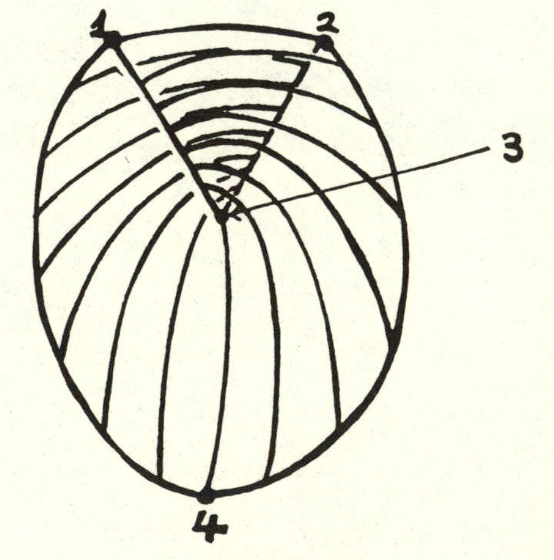

<u>Usual Mobius</u> <u>Twist Over</u>

Straighten 1-2-3, and begin to smooth.

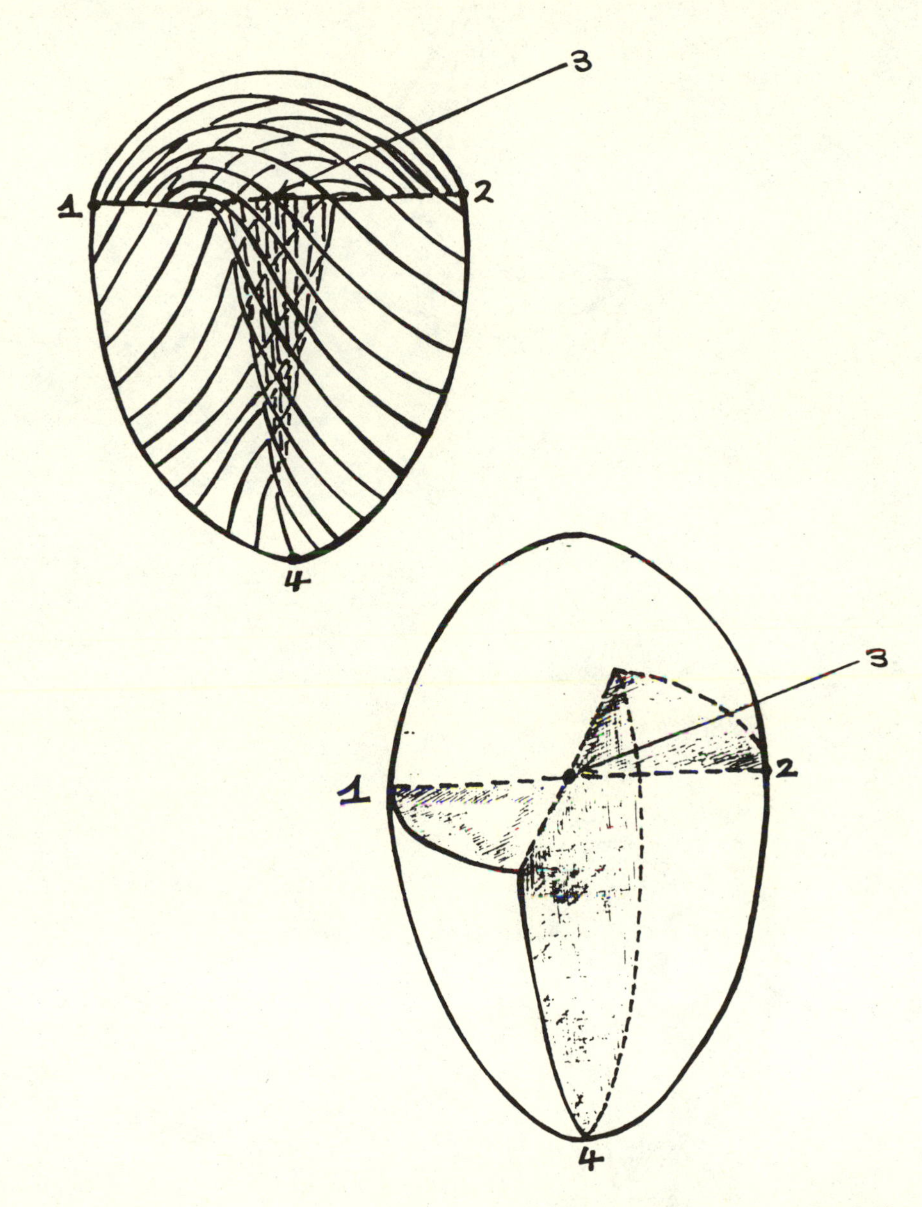

somewhat more rounded

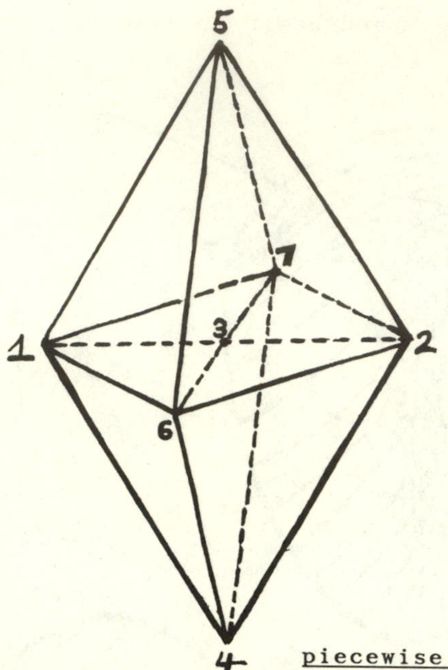

piecewise linear

Here the simplices {156, 652, 257, 751, 136, 237, 467, 462, 417} form the Mobius Band. Edges 14, 24, 23, 31 form the boundary.

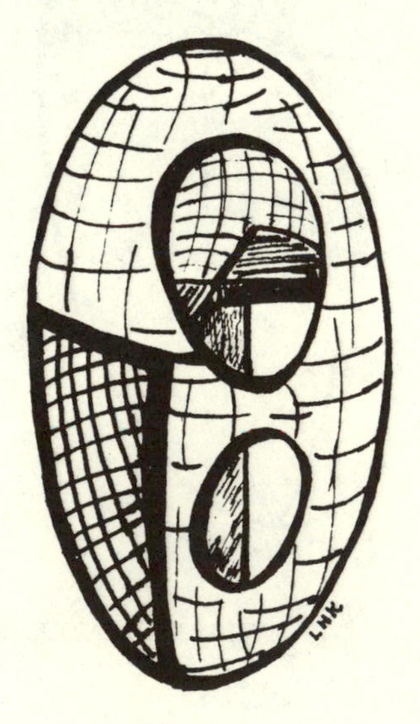

The Goblet of Mobius
(with two (round) windows)

Another possibility for representation is the use of a computer to produce drawings of the stereographic projection of the following embedding of the Mobius band in $S^3 = \{(z_1, z_2) \mid |z_1|^2 + |z_2|^2 = 1\}$: (This has been done by L. Siebenmann, Dan Asimov, the author,...)

$$[0, 2\pi] \times \left[-\frac{\pi}{2}, \frac{\pi}{2}\right] \xrightarrow{M} S^3$$

$$M(\theta, \phi) = (\cos(\phi)e^{i\theta},\ \sin(\phi)e^{i\theta/2})$$

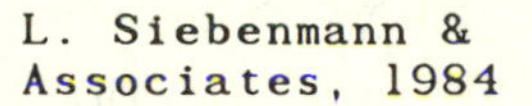

L. Siebenmann & Associates, 1984

§18. *THE GENERALIZED POLYNOMIAL*

An extraordinary breakthrough occurred in knot theory during the Summer of 1984 (just after the course on which these notes are based, and during the time they were being prepared). Vaughan Jones [JO1] discovered a new polynomial invariant of oriented knots and links that satisfied a set of axioms as follows:

AXIOMS FOR JONES' POLYNOMIAL

1. *To each oriented knot or link* K *there is associated a Laurent polynomial* $V_K(t) \in Z[t,t^{-1}]$ *such that* $V_K = V_{K'}$ *whenever* K *and* K′ *are ambient isotopic.*
2. $V \quad = 1$.
3. $t^{-1}\nabla_K - t\,V_{\overline{K}} = (\sqrt{t} - 1/\sqrt{t})V_L$ *whenever* K, $\overline{K}$ *and* L *are related as*

Another "Conway polynomial"! For knot theorists this was like the appearance of a new star in the sky. And Jones' work was stranger than that. He obtained his polynomial via a construction of new representations of the classical braid group into operator algebras previously used mainly in quantum mechanics and statistical physics! Furthermore, this new polynomial has the incredible ability to distinguish many <u>knots</u> (including the trefoil) from their mirror images. Thus it has a power that no simple invariant had previously assumed.

It is beyond the capability of these notes to go into Jones' representation theory. However, the story does not end there. Many people heard Jones' initial lectures in July of 1984 (the present author alas not among them) and among these a number saw clearly to a generalization embracing both the Conway and Jones polynomials. Thus the <u>generalized polynomial</u>, G_K, was brought forth independently by: Ken Millet and Raymond Lickorish, Jim Hoste, Peter Freyd and David Yetter, Adrian Ocneanu (see [HOMFLY]). Of these people, all did their work by inductive technique on diagrams or braid diagrams (in the case of Freyd-Yetter) except Ocneanu who generalized the braid-representation theory. The polynomial was also independently discovered by Jozef H. Przytycki and Pawel Traczyk [PR].

The generalization comes about as follows:

$$\text{Conway} \Longrightarrow \text{(A)} \quad \nabla_K - \nabla_{\overline{K}} = z\nabla_L$$

$$\text{Jones} \Longrightarrow \text{(B)} \quad t^{-1}V_K - t\,V_{\overline{K}} = \left(\sqrt{t} - \frac{1}{\sqrt{t}}\right)V_L.$$

(For yet another generalized polynomial due to the author, and further discussion of the Jones polynomial, see the Appendix to these notes.)

Question: For what "arbitrary" coefficients a,b,c does the relation

$$aG_K + bG_{\overline{K}} = cG_L$$

give an invariant? And incredibly, it always does. After normalization the situation can be expressed by the

AXIOMS FOR GENERALIZED POLYNOMIAL.

1. *To each oriented knot or link there is associated a Laurent polynomial in two variables*

$$G_K(\alpha, z) \in \mathbb{Z}[\alpha, \alpha^{-1}, z, z^{-1}]$$

If K *and* $\overline{K}$ *are ambient isotopic, then* $G_K = G_{\overline{K}}$.

2. $G_{\bigcirc} = 1$.

3. $\alpha G_{\times} - \alpha^{-1} G_{\times} = z G_{\rightrightarrows}$.

Thus: (i) $\alpha = 1 \Longrightarrow G_K = \nabla_K$, the Conway polynomial.

(ii) $z = \left[\frac{1}{\sqrt{\alpha}} - \sqrt{\alpha}\right] \Longrightarrow G_K = V_K(\alpha)$, the Jones polynomial.

Other authors use different lettering. Thus $\alpha = \ell^{-1}$, $z = -m$ gives the polynomial of Lickorish and Millet.

What is going on here?

There are many questions to ask about these new invariants. I take a step in the next two sections toward a geometric interpretation. What I show is that a version of the generalized polynomial is an ambient isotopy invariant of knotted, twisted bands in three-dimensional space.

The new variable measures twisting of bands. Thus, we shall (by Section 20) construct a B-polynomial whose axioms are:

AXIOMS FOR THE B-POLYNOMIAL

1. *Let* K,K′ *be oriented links whose components are (knotted) oriented twisted bands. (Sometimes called* *__framed__ __links__.) Let* $\simeq$ *denote ambient isotopy of bands in three-dimensional space. Then there is a Laurent polynomial* $B_K(\alpha,z) \in Z[\alpha,\alpha^{-1},z,z^{-1}]$ *such that* $B_K = B_{K'}$ *whenever* $K \simeq K'$.

2. 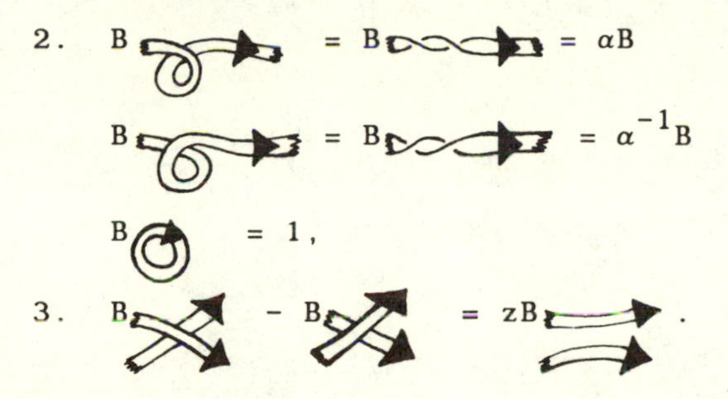

Note that here the exchange identity is in standard form (no new variables). __The new variable measures twisting in the bands.__

By using topological script and the concept of "regular isotopy" (see Section 18) the next two sections draw the connection between the B-polynomial and the generalized polynomial.

For the rest of this section we go through a series of sample B-polynomial calculations:

1. B $= \alpha$B $= \alpha \cdot 1 = \alpha$

   and (similarly) B $= \alpha^{-1}$.

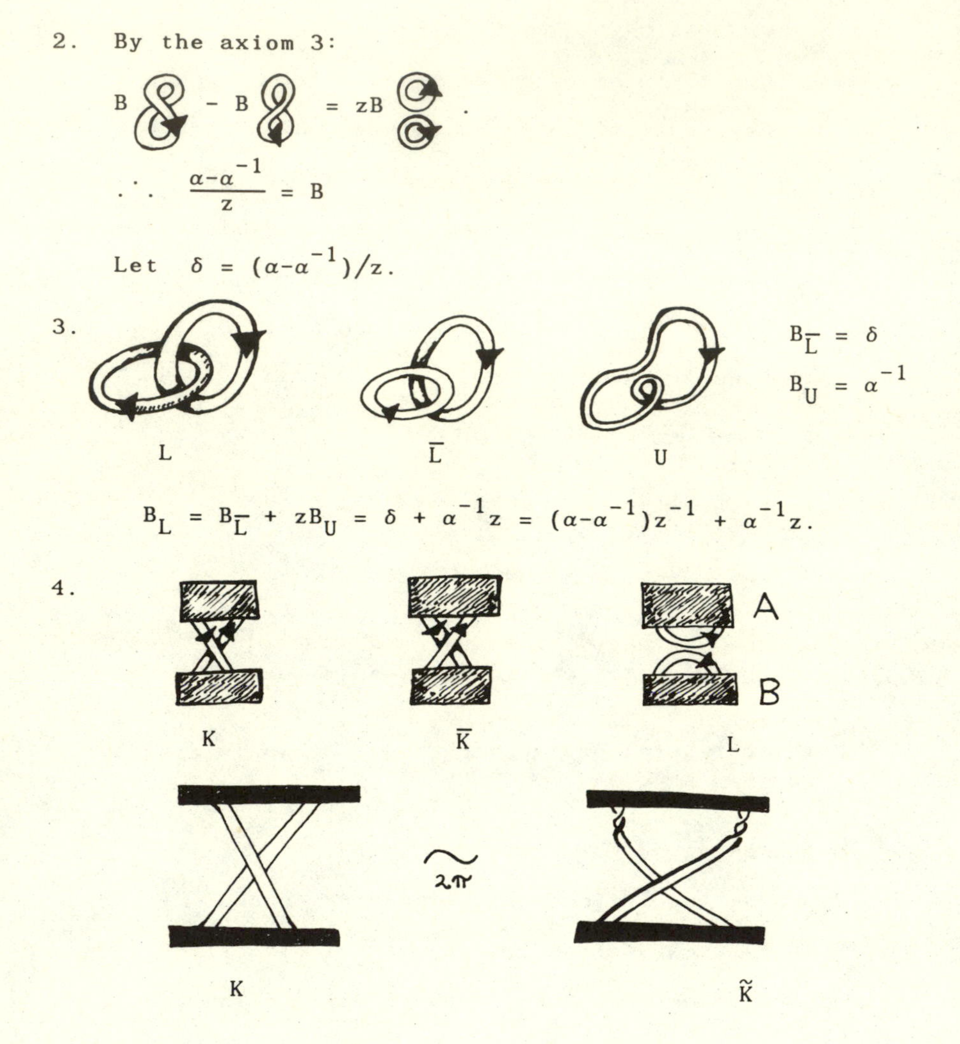

2. By the axiom 3:

$$B(\text{figure}) - B(\text{figure}) = zB(\text{figure}).$$

$$\therefore \quad \frac{\alpha-\alpha^{-1}}{z} = B$$

Let $\delta = (\alpha-\alpha^{-1})/z$.

3. L $\qquad \bar{L} \qquad$ U $\qquad B_{\bar{L}} = \delta$, $\quad B_U = \alpha^{-1}$

$$B_L = B_{\bar{L}} + zB_U = \delta + \alpha^{-1}z = (\alpha-\alpha^{-1})z^{-1} + \alpha^{-1}z.$$

4.

$$\therefore \quad B_K = B_{\tilde{K}} = \alpha^2 B_{\bar{K}}$$

$$\therefore \quad B_K - \alpha^{-2}B_K = zB_L$$

$$(1-\alpha^{-2})B_K = zB_L.$$

It is easy to see by induction that $B_L = \delta B_A B_B$ $(\delta = (\alpha-\alpha^{-1})/z)$, where $L = A \cup B$ is a split link.

Hence
$$B_K = \left(\frac{z\alpha}{\alpha-\alpha^{-1}}\right)\left(\frac{\alpha-\alpha^{-1}}{z}\right)B_A B_B$$

$$B_k = \alpha B_A B_B$$

and
$$B_{\overline{K}} = \alpha^{-1} B_A B_B .$$

A similar argument shows that a straight "connected sum" gives the product of the polynomials:

5.

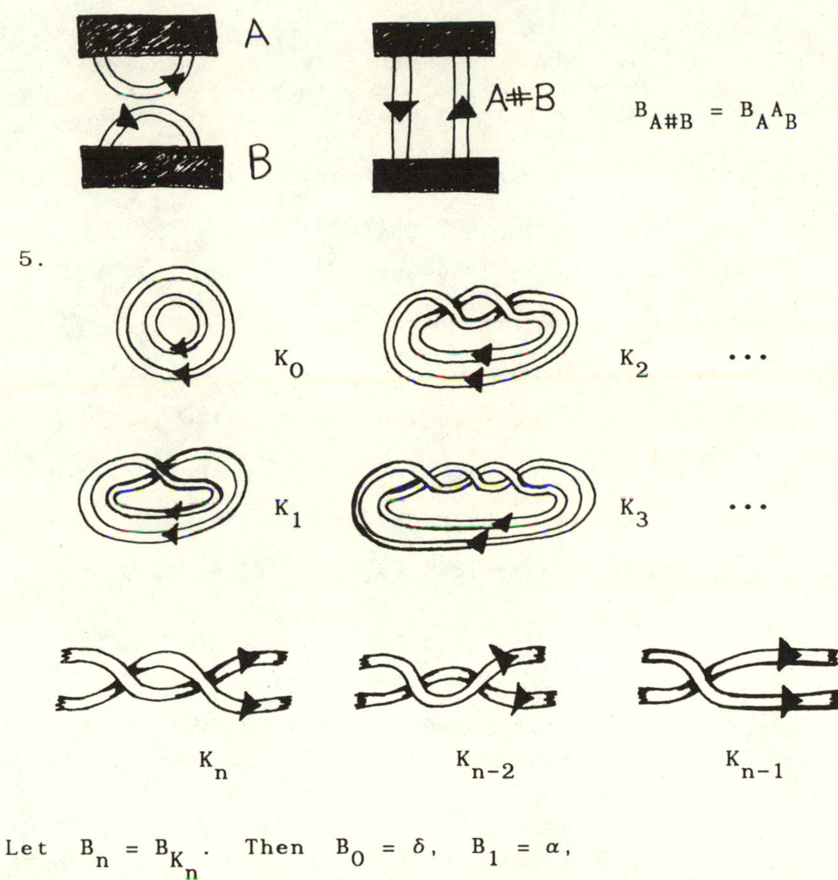

Let $B_n = B_{K_n}$. Then $B_0 = \delta$, $B_1 = \alpha$,

$$B_n = B_{n-2} + zB_{n-1}$$

$B_0 = \delta \qquad = z^{-1}(\alpha-\alpha^{-1})$

$B_1 = \alpha \qquad = \alpha$

$B_2 = z\alpha + \delta \qquad = z\alpha + z^{-1}(\alpha-\alpha^{-1})$

$B_3 = z^2\alpha + z\delta + \alpha \qquad = z^2\alpha + (2\alpha-\alpha^{-1})$

$B_4 = z^3\alpha + z^2\delta + 2z\alpha + \delta = z^3\alpha + z(3\alpha-\alpha^{-1}) + z^{-1}(\alpha-\alpha^{-1})$

....

6. It follows by induction that if $K^!$ is obtained from K by reversing all crossings, then $B_{K^!}(\alpha,z) = B_K(-\alpha^{-1},z)$. Since δ is invariant under $\alpha \longrightarrow -\alpha^{-1}$ it is useful to leave the polynomials in the form of functions of z,α,δ. Thus we see at once that $B_3(\alpha,z)$ is not equal to $B_3(-\alpha^{-1},z)$.

This should give the flavor of calculating the B-polynomial for twisted bands. The next two sections delineate the connection with the generalized polynomial. The link goes by way of topological script. If we denote

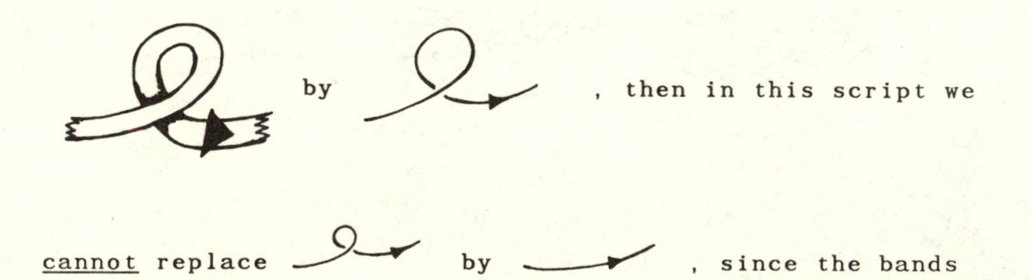

by , then in this script we

cannot replace by , since the bands

 and are not ambient isotopic.

Consequently, we are led to consider diagram moves without

the type I replacement.

§19. *THE GENERALIZED POLYNOMIAL AND REGULAR ISOTOPY*

Recall the crossing signs:

$$\epsilon[\ \ ] = +1$$

$$\epsilon[\ \ ] = -1.$$

Define the <u>writhe</u>, $w(K)$, of a diagram K to be the sum of all of its crossing signs.

$$w(K) = \sum_{p} \epsilon(p)$$

where p runs over all crossings in the diagram K.

Thus the trefoil T and its mirror image $T^!$ have writhe of $+3$ and -3, respectively.

$w(T) = +3$ $\qquad$ $w(T^!) = -3$

In general, a knot and its mirror image (obtained by switching all the crossings) have writhe of opposite sign.

If the writhe were an invariant of ambient isotopy then its calculation would be an excellent method for distinguishing mirror images. Writhe *is* an invariant of *regular isotopy* (denoted $\approx$). Two diagrams are said to be regularly isotopic if one can be obtained from the other by a sequence of Reidemeister moves of type II or type III.

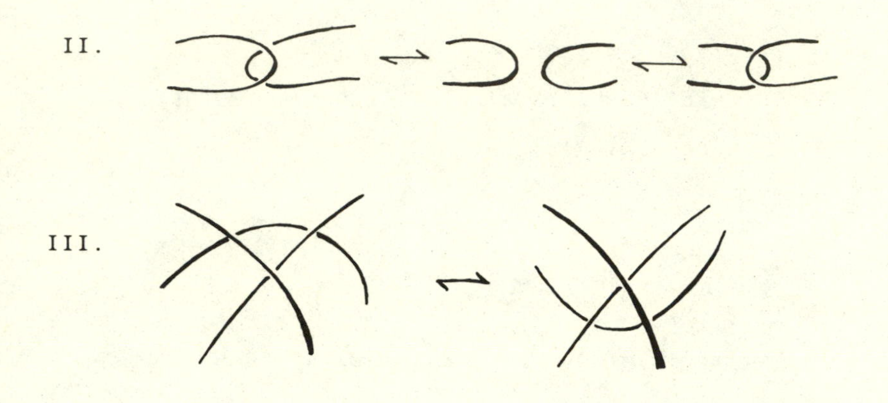

Generators of Regular Isotopy $\approx$

Thus the writhe is an invariant in the "flat" knot theory of regular isotopy. And any diagram with nonzero writhe is not regularly isotopic to its mirror image. Unfortunately (or fortunately (!) if you love deep and simply stated problems) the ambient isotopy problem for mirror images is not so simple. Perhaps some combination of moves II, III and also I will turn a trefoil into its mirror image?

In fact, this is not so! The trefoil is not ambient

isotopic to its mirror image. (We shall prove this again later in the book by signature methods.) Here we construct an invariant that I call the R–polynomial, R_K. The R-polynomial is an invariant of regular isotopy, and it is related to the <u>generalized</u> <u>polynomial</u> G_K by the formula

$$G_K = \alpha^{-w(K)} R_K$$

where $w(K)$ is the writhe.

AXIOMS FOR R_K

1. *To each oriented knot or link there is associated a Laurent polynomial in two variables* α, z. The *polynomial is denoted*

   $$R_K(\alpha, z) \in Z[\alpha, \alpha^{-1}, z, z^{-1}].$$

   It is an invariant of regular isotopy:

   $$K \approx K' \Longrightarrow R_K = R_{K'}.$$

2. $R = R = 1,$

   $R = \alpha R,$

   $R = \alpha^{-1} R.$

3. $R - R = zR$

Remark: As with the Conway polynomial, these diagrammatic abbreviations refer to larger diagrams in which the indicated patterns are embedded.

Unfolding Axiom 2, we see that for a curly unknot diagram such as

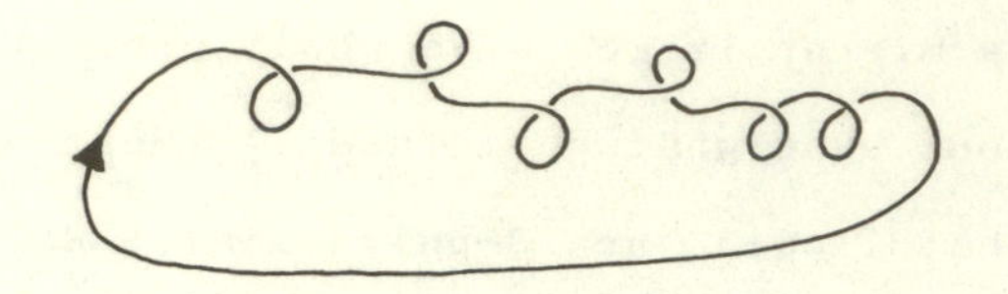

the R-polynomial returns the writhe $w(U)$ in the form $R_U = \alpha^{w(U)}$.

The third axiom looks just like the exchange identity for the Conway polynomial. But R is <u>not</u> the Conway polynomial. Not at all! View the following calculation:

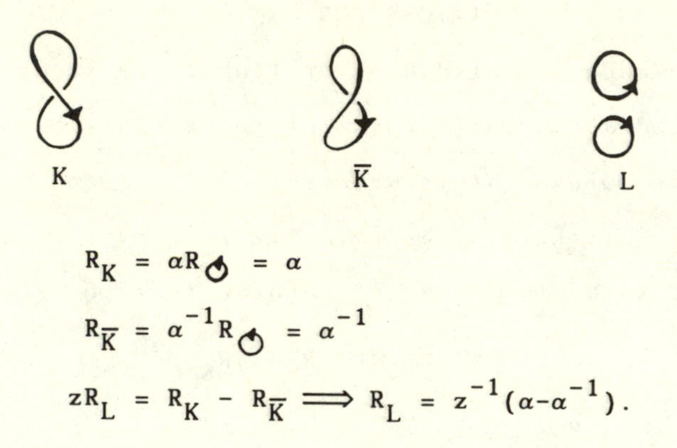

$$R_K = \alpha R_{\bigcirc} = \alpha$$

$$R_{\overline{K}} = \alpha^{-1} R_{\bigcirc} = \alpha^{-1}$$

$$zR_L = R_K - R_{\overline{K}} \Longrightarrow R_L = z^{-1}(\alpha - \alpha^{-1}).$$

We see that with R, an unlink can receive a nonzero polynomial. In fact, it is useful to record $\delta = z^{-1}(\alpha - \alpha^{-1})$, the value of R on a split unlink of two components. [As the reader can see, R is a script-version of B of Section 18.]

<u>Exercise</u>. Show that if $U_n = \bigcirc \bigcirc \bigcirc \cdots \bigcirc$ denotes a split unlink of n-components then $R_{U_n} = \delta^{n-1}$.

<u>Exercise</u>. Show that $R_{K^!}(\alpha, z) = R_K(-\alpha^{-1}, z)$ where $K^!$ is the mirror-image diagram that is obtained from K by measuring all the crossings.

The <u>generalized</u> <u>polynomial</u> $G_K(\alpha, z)$ is defined by the equation

$$G_K = \alpha^{-w(K)} R_L .$$

Since $R_{\circlearrowleft\!\rightarrow} = \alpha R_{\rightarrow}$ and $R_{\circlearrowright\!\rightarrow} = \alpha^{-1} R_{\rightarrow}$, it follows at once that G_K is invariant under moves I, II and III. Hence the generalized polynomial is an invariant of ambient isotopy. [We beg pardon for reversing historical and logical order! G came first, and we do not prove here that R is consistent. See [K8], or start with G and define R by $R = \alpha^{w(K)} G$ for a logical path.]

<u>Exercise</u>. Show that the generalized polynomial satisfies the exchange identity

$$\alpha G_{\times} - \alpha^{-1} G_{\times} = z G_{\rightrightarrows} .$$

Show that $G_{K^!}(\alpha, z) = G_K(-\alpha^{-1}, z)$.

The time has come to compute R and G for the trefoil. The calculation exactly parallels the familiar calculation for ∇_K:

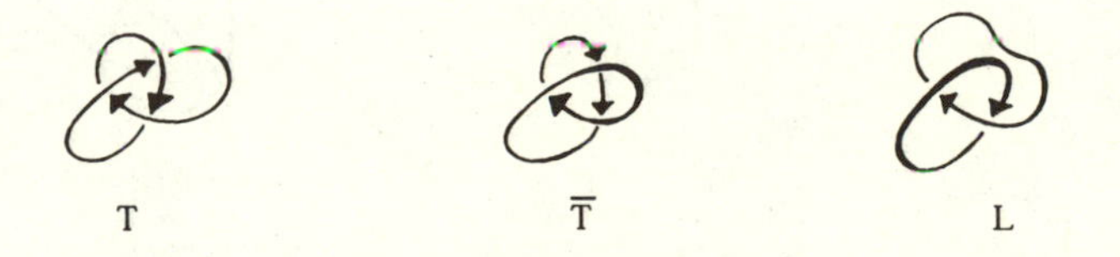

T $\qquad$ $\overline{T}$ $\qquad$ L

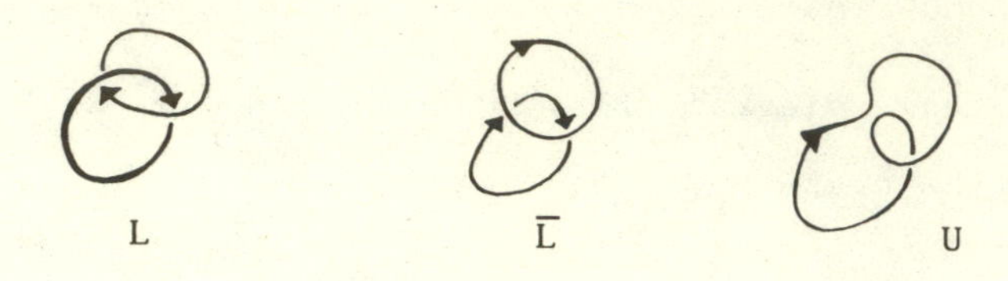

Thus $T = \overline{T} \oplus L$, $L = \overline{L} \oplus U$,

$$\therefore \quad T = \overline{T} \oplus (\overline{L} \oplus U),$$

$$\therefore \quad R_T = R_{\overline{T}} + z(R_{\overline{L}} + zR_U).$$

Since $R_{\overline{T}} = \alpha$, $R_{\overline{L}} = \delta (= z^{-1}(\alpha - \alpha^{-1}))$ and $R_U = \alpha$, we conclude that

$$R_T = \alpha + z(\delta + z\alpha) = \alpha + z\delta + z^2\alpha$$

$$= \alpha + \alpha - \alpha^{-1} + z^2\alpha,$$

$$\therefore \quad R_T = (2\alpha - \alpha^{-1}) + z^2\alpha$$

and $$G_T = \alpha^{-w(T)} R_T = \alpha^{-3} R_T$$

$$\therefore \quad G_T = (2\alpha^{-2} - \alpha^{-4}) + z^2\alpha^{-2}.$$

Note that G_T is not invariant under the substitution $\alpha \longmapsto -\alpha^{-1}$. *This proves that the trefoil is not equivalent to its mirror image.*

Here is another sample calculation. This time for the figure-eight knot K:

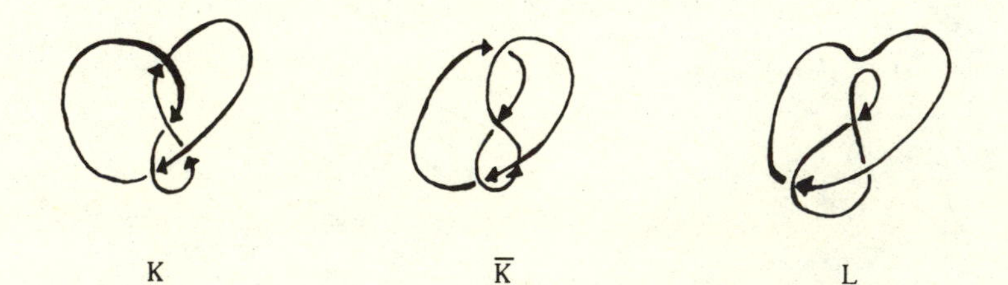

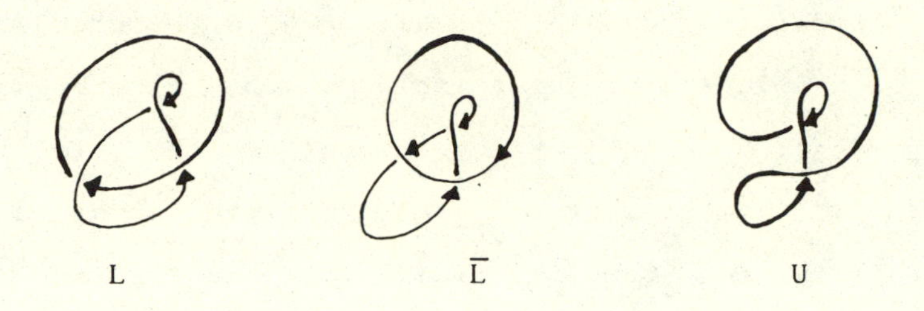

Here $K = \overline{K} \oplus L'$ and $L - \overline{L} \ominus U$. Hence $K = \overline{K} \oplus (\overline{L} \ominus U)$ with

$$R_{\overline{K}} = \alpha^{-2}$$

$$R_{\overline{L}} = a\delta$$

$$R_{\overline{U}} = \alpha\alpha^{-1} = 1.$$

Thus

$$R_K = R_{\overline{K}} + zR_L - z^2 R_U$$

$$= \alpha^{-2} + \alpha(\alpha - \alpha^{-1}) - z^2$$

$$\therefore \quad R_K = (\alpha^{-2} + \alpha^2) - z^2.$$

Note that $w(K) = 0$. Hence $R_K = G_K$ for this diagram of the figure-eight knot. Also $R_K ! = R_K$. Since the figure-eight knot is ambient isotopic to its mirror image, we knew this would happen. Note that $R_K(\alpha, z) = R_K(-\alpha^{-1}, z)$.

What may not be so obvious is that <u>the figure-eight knot is regularly isotopic to its mirror image</u>. In fact, we shall prove the following

MIRROR THEOREM. *Let* K *be a knot diagram with zero writhe. Suppose that* K *is ambient isotopic to* $K^!$. *Then the diagrams* K *and* $K^!$ *are regularly isotopic.*

In order to prove this result we need to introduce yet another invariant of regular isotopy. This is the <u>Whitney degree</u> $d(K)$. The Whitney degree measures the total turn of the unit tangent vector to the underlying plane curve of the knot diagram. Combinatorially it is defined as follows:

1. d( ) = -1.

2. d( ) = +1

3. $d(X \cup Y) = d(X) + d(Y)$ if $X \cup Y$ is a disjoint union of collections of curves in the plane.

4. d( ) = d( ) = d( ).

By splicing all the crossings in the diagram, we obtain a disjoint collection of circles (these are called the Seifert circuits of the diagram). The Whitney degree is then the sum of ±1 for each Seifert circuit.

For example, in the case of the trefoil:

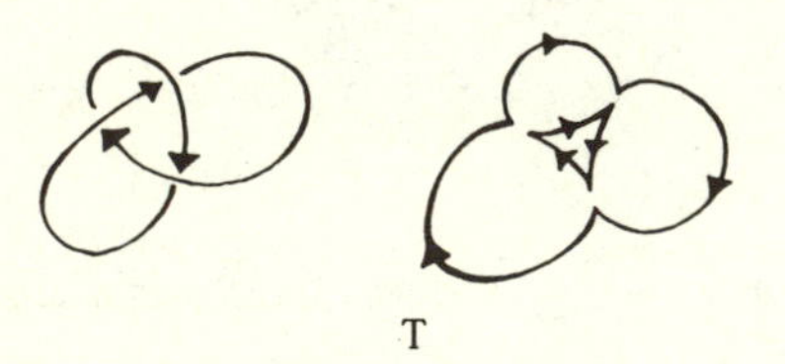

$\Longrightarrow d(T) = -2.$

and for the figure eight:

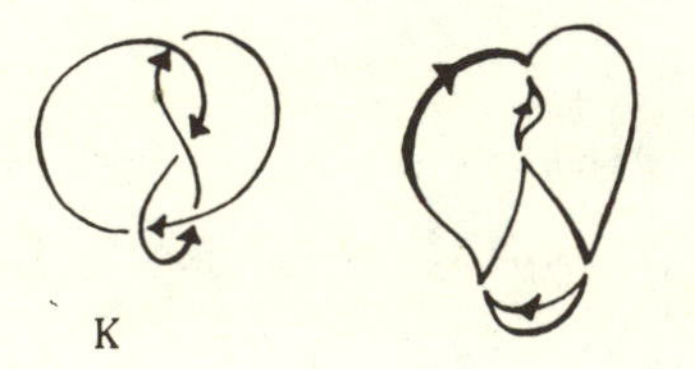

$\Longrightarrow d(K) = -1.$

Since the Whitney degree does not depend upon over-crossings or under-crossings it actually measures a property of the underlying plane curve immersion. In fact, Whitney and Graustein [W1] proved that *two immersed curves in the plane are regularly homotopic if and only if they have the same Whitney degree*.

We will not explain *regular homotopy* here except to state that it is combinatorially equivalent to the relation generated by the *projections* of the moves II and III. That is, by

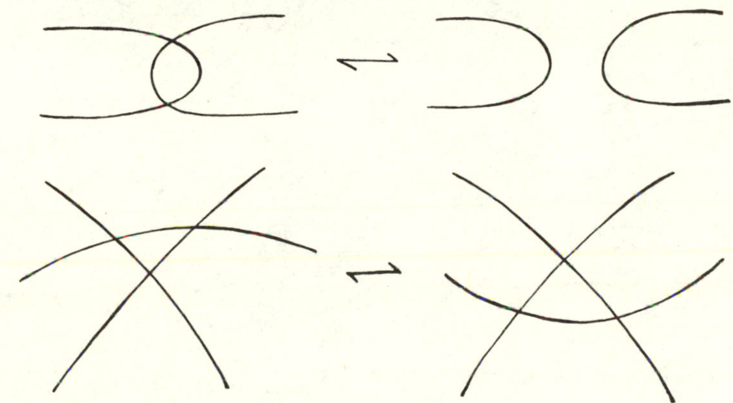

and

It is a nice exercise to prove that the Whitney degree is an invariant of this relation (see [K1]). The basic move underlying the Whitney-Graustein Theorem is the Whitney Trick (this the reader has already encountered in Section 3 of this chapter!):

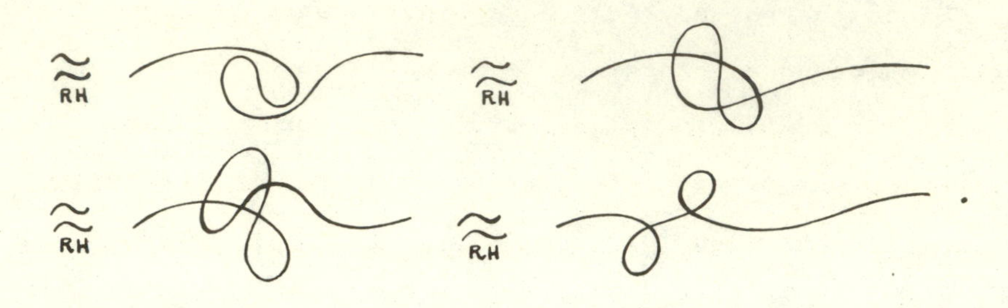

In order to generalize the Whitney trick we shall include the crossings and create a regular isotopy:

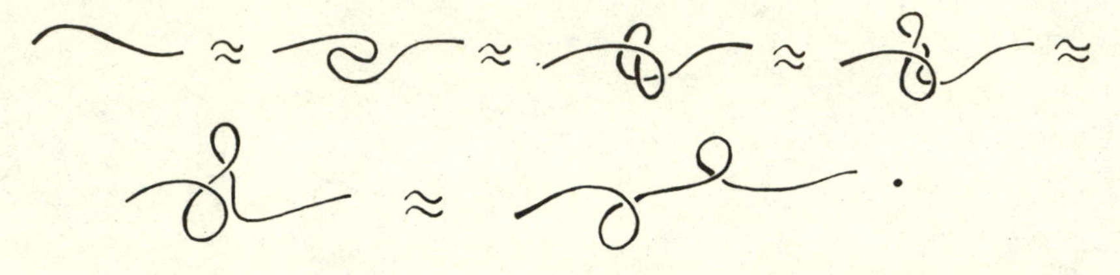

Thus we have the fundamental curl-cancelling regular isotopy:

On the other hand is not regularly isotopic to . (Prove this by using the regular isotopy invariance of the writhe.) As a result, one can prove the following generalization of the Whitney-Graustein Theorem (See [W1], [TR].):

PROPOSITION. *Let* K *and* K′ *be knot diagrams each ambient isotopic to the unknot. Then* K *and* K′ *are regularly isotopic if and only if they have the same writhe and the same Whitney degree:* $K \approx K' \Leftrightarrow w(K) = w(K')$ *and* $d(K) = d(K')$.

We omit the proof here, but remark that the result is obtained by a regular isotopy to a normal form consisting of a string of curls such as

and the appropriate cancellations using the Whitney trick when it is appropriate.

Now it is time to prove the

MIRROR THEOREM. *Let* K *be a knot diagram with zero writhe. Then* K *is ambient isotopic to* $K^!$ *if and only if* K *is regularly isotopic to* $K^!$.

Proof: Since regular isotopy is an ambient isotopy, it suffices to prove that $K \approx K^!$ given that $K \sim K^!$. Assuming that $K \sim K^!$ means that K is regularly isotopic to the connected sum of $K^!$ and C where C is an unknot diagram (appropriately curly). Thus we have

$$K \approx K^! \# C.$$

(<u>Exercise</u>: Prove this last assertion.)

Since the Whitney degree and the writhe are each additive under connected sums (exercise) we see that

$$w(C) = 0 \quad (\text{because} \quad w(K) = w(K^!) = 0)$$

and

$$d(C) = 0 \quad (\text{because} \quad d(K) = d(K^!)).$$

By the proposition it then follows that $C \approx 0$ and hence $K \approx K^!$. This completes the proof. ∎

The Mirror Theorem shows that the problem of distinguishing knots from their (oriented) mirror images is actually a problem in the regular isotopy category.

Remark: We have not proved the consistency of the axioms for G (or R). The reader is referred to the papers listed at the beginning of this section for discussions of G, and to [K8] for a proof for the L-polynomial discussed in our appendix.

We conclude with a picture of a regular isotopy between the figure-eight and its mirror image.

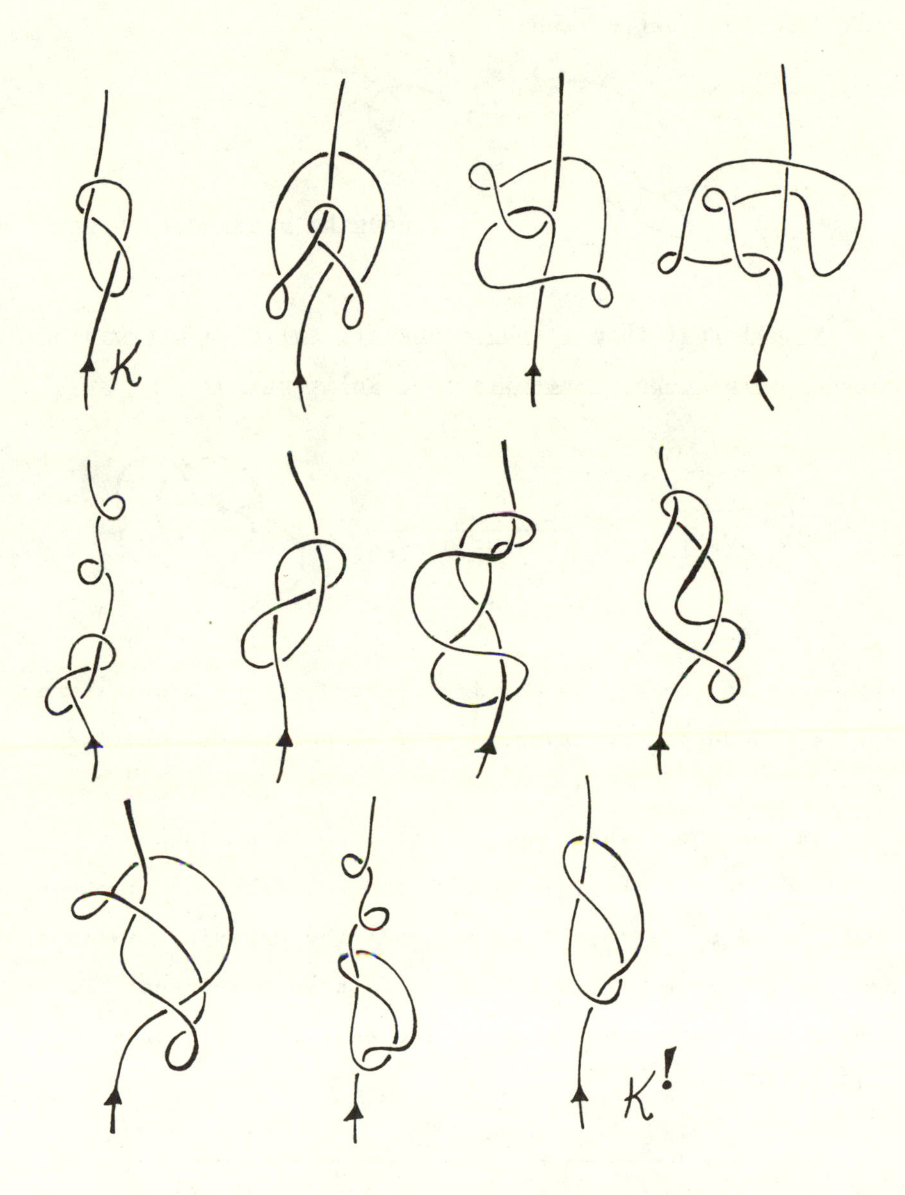

Regular Isotopy of Figure Eight and its Mirror Image

Remark on Well-Definition:

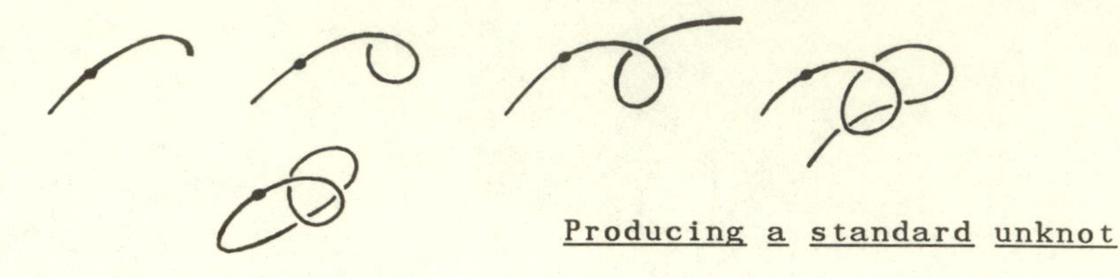

<u>Producing a standard unknot</u>

<u>Recall that in a standard unknot, splicing a crossing nearest to base-point results in a split unlink</u>.

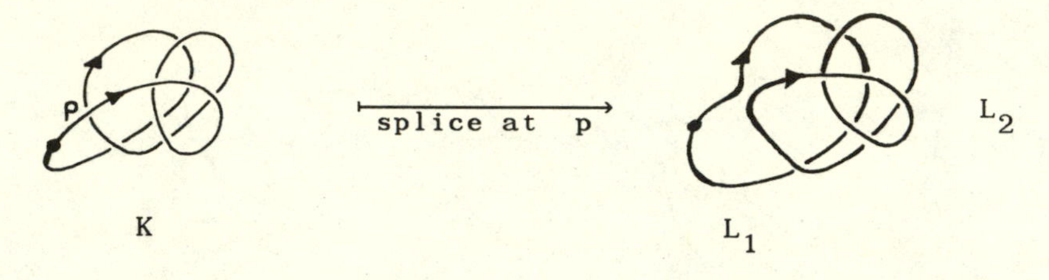

This fact is the key to any inductive argument proving the well-definedness and invariance of either the generalized polynomial, or the R-polynomial.

To see the issue, suppose that $\tilde{K}$ is an unknot obtained from K by switching crossings labelled $1, \cdots, n$. Thus $\tilde{K} = S_n S_{n-1} \cdots S_1 K$. Assume that the crossings switched are the difference between K and a standard unknot $\tilde{K}$. Assume that n is adjacent to the base-point as shown below:

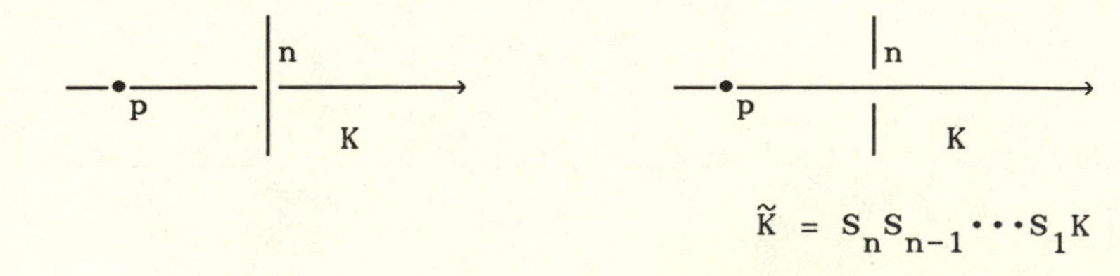

$$\tilde{K} = S_n S_{n-1} \cdots S_1 K$$

Then $R_K = R_{\tilde{K}} + z \sum_{i=1}^{n} \epsilon_n(K) R E_i S_{i-1} \cdots S_1 K.$

(As in Chapter 3 of these notes, there is another basic case. It can be handled directly or by first proving a lemma about invariance under cyclic permutations of switching elements.)

If we slide the base point through the crossing we get

[Diagrams: a vertical strand crossed by a horizontal strand labelled n, base point q, labelled K (left) and $\tilde{K}'$ (right)]

$$\tilde{K}' = S_{n-1} \cdots S_1 K.$$

Thus $R_K = R_{\tilde{K}'} + z \sum_{i=1}^{n-1} \epsilon_i(K) R\ E_i S_{i-1} \cdots S_1 K.$ The right-hand sums from these two calculations had better match!

Their difference Λ is

$$\Lambda = R_{\tilde{K}} - R_{\tilde{K}'} + z\ \epsilon_n(K) R\ E_n S_{n-1} \cdots S_1 K.$$

Let $\epsilon = \epsilon_n(K)$. We know that $L = E_n S_{n-1} \cdots S_1 K = E_n \tilde{K}$ is a split unlink of writhe (say) α^p. Then

$$w(\tilde{K}) = \alpha^{p-\epsilon}$$

$$w(\tilde{K}') = \alpha^{p+\epsilon}$$

$$R_L = \alpha^p \delta.$$

$$\therefore \quad \Lambda = \alpha^{p-\epsilon} - \alpha^{p+\epsilon} + \epsilon\alpha^p(\alpha - \alpha^{-1})$$

$$\therefore \quad \Lambda = 0.$$

This miracle makes the generalized polynomial work.

<u>Exercise</u>. Show that the knot 9_{42} (below) has generalized polynomial $G_K = (2\bar{\alpha}^2-3+2\alpha^2) + (\bar{\alpha}^2-4+\alpha^2)z^2 - z^4$. Since $G_K(\alpha,z) = G_K(-\alpha^{-1},z)$ we see that K might be amphicheiral. However, this is not the case. In Chapter 7 we will see that the nonvanishing of the signature of K (= 9_{42}) prevents it from being equivalent to its mirror image. The knot 9_{42} is the first nonamphicheiral knot that fails to be distinguished from its mirror image by the generalized polynomial (<u>Millet's</u> <u>Example</u>).

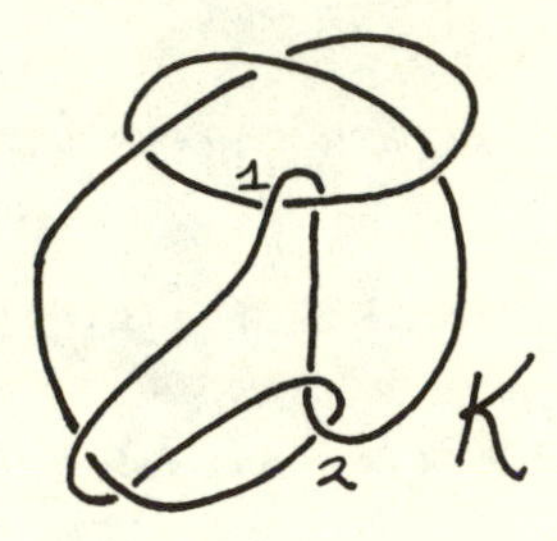

(K = 9_{42})

[*Hint:* Use crossings 1 and 2 for switching and splicing.]

§20. *TWISTED BANDS (AGAIN)*

The regular isotopy of the figure eight knot with its mirror image can be viewed as topological script for an ambient isotopy of a knotted band. That is, we have actually shown that

and

are ambient isotopic bands!

Another way to view this interpretation of knotted twisted bands is to consider a knot formed on a rubber tube so that twisting causes tension along the tube. If we put a knot that is equivalent to its mirror image on the rope so that the tube is relaxed, then it will also be relaxed after we deform the mirror image. This is the physical content of the mirror theorem.

It is very illuminating to perform this experiment with the figure eight knot. Tension appears in the intermediate stages and relaxes away as we get the mirror image. Energetically speaking, the knot and its mirror image are separated by the walls of their individual potential wells.

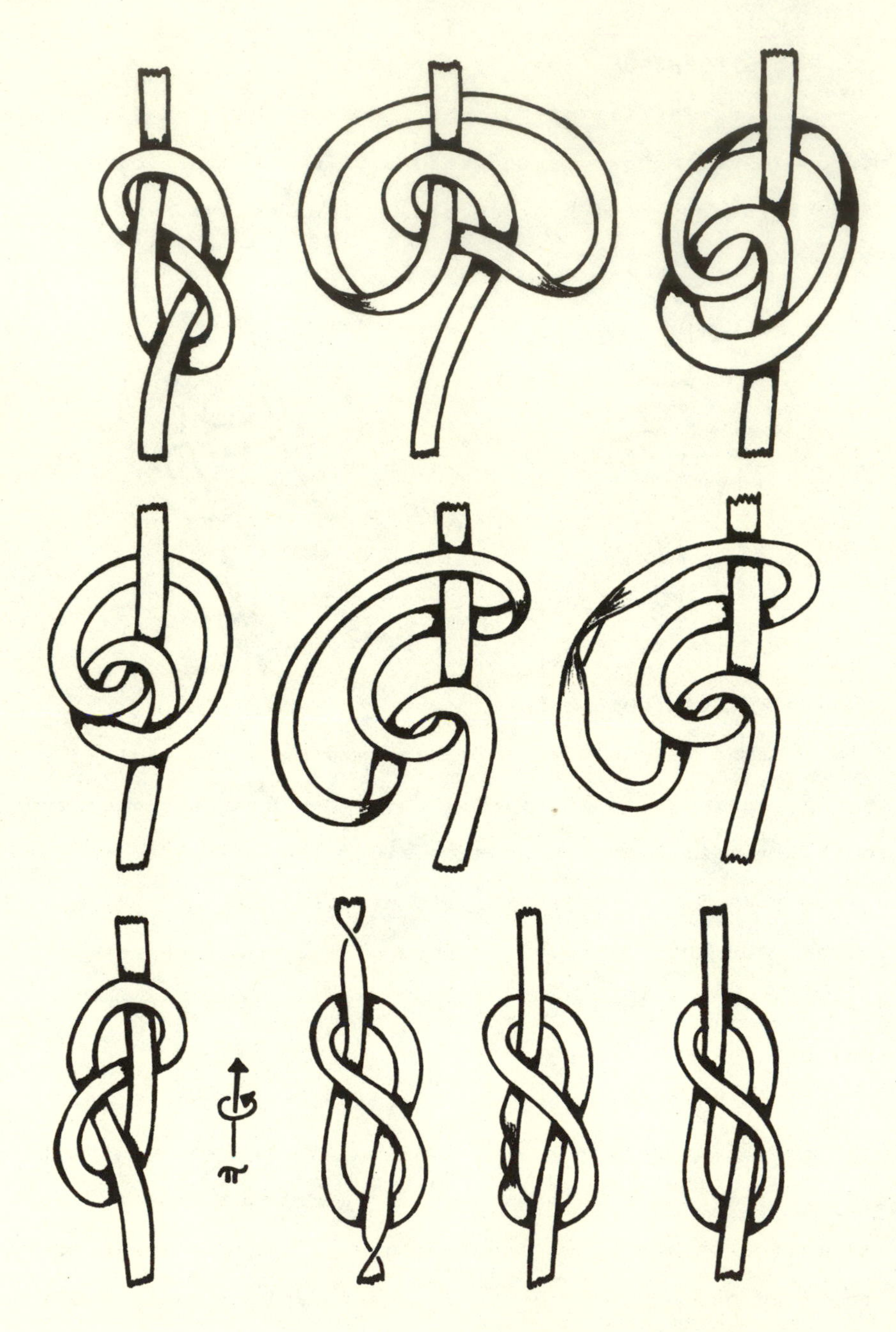
π

VII

SPANNING SURFACES AND THE SEIFERT PAIRING

Let's begin by determining the genus of the Seifert surface. Recall that the Seifert surface is a surface obtained by Seifert's algorithm from a knot or link diagram (see Chapter 5.) Also, if F is an orientable surface, then the genus of F, $g(F)$, is given by the formula

$$2g(F) = \rho(F) - \mu(F) + 1$$

where $\rho(F)$ is the rank of the first homology group $H_1(F)$, and $\mu(F)$ is the number of boundary components of F. (We assume that $\mu(F) \geq 1$.) In other words, the genus of F is the number of handles in the standard form for F' where F' is obtained from F by adding disks to all the boundary components.

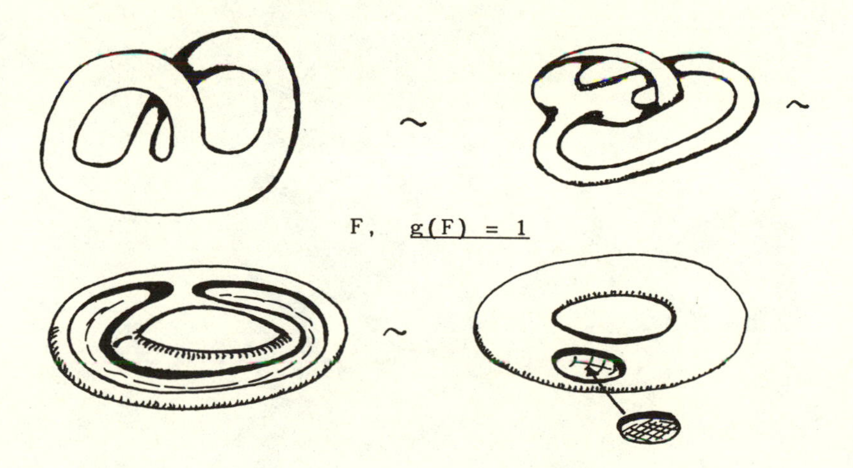

F, $g(F) = 1$

For one boundary component, $2g(F) = \rho(F)$.

DEFINITION 7.1. *Let* K *be an oriented knot or link. Then the* genus of K, $g(K)$, *is the minimal value of* $g(F)$ *among all connected, oriented surfaces* $F \subset S^3$ *that span* K.

Similarly, the rank of K, $\rho(K)$, *is the minimum value of* $\rho(F)$ *among connected, oriented spanning surfaces. By our formula, we have* $2g(K) = \rho(K) - \mu(K) + 1$ *where* $\mu(K)$ *is the number of boundary components of* K.

Example: Any knot K that is knotted and bounds a surface of genus 1 must have $g(K) = 1$ (since if it bounded a disk it would be unknotted). Thus the genus of the trefoil knot is 1.

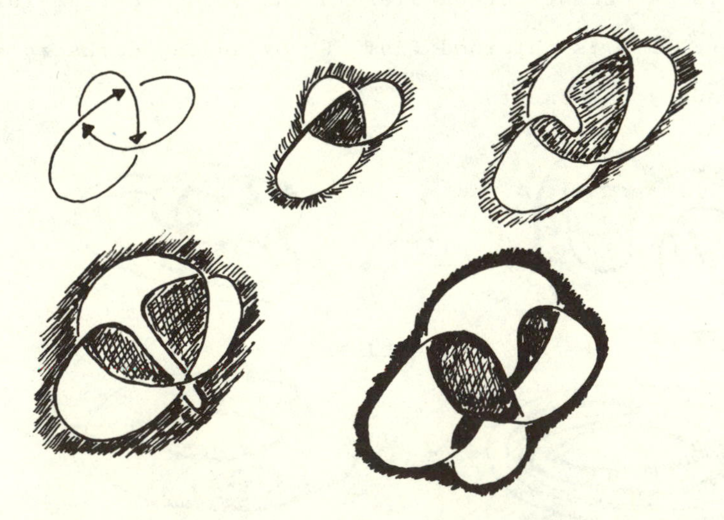

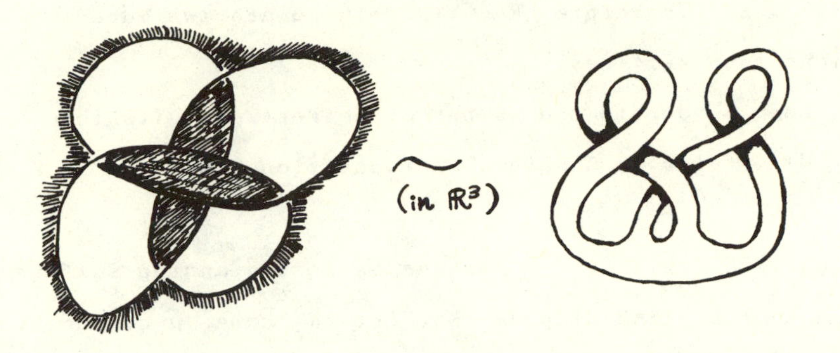

Let K be a knot (or link) diagram and let U be its underlying universe (the planar graph).

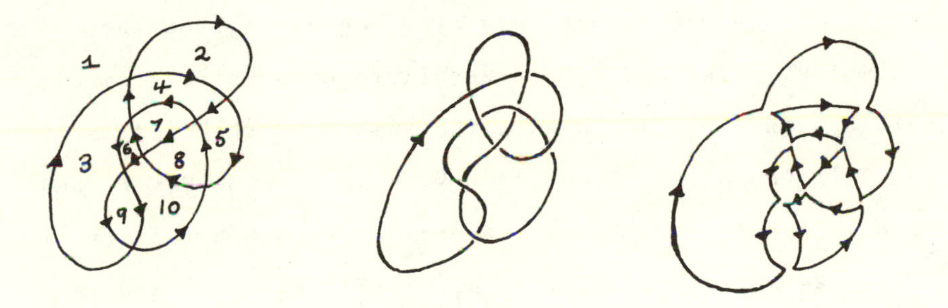

U K

<u>R = 10, V = 8, E = 16</u> <u>S = 3</u>

U is a planar graph with <u>R regions</u>, <u>E edges</u>, and <u>V vertices</u>. By Euler's Formula, we have V-E+R = 2. Since 4 edges are incident to each vertex, we also have 4V = 2E or $2V = E$. Therefore $R = V+2$. There are two more regions than vertices.

Let S denote the number of Seifert circuits for K (or U) (refer to Chapter 5, Proposition 5.8).

PROPOSITION 7.2. *Let* F *be the Seifert spanning surface for a knot or link diagram* K. *Let* K *have* μ *components,* S *Seifert circles and* R *regions. Then the rank and genus of* F *are given by the formulas*

$$\rho(F) = R-S-1$$

$$g(F) = \frac{1}{2}(R-S-\mu).$$

Proof: F is, up to homotopy type, obtained from the 1-complex U by adding one 2-cell to each Seifert circle. Let e_k $(k = 0,1,2)$ denote the number of k-cells in the resulting complex. Then $e_0 = V$, $e_1 = E$, $e_2 = S$. We have $e_0-e_1+e_2 = \chi(F) = \rho_0-\rho_1+\rho_2$ where $\rho_k = \text{rank } H_k(F)$ $(k = 0, 1,2)$. We know that $\rho_0 = 1$, $\rho_1 = \rho(F)$, $\rho_2 = 0$. And we know $V-E+S = V-2V+S = -V+S = 2-R+S$. Therefore $2-R+S = 1-\rho(F)$. ∎

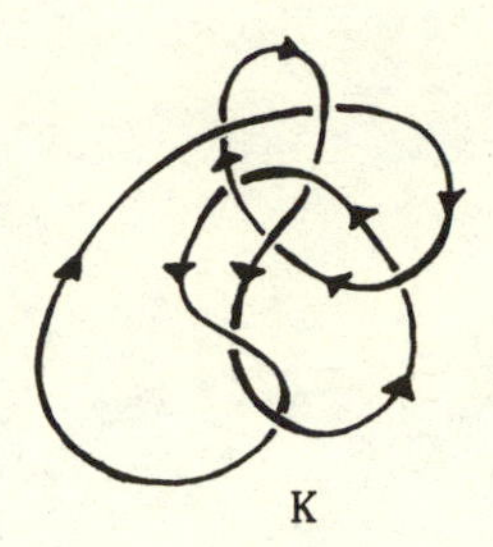

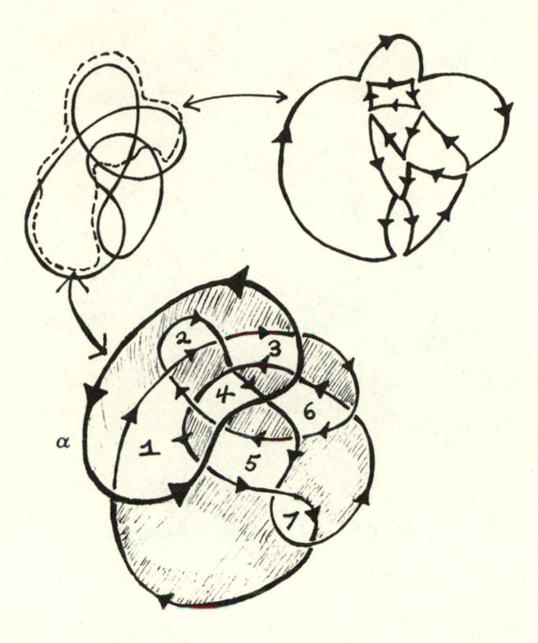

Seifert surface for K is obtained by adding a disk to the curve α.

In the figure above, I have indicated a method for understanding the Seifert surface for a knot or link, even when there are type II circuits present. (A type II Seifert circuit divides the plane into two regions, each containing Seifert circuits.)

We know that we are supposed to add a disk to the type II circuit, but in the usual drawing, the neighborhood of the boundary of this disk has an unclear structure. My solution is to draw a "tracer circuit" α corresponding to the given type II circuit. This tracer is drawn as follows:

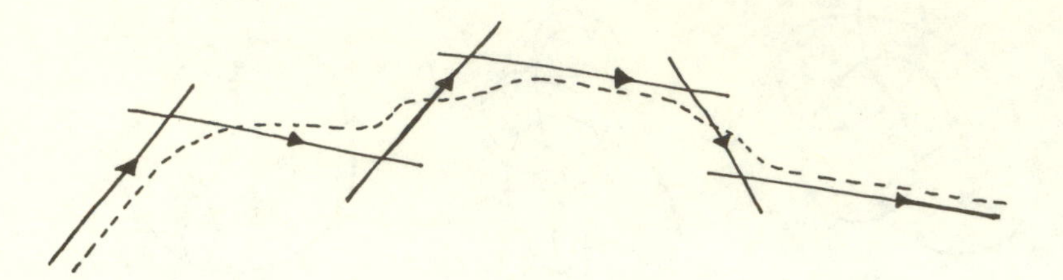

Tracer Circuit

Thus, it follows the type II circuit, crossing it whenever necessary. If we draw the tracer on the knot diagram as a new component α that over-passes the diagram, then the regions between the tracer and the old type II circuit become bounded by type I circuits. The disks filling in these regions draw out a picture of the boundary of a neighborhood of the disk added to the original type I.

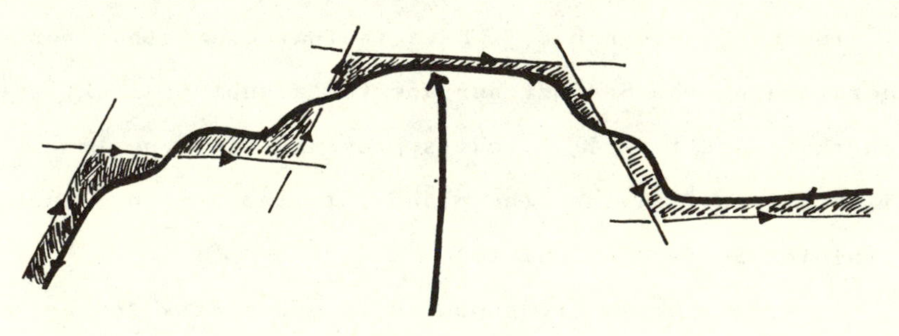

Tracer α

The new component α is oriented in the opposite direction to the type II circuit it traces.

Thus we have proved:

PROPOSITION 7.3. *Let* K *be an oriented knot or link diagram. Let* F_K *be the Seifert surface for* K. *Let* K' *be the diagram obtained from* K *by adding disjoint overpassing tracer circuits for each type* II *Seifert circuit in* K. *Let these tracer circuits be labelled* $\alpha_1, \alpha_2, \cdots, \alpha_k$. *Then*

(i) K' *has only type* I *Seifert circuits and*

(ii) F_K *is ambient isotopic to* $F_{K'} \cup D_1 \cup \cdots \cup D_k$ *where* D_i $(i = 1, \cdots, k)$ *is a disk with boundary* α_i. *And* $D_i \cap D_j = \emptyset$ *for* $i \neq j$.

This method of representing the Seifert surface will be particularly useful for certain calculations later on. Here is a fundamental problem about the Seifert surfaces: *Does every knot (link) achieve its minimal genus on a Seifert surface?* This is true for some classes of knots and links as we shall see shortly. It is false in general, by an application of the generalized polynomial. See the paper [M] by Hugh Morton.

While we're at it, let's remark on how to see a homology basis for $H_1(F_k)$ by looking at the tracer surface $F_{K'}$. Now $F_{K'}$ is a *planar surface*. That is, it is obtained as a checkerboard pattern from the diagram K'. It only leaves the plane at the twists

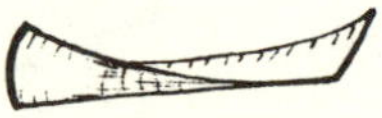

Consequently, $H_1(F_{K'})$ is generated by cycles

$\{c \mid c$ encircles a white region in the diagram$\}$.

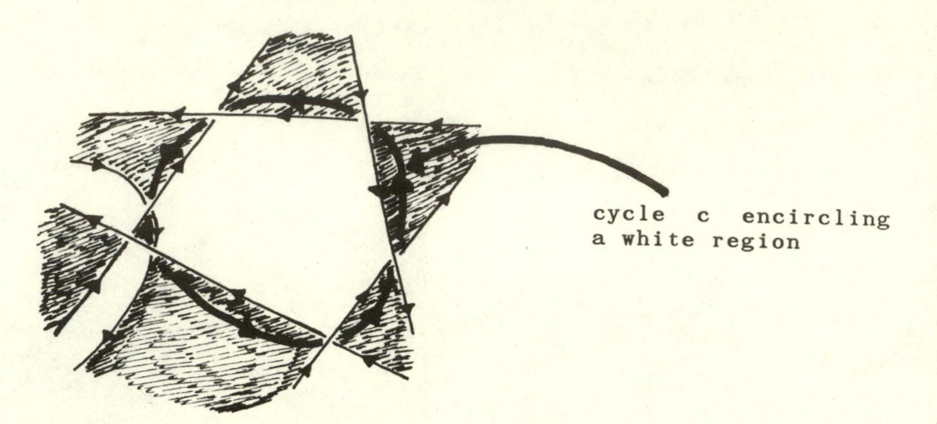

We orient c compatibly with the planar orientation ↺

Since $F_{K'}$ has the homotopy type of the plane punctured by the white regions, we see that $\text{rank } H_1(F_{K'})$ $= \#(\text{white regions})-1$. In counting, count all the bounded white regions. Then to obtain $H_1(F_K)$, note that $\text{rank } H_1(F_K) = \text{rank } H_1(F_{K'})-k$ where k is the number of tracer circles. For example, in the figure on p. 185, we have by this account $\rho(F_K) = \rho(F_{K'})-1$ and $\rho(F_{K'}) = 7$. Therefore $\rho(F_K) = 6$. Note that this is in accord with the formula of Proposition 7.2. In fact, $H_1(F_K)$ has as basis the cycles $\{c_1,c_2,c_3,c_4,c_5,c_6,c_7\}$. We have added c_4 since $c_1+c_2+c_3+c_4 \approx \alpha$ ($\approx$ denotes homology of cycles) and α bounds a disk in F_K.

<u>Exercise</u>. Explain how to obtain a basis for $H_1(F_K)$ in the general case of k tracer circuits $\alpha_1,\alpha_2,\cdots,\alpha_k$.

Give a procedure for deciding which white cycles to retain or throw away.

SEIFERT PAIRING

We now define an algebraic method for measuring the embedding of an oriented surface $F \subset S^3$. Given $F \subset S^3$, and a cycle a on F, let a^* denote the result of pushing a a very small amount into S^3-F along the positive normal direction to F. Using this, we define the Seifert pairing $\theta : H_1(F) \times H_1(F) \longrightarrow Z$ by the formula $\theta(a,b) = lk(a^*,b)$. This is a well-defined, bilinear pairing. It is an invariant of the ambient isotopy class of the embedding $F \subset S^3$.

Seifert invented a version of this pairing in [S]. He used it to investigate branched covering spaces. It has since proved to be extraordinarily useful in both classical and higher-dimensional knot theory.

Example 7.4:

θ	a	b
a	-1	1
b	0	-1

The surface F is oriented so that the positive normal points out of the page, toward the reader. For the self-linking $\theta(a,a) = lk(a^*,a)$, a^* may be represented by a parallel copy of a along the surface. Thus $\theta(a,a)$ can be computed from a disk with bands, by counting curls with sign.

Example 7.5:

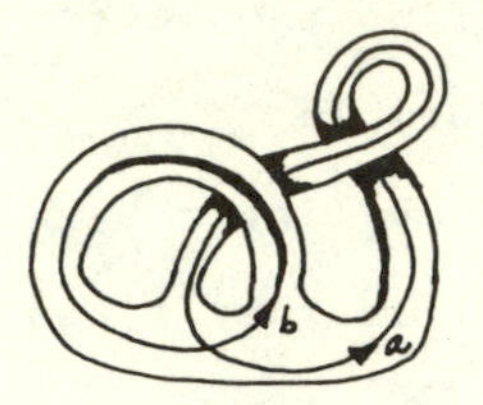

θ	a	b
a	−1	1
b	0	0

Note:

θ	a+b	b
a+b	0	1
b	0	0

$$\theta(a+b,a+b) = \theta(a,a)+\theta(b,a)+\theta(b,b) = 0$$

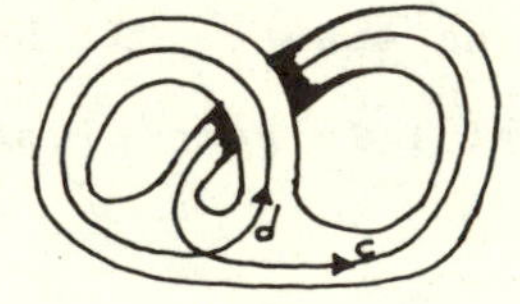

θ'	c	d
c	0	1
d	0	0

Thus these pairings are isomorphic. In fact, these two embeddings are isotopic:

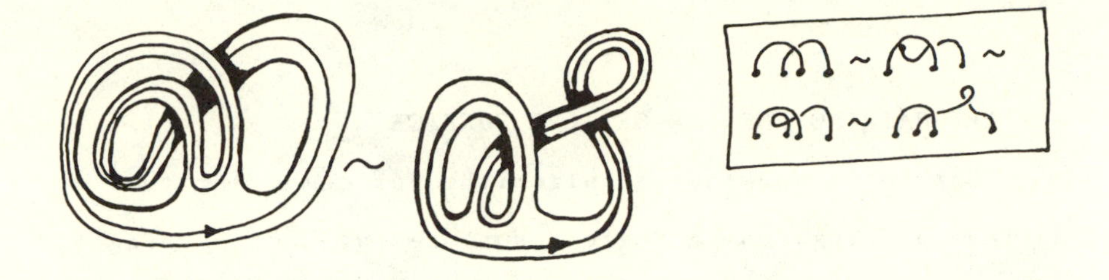

We can, if we want to do it, indicate a banded surface entirely in topological script. Thus

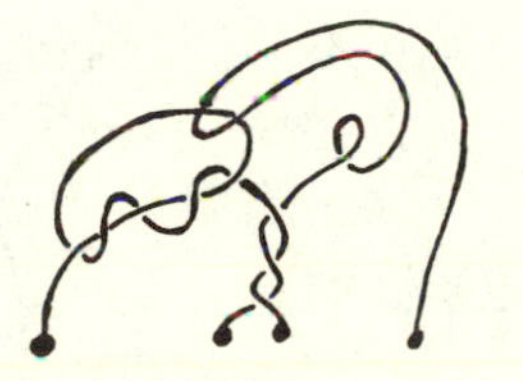

represents the surface:

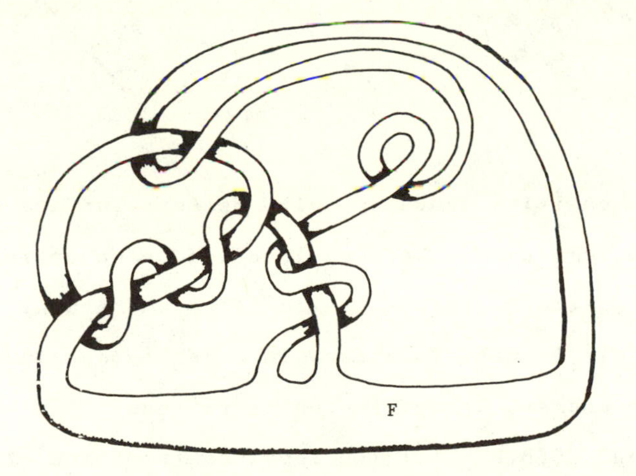

Exercise. Determine the Seifert pairing for this surface F.

SEIFERT PAIRING FOR THE SEIFERT SURFACE

Now let's work out an algorithm for computing the Seifert pairing from a Seifert surface (without pushing it into band-form). Recall that $H_1(F_K)$ is generated by the white cycles. (These are circles encircling white regions in $F_{K'}$.) Thus we must determine how each crossing in the diagram contributes to the Seifert linking number $\theta(a,b)$.

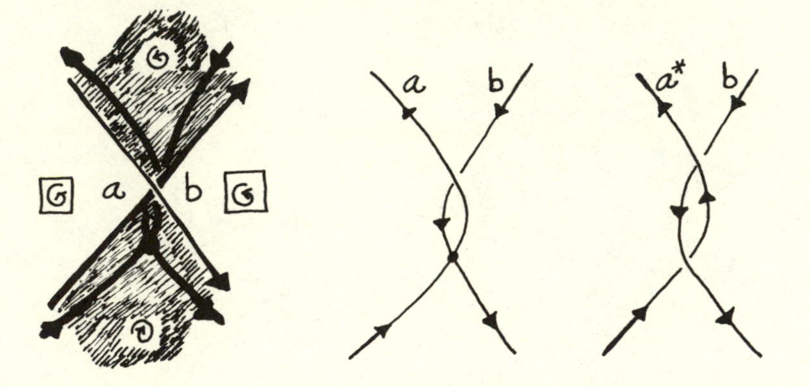

Here is a positive crossing, with Seifert surface shaded, and white regions a and b labelled. The cycles corresponding to these regions are labelled and drawn. Note that the cycles must intersect in order to continue following their courses around the white regions. Let's write $\theta(a,b)$ and $\theta(b,a)$ for the local contribution of this crossing. Then

$$\theta(a,b) = +1$$

$$\theta(b,a) = 0.$$

Note that $a \cdot b = +1$ also, where $x \cdot y$ denotes intersection number of cycles on the surface. (The signs reverse for a negative crossing.)

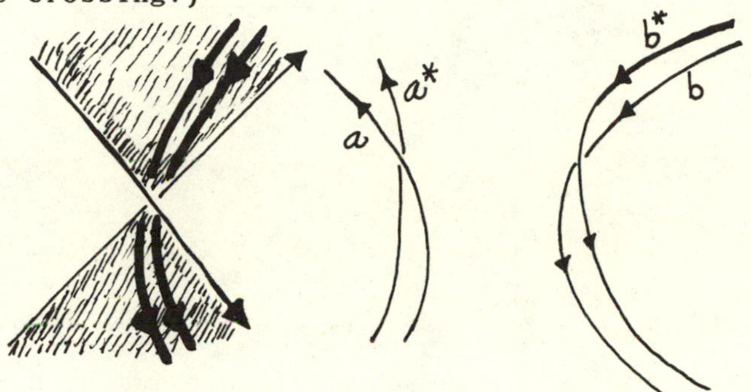

The self-linking contribution is $\theta(a,b) = -\frac{1}{2} = \theta(b,b)$. (*Note:* The cycles bounding white regions are all oriented compatibly with an orientation for the white region itself.)

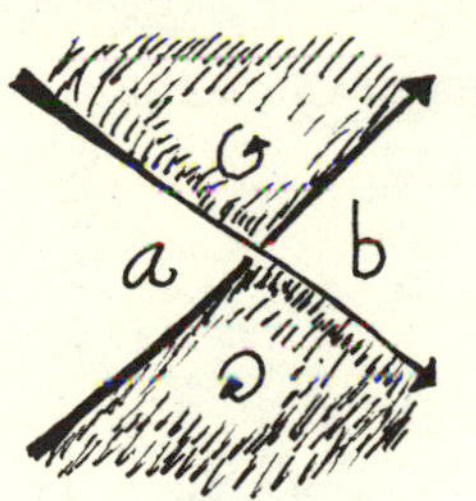

$$\begin{cases} \theta(a,b) = +1 \\ \theta(b,a) = 0 \\ \theta(a,a) = \theta(b,b) = -1/2 \end{cases}$$

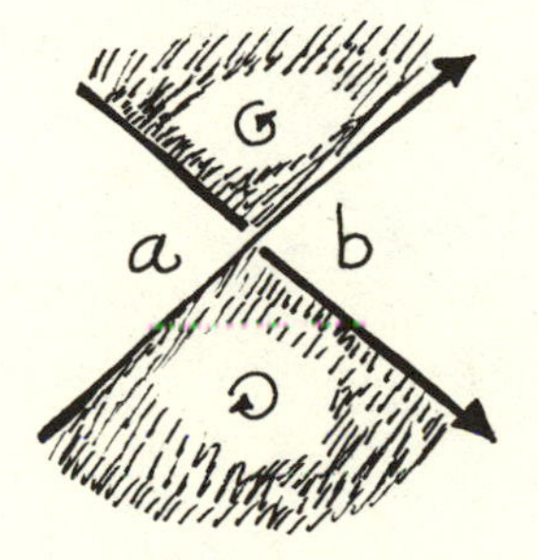

$$\begin{cases} \theta(a,b) = 0 \\ \theta(b,a) = -1 \\ \theta(a,a) = \theta(b,b) = +1/2 \end{cases}$$

For example:

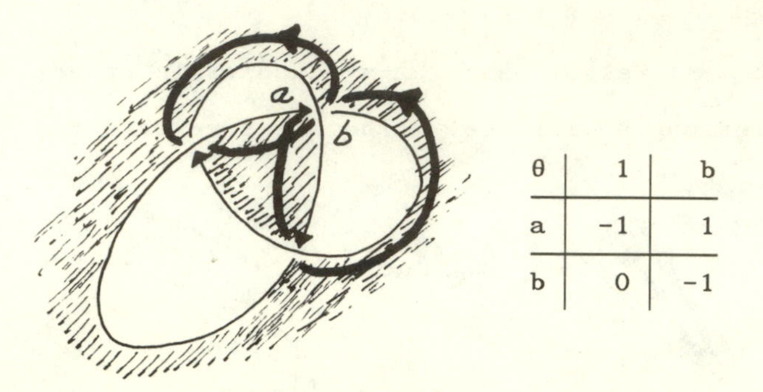

θ	1	b
a	-1	1
b	0	-1

Here a and b interact at only one crossing. But we look at two crossings to compute $\theta(a,b)$ and $\theta(b,b)$.

<u>Exercise</u>. Compute the Seifert pairing for F_K of Figure 7.1.

<u>Exercise</u>. Let $x \cdot y$ denote intersection number on the surface F. Show that for all $x,y \in H_1(F)$,

$$\theta(x,y)-\theta(y,x) = x \cdot y.$$

Hint: Do it for Seifert surface first. Then try the general case. To do the general case it helps to have the following description of linking numbers: Let $\alpha,\beta \subset S^3$ be two disjoint oriented curves. Let B be an oriented surface bounding β. Isotope α so that α intersects B transversally. Then $lk(\alpha,\beta) = \alpha \cdot B$.

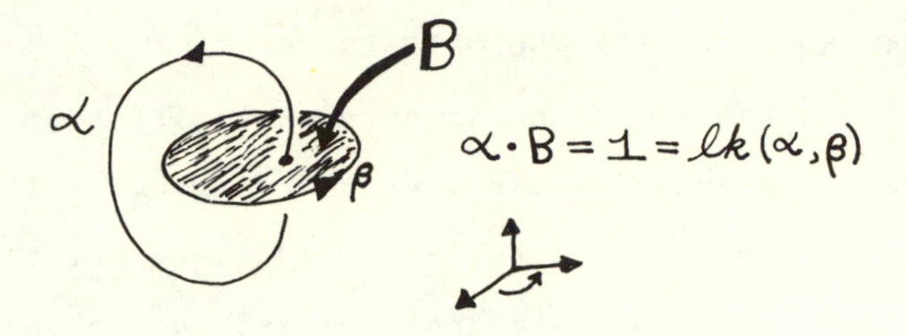

[Why is this independent of the choice of B?]

<u>Exercise</u>. Prove, using Seifert (or spanning) surfaces, that this description of linking implies our original description.

Now return to the formula $\theta(x,y)-\theta(y,x) = x \cdot y$, contemplate

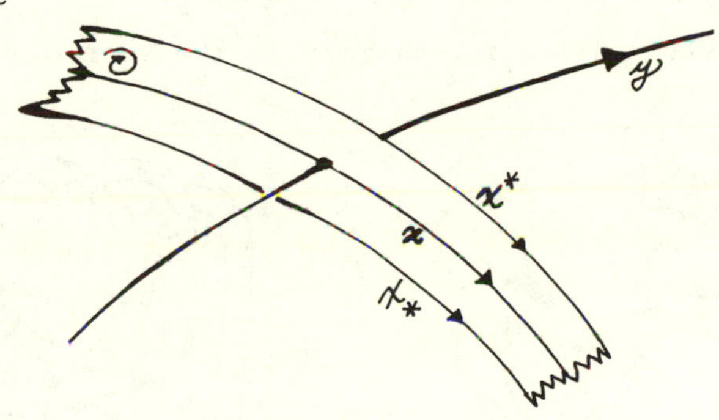

$$\partial B = \text{boundary of } B = x^{*}-x_{*}$$

$$\begin{aligned}\theta(x,y)-\theta(y,x) &= lk(x^{*},y)-lk(y^{*},x)\\ &= lk(y,x^{*})-lk(y^{*},x)\\ &= lk(y,x^{*})-lk(y,x_{*})\\ &= lk(y,x^{*}-x_{*})\\ &= y\cdot B\\ &= x\cdot y.\end{aligned}$$

DIFFERENT SURFACES FOR ISOTOPIC KNOTS

A given knot or link can have many different spanning surfaces. For example, two isotopic diagrams will have rather different Seifert surfaces. How are all the different surfaces spanning a knot related to one another?

The answer is, in principle, surprisingly simple. Consider the following way to complicate a spanning surface:

1) Cut out two discs, D_1, D_2.
2) Take a tube $S^1 \times I$ and embed it in S^3 disjointly from the surface, but with the tube boundary attached to ∂D_1 and ∂D_2.

This is called doing a 1-surgery to the surface.

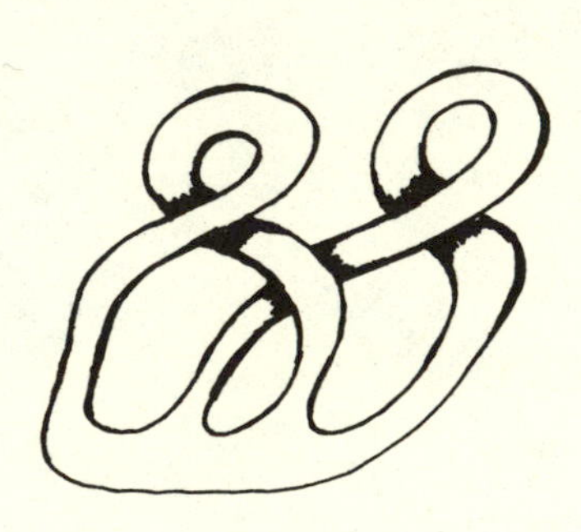

F

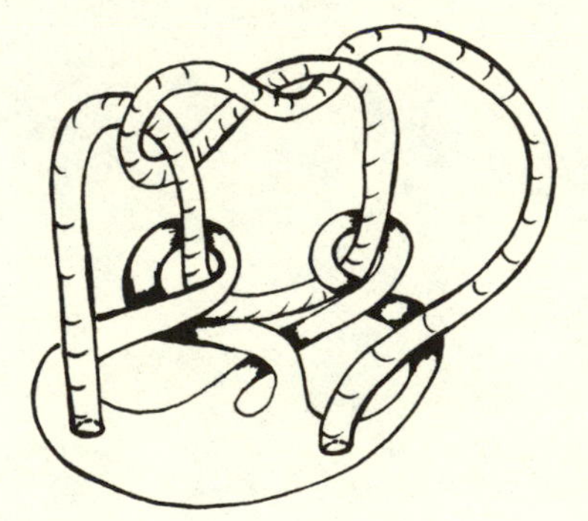

F after surgery

The reverse operation consists in finding a curve α on F such that α bounds a disk S^3-F. Then cut out $\alpha \times I$ from F and cap off with two D^2's.

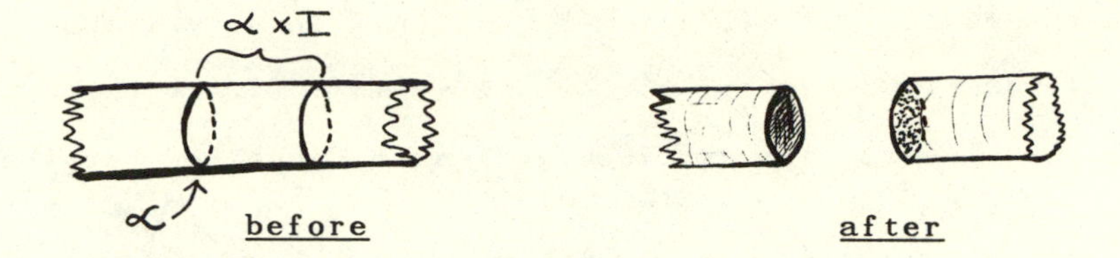

This is a <u>0-surgery</u>. It simplifies the surface (i.e., reduces genus).

These two surgery operations give us different surfaces with the same boundary.

DEFINITION 7.6. *Let* F *and* F′ *be oriented surfaces with boundary that are embedded in* S^3. *We say that* F *and* F′ *are* <u>*S-equivalent*</u> ($F \underset{S}{\approx} F'$) *if* F′ *may be obtained from* F *by a combinations of* 0-*surgery*, 1-*surgery and ambient isotopy*.

THEOREM 7.7 [L1]. *Let* F *and* F′ *be connected, oriented spanning surfaces for ambient isotopic links* $L, L' \subset S^3$. *Then* F *and* F′ *are* S-*equivalent*.

Proof sketch: Let $X = S^3 \times I$ and suppose that $\alpha: S^1 \times I \longrightarrow S^3$ is the ambient isotopy from $L = \alpha(S^1 \times 0)$ to $L' = \alpha(S^1 \times 1)$. Then we get an embedding of an annulus in X via $\hat{\alpha} : S^1 \times I \longrightarrow X$, $\hat{\alpha}(\lambda, t) = (\alpha(\lambda, t), t)$. If we form $M = (F \times 0) \cup \hat{\alpha}(S^1 \times I) \cup (F^1 \times 1)$, then this is a closed surface embedded in $S^3 \times I$. One then shows that $M = \partial W$ where W is a 3-manifold embedded in $S^3 \times I$. W can be arranged so that $(S^3 \times t) \cap W$ has only Morse critical points of type $x^2+y^2-z^2$ or $-z^2-y^2+z^2$. These correspond to the 0-surgeries and 1-surgeries we described earlier.

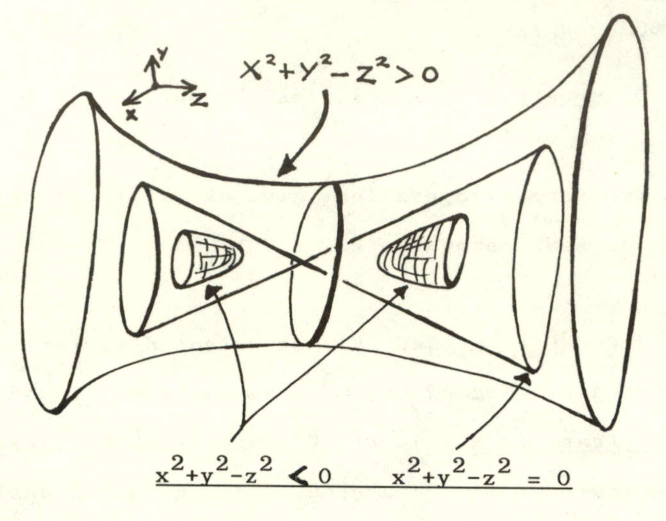

Remark: It may be of interest to look directly at the S-equivalences between Seifert surfaces for diagrams that are related by Reidemeister moves. For example,  is obtained from by the surgery

Now consider the Seifert pairings for S-equivalent surfaces. Suppose that F' is obtained from F by adding a tube. Then $H_1(F') \cong H_1(F) \oplus Z \oplus Z$ where these two

extra factors are generated by a <u>meridian for the tube</u> a, and an element <u>b</u> that passes once along the tube oriented so that $a \cdot b = 1$.

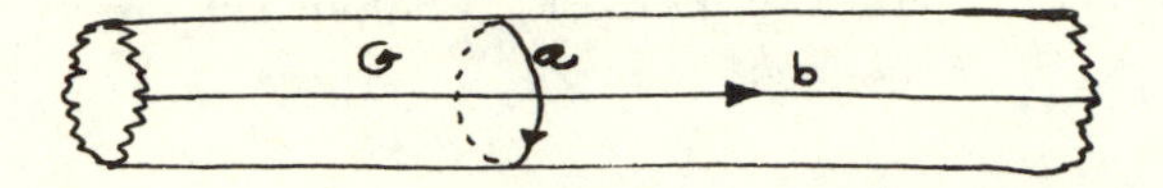

We then have $\theta(a,a) = 0$, $\theta(a,b) = 1$, $\theta(b,a) = 0$ and $\theta(a,x) = \theta(x,a) = 0$ for all $x \in H_1(F)$. Let θ_0 denote the Seifert pairing for F. Then we have

$$\theta = \begin{array}{c|c|c|c|} \cline{2-4} & \theta_0 & 0 & \alpha \\ \cline{2-4} a & 0 & 0 & 1 \\ \cline{2-4} b & \beta & 0 & n \\ \cline{2-4} \end{array}$$

where β is a row vector, and α is a column vector. Because of the row $(\overline{0},0,1)$, θ becomes on change of basis

$$\left[\begin{array}{c|cc} \theta_0 & 0 & 0 \\ \hline 0 & 0 & 1 \\ \beta & 0 & 0 \end{array}\right].$$

An enlargement of this kind is called an S-equivalence. More generally, two matrices θ and ψ are said to be S-equivalent if ψ can be obtained from θ by a combination of congruence ($\theta \longrightarrow P\,\theta\,P'$ where P' is the transpose of P, P invertible over Z. This corresponds to basis change.) and enlargements and contractions (reverse of enlargement) as above. If θ and ψ are S-equivalent, we write $\theta \underset{S}{\approx} \psi$.

COROLLARY 7.8. *Let K and K′ be ambient isotopic knots or links with connected spanning surfaces F (for K) and F′ (for K′). Let θ be the Seifert pairing for F and ψ be the Seifert pairing for F′. Then θ and ψ are S-equivalent.*

INVARIANTS OF S-EQUIVALENCE

DEFINITION 7.9. *Let F be a connected spanning surface for the knot or link K and θ the Seifert pairing for F. Define*

(i) *The determinant of K*, $D(K) = D(\theta+\theta')$ *where D denotes determinant.*

(ii) *The potential function of K*, $\Omega_K(t) \in Z[t^{-1},t]$ *by the formula* $\Omega_K(t) = D(t^{-1}\theta - t\theta')$.

(iii) *The signature of K*, $\sigma(K) \in Z$, *by* $\sigma(K) = \mathrm{Sign}(\theta+\theta')$ *where* Sign *denotes the signature of this matrix.* (See definition below.)

Of course the gadgets produced in this definition are not going to change under S-equivalence! Hence they will be invariants of K.

For example, if $\theta = \left[\begin{array}{c|cc} \theta_0 & 0 & 0 \\ \hline 0 & 0 & 1 \\ \alpha & 0 & 0 \end{array}\right]$ then $\theta+\theta' =$

$\left[\begin{array}{c|cc} \theta_0+\theta_0' & 0 & \alpha' \\ \hline 0 & 0 & 1 \\ \alpha & 1 & 0 \end{array}\right]$ and $D(t^{-1}\theta - t\theta') = D(t^{-1}\theta_0 - t\theta_0')$ because

$$D\begin{bmatrix} 0 & t^{-1} \\ -t & 0 \end{bmatrix} = 1.$$

For the signature, recall that a symmetric matrix M over Z can be diagonalized through congruence over Q (the rationals) or over $\mathbb{R}$. Let e_+ denote the number of positive diagonal entries, and e_- the number of negative diagonal entries. The signature, $\mathrm{Sign}(M)$, is defined by the formula $\mathrm{Sign}(M) = e_+ - e_-$. It is an invariant of the congruence class of M. (See [HNK].) Note in particular, that $\mathrm{Sign}\begin{bmatrix} 0 & 1 \\ 1 & 0 \end{bmatrix} = 0$. From this it follows that $\mathrm{Sign}(\theta+\theta')$ is an invariant of its S-equivalence class, hence an invariant of K. We shall also show that $\sigma(K)$ is an invariant of concordance.

The potential function provides a model for the Conway polynomial:

THEOREM 7.10.

(i) *If* K *and* K' *are ambient isotopic oriented links, then* $\Omega_K(t) = \Omega_{K'}(t)$.

(ii) *If* $K \sim 0$, *then* $\Omega_K(t) = 1$.

(iii) *If links* K, $\overline{K}$ *and* L *are related as below, then* $\Omega_K - \Omega_{\overline{K}} = (t-t^{-1})\Omega_L$.

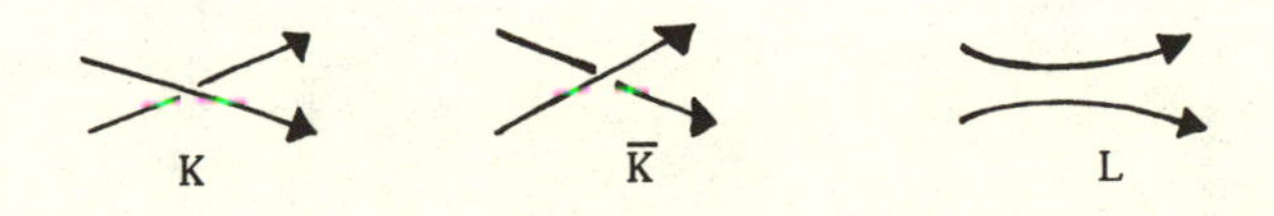

Proof: We have already proved (i) and (ii). Note that $\Omega_K = 0$ if K is a split link. To see this, choose disjoint spanning surfaces for two pieces of the link, and connect these by a tube to form a connected spanning surface F.

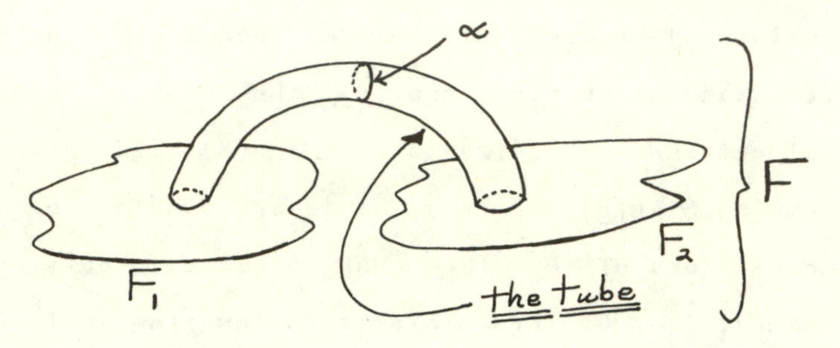

If α is a meridian of this type, then

$$H_1(F) \cong H_1(F_1) \oplus H_1(F_2) \oplus Z$$

where α generates the extra copy of Z. Since $\theta(\alpha, x) = \theta(x, \alpha) = 0 \quad \forall\, x \in H_1(F)$, it follows that $\Omega_K(t) = 0$.

We use this discussion as follows. Consider Seifert surfaces for K, $\overline{K}$ and L. Locally, they appear as

F_K $\qquad$ $F_{\overline{K}}$ $\qquad$ F_L

We see that $H_1(F_K)$ and $H_1(F_{\overline{K}})$ will have one more homology generator than F_L, unless it should happen that L is a split diagram. But in this case $\Omega_L = 0$ while F_K and $F_{\overline{K}}$ are isotopic by a 2π twist. Thus $\Omega_K - \Omega_{\overline{K}} = 0 = \Omega_L$, proving (iii).

If L is not a split diagram, then the extra generator may be represented as $\underline{a}$ on F_K and $\underline{a'}$ on $F_{\overline{K}}$. We see that $\theta(a',a') = \theta(a,a)+1$. Hence $\theta_K = \left[\begin{array}{c|c} n & \beta \\ \hline \alpha & \theta_L \end{array}\right]$, $\theta_{\overline{K}} = \left[\begin{array}{c|c} n+1 & \beta \\ \hline \alpha & \theta_L \end{array}\right]$, with appropriate choice of bases. It is now a straightforward determinant calculation to show that $\Omega_K - \Omega_{\overline{K}} = (t-t^{-1})\Omega_L$. ∎

Remark: By our axiomatics, it follows that the Conway polynomial and our potential function are related by the substitution $z = t-1/t$. Thus $\Omega_K(t) = \nabla_K(t-1/t)$. It is amusing to solve the reverse. Then

$$t = z+1/t.$$

Hence

$$t = z + \cfrac{1}{z + \cfrac{1}{z + \cdots}}.$$

Using the notation $[z+1]\circlearrowleft$ for the continued fraction $z + \cfrac{1}{z + \cfrac{1}{z + \cdots}}$, we have $\nabla_K(z) = \Omega_K([z+1]\circlearrowleft)$. In particular,

$$\nabla_K(1) = \Omega_K\left[\frac{1+\sqrt{5}}{2}\right].$$

We shall return to this subject!

Example: Let T be a trefoil with $\theta = \begin{bmatrix} -1 & 1 \\ 0 & -1 \end{bmatrix}$. Then

$$\Omega_T = D\begin{bmatrix} -t^{-1}+t & t^{-1} \\ -t & -t^{-1}+t \end{bmatrix} = (t-t^{-1})^2+1 = z^2+1.$$

This agrees with our previous calculations.

Example: Given a knot K, let K denote the numerator of the tangle obtained by running a parallel copy of K with opposite orientation. K is a link of two components.

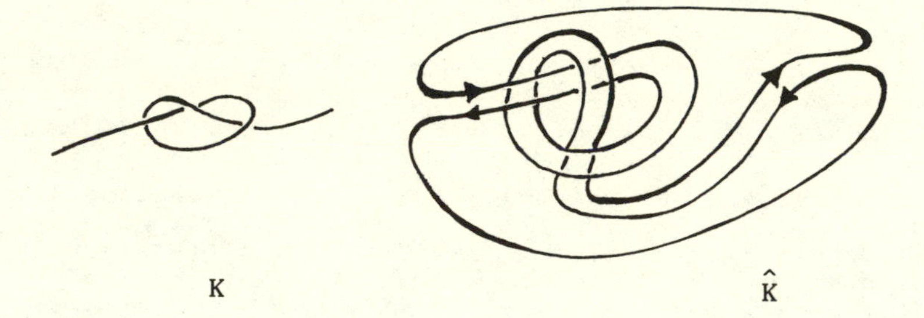

Since K has a spanning surface that is an annulus, we see that $\theta = [-\mathrm{lk}(\hat{K})]$ is a Seifert matrix for $\hat{K}$. Therefore $\Omega_K = (t^{-1}-t)(-\mathrm{lk}(\hat{K}))$ and hence $\underline{\nabla_K = \mathrm{lk}(\hat{K})z}$. Apparently, in this case the Conway polynomial is much easier to compute using the Seifert pairing. (Compare this discussion with the last exercise of Chapter IV of these notes.)

TRANSLATING ∇ *AND* Ω.

Note that $\Omega_K(t) = D(t^{-1}\theta - t\theta')$. Therefore

$$\Omega_K(t^{-1}) = D(t\theta - t^{-1}\theta')$$
$$= D(t\theta' - t^{-1}\theta)$$
$$\therefore \quad \Omega_K(t^{-1}) = D(-(t^{-1}\theta - t\theta')).$$

Since θ is $2g \times 2g$ for knots, $(2g+1)\times(2g+1)$ for 2-component links, we conclude that $\Omega_K(t^{-1}) = (-1)^{\mu+1}\Omega_K(t)$ where μ is the number of components of K.

To obtain a practical method of translation between Ω_K and ∇_K, we need to write $t^n+(-1)^n t^{-n} = T_n$ in terms of $z = t-t^{-1}$. Look at the pattern:

$$t^2+t^{-2} = (t-t^{-1})^2+2 = z^2+2$$

$$t^3-t^{-3} = (t-t^{-1})^3+3t-3t^{-1} = z^3+3z.$$

Exercise. Let $T_n = t^n+(-1)^n t^{-n}$ and $z = t-t^{-1}$. Show that $T_{n+2} = zT_{n+1}+T_n$ for $n \geq 0$.

$$t-t^{-1} = z$$
$$t^2+t^{-2} = z^2+2$$
$$t^3-t^{-3} = z^3+3z$$
$$t^4+t^{-4} = z^4+4z^2+2$$
$$t^5-t^{-5} = z^5+5z^3+5z$$
$$t^6+t^{-6} = z^6+6z^4+9z^2+2.$$

Show that the coefficient of z^2 in $t^{2n}+t^{-2n}$ is n^2.

We can use this exercise to obtain a curious formula for the second Conway coefficient $a_2(K)$. For let K be a knot. Then K has potential function in the form

$\Omega_K(t) = b_0+b_1(t^2+t^{-2})+b_2(t^4-t^{-4})+\cdots+b_n(t^{2n}+t^{-2n})$. It follows from our exercise that

$$a_2(K) = b_1+4b_2+9b_3+16b_4+\cdots+n^2b_n.$$

Exercise. Compute Seifert pairing, determinant, potential function and signature for the torus knots and links of type $(2,n)$.

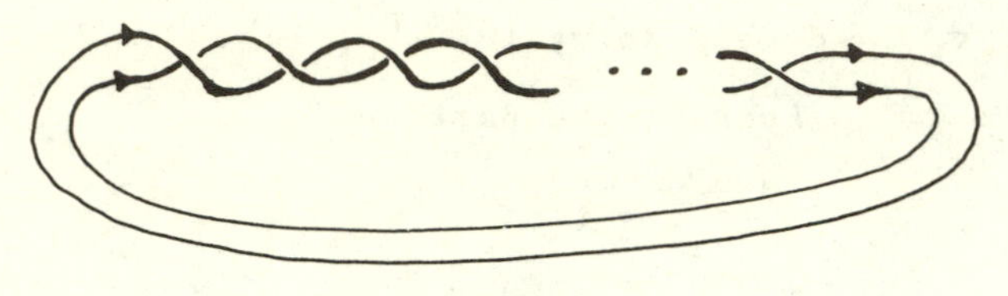

n Crossings

Exercise. Prove that $\sigma(K^!) = -\sigma(K)$ when K is a knot and $K^!$ is its mirror image. Calculate $\sigma(T)$ and thereby show that $T =$ and $T^! =$ are not ambient isotopic.

Exercise. Prove that for knots K, K',

$$\sigma(K\#K') = \sigma(K)+\sigma(K').$$

Use this exercise and the previous exercise to distinguish the granny and the square knot.

square granny

<u>Exercise</u>. Choose a knot or link and compute everything you can.

<u>Exercise</u>. Let K be a knot. Show that $\nabla_K(2i)/|\nabla_K(2i)| = i^{\sigma(K)}$. Use this in conjunction with the (easily proved) fact $\sigma_{K-} \leq \sigma_{K+} \leq 2 + \sigma_{K-}$ [K+ K-] to show how to inductively calculate knot signatures using a skein decomposition (see [C1], [G1]).

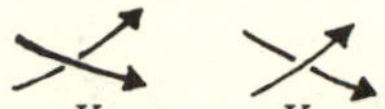

Apply this method to the knot 9_{42} (see the end of Section 19 of Chapter VI in these notes) to show that 9_{42} has signature 2. This completes our earlier assertion that 9_{42} is not amphicheiral.

VIII

RIBBONS AND SLICES

First a lemma about 3-manifolds:

LEMMA 8.1. *Let* M^3 *be a compact, connected orientable 3-manifold with boundary. Denote the boundary by* $\partial M^3 = N^2$. *Let* $j : N \longrightarrow M$ *be the inclusion, and let* $H_* = H_*(\ ;Q)$ *denote homology with rational coefficients. Let* $K = \mathrm{Kernel}(j_* : H_1(N) \longrightarrow H_1(M))$. *Then, as vector spaces over* Q, $\dim K = \left[\frac{1}{2}\right] \dim H_1(N)$.

Proof. Look at the homology exact sequence for the pair (M,N):

$$0 \longrightarrow \overset{1}{H_3(M,\partial M)} \longrightarrow \overset{a}{H_2(\partial M)} \longrightarrow \overset{b}{H_2(M)} \longrightarrow \overset{c}{H_2(M,\partial M)} \longrightarrow$$

$$\begin{array}{ccc} & & 0 \\ k & & \nearrow \\ & K & \\ \nearrow & & \searrow \end{array}$$

$$\underset{d}{H_1(\partial M)} \longrightarrow \underset{c}{H_1(M)} \longrightarrow \underset{b}{H_1(M,\partial M)} \longrightarrow \underset{a}{H_0(\partial M)} \longrightarrow \underset{1}{H_0(M)} \longrightarrow 0.$$

We have denoted dimensions by $a,b,c,d,1,k$, and used Poincaré-Lefschetz Duality $(H_1(M,\partial M) \cong H_2(M)$; $\therefore \dim H_1(M,\partial M) = \dim H_2(M)$, etc.). Exactness of the sequence implies that

$$1 - a + b - c + d - c + b - a + 1 = 0$$

$$\therefore \quad 2 - 2a + 2b - 2c + d = 0$$

$$\therefore \quad d = 2(c-b+a-1).$$

But $k = \dim K = d - c + b - a + 1 = d - (c-b+a-1)$. Hence, $k = d/2$ as desired.

Thus half the cycles die into the interior of M. The typical case is a handlebody with the meridians representing K:

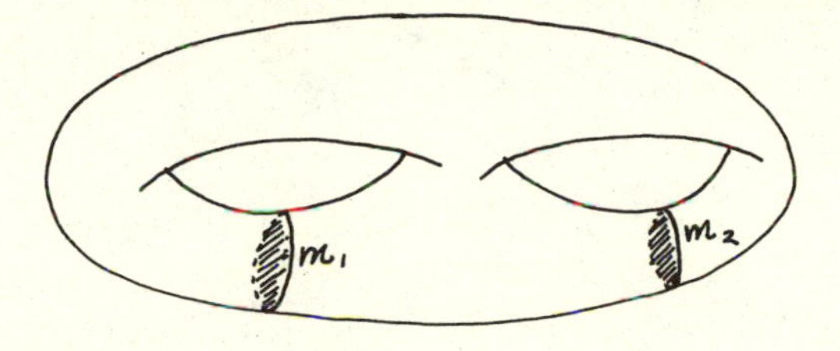

The lemma is useful in conjunction with a 4-dimensional interpretation of linking numbers. Let $\alpha, \beta \subset S^3$ be disjoint oriented curves. We have interpreted $\ell k(\alpha,\beta)$ as $\ell k(\alpha,\beta) = \alpha \cdot B$ where B represents a surface with boundary $\partial B = \beta$. Here is a more symmetrical way: Regard $S^3 = \partial D^4$. Choose surfaces A and B embedded in D^4 transverse to the boundary such that $\partial A = \alpha$, $\partial B = \beta$. Orient A and B so that they induce the given orientations on α and β. Then $\ell k(\alpha,\beta) = A \cdot B$ where this intersection number is taken in D^4.

Picture reduced one dimension:

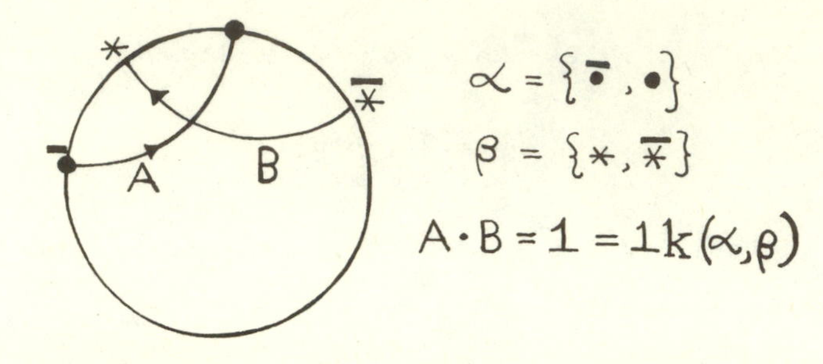

Example: [link] $_L$. Show that the components of this link bound disks in D^4 that intersect transversely in one point.

Solution: $S^3 \cong \partial(D^2 \times D^2)$, $L = (S^1 \times 0) \cup (0 \times S^1) \subset S^3$.
$\alpha = \partial(D^2 \times 0)$, $\beta = \partial(0 \times D^2)$ and $(D^2 \times 0) \cap (0 \times D^2) = \{(0,0)\}$.

THEOREM 8.2. *Let* $K \subset S^3$ *be a slice knot, then there exists a submodule* $W \subset H_1(F)$ *(*F *any spanning surface for* K, $F \subset S^3$*) of half the rank of* $H_1(F)$ *such that the Seifert pairing vanishes identically on* W.

Proof: By the method of Theorem 7.7, there exists a 3-manifold $M \subset D^4$ such that $\partial M = D \cup F$ where D is a slice disk for K embedded in D^4 and F is the given spanning surface. Apply Lemma 8.1 and the fact that $H_1(\partial M) = H_1(F)$ to deduce the existence of a submodule $W \subset H_1(F)$ of rank $= g = \frac{1}{2}$ rank $H_1(F)$ and $W = \mathrm{Ker}(H_1(F) \to H_1(M))$. We claim that this W is a vanishing subspace for θ: Let $\alpha, \beta \in W$

(regard these as representative cycles). Then there exist surfaces $A, B \subset M$ with $\partial A = \alpha$, $\partial B = \beta$. Furthermore the operation $x \longrightarrow x^{*}$, translating $F \longrightarrow S^3 - F$ along the positive normal, extends to $M \longrightarrow D^4 - M$. Hence $\partial A^{*} = \alpha^{*}$. Hence $\theta(\alpha, \beta) = \ell k(\alpha^{*}, \beta) = A^{*} \cdot B = 0$. This completes the proof. ■

THEOREM 8.3. *Let* $K \subset S^3$ *be a slice knot. Then*

(i) *The signature* $\sigma(K) = \mathrm{Sign}(\theta + \theta')$ *vanishes:* $\sigma(K) = 0$.

(ii) *The potential function* $\Omega_K(t)$ *has the form* $\Omega_K(t) = f(t)f(t^{-1})$ *where* $f(t) \in Z[t^{-1}, t]$.

Proof: Let F be any spanning surface for K, $F \subset S^3$. Let $\theta : H_1(F) \times H_1(F) \longrightarrow Z$ be the Seifert pairing for F. We then know that $\theta = \begin{bmatrix} 0 & X \\ Y & T \end{bmatrix}$ in matrix form. The 0 block is $g \times g$ and corresponds to the subspace W of Theorem 8.2. Thus $M = \theta + \theta' = \begin{bmatrix} 0 & L \\ L' & S \end{bmatrix}$ where $L = X + Y'$ and $S = T + T'$. $\sigma(K) = \mathrm{Sign}(M)$ and we wish to show that this signature vanishes. Note that $M \equiv \theta - \theta' \pmod 2$ and that $|\theta - \theta'| \neq 0$ since this is the determinant of the intersection form on F. Consequently $|M| \neq 0$ and therefore $|L| \neq 0$. Hence L is invertible (over Q).

[<u>Notation</u>: $\bar{A} = A^{-1}$.] Let $P = \begin{bmatrix} \bar{L} & 0 \\ Z\bar{L} & -I \end{bmatrix}$ and verify that $PMP' = \begin{bmatrix} 0 & -I \\ -I & 0 \end{bmatrix}$. Conclude that $\sigma(K) = \sigma(M) = \sigma(PMP') = 0$.

To see the form of the potential function, note that

$$t^{-1}\theta - t\theta' = \begin{bmatrix} 0 & t^{-1}X - tY' \\ t^{-1}Y - tX' & t^{-1}Z - tZ' \end{bmatrix}.$$

Hence

$$\Omega_K(t) = (-1)^g |t^{-1}X - tY'||t^{-1}Y - tX'|$$

$$= |t^{-1}X - tY'||tX - t^{-1}Y'|$$

$$\therefore \quad \Omega_K(t) = f(t)f(t^{-1})$$

where $f(t) = |t^{-1}X - tY'|$. This completes the proof. ■

Example: K has $\theta = \begin{bmatrix} -1 & 1 \\ 0 & -1 \end{bmatrix}$. $\theta + \theta' = \begin{bmatrix} -2 & 1 \\ 1 & -2 \end{bmatrix} = M$.

$$\begin{bmatrix} -2 & 1 \\ 1 & -2 \end{bmatrix} \xrightarrow[\text{operation}]{\text{column}} \begin{bmatrix} -2 & 0 \\ 1 & -3/2 \end{bmatrix} \xrightarrow[\text{operation}]{\text{row}} \begin{bmatrix} -2 & 0 \\ 0 & -3/2 \end{bmatrix}$$

$\Longrightarrow \sigma(K) = -2$.

Example: In the case of a ribbon knot we can construct a spanning surface on which the half-rank vanishing submodule for θ is geometrically obvious. Start with the singular ribbon and replace each ribbon singularity by a piece of nonsingular surface as follows:

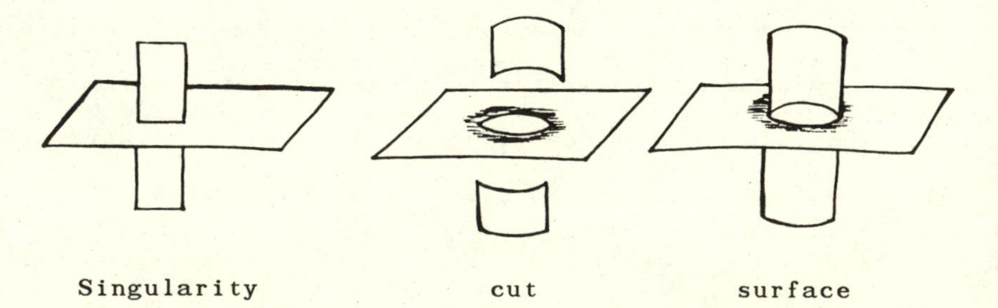

Singularity cut surface

Cut along the ribbon singularity. This produces a hole in

one ribbon, and two ribbon-ends. Paste the edge of one end to one half of the hole-boundary, and paste the edge of the other to the other half of the hole-boundary (as illustrated above).

In the resulting surface F there will be a set of cycles $\lambda_1,\cdots,\lambda_g$, one for each hole (and encircling it as shown above). Then $g(F) = g$ and $\{\lambda_1,\cdots,\lambda_g\}$ generates a submodule of $H_1(F)$ on which θ vanishes.

Note that this geometry gives us a direct proof for ribbons that the potential has form $f(t)f(t^{-1})$.

<u>Exercise</u>. Given a knot K, let $\overline{K} = -K^!$. That is, $\overline{K}$ is K mirror-imaged and with reversed orientation. We know from Chapter V that $K \# K^!$ is a ribbon knot. Think about this.

<u>Exercise</u>. Formalize the details for obtaining an orientable spanning surface for a ribbon knot, as described above. On the other hand, you can show that if F is a spanning surface for K with Seifert pairing (matrix) θ then it is easy to construct a surface for $\overline{K}$ with Seifert pairing $-\theta$. (Do this!) Hence $K \# \overline{K}$ bounds a surface with Seifert pairing $\begin{bmatrix}\theta & 0\\ 0 & -\theta\end{bmatrix} = \tilde{\theta}$. From this we see directly that $\Omega_{K\#\overline{K}} = \Omega_K(t)\Omega_K(t^{-1})$, and this satisfies the form demanded by the ribbon knot.

(i) Locate the half-rank submodule on which $\tilde{\theta}$ vanishes. Interpret this geometrically in terms of the connected sum.

(ii) One way of looking at the result K ribbon $\Rightarrow$ $\Omega_K(t) = f(t)f(t^{-1})$, is to see that it says that in some algebraic sense K looks like a connected sum with a mirror image. Investigate this concept.

We now apply 8.3 to some examples.

n curls F_n.

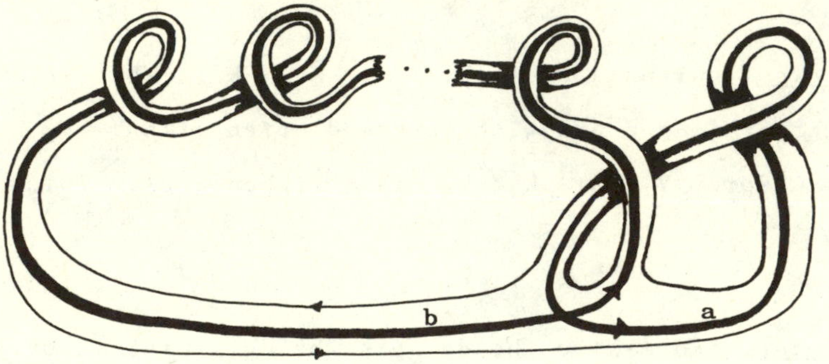

a

b

Let K_n be the boundary of F_n.

K_1 = figure eight knot

K_2 = stevedore's knot (which we know is slice) (see Sec. 5)

On F_n we have $\theta(a,a) = -1$ $\theta(a,b) = 1$

$\theta(b,b) = n$ $\theta(b,a) = 0$.

Thus F_n has Seifert matrix $\begin{bmatrix} -1 & 1 \\ 0 & n \end{bmatrix}$. Whence

$$\Omega_n = \Omega_{K_n} = \begin{vmatrix} -t^{-1}+t & t^{-1} \\ -t & nt^{-1}-nt \end{vmatrix}.$$

$$\therefore \quad \Omega_n = -n(t^{-1}-t)^2+1 \quad \text{and}$$
$$\nabla_n = -nz^2+1.$$

Thus we know already, from our results about the coefficient $a_2(K)$, that n must be even if K_n is to be slice. But Theorem 8.3 gives more information: In order for Ω_n to have the form $f(t)f(t^{-1})$ we need $f(t)$ to have degree 1. That is, $f(t) = at+bt^{-1}$ where $a,b \in Z$. Then

$$\begin{aligned} f(t)f(t^{-1}) &= (at+bt^{-1})(at^{-1}+bt) \\ &= a^2 + b^2 + ab(t^2+t^{-2}) \\ &= a^2 + b^2 + ab((t-t^{-1})^2+2) \\ &= a^2 + b^2 + ab(z^2+2). \end{aligned}$$
$$\therefore \quad f(t)f(t^{-1}) = (a+b)^2 + abz^2.$$

Thus 8.3 tells us that if K_n is slice, then $-n = ab$ where $a+b = \pm 1$. In other words, _n has the form $n = k(k+1)$_.

This rules out many even n.

Now, in fact, the work of Casson and Gordon [CG] which we will consider later in these notes, shows that _none_ of the K_n (except K_2) are slice knots. This is a deep result, and it requires much more technique. The first example is K_6.

We're going to look at some of the geometry of these examples right now. First note the following general principle. Suppose $\alpha \subset F$ is a curve embedded in a surface $F \subset S^3$, F orientable. Suppose also that $[\alpha] \in H_1(F)$ is a nonzero class.

If (i) $\theta(\alpha,\alpha) = 0$ and

(ii) $\alpha \subset S^3$ is a slice knot, then we can surger the surface along α to obtain a surface F' of lower genus embedded in D^4 with $\partial F' = \partial F$.

This surface F' is obtained by attaching a slice disk Δ to α and thickening this disk for the surgery. That is, since $\theta(\alpha,\alpha) = 0$, we can embed $\alpha \times I \longrightarrow F$ and $\Delta \times D^2 \longrightarrow D^4$ and perform embedded surgery.

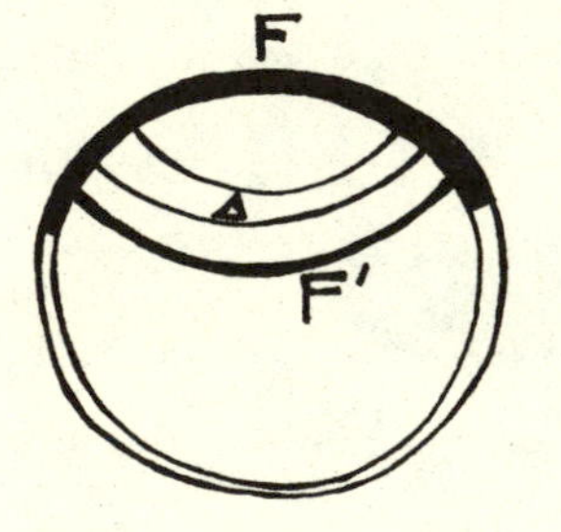

The resulting surface is best described by

(i) Using $\theta(\alpha,\alpha) = 0$, take two parallel copies of α in F.

(ii) Cut out the annulus between them.

(iii) Add slicing disks for α and α' into D^4.

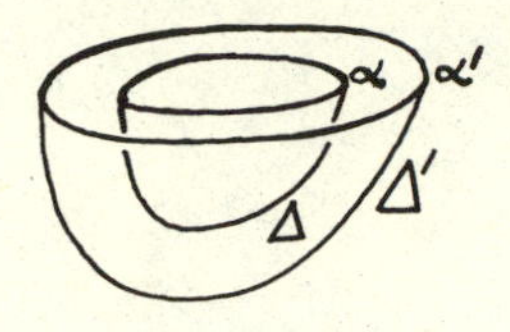

<u>Exercise</u>. If you can find on F a set of curves $\alpha_1, \cdots, \alpha_g$ representing a half-rank (rank $H_1(F) = 2g$) submodule on which θ vanishes <u>and</u> if each $\alpha_i \subset S^3$ is a slice knot <u>then</u> $K = \partial F$ is a slice knot. If, furthermore, each $\alpha_i \subset S^3$ is ribbon, then K is ribbon.

Example 8.4. Generalized Stevedore Knots. This surgery procedure is a good way to try to slice a knot. Let's see how it works for Stevedore (K_2): Here $\theta = \begin{bmatrix} -1 & 1 \\ 0 & 2 \end{bmatrix}$ with respect to the basis $\{a, b\}$. Let $\alpha = ka + \ell b$ and suppose $\theta(\alpha, \alpha) = 0$. Then

$$0 = k^2\theta(a,a) + \ell^2\theta(b,b) + k\ell(\theta(a,b) + \theta(b,a)).$$

$$\therefore \quad 0 = k^2(-1) + \ell^2(2) + k\ell$$

$$k^2 - k\ell - 2\ell^2 = 0$$

$$k = (\ell \pm \sqrt{\ell^2 + 8\ell^2})/2$$

$$k = (\ell \pm 3\ell)/2$$

$$k = 2\ell, \quad -\ell$$

$$\therefore \quad \alpha = ka + \ell b = \ell(2a+b), \ \ell(-a+b).$$

This tells us to try embedded representatives for $2a+b$ or $-a+b$.

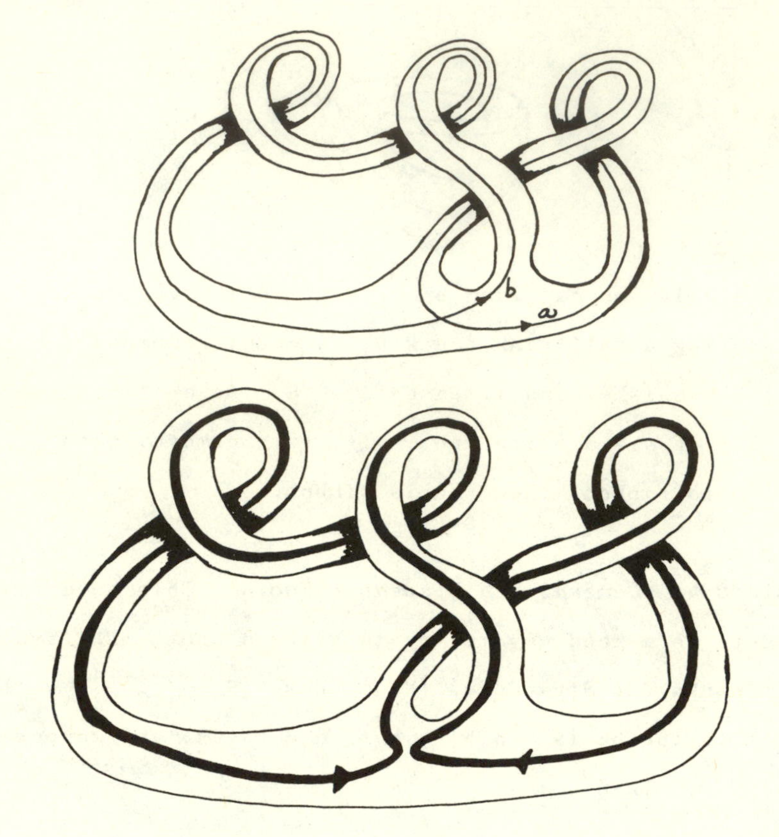

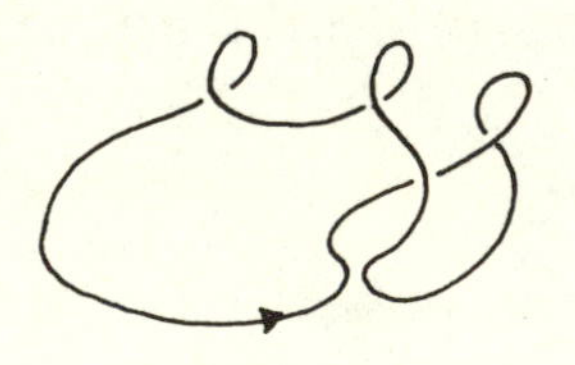

-a + b is an unknotted curve. Hence it will work!

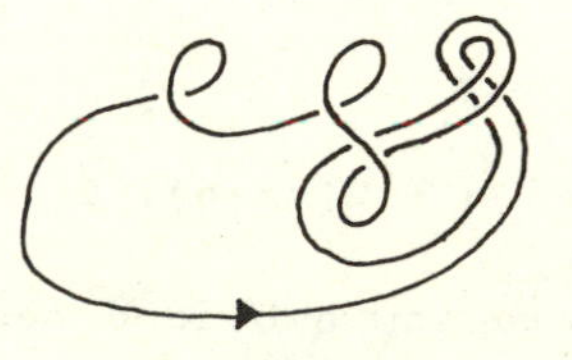

2a + b is also unknotted.

Either of these curves can be surgered to produce a ribbon surface for K_2.

<u>Exercise</u>. Draw pictures of the results of these surgeries. Let's set up the exercise for <u>$-a+b$</u>. First cut out an annulus $(-a+b)\times I$:

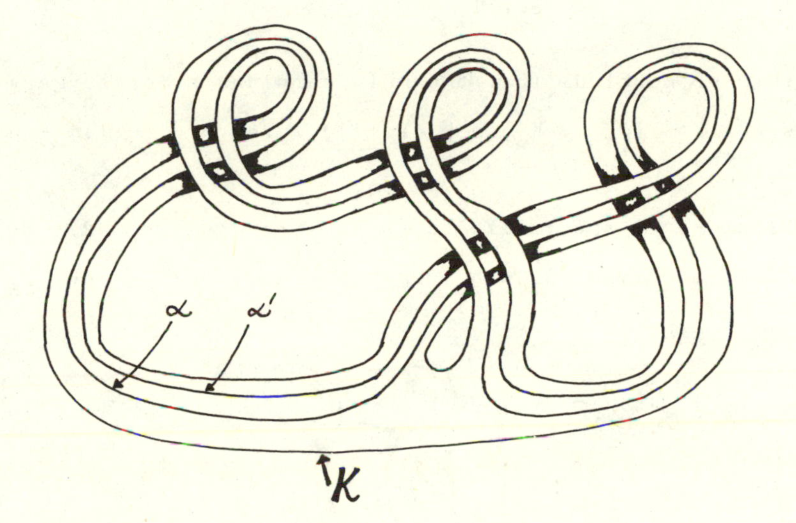

In principle you can now insert disks with boundaries α and α' introducing only ribbon singularities in the process. Enjoy!.

What goes wrong with the surgery process for K_n, $n > 2$? Well, let's look. The Seifert matrix is $\begin{bmatrix} -1 & 1 \\ 0 & n \end{bmatrix}$. If $\alpha = ka+\ell b$ and $\theta(\alpha,\alpha) = 0$, then

$$0 = k^2\theta(a,a) + \ell^2\theta(b,b) + k\ell(\theta(a,b) + \theta(b,a))$$

$$0 = -k^2 + n\ell^2 + k\ell.$$

Now we may as well assume $n = e(e+1)$, since this is necessary for slice. Then

$$k^2 - \ell k - e(e+1)\ell^2 = 0$$

$$\begin{aligned} k &= (\ell \pm \sqrt{\ell^2 + 4e(e+1)\ell^2})/2 \\ &= (\ell \pm (2e+1)\ell)/2 \\ &= (e+1)\ell, \quad -e\ell. \end{aligned}$$

From this we conclude that the only candidates for surgery curves on F_n $(\partial F_n = K_n,\ n = e(e+1))$ are $\alpha = (e+1)\underline{a} + \underline{b}$ or $\alpha = -e\underline{a} + \underline{b}$.

First we'll specialize to $n = 6$ so that $e = 2$. So consider the curves $\alpha_1 = \underline{b} - 2\underline{a}$ and $\alpha_2 = \underline{b} + 3\underline{a}$ on F_6:

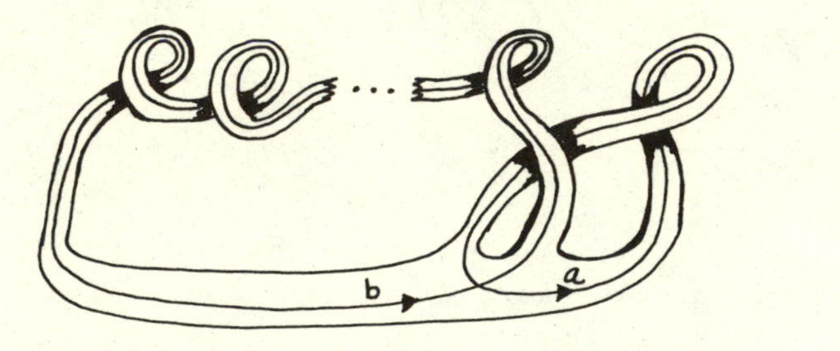

F_6

$\alpha_1 = \underline{b}-2\underline{a}$

Thus α_1 is a trefoil and not slice.

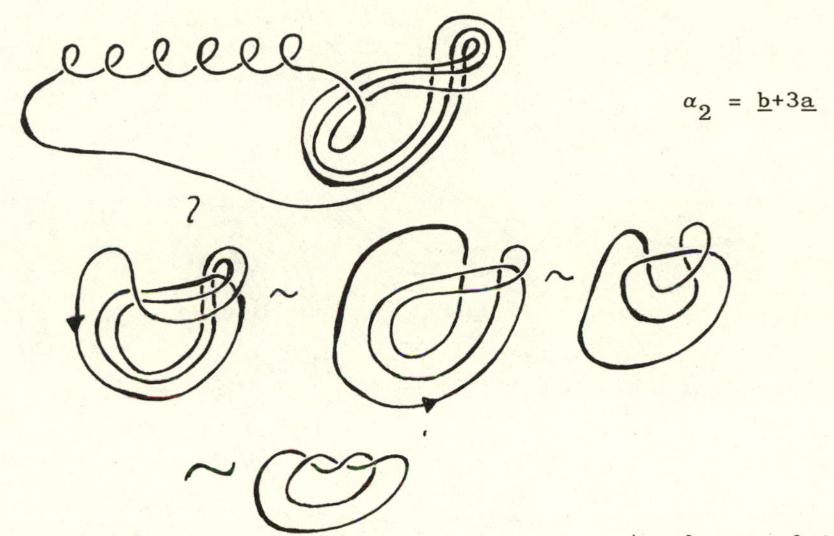

Another trefoil.

The result is that no surgery curves are available on F_6. This does not prove that K_6 is not slice, but it certainly underscores the difficulty in slicing it!

As we have mentioned, the knots K_n are not slice ([CG]). The attempt at surgery on F_n yields two possible curves whenever $n = e(e+1)$ but these curves are not slice knots. We <u>will</u> prove this fairly soon! Let's get a picture (for the record) of these curves: They are of the form $m\underline{a} + \underline{b}$ where $m = -e$ or $(e+1)$.

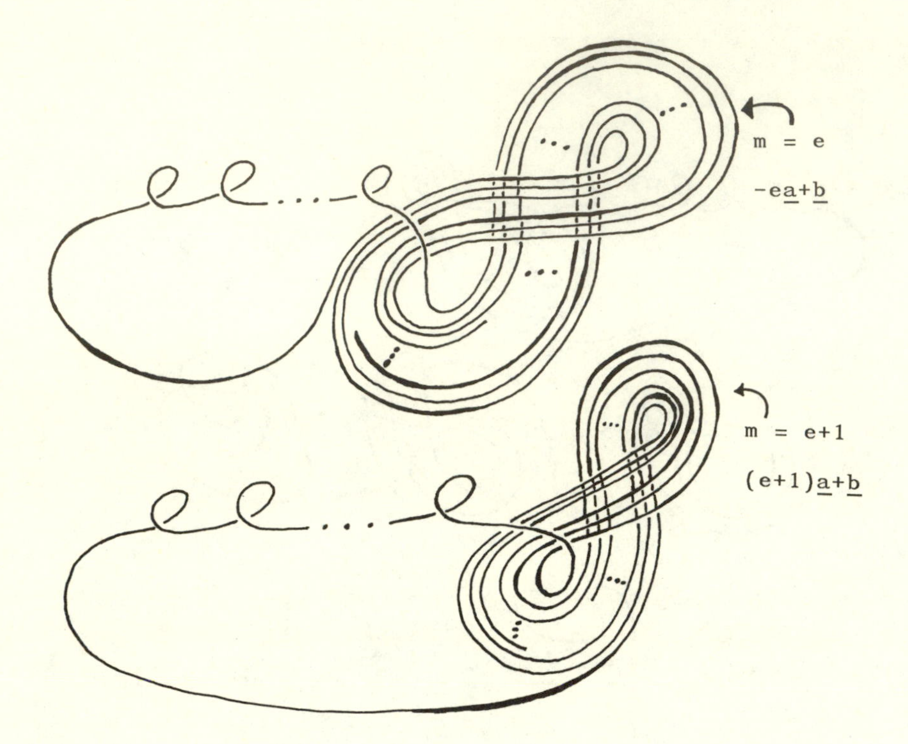

Thus these knots are of the form:

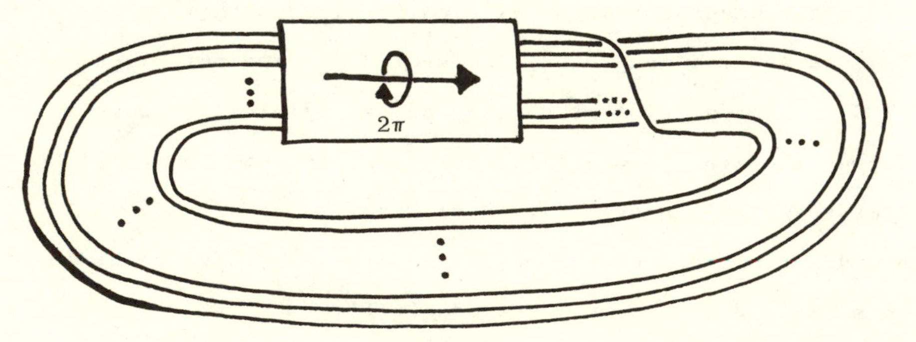

$\underline{m}$ strands

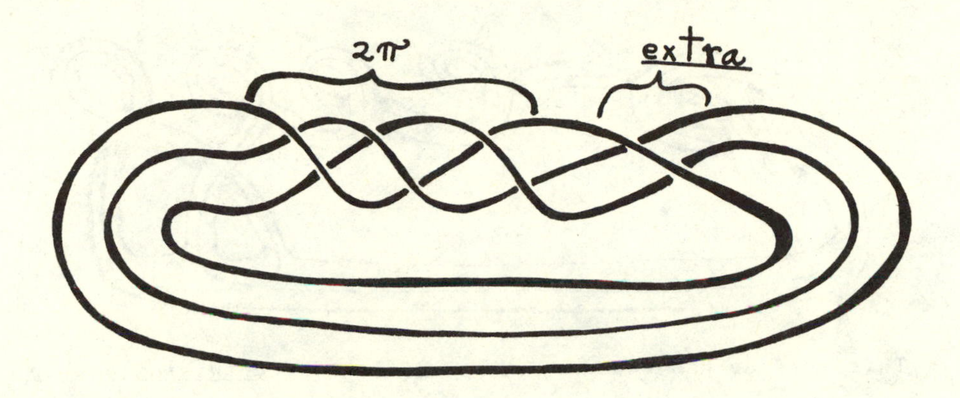

This extra twist can also be in the opposite direction. This shows that we are looking at torus knots of type $(m, m\pm 1)$. Hence, in fact at torus knots of type $(k, k+1)$.

To see this, note that a 2π-twist on a cable of n-strands can be accomplished by a series of $(2\pi/n)$-twists. Thus for $m = 3$:

Signature calculations will show that these torus knots are not slice.

Example 8.5. The Untwisted Double.

This surface has boundary $\partial F = D(T)$, the <u>untwisted</u> <u>double</u> of the trefoil

T  An equivalent way to describe $D(T)$ is to take two parallel copies of T, add twists to obtain zero linking number, then change the parallel strands to a clasp to obtain a knot.

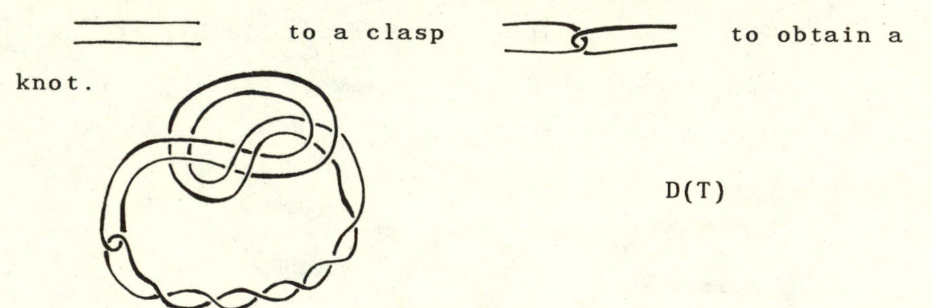

D(T)

There are two choices for the clasp: We will refer to either as the untwisted double unless some need to distinguish them arises.

The operation of forming the untwisted double of any

knot K always produces a knot that looks algebraically slice.

DEFINITION 8.6. *A knot* K *is said to be* *<u>algebraically slice</u>* *if given a surface* F *bounding* K *(connected, oriented, embedded in* S^3*) then the Seifert pairing* $\theta : H_1(F) \times H_1(F) \longrightarrow Z$ *has a* $\frac{1}{2}$ *rank vanishing submodule.*

Thus we have shown that the generalized Stevedore knots K_n are algebraically slice whenever n is of the form $n = e(e+1)$. Clearly, for any algebraically slice knot, there is a nice collection of candidates for doing surgery to locate a slice disk. But, as we have seen, none of these candidates may be slice themselves.

In the case of the untwisted double D(K) we find that the surgery candidate is K itself! Thus we can prove

<u>PROPOSITION</u> <u>8.7</u>. *If* K *is slice (ribbon), then the untwisted double* D(K) *is also slice (ribbon). For any* K, *the Conway polynomial of* D(K) *is equal to* 1, *and* D(K) *is algebraically slice.*

Proof: Just as with the surface F depicted at the beginning of Example 8.5, the Seifert pairing for D(K) (any knot K) has the form $\theta = \begin{bmatrix} -1 & 1 \\ 0 & 0 \end{bmatrix}$ with respect to the basis $\{a,b\}$. It follows at once that $\Omega_{DK} = 1$, hence

$\nabla_{DK} = 1$. The only possible surgery curves are b and a+b. The embedded representatives of these form the original knot K. Thus if K is slice, then D(K) is also slice. ■

Remark: Michael Freedman [F] has shown that D(K) is topologically slice for any K. On the other hand, Robert Gompf in his thesis [GO] shows that the double of the trefoil is not differentiably slice.

Here are a couple of conjectures about the general situation: Let F be a spanning surface in S^3. Call an embedded curve $\alpha \subset F$ good if $[\alpha] \in H_1(F)$ is a nonzero class, and $\theta(\alpha,\alpha) = 0$ where θ is the Seifert pairing for F. Recall that we have the mod-2 invariant $A(K) \equiv a_2(K) \pmod 2$ for knots K. [I will call this the Arf invariant, anticipating Chapter IX.]

STRONG CONJECTURE. Let F be any oriented connected spanning surface in S^3 for a slice knot $K \subset S^3$. Then there exists a good curve $\alpha \subset F$ such that $\alpha \subset S^3$ is a slice knot.

WEAK CONJECTURE. Let F be any oriented, connected spanning surface in S^3 for a slice knot $K \subset S^3$. Then there exists a good curve $\alpha \subset F$ such that $\alpha \subset S^3$ has vanishing Arf invariant, $A(\alpha) = 0$.

Remark. We showed that ribbon knots have vanishing Arf. In Chapter X we will prove that any slice knot also has Arf = 0. Thus the <u>weak conjecture</u> says that some good curve looks slice to the eyes of the weakest available invariant. Note that even the weak conjecture implies that D[ ] is not slice.

Example 8.8. A Twisted Double.

<u>K is ribbon</u> (Casson's example).

θ	a	b
a	-1	1
b	0	6

If $\alpha = ka + \ell b$ and $\theta(\,,\alpha) = 0$, then
$k^2(-1) + \ell^2(6) + k\ell = 0$
$\Longrightarrow k = 3\ell$ or $k = -2\ell$.

Let $\alpha = \underline{b} + 3\underline{a}$.

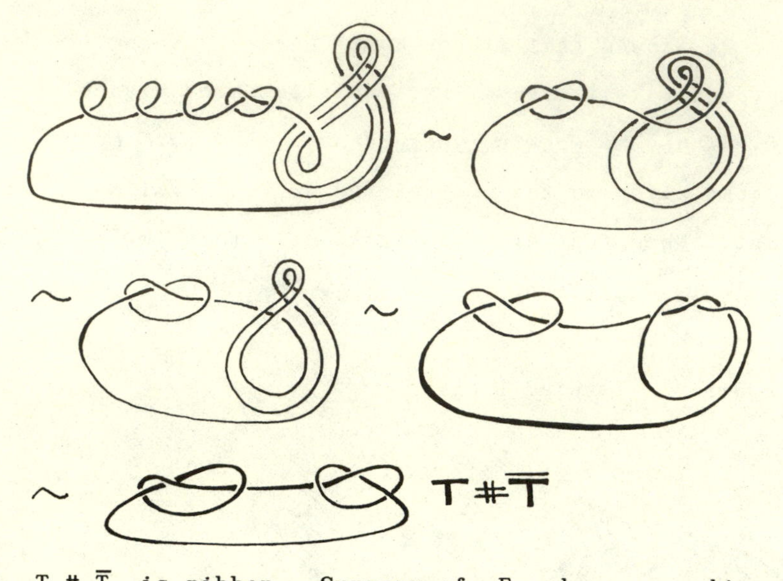

$T \# \overline{T}$ is ribbon. Surgery of F along α slices K.

[See Chapter VI, Section 13 for a detailed rendering of the slicing disk.]

IX

THE ALEXANDER POLYNOMIAL AND BRANCHED COVERINGS

Let $K \subset S^3$ be an oriented knot or link, and $F \subset S^3$ a connected oriented spanning surface for K. Let $\theta : H_1(F) \times H_1(F) \longrightarrow Z$ be the Seifert pairing.

DEFINITION 9.1. *Two polynomials* $f(t), g(t) \in Z[t]$ *are said to be* balanced *(written* $f \doteq g$*) if there is a non-negative integer* n *such that* $\pm t^n f(t) = g(t)$ *or* $\pm t^n g(t) = f(t)$. *This definition is also extended to rational functions. Thus* $t + \frac{1}{t}$ *and* t^2+1 *are balanced and we write* $t + \frac{1}{t} \doteq t^2+1$.

DEFINITION 9.2. *Let* K, F, θ *be as above. The* Alexander polynomial, $\Delta_K(t)$, *is the balance class of the polynomial* $\Delta_K(t) = D(\theta - t\theta')$. *(It follows from S-equivalence (Chapter VII) that this determinant is well defined on isotopy classes of knots and links up to multiplication by factors of the form* $\pm t^n$.*)*

The Alexander polynomial precedes the potential function (see Definition 7.9) historically. The reasons for this go back to the evolution of invariants. We will explain the geometrical meaning of $\theta - t\theta'$ shortly. First some calculations of Alexander polynomials.

Example: K, $\theta = \begin{bmatrix} -1 & 1 \\ 0 & -1 \end{bmatrix}$. (See Example 7.4.)

$$\Delta_K = \left| \begin{bmatrix} -1+t & 1 \\ -t & -1+t \end{bmatrix} \right| = t^2+1-2t+t = t^2-t+1.$$

PROPOSITION 9.3. *The Alexander polynomial, Conway polynomial and the potential function are related by the following formulas:*

$$\Delta_K(t^2) \doteq \Omega_K(t)$$

$$\therefore \quad \Delta_K(t^2) \doteq \nabla_K(t-t^1)$$

$$\Delta_K(t) \doteq \nabla_K(\sqrt{t} - 1/\sqrt{t}).$$

Proof:

$$\begin{aligned} \Delta_K(t^2) &= D(\theta - t^2\theta') \\ &= D(t^1(t^{-1}\theta - t\theta')) \\ &= t^n D(t^{-1}\theta - t\theta') \\ &= t^n \Omega_K(t) \end{aligned}$$

$$\therefore \quad \Delta_K(t^2) \doteq \Omega_K(t).$$

Since $\nabla_K(t-t^{-1}) = \Omega(t)$, the other identities follow at once.

Example: For , $\nabla_K = 1 + z^2$.

$$1+z^2 = 1+(\sqrt{t}-1/\sqrt{t})^2 = 1+t-2+t^{-1}$$
$$= t-1+t^{-1}$$
$$\doteq t^2-t+1 = \Delta_K(t).$$

The extra information carried by the sign of the potential function (or Conway polynomial) is not present with the Alexander polynomial. But the Alexander polynomial is very closely connected with the geometry of this situation. To see this we first reinterpret the Seifert pairing.

θ EXPRESSES THE ALEXANDER DUALITY OF F AND S^3-F.

It follows from Alexander Duality [G3] that $H_1(F) \cong H_1(S^3-F)$. In fact, this duality is expressed by the nonsingular pairing $\ell k : H_1(F) \times H_1(S^3-F) \longrightarrow Z$ (ℓk denotes linking number).

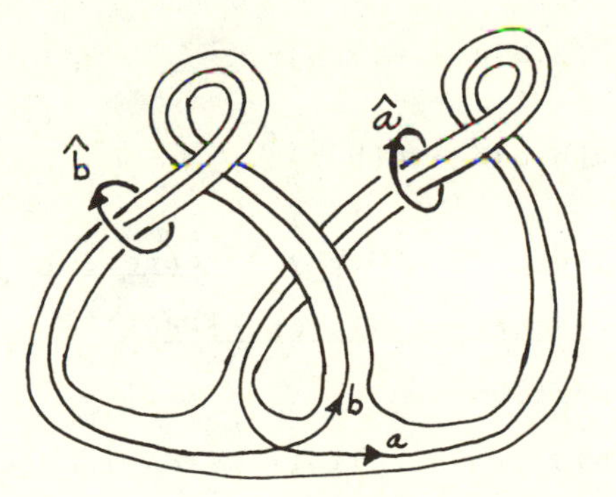

θ	a	b
a	−1	1
b	0	−1

Here, for example, is a spanning surface F for the

trefoil. $\{a,b\}$ is a basis for $H_1(F)$ and $\{\hat{a},\hat{b}\}$ is a basis for $H_1(S^3-F)$ that is dual to $\{a,b\}$ in the sense that the linking matrix between these bases is an identity matrix:

ℓk	$\hat{a}$	$\hat{b}$
a	1	0
b	0	1

We have used the map $i : F \longrightarrow S^3-F$, $i(x) = x^* =$ result of pushing x along the positive normal to F. This map induces a corresponding map on homology $i: H_1(F) \longrightarrow H_1(S^3-F)$. Here we see that $i(a) = -\hat{a}+\hat{b}$. For if $a^* = ia = x\hat{a}+y\hat{b}$ then

$$-1 = \theta(a,a) = \ell k(a^*,a) = \ell k(x\hat{a}+y\hat{b},a) = x$$

$$1 = \theta(a,b) = \ell k(a^*,b) = \ell k(x\hat{a}+y\hat{b},b) = y.$$

Similarly, $i(b) = -\hat{b}$.

Thus we see $i(a) = \theta(a,a)\hat{a} + \theta(a,b)\hat{b}$,

$$i(b) = \theta(b,a)\hat{a} + \theta(b,b)\hat{b}.$$

In other words, <u>the matrix of θ'</u> (Sorry!) <u>represents the map $i : F \longrightarrow S^3-F$ with respect to Alexander-dual bases</u>. (<u>Exercise</u>. Prove this assertion.)

Another way to put this is as follows: Imagine <u>cutting S^3 along F</u>. The result is a 3-dimensional manifold W whose boundary consists in two copies of F that we denote F_+ and F_-. We have $F_+ \cap F_- = K$. The two copies meet along K. Here is a picture 1-dimension down:

K is 2 points in S^2.

F is an arc between them.

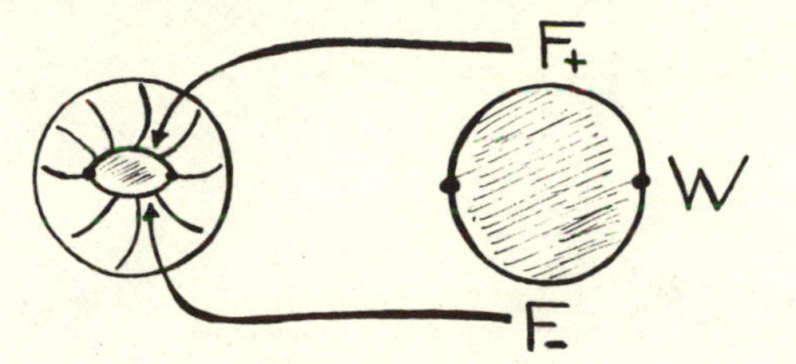

In this case, the manifold W obtained by cutting along F is homeomorphic to a disk.

In the 3-dimensional case, we might sketch

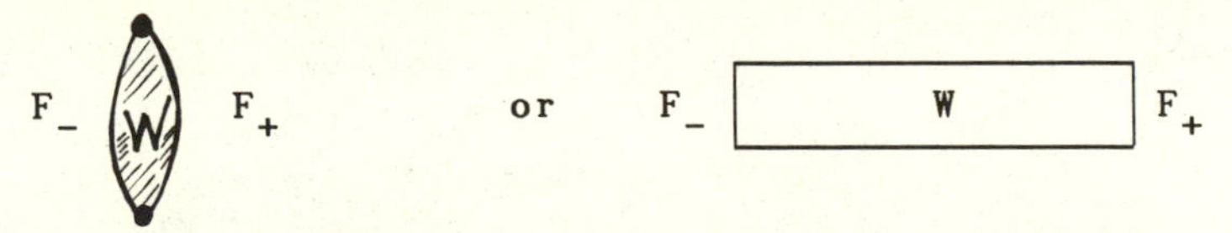

but this is purely schematic. *To visualize W, look at the spanning surface F and imagine that you are not allowed to cross F when you meet it transversely at an interior point*.

As we have remarked, the Seifert pairing gives an algebraic picture of how the boundary pieces of W fit into $H_1(W)$.

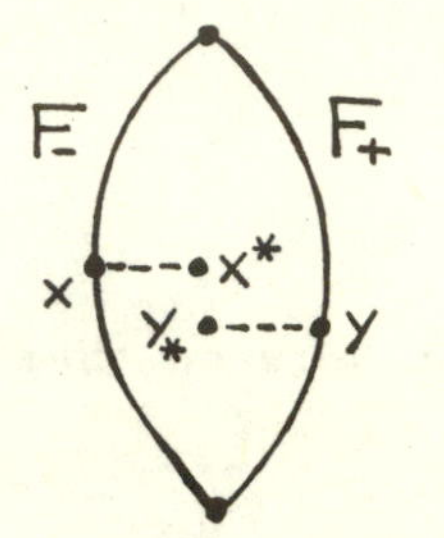

Our convention:

Inclusion of F_- corresponds to x^*.
Inclusion of F_+ corresponds to x_*.

Therefore, on the level of homology we have:

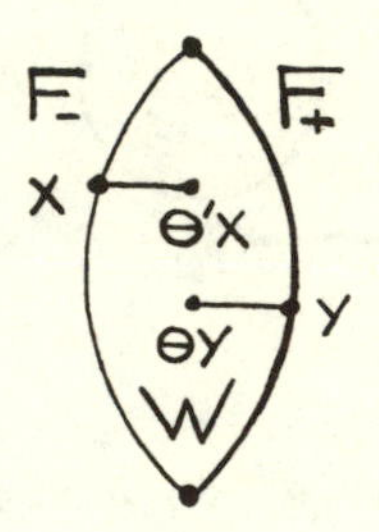

That is, if i_-, i_+ are the respective inclusions of F_- and F_+ into W and if $\mathcal{B}_-$, $\mathcal{B}_+$, $\mathcal{B}$ are bases for $H_1(F_-)$, $H_1(F_+)$, $H_1(W)$ so that $\mathcal{B}_-$, $\mathcal{B}$, and $\mathcal{B}_+$, $\mathcal{B}$ are each dual pairs, then θ' represents i_- on homology and its transpose θ represents i_+.

Seifert's beautiful idea was that W can be used as the building block for closed 3-manifolds whose homeomorphism class is an invariant of K. These are cyclic branched covers of S^3 branched along K, and also the infinite cyclic cover of the complement of K. Let's proceed by examples:

Example 9.4. The 2-fold Branched Cover. Take two copies of W: Call them W and tW where <u>t is a formal symbol</u> such that $t^2 = 1$ and if $x \in W$ then tx is the corresponding point in tW. Identify ∂W with ∂tW so that

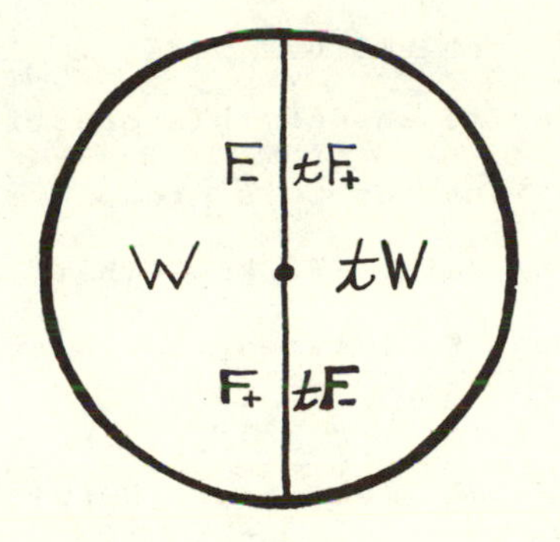

F_- is identified with tF_+ and tF_- is identified with F_+. (Thus t looks like rotation by π in our schematic.) Let $M = W \cup tW$. By construction, t becomes an order-2 homeomorphism $t : M \longrightarrow M$. The set of points fixed by t is precisely the knot $K = F_+ \cap F_-$, and $M/t \cong W/(F_- \equiv F_+) \cong S^3$. Thus there is a projection

$\pi : M \longrightarrow S^3$ such that $\begin{array}{ccc} M & \xrightarrow{t} & M \\ & \searrow^{\pi} \quad \swarrow_{\pi} & \\ & S^3 & \end{array}$ commutes and the fixed

point set of t projects to the knot or link K. M is the 2-fold cyclic branched cover of S^3 along K.

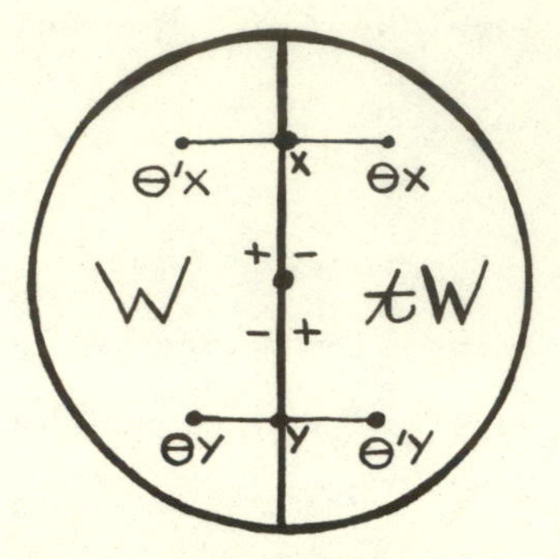

What is $H_1(M)$? Assume now that K is a knot (one component). In order to answer this question we would like generators and relations for this group. The relations come from the pasting data. We know that $H_1(W) \cong H_1(F)$, but how are cycles on W related to cycles on ∂W?

LEMMA 9.5. *If* α *is an element of* $H_1(W)$ *then* α *is homologous to a difference* i_-x-i_+x *where* x *belongs to* $H_1(F)$ *and* $i_\pm$ *denotes the identification of* F *with* $F_\pm$. *In fact, the map* $f : H_1(F) \longrightarrow H_1(W)$, $f(x) = i_-x-i_+x$ *is an isomorphism.*

Proof: Regarded as a cycle on $S^3 = W/\sim$, α is null-homologous. Thus $\alpha = \partial B$, B a 2-chain on S^3. Cutting S^3 along F, cuts B along a relative 1-cycle x giving $\alpha \approx i_-x-i_+x$ on W. Use the Mayer-Vietoris sequence [G3] for a more algebraic treatment.

Remark: Note that Lemma 9.5 is false for links. It works

for knots because there is a single boundary component to the surface F, along which we can make a path to create the desired relative cycle. For example:

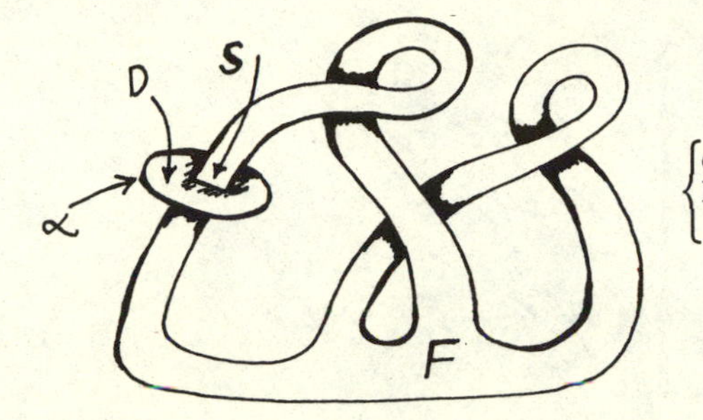

$\begin{cases} \alpha \text{ bounds the disk } D. \\ D \text{ intersects } F \text{ along an arc } s. \end{cases}$

We create a relative cycle from s by joining its end points with a path along the boundary of F:

This is the x of Lemma 9.5.

Thus we have $\alpha \approx i_- x - i_+ x$. Note that x itself is homologous to the curve shown below:

From this lemma we know that every element of $H_1(W)$ can be represented through its boundary. For example, there must exist a mapping $\Gamma : H_1(F) \longrightarrow H_1(F)$ so that $i_+ = (i_+ - i_-) \circ \Gamma$.

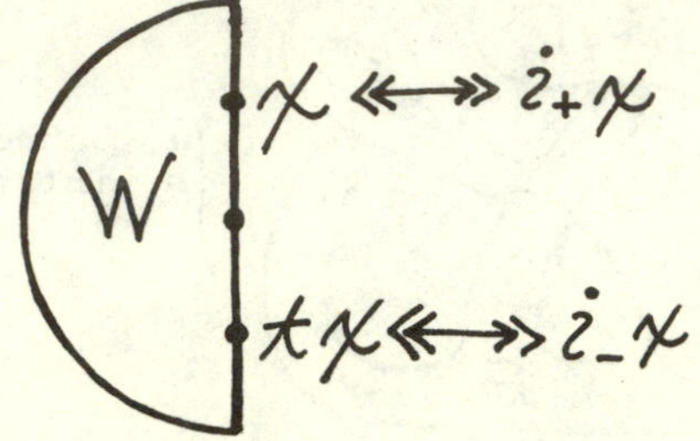

In the t-notation this reads as a relation: $x = \Gamma x - \Gamma tx$ or $\underline{t\Gamma x = (\Gamma - I)x.}$ This is the basic relation on $H_1(F) \oplus H_1(tF)$ reducing it to $H_1(W)$. Formally, we have

LEMMA 9.6. *The following sequence is exact:*

$$0 \longrightarrow H_1(F) \xrightarrow{g} H_1(F) \oplus H_1(F) \xrightarrow{i_- + i_+} H_1(X) \longrightarrow 0$$

where $g(x) = (\Gamma x, (I - \Gamma)x)$.

Proof: Exercise.

Now what is Γ?

LEMMA 9.7. *Let* $S : H_1(F) \times H_1(F) \longrightarrow Z$ *be the intersection pairing, and* θ *the Seifert pairing. Then* $\theta(x,y) = S(x,\Gamma y)$ $\forall$ $x,y \in H_1(F)$. *Matrix-wise this says*

$\theta = S\Gamma$ *(for knots, S is invertible).*

Proof:

$$\begin{aligned} \theta(x,y) &= \ell k(x^*,y) = \ell k(x,y_*) = \ell k(x,i_+y) \\ &= \ell k(x,i_+\Gamma y - i_-\Gamma y) \\ &= \ell k(x,i_+\Gamma y) - \ell k(x,i_-\Gamma y) \\ &= \theta(x,\Gamma y) - \theta(\Gamma y,x) \\ \theta(x,y) &= S(x,\Gamma y) \end{aligned}$$

(since $\theta(a,b) - \theta(b,a) = a \cdot b = S(a,b)$). ∎

Note: $S(I-\Gamma) = \theta-\theta'-S\Gamma = \theta-\theta'-\theta = -\theta'$. It was by this route that Seifert [S1] discovered the Seifert pairing.

Now let's go back to $H_1(M)$ where M is the 2-fold branched cover.

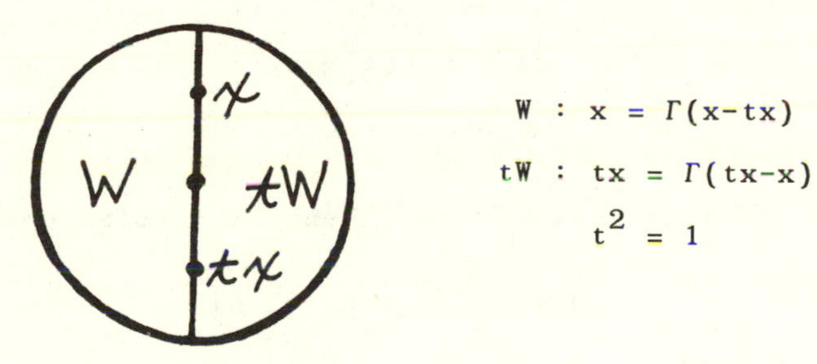

The relations on $H_1(M)$ are then as above. Thus $x+tx = 0$ and the relations reduce to $x = 2\Gamma x$, or

$$(\Gamma-I)x+\Gamma x = 0.$$

Remember that $S\Gamma = \theta$, $S(\Gamma-I) = \theta'$. Thus if the branch set is a knot so that S is invertible, then these relations become $(\theta+\theta')x = 0$. In other words, $\underline{\theta+\theta'}$ $\underline{\text{is a}}$

<u>relation matrix for the 2-fold branched covering along a knot</u>.

In the case of the trefoil, $\theta = \begin{bmatrix} -1 & 1 \\ 0 & -1 \end{bmatrix}$, $\theta+\theta' = \begin{bmatrix} -2 & 1 \\ 1 & -2 \end{bmatrix}$. Integer invertible row and column operations give us: $\begin{bmatrix} -2 & 1 \\ 1 & -2 \end{bmatrix} \to \begin{bmatrix} 0 & -3 \\ 1 & -2 \end{bmatrix} \to \begin{bmatrix} 0 & -3 \\ 1 & 0 \end{bmatrix}$. Hence $H_1(M_2($  $)) = Z_3$.

Example 9.8. Higher Order Branched Covers. Let $M_n(K)$ denote the n-fold cyclic cover of S^3 branched along a knot K. Seifert's method generalizes. For example, to compute $H_1(M_3)$ we have the relations

$$\left\{\begin{array}{l} x = \Gamma(x-tx) \\ tx = \Gamma(tx-t^2x) \\ t^2x = \Gamma(t^2x-x) \\ t^3x = x \end{array}\right\}$$

Thus we see that $1+t+t^2 = 0$. Hence the relations become

$$\left\{\begin{array}{l} (\Gamma-I)x - \Gamma tx \quad = 0 \\ (\Gamma-I)tx - \Gamma t^2x = 0 \\ t^2x = -x - t^2 \end{array}\right\}$$

In matrix block form this becomes:

$$\begin{bmatrix} \Gamma-I & -\Gamma \\ +\Gamma & (\Gamma-I)+\Gamma \end{bmatrix} \to \begin{bmatrix} -I & -3\Gamma+I \\ \Gamma & 2\Gamma-I \end{bmatrix} \to \begin{bmatrix} +I & 3\Gamma-I \\ 0 & -3\Gamma^2+3\Gamma-I \end{bmatrix}.$$

Hence $-3\Gamma^2+3\Gamma-I = (\Gamma-I)^3-\Gamma^3$ is a relation matrix for M_3. Seifert proved:

THEOREM 9.9. *Let* Γ *be the matrix defined above so that* $S\Gamma = \theta$ *where* θ *is the Seifert pairing matrix and* S *is the intersection matrix on a spanning surface* F *for a knot* K. *Then* $(\Gamma-I)^g-\Gamma^g$ *is a relation matrix for* $H_1(M_g(K))$, *the first homology group of the* g-*fold cyclic branched cover of* S^3 *along* K.

Example: For the trefoil we have $S = \begin{bmatrix} 0 & 1 \\ -1 & 0 \end{bmatrix}$, $\theta = \begin{bmatrix} -1 & 1 \\ 0 & -1 \end{bmatrix}$.

$$\text{So} \quad \Gamma = S^{-1}\theta = \begin{bmatrix} 0 & -1 \\ 1 & 0 \end{bmatrix}\begin{bmatrix} -1 & 1 \\ 0 & -1 \end{bmatrix} = \begin{bmatrix} 0 & 1 \\ -1 & 1 \end{bmatrix}$$

$$\Gamma - I = \begin{bmatrix} -1 & 1 \\ -1 & 0 \end{bmatrix}.$$

$$\Gamma^2 = \begin{bmatrix} 0 & 1 \\ -1 & 1 \end{bmatrix}\begin{bmatrix} 0 & 1 \\ -1 & 1 \end{bmatrix} = \begin{bmatrix} -1 & 1 \\ -1 & 0 \end{bmatrix}$$

$$\Gamma^3 = \begin{bmatrix} 0 & 1 \\ -1 & 1 \end{bmatrix}\begin{bmatrix} -1 & 1 \\ -1 & 0 \end{bmatrix} = \begin{bmatrix} -1 & 0 \\ 0 & -1 \end{bmatrix} = -I$$

$$(\Gamma-I)^2 = \begin{bmatrix} -1 & 1 \\ -1 & 0 \end{bmatrix}\begin{bmatrix} -1 & 1 \\ -1 & 0 \end{bmatrix} = \begin{bmatrix} 0 & -1 \\ 1 & -1 \end{bmatrix}$$

$$(\Gamma-I)^3 = \begin{bmatrix} -1 & 1 \\ -1 & 0 \end{bmatrix}\begin{bmatrix} 0 & -1 \\ 1 & -1 \end{bmatrix} = \begin{bmatrix} 1 & 0 \\ 0 & 1 \end{bmatrix} = I.$$

This implies a <u>periodicity</u>: $H_1(M_{g+6}(K)) \cong H_1(M_g(K))$ for $K =$  . These periodicities were first noticed by Antonio Plans [P]. There is also a paper on this by R. H. Fox [FO]. As we shall see later on, these periodicites are related to a geometric periodicity in the fiber structure of S^3-K, and they have implications for higher dimensional phenomena.

<u>Exercise</u>. (i) Actually compute $H_1(M_g(\ ))$ for all g.

(ii) Compute $H_1(M_g(K))$ when K is the figure-eight knot. (iii) Prove Theorem 9.9.

Example 9.10. The Infinite Cyclic Covering. The infinite cyclic covering space $X_\infty \xrightarrow{\pi} S^3-K$ is a covering space of the knot complement that covers all of the $M_g(K)-K$ (remove the branch set). We form it just as we formed $M_g(K)$ except that K is removed from W and $g = \infty$. Thus we diagram W as

F_- [W] F_+

thinking of this as the original W minus a tubular neighborhood of K. Let X denote $\overline{W}$. We can assume that X

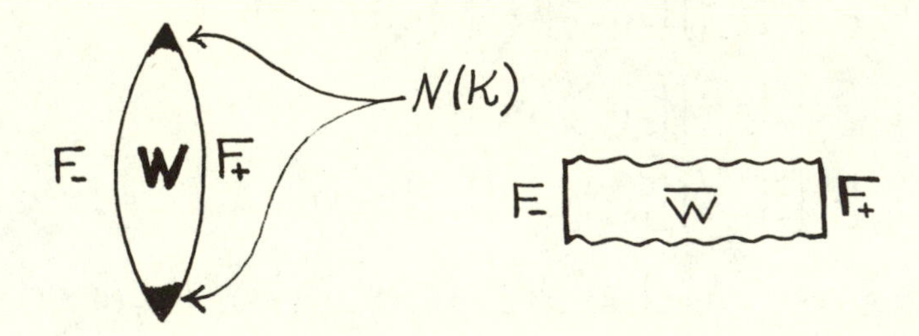

is closed and that $X/(F_- \equiv F_+) = E_K$ where E_K denotes S^3-Interior(N(K)), N(K) a tubular neighborhood of K.

Then X_∞ has diagram

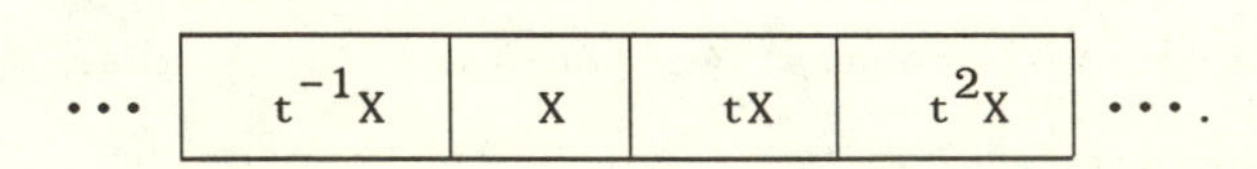

The homology of X_∞ is generated by $\{H_1(t^nF) \mid n \in Z\}$.

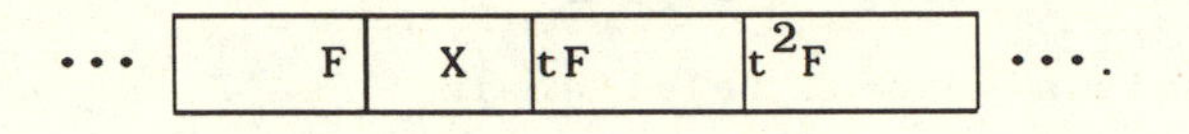

On X we have as before the relation $(\Gamma-I)x-t\Gamma x = 0$ $\forall\ x \in H_1(F)$. $S\Gamma = \theta$, $S(\Gamma-I) = \theta'$.

Hence $S(\Gamma-I)x-tS\Gamma x = 0$ implies that $\theta'x-t\theta x = 0$ for all x in $H_1(F)$.

Therefore we have the string of relations:

$$
\begin{array}{rcl}
\vdots & & \vdots \\
t^{-2}X & : & t^{-2}\theta'-t^{-1}\theta \\
t^{-1}X & : & t^{-1}\theta'-\theta \\
X & : & \theta'-t\theta \\
tX & : & t\theta'-t^{2}\theta \\
t^{2}X & : & t^{2}\theta'-t^{3}\theta \\
\vdots & & \vdots
\end{array}
$$

And these form a complete set of relations for $H_1(X_\infty)$. A good way to say this is to let $R = Z[t,t^{-1}]$ and to regard $H_1(X_\infty)$ as an R-module. Then <u>as an R-module</u> $H_1(X_\infty)$ <u>has relation matrix</u> $\theta'-t\theta$.

Note that <u>the Alexander polynomial</u> $\Delta_K(t) = D(\theta-t\theta')$ $\doteq D(\theta'-t\theta)$. Hence $\Delta_K(t)\alpha = 0$ for all $\alpha \in H_1(X_\infty)$. This is <u>a geometric interpretation of the Alexander polynomial</u>, and it can be used to define it.

Notice that this says that if $\alpha \in H_1(X_\infty($ ... $))$, then $t^2\alpha - t\alpha + \alpha = 0$, $t^2\alpha = t\alpha - \alpha$. The Alexander polynomial is a general relation in

$$H_1(\text{Infinite Cyclic Cover}).$$

Perhaps you would like to see how this happens more geometrically: <u>Here is Dale Rolfson's method [R2] for looking at</u> X_∞:

(1) Surger S^3-K to S^3-K' where K' is unknotted.

(2) Form the infinite cyclic cover of S^3-K' $\cong D^2 \times S^1$, and so get $D^2 \times \mathbb{R} \longrightarrow S^3-K'$.

(3) Re-surger $D^2 \times \mathbb{R}$ to get X_∞.

(4) Use this model to look at $H_1(X_\infty)$.

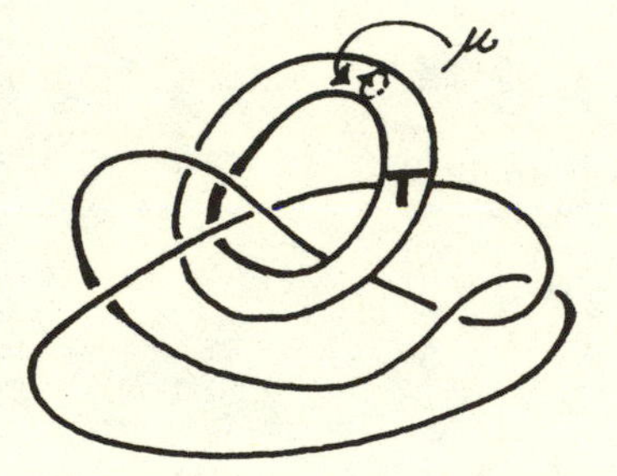

K = figure-eight knot

The crossing surrounded by the solid torus T may be changed by a homeomorphism $h : S^3-\mathring{T} \longrightarrow S^3-T$ ($\mathring{A}$ = Interior A) to yield:

The infinite cyclic cover of $S^3-h(K)$ is

$$p : \mathbb{R}^1 \times \mathbb{R}^2 \longrightarrow S^1 \times \mathbb{R}^2 \cong S^3-h(K).$$

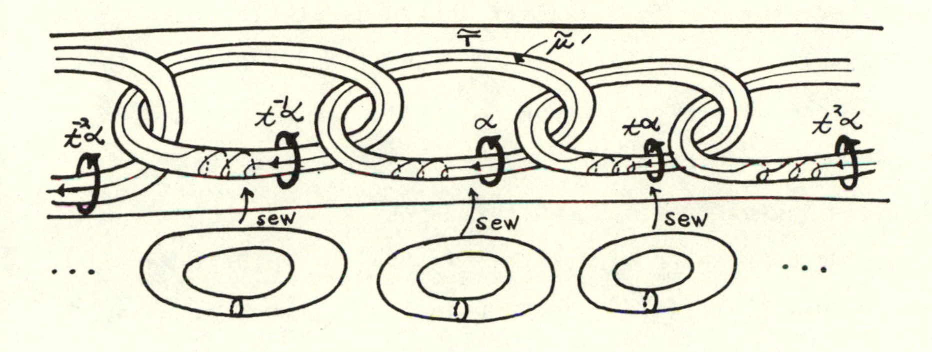

X_∞ is obtained by removing each t^nT and replacing it with the meridian along $t^n\mu'$. Sewing in T kills $\mu' = t\alpha-3\alpha+t^{-1}\alpha$. $H_1(X_\infty) = (\alpha\,|\,(t-3+t^{-1})\alpha = 0)$ as a $Z[t,t^{-1}]$ module. $\Delta(t) = t-3+t^{-1} \doteq t^2 - 3t+1$, the Alexander polynomial of the figure-eight knot.

ALEXANDER'S METHOD

Example 9.11. It was by way of the geometry of coverings and particularly the infinite cyclic covering that people realized that the Alexander polynomial could be extracted from the fundamental group of the knot complement. This comes about as follows: By construction, $\pi_1(X_\infty) \cong G'$ where G' denotes the commutator subgroup of $G = \pi_1(S^3-K)$. Thus $H_1(X_\infty) \cong G'/G''$, and the action of $Z[t,t^{-1}]$ on

$H_1(X_\infty)$ corresponds algebraically to $t \cdot g = sgs^{-1}$ where $s \in \pi_1(S^3-K)$ is a chosen element having linking number $+1$ with K. By definition, $\Delta_K(t)$ is the balance class (i.e., defined up to $\pm t^n$) of the ideal of elements $\rho \in Z[t,t^{-1}]$ such that $\rho g = 0$ for all $g \in G'/G''$. This can be computed purely group theoretically by using a standard presentation of $\pi_1(S^3-K)$. [One says that $\Delta_K(t) \doteq$ the order of G'/G'' over $Z[t,t^{-1}]$.]

There are many algorithms of this sort. Here we will sketch one that yields the computational method given by Alexander in his original paper [A1]: If $K \subset S^3$ is a knot and $G = \pi_1(S^3-K)$, then $G/G' \cong Z$ and the map $f : G \longrightarrow Z$ is given by $f(\alpha) = \ell k(\alpha,K)$. [In the case of oriented links, everything works in similar fashion to give a map $G \longrightarrow Z$ even though this is not the abelianization.] Thus the action of $R = Z[t,t^{-1}]$ can be obtained by finding $s \in G$ with $f(s) = 1$.

Choose a presentation for the fundamental group $G = (g_0,\cdots,g_n | r_1,\cdots,r_n)$ with one more generator than there are relations. (For example, the Dehn presentation—we will use it in the next paragraph.) We can choose $s = g_0$ and rewrite generators and relations so that $f(g_1) = \cdots = f(g_n) = 0$ (replacing g_k by $s^{i_k} g_k$ when necessary). Then $g_1,\cdots,g_n \in G'$ and one can show that $G' = (g_1,\cdots,g_n | r_i = 1,\ i = 1,\cdots,n)$ is a presentation of G' as an R-module. Now abelianize G', and look for the relations.

To make this concrete we use the Dehn presentation of $\pi_1(S^3-K) = G$.

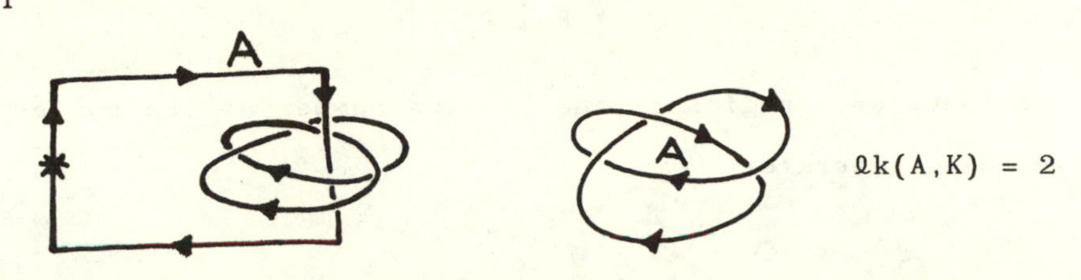

In the Dehn presentation, generators of $\pi_1(S^3-K)$ are in 1-1 correspondence with all-but-one of the regions of the knot diagram. We let the base-point, *, live in the unbounded region. Each of the other regions becomes an element of $\pi_1(S^3-K)$ by taking a path through it as illustrated above.

Each crossing gives rise to a relation:

A D
B C

$$AB^{-1}CD^{-1} = 1.$$

<u>Exercise</u>. Draw a picture of this relation.

The linking number of a generator with the knot is computed by a method of indexing the regions of the diagram with integers:

(1) Index (outer unbounded region) = 0.

(2) Relative indices across an oriented edge form

this pattern.

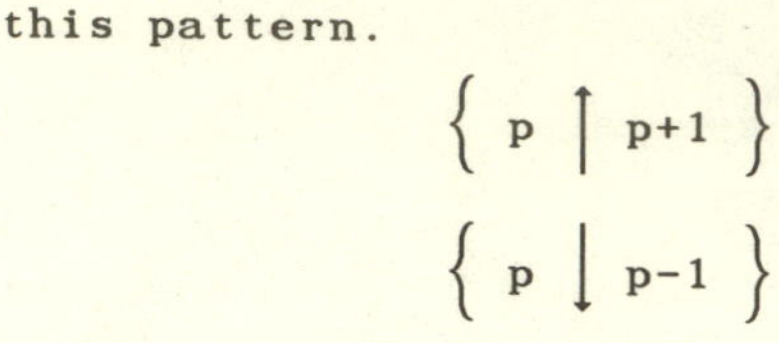

The index of a region is the linking number of the corresponding generator.

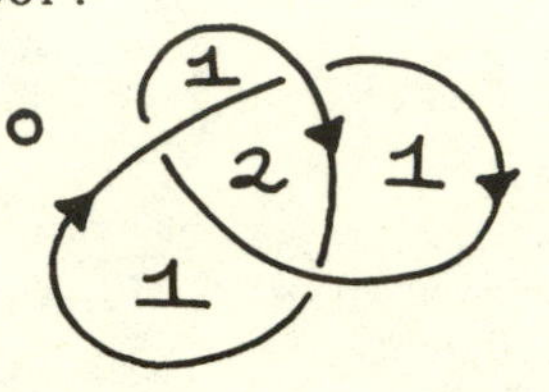

Let's use this format to get a presentation of G'/G''. First $G = (s, h_1, \cdots, h_n \mid r_1, \cdots, r_n)$ where s corresponds to a region adjacent to the unbounded region.

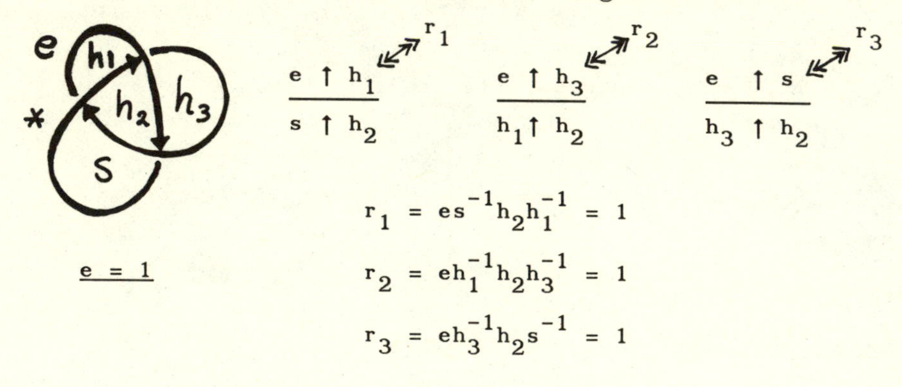

$$r_1 = es^{-1}h_2h_1^{-1} = 1$$

$$r_2 = eh_1^{-1}h_2h_3^{-1} = 1$$

$$r_3 = eh_3^{-1}h_2s^{-1} = 1$$

$$G = (s, h_1, h_2, h_3 \mid r_1 = 1,\ r_2 = 1,\ r_3 = 1)$$

But now we want to let $g_k = s^{-i_k}(h_k)$ so that $\ell k(g_k, K) = 0$. Thus $i_k = \text{Index}(h_k)$. And we have to rewrite the relations in terms of this new basis:

Here is one case. We leave the other as an exercise.

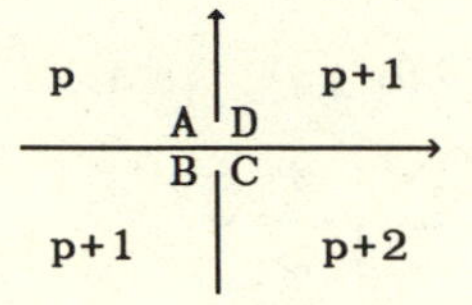

With these orientations, the indices are p, $p+1$, $p+2$ as indicated.

$$r = A\bar{B}C\bar{D} \quad \text{(using bar } (\bar{\ }) \text{ for } (\ )^{-1}\text{)}.$$

Let
$$a = s^{-p}A \qquad c = s^{-p-2}C$$
$$b = s^{-p-1}B \qquad d = s^{-p-1}D.$$

Then
$$A = s^{p}a \qquad C = s^{p+2}c$$
$$B = s^{p+1}b \qquad D = s^{p+1}d$$

$$r = (s^{p}a)(\overline{s^{p+1}b})(s^{p+2}c)(\overline{s^{p+1}d})$$
$$= (s^{p}a)(\bar{b}s^{-p-1})(s^{p+2}c)(\bar{d}s^{-p-1})$$
$$r = (s^{p}as^{-p})(s^{p}\bar{b}s^{-p})(s^{p+1}cs^{-p-1})(s^{p+1}\bar{d}s^{-p-1}).$$

This is now written correctly as an element of G' as an R-module, where $ta = sas^{-1}$. In $H = G'/G''$ (the abelianization), this relation becomes:

$$t^{p}a - t^{p}b + t^{p+1}c - t^{p+1}d = 0$$

or equivalently

$$\boxed{a - b + tc - td = 0}$$

1	$-t$
-1	t

This "code" may help remember how to write the relation.

Thus, for the trefoil, we get

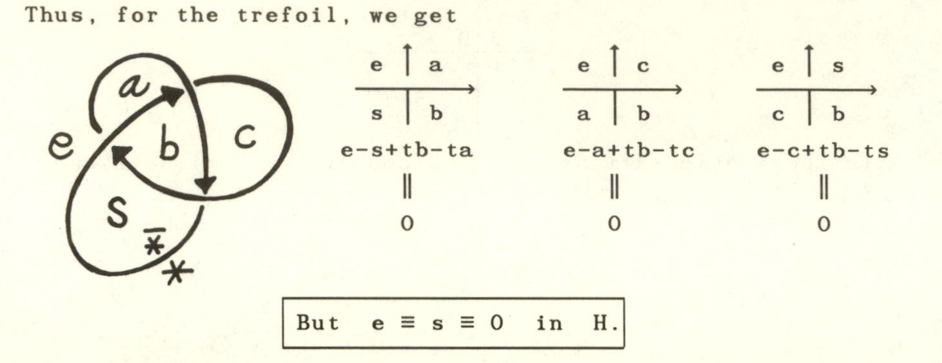

But $e \equiv s \equiv 0$ in H.

Thus the relations become $tb-ta = 0$, $-a+tb-tc = 0$, $-c+tb = 0$,

<u>or</u>
$$\begin{cases} b-a = 0 \\ -a+tb-tc = 0 \\ -c+tb = 0 \end{cases}$$

<u>or</u>
$$\begin{cases} -a+ta-tc = 0 \\ c = tb = ta \end{cases}$$

<u>or</u>
$$\begin{cases} -a+ta-t^2a = 0 \\ \hline (t^2-t+1)a = 0 \end{cases}$$

And so $\Delta_K(t) \doteq t^2-t+1$. Not a surprise by now.

Using Alexander's formalism, you can find $\Delta_K(t)$ directly by taking a determinant of the $n \times n$ relation matrix. Thus here we have:

	e	s	a	b	c	
1st crossing	1	-1	-t	t	0	$e-s+tb-ta = 0$
2nd crossing	1	0	-1	t	-t	$e-a+tb-tc = 0$
3rd crossing	1	-t	0	t	-1	$e-c+tb-ts = 0$

Let $M = \begin{bmatrix} -t & t & 0 \\ -1 & t & -t \\ 0 & t & -1 \end{bmatrix}$ (deleting the columns corresponding to e and s). Then M is a relation matrix for $H = G'/G''$ as a $Z[t,t^{-1}]$-module, and $\Delta_K(t) \doteq D(M)$.

Remark: Alexander begins his paper [A1] by giving this formula as the definition! If you read diligently, there are hints about fundamental group and covering spaces in the last two pages of the paper. I had the pleasure of discovering that there is a whole world of combinatorics related to this version of $\Delta_K(t)$. And it yields another model of the Conway axioms. For more, read [K1]. The discovery of the first generalized polynomial by Jones, Ocneanu, Lickorish, Millet, Hoste, Freyd, Yetter, Przytycki and Traczyk (!) may be regarded as a remarkable confirmation of Alexander's intuition in formulating a combinatorial approach. (See these notes, Chapter VI, sections 18, 19, 20, and the Appendix for more about generalized polynomials.)

X

THE ALEXANDER POLYNOMIAL AND THE ARF INVARIANT

Recall that we have defined, for a knot K, the invariant $A(K) \in Z_2$ via $A(K) \equiv a_2(K)$ (modulo-2) where $a_2(K)$ is the second Conway coefficient. And we showed (Chapter V) that $A(K) = 0$ for ribbon knots. In this chapter we will show that $A(K)$ is identical with the Arf invariant, $ARF(K)$, which is the Arf invariant of a mod-2 quadratic form related to K.

MOD-2 QUADRATIC FORMS

First recall that a mod-2 quadratic form q is a mapping $q : V \longrightarrow Z_2$ where V is a Z_2-vector space such that V has a bilinear symmetric pairing $\langle\ ,\ \rangle : V \times V \longrightarrow Z_2$. The mapping q must satisfy the following property:

$$(*) \qquad q(x+y) = q(x)+q(y)+\langle x,y\rangle \quad \text{for all} \quad x,y \in V.$$

Remark: Over a field of characteristic $\neq 2$ quadratic forms and symmetric bilinear forms are in 1-1 correspondence. Thus if $[\ ,\ ] : W \times W \longrightarrow F$ is a symmetric bilinear form, and char $F \neq 2$, then we can define $Q(x) = [x,x]/2$ and obtain:

$$Q(x,y) = \frac{1}{2}([x+y,\ x+y])$$
$$= \frac{1}{2}([x,x]+2[x,y]+[y,y])$$
$$\therefore\ Q(x,y) = Q(x) + Q(y) + [x,y].$$

In characteristic 2 the situation is subtler, and more than one quadratic form may correspond to a given bilinear form.

Classically, a quadratic form in two variables looks like a quadratic polynomial,

Beware the change of variables.

$Q(x,y) = ax^2+bxy+cy^2$ and if char $\neq 2$ then we can write $ax^2+bxy+cy^2 = (x,y)\begin{bmatrix} a & b/2 \\ b/2 & c \end{bmatrix}\begin{bmatrix} x \\ y \end{bmatrix}$ and classify the form $ax^2+bxy+cy^2$ by analyzing the congruence class of the matrix $\begin{bmatrix} a & b/2 \\ b/2 & c \end{bmatrix}$.

In characteristic = 2, there is still a symmetric bilinear form associated with a quadratic polynomial, but now it occurs because $2 = 0$: If $Q(x,y) = ax^2+bxy+cy^2$, let $v = (x,y)$, $v_1 = (x_1,y_1)$, $v_2 = (x_2,y_2)$. Then

$$Q(v_1+v_2) = a(x_1+x_2)^2 + b(x_1+x_2)(y_1+y_2) + c(y_1+y_2)^2$$
$$= ax_1^2 + ax_2^2 + b(x_1y_1+x_2y_2+x_1y_2+x_2y_1) + cy_1^2 + cy_2^2$$
$$= Q(v_1) + Q(v_2) + b(x_1y_2+x_2y_1)$$
$$= Q(v_1) + Q(v_2) + v_1\begin{bmatrix} 0 & b \\ b & 0 \end{bmatrix}v_2'.$$

The associated symmetric bilinear form has matrix $\begin{bmatrix} 0 & b \\ b & 0 \end{bmatrix} = b\begin{bmatrix} 0 & 1 \\ 1 & 0 \end{bmatrix}$. This should remind us of the mod-2 intersection form on the (punctured) torus:

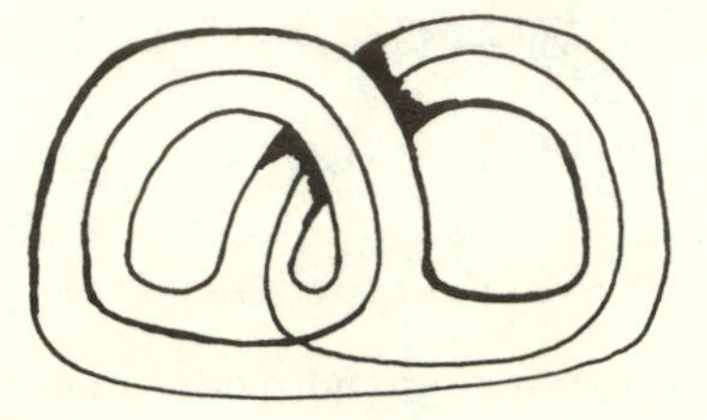

DEFINITION 10.1. *Let* $K \subset S^3$ *be a knot and* F *a connected oriented spanning surface for* K *with Seifert pairing* $\theta : H_1(F) \times H_1(F) \longrightarrow Z$. *Let* $V = H_1(F) \otimes Z_2$, $\overline{\theta} = \theta$ *on* V, *and let* $\langle\ ,\ \rangle$ *denote the* mod-2 *reduction of the intersection form* S *on* $H_1(F)$. *The* <u>mod-2</u> <u>*quadratic form of*</u> F *is then defined by* $q(x) = \overline{\theta}(x,x)$ for all $x \in V$.

Note that

$$\begin{aligned} q(x+y) &= \overline{\theta}(x+y,\ x+y) \\ &= \overline{\theta}(x,x) + \overline{\theta}(y,y) + \overline{\theta}(x,y) + \overline{\theta}(y,x) \\ &\equiv q(x) + q(y) + (\theta(x,y)-\theta(y,x)) \pmod 2 \\ &\equiv q(x) + q(y) + S(x,y) \pmod 2 \end{aligned}$$

$$\therefore\ q(x+y) = q(x) + q(y) + \langle x,y\rangle .$$

Thus the Seifert pairing produces a mod-2 quadratic form that is naturally associated with any spanning surface. We see that with respect to the standard basis (symplectic basis) for the surface it is easy to write the quadratic polynomial that corresponds to the form. Thus:

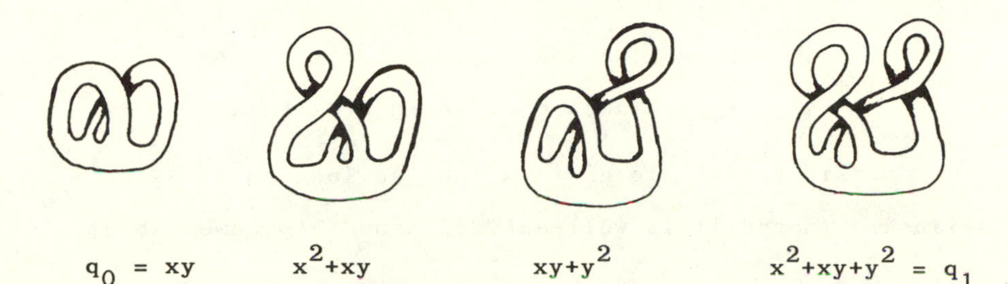

$q_0 = xy$ x^2+xy $xy+y^2$ $x^2+xy+y^2 = q_1$

We know that the first three surfaces are isotopic, hence the forms xy, x^2+xy and $xy+y^2$ must be isomorphic! Indeed, this is the case. For example $x^2+xy = x(x+y)$ and so is isomorphic to xy via the change of basis $x' = x$, $y' = x+y$.

These four forms are nondegenerate in the sense that the associated bilinear form is nondegenerate. Here it is in matrix form $\begin{bmatrix} 0 & 1 \\ 1 & 0 \end{bmatrix}$. Nondegeneracy of $\langle\ ,\ \rangle$ means that the matrix of $\langle\ ,\ \rangle$ is nonsingular.

In fact, we have just shown that there are at most two isomorphism classes of nondegenerate dimension-two mod-2 forms: $q_0 = xy$ and $q_1 = x^2+xy+y^2$. It is easy to see that q_0 and q_1 are not isomorphic. For, if $V = Z_2 \times Z_2$ then q_0 takes a majority of elements to 0, while q_1 takes a majority of elements to 1. Thus we have classified rank-2 forms over Z_2.

DEFINITION 10.2. *Let* V *be a finite dimensional vector space over* Z_2 *and* $q : V \longrightarrow Z_2$ *a nondegenerate quadratic form. The* <u>*Arf invariant*</u> $\mathrm{ARF}(q) \in Z_2$ *is defined by the formula*

$$\mathrm{ARF}(q) = \begin{cases} 0 & \text{if } q \text{ takes a majority of elements to } 0 \\ 1 & \text{if } q \text{ takes a majority of elements to } 1. \end{cases}$$

Certainly ARF is an invariant so long as it is well-defined. Indeed it is well-defined, and this comes about as follows:

(i) Symmetric bilinear forms over Z_2 are all (when nondegenerate) sums of forms of type $\begin{bmatrix} 0 & 1 \\ 1 & 0 \end{bmatrix}$. That is, there is a <u>symplectic basis</u> $\{a_1, \cdots, a_g, b_1, \cdots, b_g\}$ for V such that $\langle a_i, b_j\rangle = \delta_{ij}$, $\langle a_i, a_j\rangle = \langle b_i, b_j\rangle = 0$ for all i and j. This, of course, is given geometrically in our Seifert form case.

(ii) It follows from (i) that any nondegenerate mod-2 quadratic form is a direct sum of the two-dimensional forms. Hence it is a direct sum involving q_0 and q_1.

(iii) $q_1 \oplus q_1 \cong q_0 \oplus q_0$. This is the basic fact. You can prove it by a basis-change, or you can see it geometrically by taking the connected sum

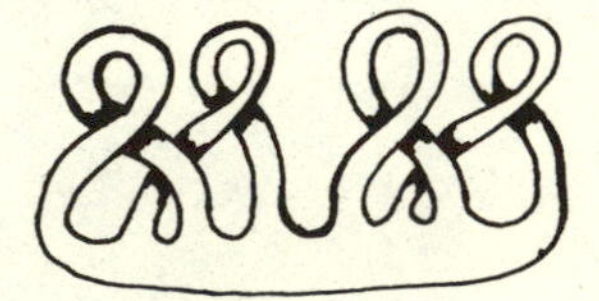

which has the form $q_1 \oplus q_1$ and find the basis change by topological script! Here we can use <u>mod-2 script in the plane so that</u>

These modifications do not change the mod-2 quadratic form of the corresponding surface.

You may also think of these script moves as equivalent to consequences of

performed on the bands (compare with pass-equivalence, Chapter V). For then

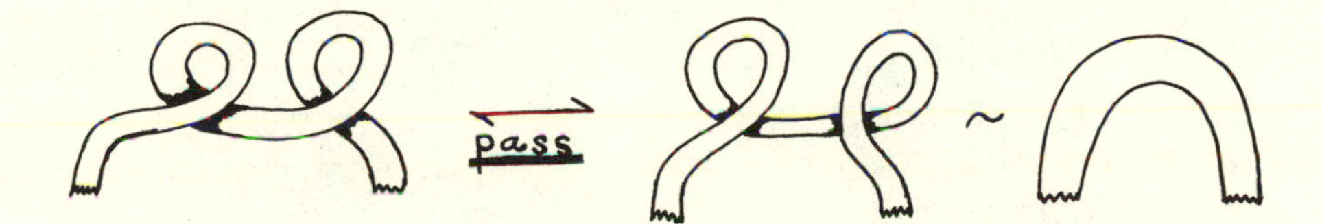

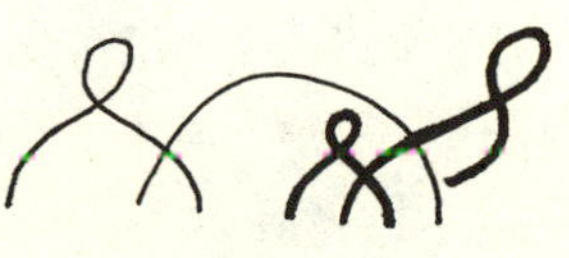

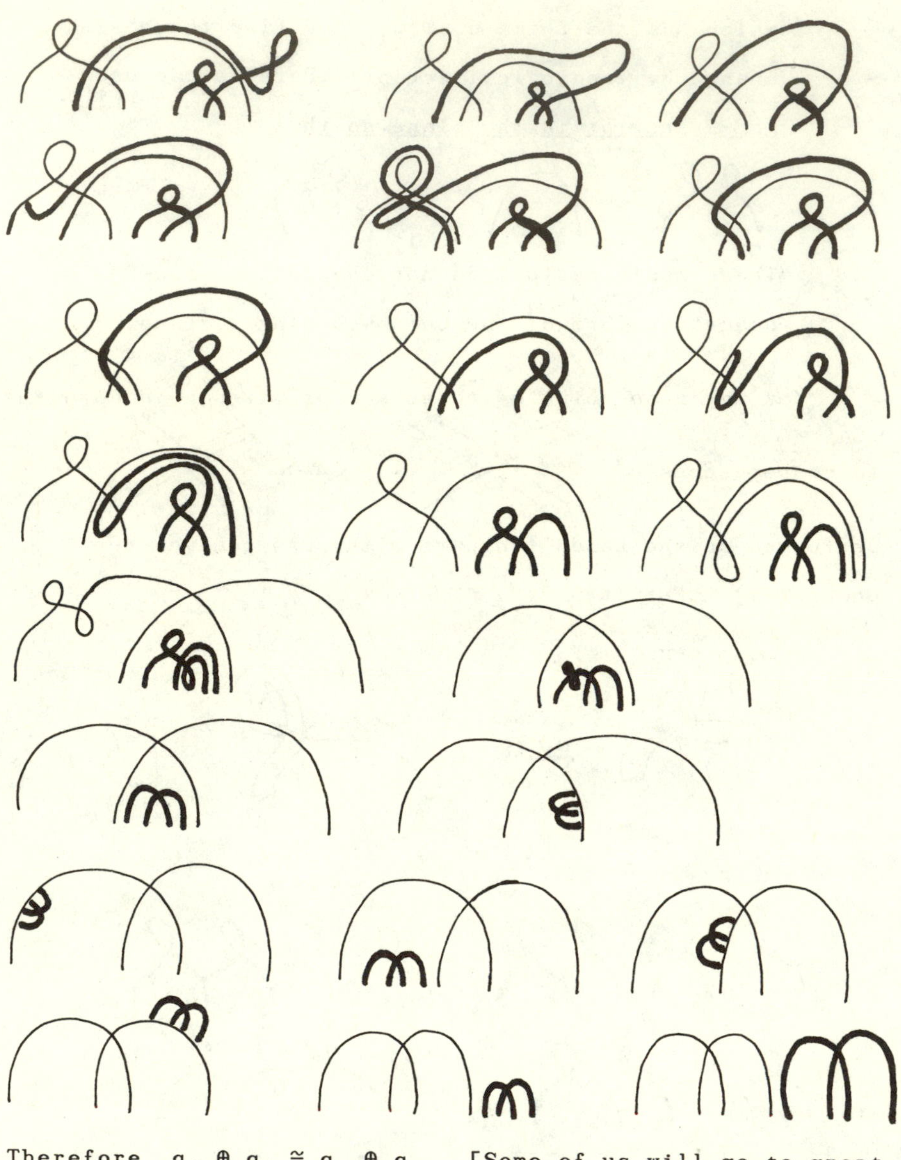

Therefore $q_1 \oplus q_1 \cong q_0 \oplus q_0$. [Some of us will go to great lengths to avoid a little algebra.]

As a result, any mod-2 form is of the form

$$q_0 \oplus \cdots \oplus q_0 = \emptyset_0 \quad \text{or}$$

$$q_1 \oplus q_0 \oplus \cdots \oplus q_0 = \emptyset_1 .$$

It is then a counting matter to see that $ARF(\emptyset_0) = 0$ and $ARF(\emptyset_1) = 1$. Thus, we have classified all nondegenerate mod-2 quadratic forms, and shown the utility of the ARF invariant in the process.

(iv) It follows from what we have said, that $q \oplus q'$ has an Arf invariant whenever q and q' have Arf invariants. Furthermore,

$$ARF(q \oplus q') = ARF(q) \oplus Arf(q').$$

(v) It can be shown that (do it!) if $\{a_1, \cdots, a_g, b_1, \cdots, b_g\}$ is a symplectic basis for V, $q : V \longrightarrow Z_2$ a mod-2 quadratic form, then $ARF(q) = \sum_{k=1}^{g} q(a_k)q(b_k)$. This gives an explicit formula for ARF.

Let $K \subset S^3$ be a knot. We now define $ARF(K) \in Z_2$ by the formula $ARF(K) = ARF(q)$ where q is the mod-2 quadratic form of any spanning surface for K. We leave it as an exercise in s-equivalence to see that this is an invariant of K.

THEOREM 10.3. *If knots* K and $\overline{K}$ *are related by one crossing change, and* L *is the 2-component link obtained by splicing this crossing, then*

$$\mathrm{ARF}(K) - \mathrm{ARF}(\overline{K}) = \ell k(L).$$

K $\qquad$ $\overline{K}$ $\qquad$ L

COROLLARY. *Let* $A(K)$ *be the* mod-2 *reduction of the second Conway coefficient* $a_2(K)$. *Then* $A(K) = \mathrm{ARF}(K)$.

Proof: <u>Exercise</u>.

Proof of Theorem. Also an exercise. Compare on a spanning surface with the curve α depicted to the left as part of a symplectic basis. Note that you can assume that this appears as part of a band, and that the the dual curve β is on another band so that the simplest picture gives:

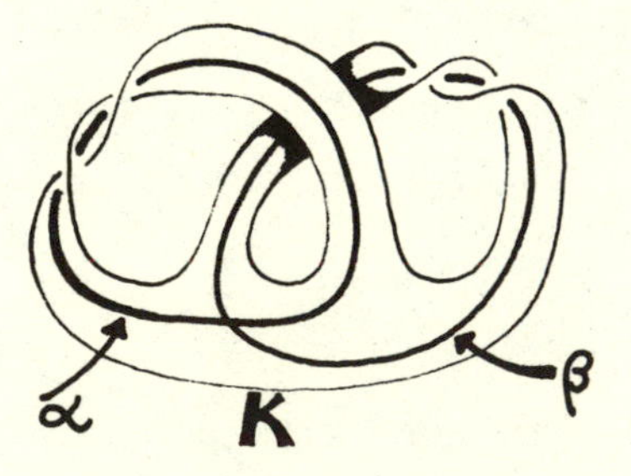

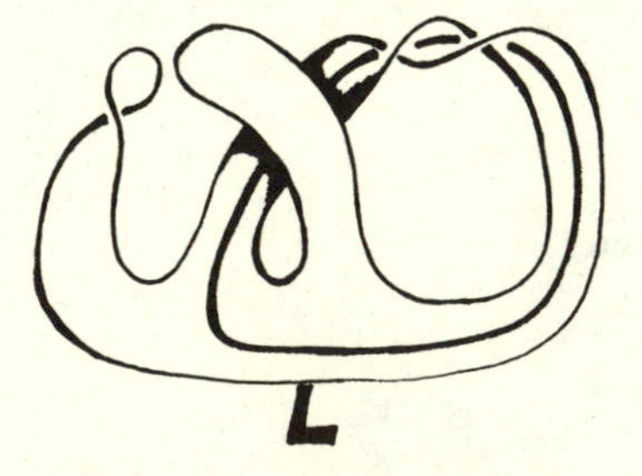

THEOREM 10.4 (Levine [L2]). *Let* $K \subset S^3$ *be a knot. Let* $\Delta_K(t)$ *be the Alexander polynomial for* K. *Then*

$$\mathrm{ARF}(K) = 0 \Longleftrightarrow \Delta_K(-1) \equiv \pm 1 \quad (\text{modulo } 8)$$

$$\mathrm{ARF}(K) = 1 \Longleftrightarrow \Delta_K(-1) \equiv \pm 3 \quad (\text{modulo } 8).$$

Proof: $\nabla_K(z)$ denotes the Conway polynomial. We know (Proposition 9.3) that

$$\nabla_K(\sqrt{t} - 1/\sqrt{t}) \doteq \Delta_K(t).$$

Hence $\nabla_K(2i) \doteq \Delta_K(-1)$ where $i = \sqrt{-1}$. Now, for a knot, $\nabla_K(z) = 1+a_2z^2+a_4z^4+\cdots$. Hence $\nabla_K(2i) \equiv 1-4a_2(K)$ (mod 8). Since $a_2(K) \equiv \mathrm{ARF}(K)$ (mod-2), the theorem follows immediately from this. ∎

Remark: $3^2 \equiv 1$ (modulo 8).

In order to get a taste of the power of Levine's result for calculating Arf invariants, we now give a brief introduction to Fox's Free Differential Calculus, and its use in computing Alexander polynomials, hence, derivatively, in computing ARF.

XI

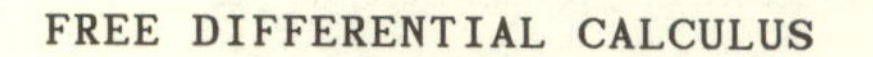

FREE DIFFERENTIAL CALCULUS

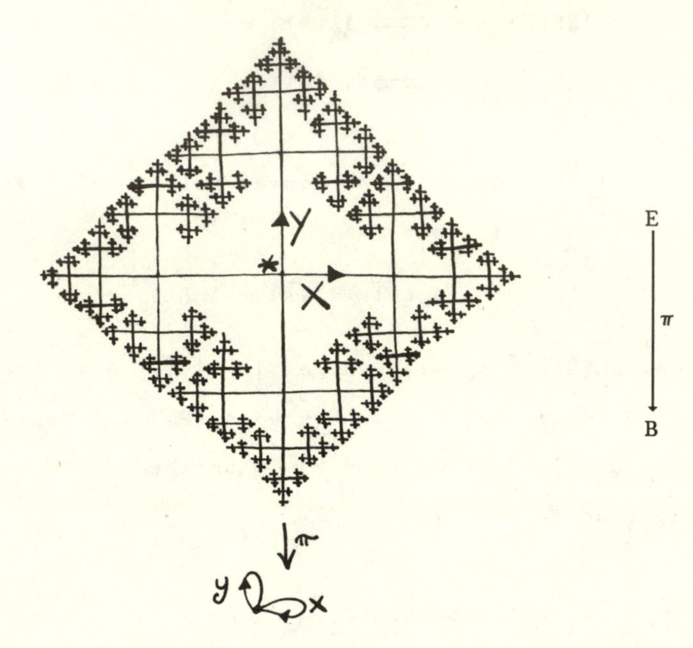

Here is a picture of the figure eight (B) and its universal covering space E. Now $\pi_1(B) = (x,y|\quad)$, the free group on two generators. Let $G = (x,y|\quad)$ and note that G is the group of automorphisms of E over B. Thus E has "generating" 1-simplices X and Y as depicted. X is the lift of x as an element in π_1 starting at $*$. Y is the lift of y. By regarding E as the set of translates of X and Y under the action of G, we can write the lift of any word $\omega \in \pi_1(B)$ as a formal sum of simplices with coefficients in G. Thus (letting ω denote the lift),

$$\widetilde{xy} = X + xY$$

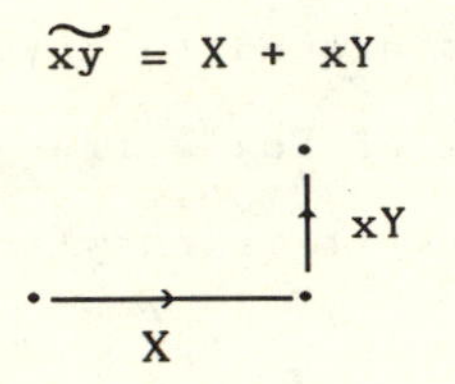

Note that we lift in lexicographic order. <u>Therefore</u>

$$\widetilde{\omega_1\omega_2} = \widetilde{\omega_1} + \omega_1\widetilde{\omega_2}.$$

Example: $x^2yx^{-1} = \omega.$

$x^2yx^{-1}X$

x^2Y

X xX

$$\begin{aligned}
\tilde{\omega} &= \tilde{x} + x(\widetilde{xyx^{-1}}) \\
&= X + x(\tilde{x} + x\widetilde{yx^{-1}}) \qquad\qquad (\widetilde{x^{-1}} = x^{-1}X) \\
&= X + xX + x^2(\tilde{y} + y\widetilde{x^{-1}}) \\
\tilde{\omega} &= X + xX + x^2Y + x^2yx^{-1}X.
\end{aligned}$$

Now collect terms.

$$\tilde{\omega} = (1 + x + x^2yx^{-1})X + (x^2)Y.$$

The coefficients belong to $\Gamma = Z[G]$, the group ring of G over the integers.

The coefficient of X is called $\partial\omega/\partial x$.

The coefficient of Y is called $\partial\omega/\partial y$.

$$\tilde{\omega} = \left[\frac{\partial\omega}{\partial x}\right]X + \left[\frac{\partial\omega}{\partial y}\right]Y.$$

These definitions extend to any number of variables, and form the beginning of Fox's free differential calculus. From this point of view it is easy to derive some rules for differentiation:

1°. $\frac{\partial(\omega\tau)}{\partial x} = \frac{\partial\omega}{\partial x} + \omega\frac{\partial\tau}{\partial x}$ (and same for $\partial/\partial y$). This is just a restatement of $\widetilde{\omega\tau} = \widetilde{\omega}+\omega\widetilde{\tau}$.

2°. Since, by definition, $\frac{\partial}{\partial x}(1) = 0$ and $1 = \omega^{-1}\omega$, we have

$$0 = \frac{\partial(1)}{\partial x} = \frac{\partial(\omega^{-1}\omega)}{\partial x} = \frac{\partial\omega^{-1}}{\partial x} + \omega^{-1}\,\frac{\partial\omega}{\partial x}.$$

Hence $\frac{\partial\omega^{-1}}{\partial x} = -\omega^{-1}\,\frac{\partial\omega}{\partial x}$.

3°. For $n > 0$, $\frac{\partial x^n}{\partial x} = 1 + x + x^2 + \cdots + x^{n-1}$. (Since $x^n = (1+x+x^2+\cdots+x^{n-1})X$.) Hence for $n < 0$

$$\frac{\partial x^n}{\partial x} = -x^n(1+x+x^2+\cdots+x^{|n|-1}).$$

The Jacobian Matrix

Here is Fox's algorithm for computing the Alexander polynomial from a presentation of $\pi_1(S^3-K)$. Let $\pi_1(S^3-K) = (x_1,\cdots,x_n|r_1,\cdots,r_m)$ be a presentation. Regard $r_1,\cdots,r_m$ as elements in the free group generated by $x_1,\cdots,x_n$. Form the <u>Jacobian matrix</u> $J = \left[\frac{\partial r_i}{\partial x_j}\right]$. Let $\phi : \pi_1(S^3-K) \longrightarrow Z = (t|\ )$ be the Abelianizing map. Let $J^\phi = \left[\frac{\partial r_i}{\partial x_j}\right]^\phi$ be the image of the Jacobian matrix under the map. Its entries are now in $Z[t,t^{-1}]$. $\underline{\Delta_K(t)}$ is, up to balance, any generator of the ideal generated by largest minors of J^ϕ. (This is a principal ideal.)

Let $K_{a,b}$ be a torus knot of type a,b. Here gcd(a,b) = 1. We can see that $\pi_1(S^3-K_{a,b}) = (\alpha,\beta \mid \alpha^a = \beta^b)$ by looking at $S^3-K_{a,b}$ as the union of pieces interior and exterior to the torus where $K_{a,b}$ lives, and using the Seifert-VanKampen Theorem [MA].

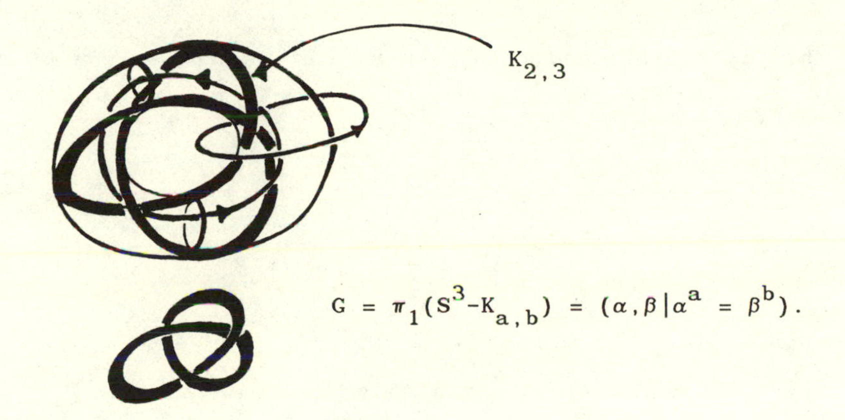

$$G = \pi_1(S^3-K_{a,b}) = (\alpha,\beta \mid \alpha^a = \beta^b).$$

It turns out that we can write a relation A = B in the form A-B and put $\left[\frac{\partial(A-B)}{\partial x_i}\right]^\phi$ into the Jacobian.

In this case the map $\phi : G \longrightarrow (t \mid \;)$ is $\phi(\alpha) = t^b$, $\phi(\beta) = t^a$ (remember gcd(a,b) = 1). Thus

$$J = \left[\frac{\partial(\alpha^a-\beta^b)}{\partial\alpha}, \frac{\partial(\alpha^a-\beta^b)}{\partial\beta}\right]$$

$$J = [1+\alpha+\alpha^2+\cdots+\alpha^{a-1}, -(1+\beta+\beta^2+\cdots+\beta^{b-1})]$$

$$\therefore \quad J^{\phi} = [1+t^b+t^{2b}+\cdots+t^{(a-1)b}, -(1+t^a+t^{2a}+\cdots+t^{a(b-1)})].$$

Now

$$1 + t^b + t^{2b} + \cdots + t^{(a-1)b} = \left[\frac{t^{ab}-1}{t^b-1}\right]$$

$$1 + t^a + t^{2a} + \cdots + t^{a(b-1)} = \left[\frac{t^{ab}-1}{t^a-1}\right]$$

$$\Delta(t) \doteq \gcd\left[\frac{t^{ab}-1}{t^b-1}, \frac{t^{ab}-1}{t^a-1}\right]$$

$$\Delta(t) \doteq \frac{(t^{ab}-1)(t-1)}{(t^a-1)(t^b-1)}.$$

This is the Alexander polynomial for the torus knot of type (a,b).

$$\Delta(t) \doteq \frac{(t^{ab}-1)(t-1)}{(t^a-1)(t^b-1)}.$$

If $\underline{a}$ and $\underline{b}$ are <u>both</u> <u>odd</u> then

$$\Delta(-1) \doteq \frac{(-2)(-2)}{(-2)(-2)} = 1.$$

Hence these knots have vanishing Arf invariant.

Suppose a is <u>odd</u> and b is <u>even</u>. Then $\Delta(-1)$ is indeterminate in this form. So apply L'Hospital's Rule.

$$\Delta(-1) \doteq \left[\frac{((ab)t^{ab-1})(t-1)+(t^{ab}-1)}{(at^{a-1})(t^b-1)+(t^a-1)(bt^{b-1})}\right] t = -1$$

$$\doteq \frac{(-ab)(-2)+0}{(-2)(-b)}.$$

$$\therefore \quad \Delta(-1) \doteq a.$$

Thus in this case $\mathrm{ARF}(K_{a,b}) = 0$ or 1 according as

$a \equiv \pm 1$ or $\pm 3 \pmod 8$.

$$\mathrm{ARF}(K_{3,2}) = 1$$
$$\mathrm{ARF}(K_{5,2}) = 1$$
$$\mathrm{ARF}(K_{7,2}) = 0$$
$$\mathrm{ARF}(K_{9,2}) = 0$$
$$\mathrm{ARF}(K_{11,2}) = 1$$
$$\mathrm{ARF}(K_{13,2}) = 1$$
$$\cdots$$

A four-fold periodicity.

<u>Exercise 11.1</u>. $K_{n,2}$ has a spanning surface of form

(1) Verify this periodicity (above) using topological script.

(2) Calculate $\nabla_K(t)$ for $K_{3,4}$ by using the Seifert pairing.

<u>One more remark</u> about Alexander Polynomial and Free Differential Calculus:

We can use the Wirtinger Presentation [F1] for $\pi_1(S^3-K)$. This associates one meridianal generator to each arc in the knot diagram, and one relation to each crossing:

$c = b^{-1}ab$,

$c = bab^{-1}$.

Let $\phi : G = \pi_1(S^3-K) \longrightarrow Z = (t| \ )$. Then ϕ(any generator in Wirtinger) = t. Each relation $c = b^{\pm 1} ab^{\mp 1} = w$ gives rise to a relation in $H_1(X_\infty)$ of the form $[c] = \left[\frac{\partial \omega}{\partial a}\right]^\phi [a] + \left[\frac{\partial \omega}{\partial b}\right]^\phi [b]$. Since $\left[\frac{\partial \omega}{\partial c}\right]^\phi = 1$ this relation corresponds to a row in the Jacobian matrix for Fox's algorithm. In the case of this presentation, the determinant of any $(n-1) \times (n-1)$ minor will produce the Alexander polynomial. (The knot has n crossings.)

<u>Exercise 11.2</u>. a) Use the notation of the above remark and show that if $w = b^{-1}ab$ then $\left[\frac{\partial \omega}{\partial a}\right]^\phi = (1-t^{-1})$ and $\left[\frac{\partial \omega}{\partial b}\right]^\phi = t^{-1}$, while if $w = bab^{-1}$ then $\left[\frac{\partial \omega}{\partial a}\right]^\phi = (1-t)$ and $\left[\frac{\partial \omega}{\partial b}\right]^\phi = t$.

b) Choose a knot and calculate its Alexander polynomial using the Wirtinger presentation.

c) Part a) of this exercise shows that if $[a]$, $[b]$, $[c]$ connote elements in $H_1(X_\infty)$ that correspond to lifts of the elements $a,b,c \in \pi_1(S^3-K)$, then the following relations ensue in $H_1(X_\infty)$ as a $Z[t,t^{-1}]$ module:

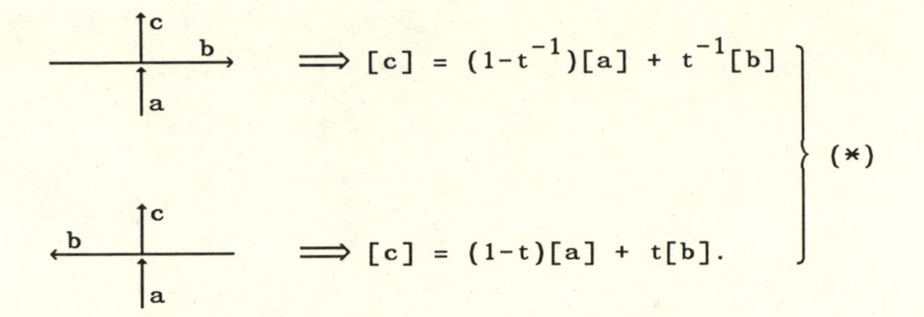

This part of the exercise asks you to compare these patterns with the patterns that arise from trying to represent the fundamental group as a group of affine transformations of the complex plane [DR]: Let

$$\mathcal{L} = \{T : \mathbb{C} \longrightarrow \mathbb{C} \mid T(z) = \alpha z+\beta, \text{ where } \alpha \text{ and } \beta \text{ are elements of } \mathbb{C}\}.$$

Call this the <u>affine group</u>. Let $G = \pi_1(S^3-K)$ with the Wirtinger presentation.

(i) Let $[\alpha,\beta]$ denote $T(z) = \alpha z+\beta$. Show that $[\alpha,\beta][\gamma,\delta] = [\alpha\gamma,\alpha\delta+\beta]$ where $[\ ][\ ]$ denotes composition of maps.

(ii) Suppose $\phi : G \longrightarrow \mathcal{L}$ is a homomorphism of groups. Show that $\phi(a) = [t,\psi(a)]$, $t \in \mathbb{C}$, for a fixed t independent of the choice of a, given that $\ell k(a,K) = +1$. Hence this holds for all the Wirtinger generators. Given an element a with $\ell k(a,K) = +1$, let $\langle a\rangle = \psi(a)$. Thus $\phi(a) = [t,\langle a\rangle]$.

Show on the Wirtinger generators,

$$c = b^{-1}ab \Rightarrow \langle c\rangle = (1-t^{-1})\langle a\rangle + t^{-1}\langle b\rangle$$

$$c = bab^{-1} \Rightarrow \langle c\rangle = (1-t)\langle a\rangle + t\langle b\rangle.$$

(iii) Show that a homomorphism such as in (ii) exists exactly for those t that are <u>roots</u> of $\Delta_K(t) = 0$.

(iv) Compute the affine representations of the trefoil and of the figure-eight knot.

XII
CYCLIC BRANCHED COVERINGS

In Chapter IX we illustrated Seifert's approach to branched covering spaces. In this chapter we turn to these spaces in a more systematic, and partially four-dimensional manner. Let K be an oriented knot or link in the oriented sphere S^3. Then there is a homomorphism $\phi : \pi_1(S^3-K) \longrightarrow Z$ defined by the equation $\phi(\alpha) = \ell k(\alpha,K)$ where this denotes the sum of the linking numbers with individual components of K. Let $\phi_b : \pi_1(S^3-K) \longrightarrow Z/nZ$ denote the composition of ϕ with the surjection $Z \longrightarrow Z/nZ$. The <u>n-fold cyclic covering of K</u> is by definition the covering space of S^3-K that corresponds to this representation.

These coverings can be described, as we have done in Chapter IX, by first cutting S^3 along a spanning surface for F, and then pasting n copies together cyclically end-to-end. Another useful description is as follows: There exists a mapping $\psi : S^3-K \longrightarrow S^1$ that induces the map ϕ on the fundamental groups. (The proof involves some obstruction theory.) This map ψ is unique up to homotopy, and it can be adjusted so that

(1) For a small tubular neighborhood of K, $N(K)$,

there is a product structure $N(K) \cong K \times D^2$ so that $\psi|\partial N(K)$ is equivalent to projection on $S^1 = \partial D^2$.

(2) There is a point $p \in S^1$ such that $\psi^{-1}(p) \cap \partial N(K)$ is $\cong$ to $K \times p$ for this $p \in \partial D^2 = S^1$. That is, $\psi^{-1}(p)$ is a parallel copy of K. And $F = \psi^{-1}(p) \cap E_K$ is a con- connected, oriented, spanning surface for this (parallel copy of) K. Here E_K denotes the closure of the exterior of this tubular neighborhood.

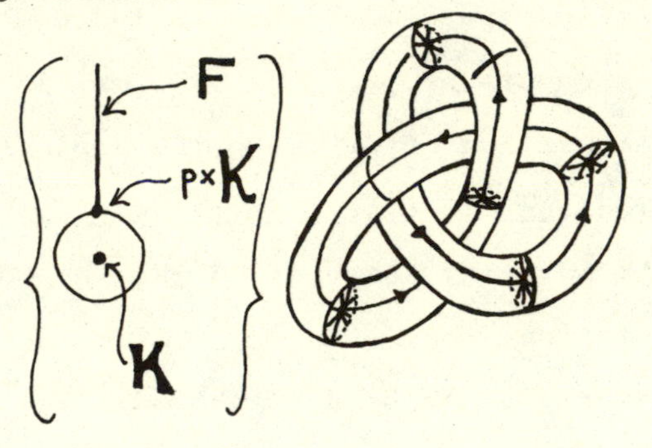

We shall call a $\psi : S^3 - K \longrightarrow S^1$ satisfying (1) and (2) above a <u>good representation for K</u>.

Some knots have especially good representation in that ψ can be a <u>fibering</u>. That is, $\psi^{-1}(p)$ gives a spanning surface for all p, and ψ^{-1}(small neighborhood(p)) $\cong$ Surface × neighborhood(p) for all p. Such a knot is called a <u>fibered knot</u>. Some examples of fiberings will appear shortly.

To return to coverings, let $\psi : S^3 - K \longrightarrow S^1$ be a good representation for K. Contemplate the following diagram:

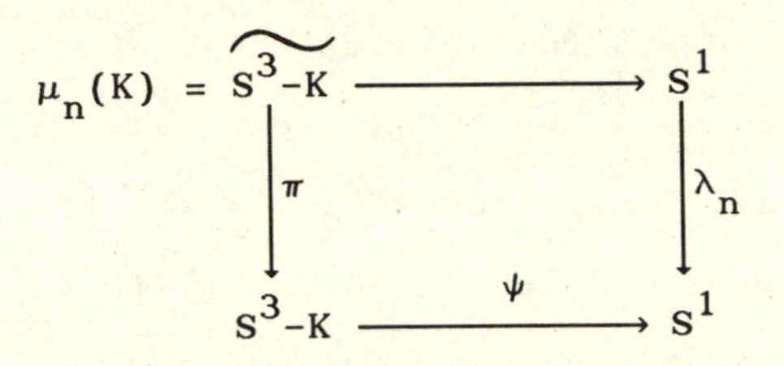

where $\lambda_n(z) = z^n$ (complex multiplication on the circle).

The meaning of the diagram is that $\mu_n(K)$ is defined to be the <u>pull-back of</u> λ_n <u>by</u> ψ. In general,

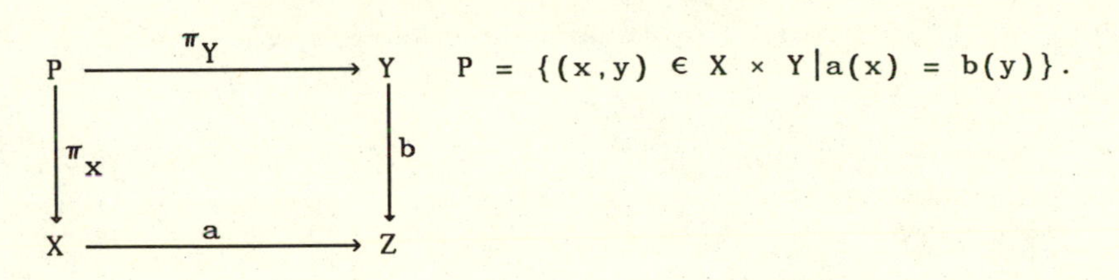

$P = \{(x,y) \in X \times Y | a(x) = b(y)\}.$

This diagram defines the pull-back (or equalizer) of two mappings.

$\widetilde{S^3 - K}$ is the n-fold cyclic cover of K. Notice that in general

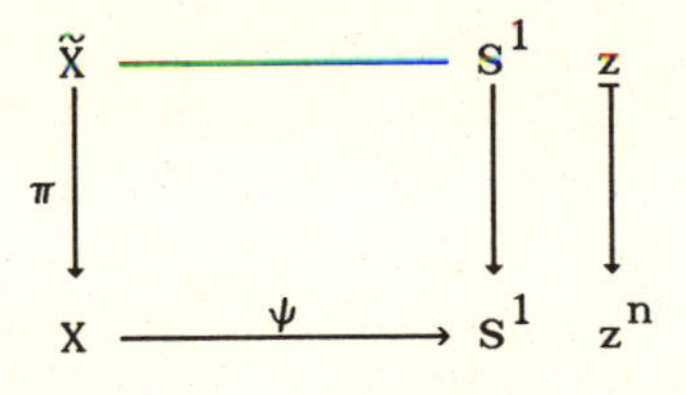

will produce an n-fold covering of X:

$$\tilde{X} = \{(x,z) \mid \psi(x) = z^n\},$$

$$\pi^{-1}(X) = \{(x,z) \mid z^n = \psi(x)\}$$

Thus we produce the n-fold covers by solving equations of the form $z^n = k$.

Branched covers follow similarly. Let E_K be the exterior of K as described above. Let $\partial E_K = K \times S^1$ via the given identification of $N(K)$ with $K \times D^2$. Then $\pi : \tilde{E}_K \longrightarrow E_K$ restricts on the boundary to $K \times S^1 \longrightarrow K \times S^1$, $(x,z) \longrightarrow (x,z^n)$. Since this extends over $K \times D^2$ (same formula), we get

$$M(K) = \tilde{E}_K \cup \widetilde{(K{\times}D^2)} \xrightarrow{\pi} E_K \cup (K{\times}D^2) = S^3.$$

This is the n-fold cyclic branched covering space, branched along K.

All we have done so far, is to redescribe our previous construction. But it is important to keep making the comparison. Thus in the covering construction, splitting along F is included in that $F = \psi^{-1}(p)$ and $S^1 \xrightarrow{\lambda_n} S^1$ can be described by splitting S^1 on a point

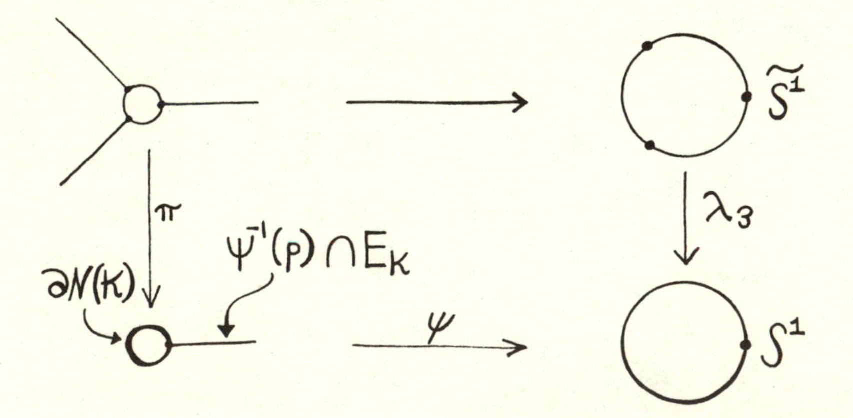

Of course, the infinite cyclic cover is obtained by using $\lambda_\infty : \mathbb{R} \longrightarrow S^1$, $\lambda_\infty(r) = e^{2\pi i r}$.

We can, if we like, use a mapping $\overline{\psi} : S^3 \longrightarrow D^2$ such that $\overline{\psi}$ is differentiably transverse to $0 \in D^2$ and $\overline{\psi}^{-1}(0) = K$. Then branched covers are directly formed as the pull-backs:

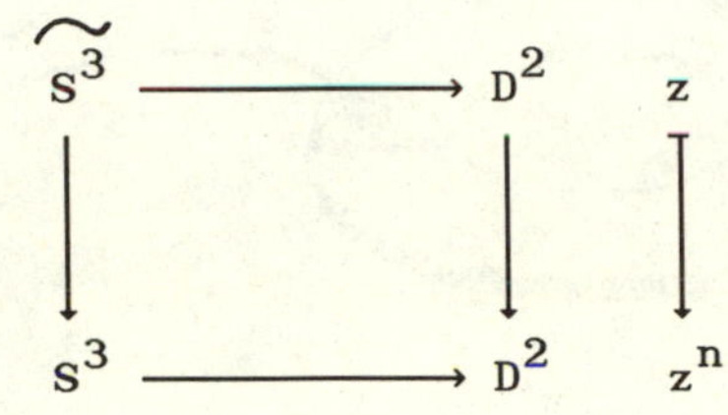

RIEMANN SURFACE DIGRESSION

Take

$$\begin{cases} K = 4 \text{ points in } S^2. \\ F = 2 \text{ intervals joining 2 pairs of points.} \end{cases}$$

Draw a picture of S^2 split along F, and a picture of the resulting 2-fold branched cover:

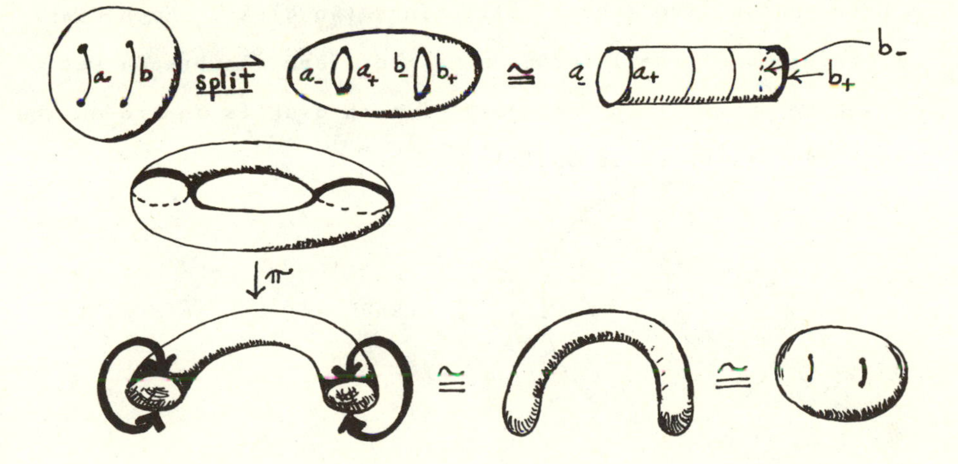

Thus the torus is the 2-fold branched cover of S^2 along 2 points.

And by the same construction, we see that the solid torus $D^2 \times S^1$ is the 2-fold branched cover of D^3 along two arcs (that meet the boundary in 4 points).

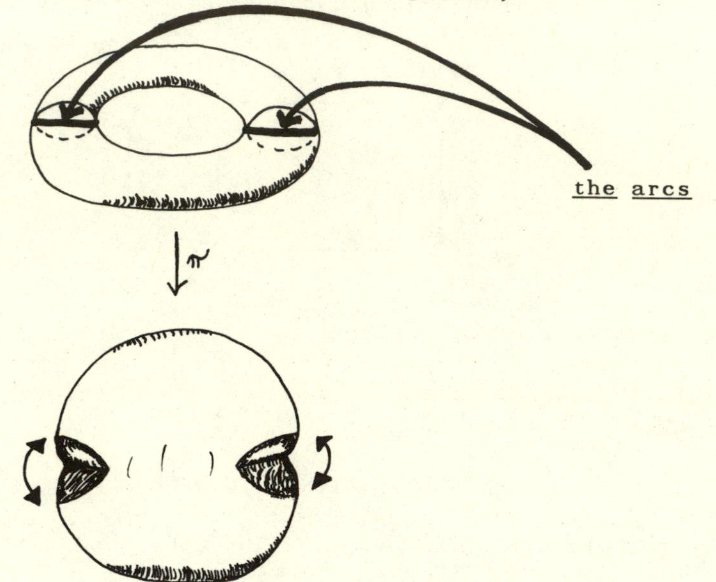

Here you see the 3-ball split along two disks. Each disk bounds one of the interior arcs that form the branch set. (Another part of the boundary of each disk is an arc on the surface of the solid ball.)

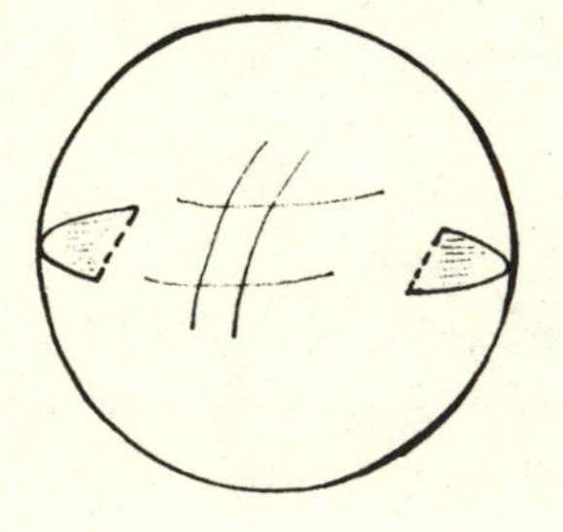

branch set in the 3-ball

This bit of geometry will be useful to us for a number of constructions, but first we generalize it! The solid torus as 2-fold branched cover of D^3 is our first example of a branched covering of a manifold with boundary.

Moving back up one dimension, consider a connected oriented surface $F \subset S^3$. Push this surface into D^4, keeping its boundary fixed in S^3.

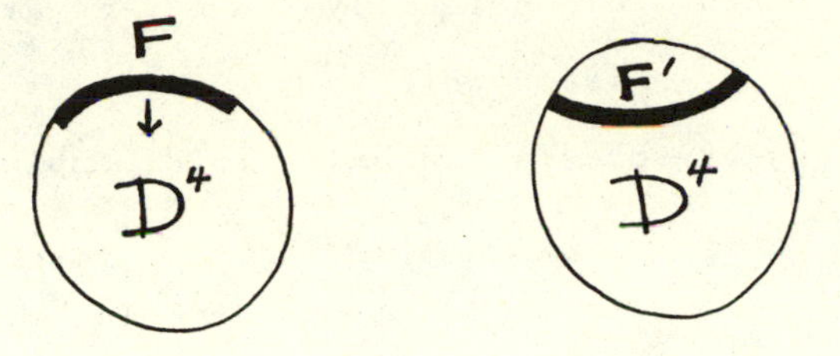

In the diagram, F' denotes the pushed-in surface. As the surface is pushed in, it traces out a 3-manifold M that is homeomorphic to $F \times I$ with $K \times I$ $(K = \partial F)$ collapsed to a single copy of K.

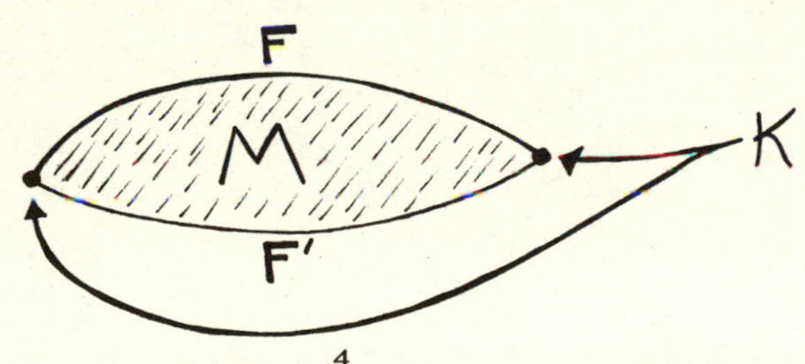

Now imagine splitting D^4 along M. The result is a 4-ball again with the surface 3-sphere now containing M_+ and M_- with $M_+ \cap M_- = F'$.

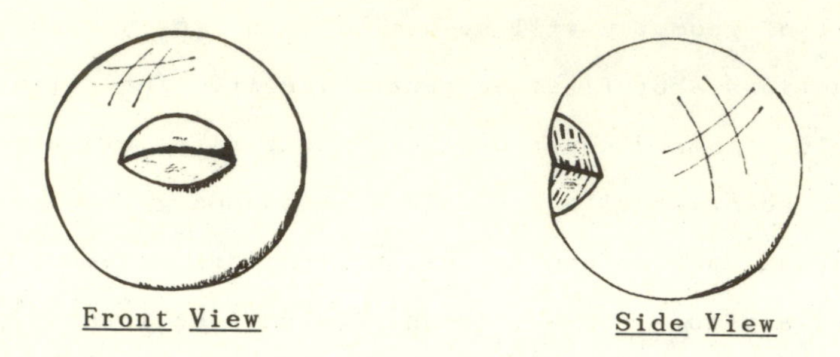

Front View Side View

Here we have drawn a dimensionally-reduced diagram. Notice that this diagram is isomorphic to our pictures for branched covers of spheres and balls in dimensions two and three.

With this model in mind, we can describe cyclic branched covers of D^4 branched along F' (the push-in of F).

Thus, if $N_a(F) = N_a$ denotes the a-fold cyclic cover of D^4, branched along F', then we make N_a from a-copies of D^4, labelled $D^4, xD^4, x^2D^4, \cdots, x^{a-1}D^4$. Here the symbol x has order $a : x^a = 1$.

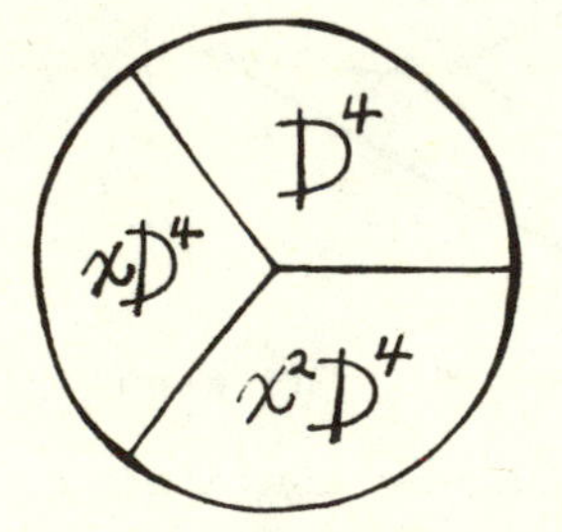

On ∂D^4 label $\begin{cases} M_+ & \text{as } xM \\ M_- & \text{as } M. \end{cases}$

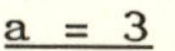

a = 3

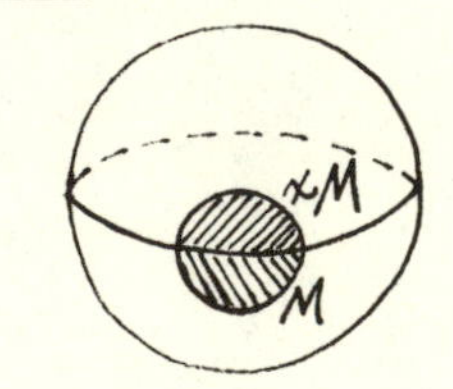

$M \cup xM \subset D^4$

$\Longrightarrow x^kM \cup x^{k+1}M \subset x^kD^4.$

Identify appropriately labelled pieces. Thus $xM = M_+ \subset D^4$ is identified with $xM = xM_- \subset xD^4$.

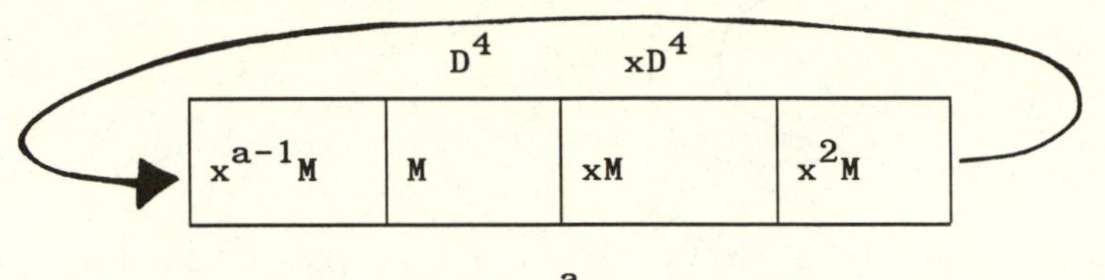

$$x^aM = M$$

It is then appropriate to identify $x: N_a \longrightarrow N_a$ with the covering translation. We have $Na/x \equiv N_a/Z_a \cong D^4$.

Of course, the boundary $M_a = \partial N_a$ is the a-fold cyclic cover of S^3, branched along K.

Example 1: Let $U \subset S^3$ be the unknot of one component U = ◯. Then the spanning surface F is a disk D and the subspace $W = M \cup xM$ consists of the lower and upper hemiballs D^3_- and D^3_+ of a 3-ball D^3 whose equator is U.

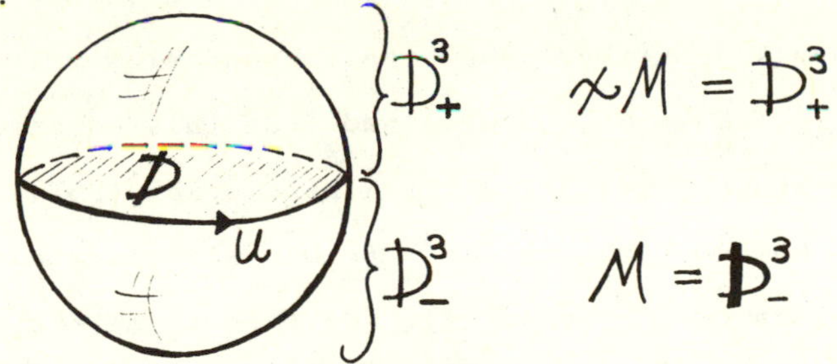

Of course, in this case $N_a \cong D^4$ for all a.

Example 2: Let L = ◯◯ be the Hopf Link. Then the

spanning surface F is an annulus:

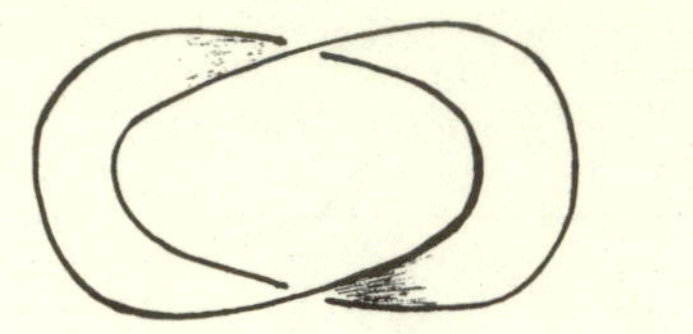

F

And $W = M \cup xM$ is a solid torus τ on whose boundary is embedded the Hopf link.

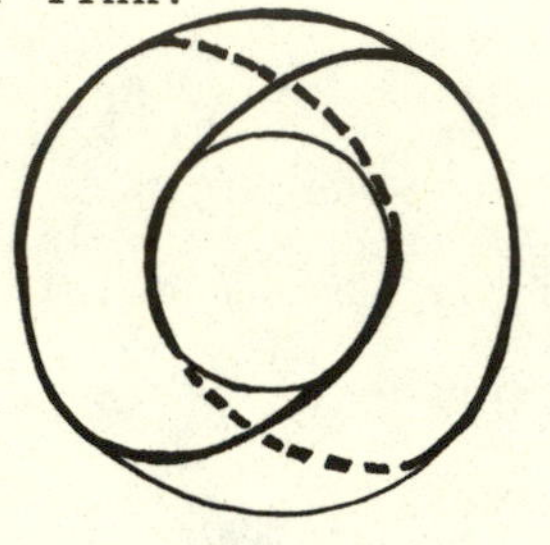

The surface of this torus is divided by L into two annuli: F and xF. M and xM meet along the central annular slice of this torus (as the page appears to cut it).

In this case we see that $M_2(L)$ can be described by taking two solid tori (the result of drilling out W from S^3) and identifying them along their boundaries via the involution $x : S^1 \times S^1 \circlearrowleft$ obtained from this construction. This shows that $M_2(L)$ is a Lens space ([ST]). In fact, it is homeomorphic to $\mathbb{RP}^3$ (= D^3 with antipodal identifications on the boundary). Do this as an exercise.

We can also see clearly the structure of $N_2(L)$. For this consists in $N_2(L) = D^4 \cup xD^4$ where the two 4-balls are pasted along W. Pasting two 4-balls along solid tori in their 3-sphere boundaries obviously produces a space with $H_2(N_2(L)) \cong Z$ generated by a suspension $\Sigma\alpha$ of the generator of the solid torus: $\Sigma\alpha = c\alpha - xc\alpha$ where $c\alpha$ denotes the result of coning α into the center of D^4.

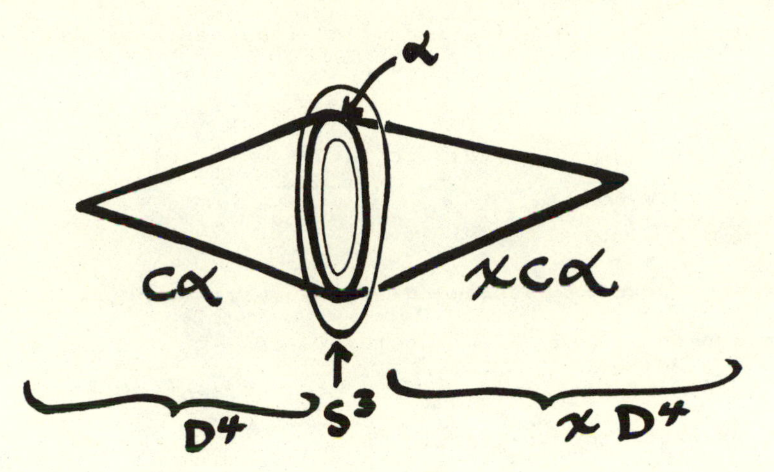

We can, if we like, choose $\alpha \subset F$ so that it generates $H_1(F)$.

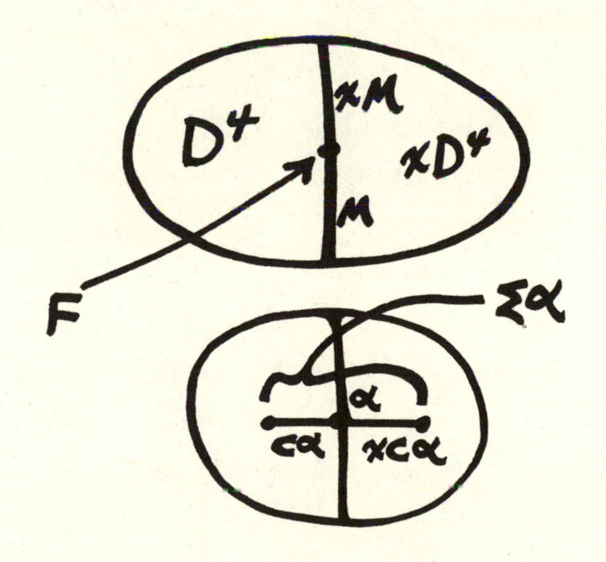

This is the sort of schematic diagram we use for this situation.

We can also see the self-intersection number of the 2-cycle $\Sigma\alpha \subset N_2(L)$. To find $(\Sigma\alpha)\cdot(\Sigma\alpha)$ deform $\Sigma\alpha$ to $d(\Sigma\alpha)$ via

$$\begin{aligned} d(\Sigma\alpha) &= d(c\alpha - xc\alpha) \\ &= c\alpha_* - xc\alpha^* \end{aligned}$$

where α_* denotes pushing α in the positive normal direction to F in S^3. The involution x exchanges pushing up and pushing down, so this is identified with $x\alpha^*$. On each side the result is

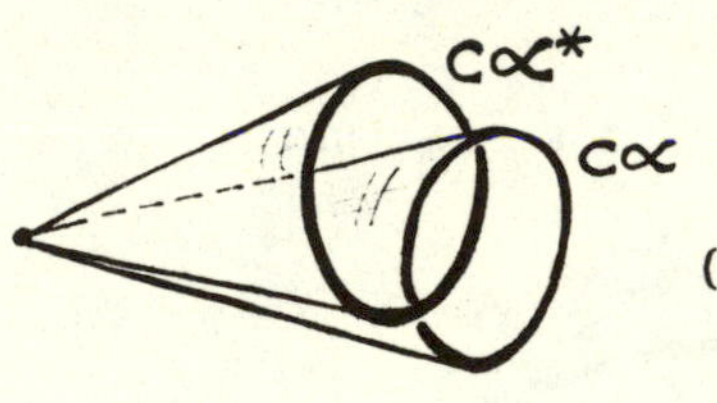

$$(c\alpha)\cdot(c\alpha_*) = \ell k(\alpha,\alpha_*) = +1$$

Thus $(\Sigma\alpha)\cdot(\Sigma\alpha) = 2$. (See [GA].)

We now generalize the geometry of this example.

LEMMA 12.1. $N_a(F)$ *has the homotopy type of* $\cup_{i=0}^{a-1} x^i CF$ *where* $CF \subset D^4$ *denotes the cone over* $F \subset S^3$ *with the apex of the cone at the center of* D^4. *Given a cycle* α *on* F, *let* $\Sigma\alpha$ *denote the cycle* $\Sigma\alpha = c\alpha - xc\alpha$. *Then* $H_1(N_a(F)) = 0$ *and* $H_3(N_a(F)) = 0$ *while*

$$H_2(N_a(F)) \cong \oplus_{k=0}^{a-2} x^k \Sigma H_1(F).$$

Proof: As in the examples,

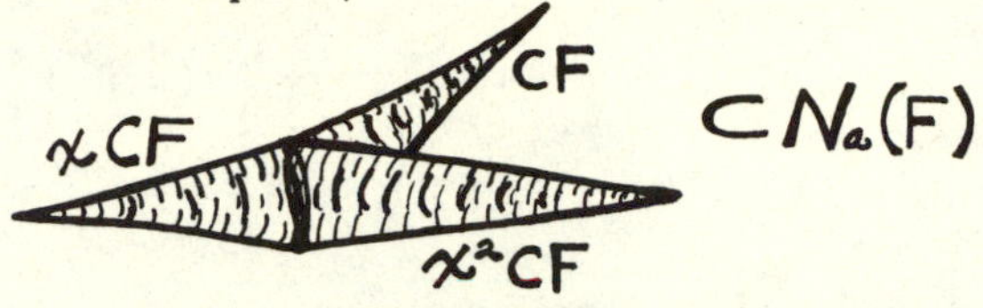

∎

Now we want to determine the intersection form $f : H_2(N_a) \times H_2(N_a) \longrightarrow Z$. (See [K4].)

THEOREM 12.2. *The intersection form* f *(above) is given by the formulas:*

$$f(x^i\Sigma\alpha, x^j\Sigma\beta) = \begin{cases} \theta(\alpha,\beta) + \theta(\beta,\alpha), & i \equiv j \pmod a \\ -\theta(\alpha,\beta), & i \equiv j-1 \pmod a \\ -\theta(\beta,\alpha), & i \equiv j+1 \pmod a \end{cases}$$

[f is zero in the other cases].

Hence f *has matrix* $\begin{bmatrix} \theta & -\theta & & & \\ \theta' & \theta & -\theta & & \\ & -\theta' & \theta & & \\ & & & \ddots & -\theta \\ & & & -\theta' & \theta \end{bmatrix}$ *where*

θ *is the Seifert form for* $F \subset S^3$, *and there are* $(a-1) \times (a-1)$ *blocks.*

Proof:

$$f(\Sigma\alpha,\Sigma\beta) = f(\Sigma\alpha, d\Sigma\beta)$$
$$= f(c\alpha - xc\alpha,\ c\beta_* - xc\beta^*)$$
$$= f(c\alpha, c\beta_*) + f(c\alpha, c\beta^*)$$
$$= \ell k(\alpha,\beta_*) + \ell k(\alpha,\beta^*)$$
$$\therefore\ f(\Sigma\alpha,\Sigma\beta) = \theta(\alpha,\beta) + \theta(\beta,\alpha).$$

Here the deformation $d\Sigma\beta$ is produced just as in the Hopf Link example. We also need $f(\Sigma\alpha, x\Sigma\beta)$ and $f(x\Sigma\alpha,\Sigma\beta)$.

$$f(\Sigma\alpha, x\Sigma\beta) = f(d\Sigma\alpha, x\Sigma\beta)$$
$$= f(c\alpha_* - xc\alpha^*,\ xc\beta - x^2 c\beta)$$
$$= -f(xc\alpha^*,\ xc\beta)$$
$$= -f(c\alpha^*, c\beta)$$
$$= -\ell k(\alpha^*,\beta)$$
$$\therefore\quad f(\Sigma\alpha, x\Sigma\beta) = -\theta(\alpha,\beta).$$

A similar calculation shows that $f(x\Sigma\alpha,\Sigma\beta) = -\theta(\beta,\alpha)$.

These calculations, and the fact that x is an isometry of f suffice to prove the theorem.

Remark: Everything can be generalized to other dimensions. We shall make use of this in Chapter XIX.

We now need some discussion of the algebraic topology of N_a and ∂N_a. Let $j : H_2(N_a) \longrightarrow H_2(N_a, \partial N_a)$ denote the mapping induced by the inclusion $N_a \subset (N_a, \partial N_a)$. This is part of the exact sequence of the pair:

$$0 \longrightarrow H_3(\partial N_a) \longrightarrow H_2(N_a) \xrightarrow{j} H_2(N_a, \partial N_a) \longrightarrow H_1(\partial N_a) \longrightarrow 0.$$

Thus a matrix for j with respect to appropriate bases will be a relation matrix for $H_1(\partial N_a)$. [Don't forget that ∂N_a is the a-fold cyclic cover of S^3 branched along K.]

LEMMA 12.3. *Let* $\mathfrak{B}$ *and* $\mathfrak{B}'$ *be bases for* $H_2(N_a)$ *and* $H_2(N_a, \partial N_a)$ *that are dual in the sense of Poincare-Lefschetz duality.* [*That is, if* $\mathfrak{B} = \{b_1, \cdots, b_n\}$ *and* $\mathfrak{B}' = \{b_1', \cdots, b_n'\}$ *and* $\langle\ ,\ \rangle : H_2(N_a) \times H_2(N_a, \partial N_a) \longrightarrow Z$ *is the (nonsingular by Poincare-Lefschetz) intersection pairing, then* $\langle b_i, b_k' \rangle = \delta_{ik}$.] *Then the matrix of* j, *with respect to* $\mathfrak{B}, \mathfrak{B}'$ *is the intersection matrix for* f *with respect to* $\mathfrak{B}$.

Proof: Let $m_{ij} = f(b_i, b_j)$. Then

$$m_{ij} = f(b_i, b_j) = f(b_i, j(b_j))$$

$$= f\left[b_i, \sum_{k=1}^{n} J_{kj} b_k\right] \quad \text{where J is the matrix of the map j}$$

$$= \sum_{k=1}^{n} J_{kj} f(b_i, b_k)$$

$$= \sum_{k=1}^{n} J_{kj} \delta_{ik} = J_{ij}.$$

$\therefore \quad M_{ij} = J_{ij}.$

Thus we now know that the matrix $\theta \otimes \Lambda_a + \theta' \otimes \Lambda_a'$ is a relation matrix for $H_1(\partial N_a)$. The matrix Λ_a is shown below.

$$\Lambda_a = \begin{bmatrix} 1 & -1 & & & \\ & 1 & -1 & & \\ & & \ddots & \ddots & \\ & & & \ddots & -1 \\ & & & & 1 \end{bmatrix} \quad (a-1) \times (a-1).$$

For example, if $a = 2$ then $\theta+\theta'$ is a relation matrix for $H_1(M_2(K))$. In the Hopf Link example, $\theta = (1)$, so (2) is the relation matrix, whence $H_1(M_2(⚭)) = Z_2$. To continue the Hopf Link example, we have that

$$\Lambda_a + \Lambda_a' = \begin{bmatrix} 2 & -1 & & & & \\ -2 & 2 & -1 & & & \\ & -1 & 2 & \ddots & & \\ & & & \ddots & \ddots & \\ & & & & & -1 \\ & & & & -1 & 2 \end{bmatrix} \quad (a-1) \times (a-1)$$

is a relation matrix for $H_1(M_a(⚭))$. Thus

$$\begin{bmatrix} 2 & -1 \\ -1 & 2 \end{bmatrix} \rightharpoonup \begin{bmatrix} 0 & 3 \\ -1 & 2 \end{bmatrix} \rightharpoonup \begin{bmatrix} 0 & 3 \\ 1 & 0 \end{bmatrix} \Rightarrow H_1(M_3(⚭)) \cong Z/3Z.$$

<u>Exercise</u>. Compute $H_1(M_a(⚭))$ for $a = 2,3,4,5,6,\cdots$.

As a next task, let's see how to compute the signature of N_a. By definition, the <u>signature of a manifold of dimension 4k is the signature of its 2k-dimensional intersection form</u>. Thus

$$\sigma(N_a) = \sigma(f) = \sigma(\theta \otimes \Lambda_a + \theta' \otimes \Lambda_a').$$

We shall proceed with a mixture of algebra and geometry. Remember that $x : H_2(N_a) \longrightarrow H_2(N_a)$ is an isometry of the form f $(f(x\alpha, x\beta) = f(\alpha,\beta)$ for all $\alpha,\beta \in H_2(N_a))$. Thus it would be helpful to decompose $H_2(N_a)$ into a sum of eigenspaces for x. Eigenspaces for different eigenvalues of an isometry of a form f are perpendicular with resepct to f.

In order to construct the eigenspaces we extend coefficients to the complex numbers $\mathbb{C}$. Then f becomes a <u>hermitian form over $\mathbb{C}$</u>. This means that $f(\lambda a, \mu b)$ $= \lambda\bar{\mu} f(a,b)$ when $\lambda, \mu \in \mathbb{C}$; and $f(a,b) = \overline{f(b,a)}$ for $a, b \in H_2(N_a;\mathbb{C})$. A hermitian form has real eigenvalues and hence a well-defined signature. The hermitian form obtained by extending scalars from a real form has the same signature as the real form (exercise).

Thinking of f over $\mathbb{C}$, construct eigenvectors as follows: Let $\omega \in \mathbb{C}$ and $e \in H_2(N_a)$. Define, <u>for $\omega^a = 1$</u>,

$$V(\omega, \epsilon) = e + x\bar{\omega}e + x^2\bar{\omega}^2 e + \cdots + x^{a-1}\bar{\omega}^{a-1} e.$$

Thus $V(\omega, e) \in H_2(N_a;\mathbb{C})$. And $V(\omega, e)$ is an eigenvector for x with eigenvalue ω.

$$\begin{aligned}
xV(\omega, e) &= x(1 + x\bar{\omega} + x^2\bar{\omega}^2 + \cdots + x^{a-1}\bar{\omega}^{a-1})e \\
&= (x + x^2\bar{\omega} + x^3\bar{\omega}^2 + \cdots + x^{a-1}\bar{\omega}^{a-2} + x^a\bar{\omega}^{a-1})e \\
&= (\bar{\omega}^{a-1} + x + x^2\bar{\omega} + x^3\bar{\omega}^2 + \cdots + x^{a-1}\bar{\omega}^{a-2})e \\
&= \omega(1 + x\bar{\omega} + x^2\bar{\omega}^2 + \cdots + x^{a-1}\bar{\omega}^{a-1})e
\end{aligned}$$

$$\therefore xV(\omega,e) = \omega V(\omega,e). \qquad (\omega^a = 1)$$

Furthermore, you can easily check that if $\{e_1, e_2, \cdots, e_n\}$ is a basis for $\Sigma H_1(F) \subset H_2(N_a(F))$, then $\{V(\omega,e_1), \cdots, V(\omega,e_n)\}$ is linearly independent whenever $\omega^a = 1$, $\omega \neq 1$. Therefore, let $V(\omega) \subset H_2(N_a; \mathbb{C})$ be the subspace $V(\omega, \Sigma H_1(F))$ for each ath root of unity:

$$V(\omega) = \{(1 + x\bar{\omega} + \cdots + x^{a-1}\bar{\omega}^{a-1})e \mid e \in \Sigma H_1(F)\}.$$

Then (by dimension count) we have that

$$H_2(N_a; \mathbb{C}) \cong \oplus_{k=1}^{a-1} V(\omega^k)$$

where $\omega = \exp(2\pi i/a)$. This is an eigenspace decomposition.

LEMMA 12.4. *The form* $f|V(\omega^k)$ has matrix

$$a((1-\omega^k)\theta + (1-\omega^k)\theta')$$

where θ *is the Seifert matrix for* $F \subset S^3$.

Proof: Let $\alpha = \Sigma X$, $\beta = \Sigma Y \in \Sigma H_1(F)$. Let ω be any a^{th} root of unity, and $f = f|V(\omega)$

$$f(V(\omega,\alpha), V(\omega,\beta)) = f\left[\sum_{i=0}^{a-1} x^i\bar{\omega}^i\alpha,\ \sum_{j=0}^{a-1} x^j\bar{\omega}^j\beta\right]$$

$$= \sum_{i=0}^{a-1} f(x\omega\alpha, x\omega\beta) + \sum_{i=0}^{a-1} f(x^i\bar{\omega}^i\alpha, x^{i+1}\bar{\omega}^{i+1}\beta)$$

$$+ \sum_{i=0}^{a-1} f(x^{i+1}\bar{\omega}^{i+1}\alpha, x^i\bar{\omega}^i\beta).$$

These are the only possible nonzero terms (using Theorem 12.2). Consequently,

$$\begin{aligned} f(V(\omega,\alpha),V(\omega,\beta)) &= \sum_{i=0}^{a-1} f(\alpha,\beta) + \sum_{i=0}^{a-1} f(\alpha,x\overline{\omega}\beta) + \sum_{i=0}^{a-1} f(x\overline{\omega}\alpha,\beta) \\ &= a[f(\alpha,\beta) + \omega f(\alpha,x\beta) + \overline{\omega}f(x\alpha,\beta)] \\ &= a[\theta(x,y) + \theta(y,x) - \omega\theta(x,y) - \overline{\omega}\theta(y,x)] \\ &= a[(1-\omega)\theta(x,y) + (1-\overline{\omega})\theta(y,x)]. \end{aligned}$$

This completes the proof. ∎

DEFINITION 12.5. *Let* $K \subset S^3$ *be an oriented knot or link,* $F \subset S^3$ *a connected oriented spanning surface for* K. *Let* θ *be the Seifert form for* F *and let* $\omega \neq 1$ *be a complex number. Define the* <u>ω-signature of K</u> *by the formula*

$$\sigma_\omega(K) = \sigma((1-\omega)\theta + (1-\overline{\omega})\theta').$$

(Compare [T].)

<u>Exercise</u>. (a) Use S-equivalence to show that $\sigma_\omega(K)$ is an invariant of the knot K.

(b) Show that $\sigma_\omega(K)$ vanishes on slice knots.

(c) Compute $\sigma_\omega(K)$ for all ω for the knot

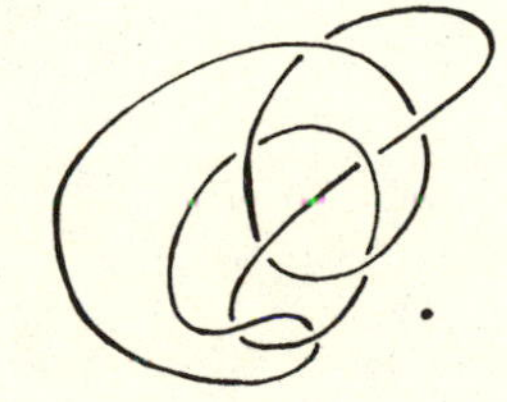.

(d) Prove that if $Te = (1+\overline{\omega}x+\cdots+\overline{\omega}^{a-1}x^{a-1})e$ then for $e \in \Sigma H_1(F)$, $Te = 0 \Rightarrow e = 0$.

We can now summarize our calculations in the

THEOREM 12.6. *Let* $N_a(F)$ *be the* a-*fold cyclic cover of* D^4 *branched along* $F' \subset D^4$ *where* F' *is the push-in of* $F \subset S^3$. F *is a connected, oriented spanning surface for a knot or link* $K \subset S^3$. *Then the signature of* $N_a(F)$ *is given by the formula*

$$\sigma(N_a(F)) = \sum_{i=0}^{a-1} \sigma_{\omega^i}(K)$$

where $\omega = \exp(2\pi i/a)$.

Proof: This follows at once from the preceding discussion.

Example: Let K be the trefoil [trefoil diagram] with Seifert form $\theta = \begin{bmatrix} -1 & 1 \\ 0 & -1 \end{bmatrix}$. Then

$$M = (1-\omega)\theta + (1-\overline{\omega})\theta' = \begin{bmatrix} \omega+\overline{\omega}-2 & 1-\omega \\ 1-\overline{\omega} & \omega+\overline{\omega}-2 \end{bmatrix}.$$

Hence

$$|M_\omega| = (\omega+\overline{\omega}-2)^2 - (1-\omega)(1-\overline{\omega})$$

$$= \omega^2 + \overline{\omega}^2 + 2 - 4\omega - 4\overline{\omega} + 4 - (1-\omega-\overline{\omega}+1)$$

$$\therefore \ |M_\omega| = \omega^2 + \overline{\omega}^2 - 3\omega - 3\overline{\omega} + 4.$$

Actually, it is better to observe that

$$|M_\omega| = |(1-\omega)(\theta-\omega\theta')| = (1-\omega)^2 \Delta_K(\bar{\omega})$$
$$= (1-\omega)^2(\bar{\omega}^2-\bar{\omega}+1).$$

Note that M_ω will be singular at roots of $\Delta_K(t) = 0$. From this we will be able to calculate $\sigma_\omega(K)$ for all ω.

Exercise: Do the calculation in this direct form.

Let's do this exercise:

$$M_\omega = \begin{bmatrix} \omega+\bar{\omega}-2 & 1-\omega \\ 1-\bar{\omega} & \omega+\bar{\omega}-2 \end{bmatrix}.$$

Given a hermitian matrix $H = \begin{bmatrix} a & z \\ \bar{z} & a \end{bmatrix}$ with $a \neq 0$ we can find a matrix H' congruent to it by (1) multiply the first row by $-\bar{z}/a$ and add it to the second row; (2) multiply the first column by $-z/a$ and add it to the second column. The result is

$$H' = \begin{bmatrix} a & 0 \\ 0 & a-z\bar{z}/a \end{bmatrix}.$$

In our case, $a = \omega+\bar{\omega}-2$ and hence $a = 0 \Rightarrow \omega+\bar{\omega} = 2$ $\Rightarrow \mathrm{Re}(\omega) = 1 \Rightarrow \omega = 1$ (since $\omega \in S^1$). Since we're only interested in σ_ω for $\omega \neq 1$, our H' applies and we have

$$a - z\bar{z}/a = (\omega+\bar{\omega}-2) - \frac{(1-\omega)(1-\bar{\omega})}{(\omega+\bar{\omega}-2)}$$

$$= (\omega+\bar{\omega}-2) - \frac{(1-\omega-\bar{\omega}+1)}{(\omega-\bar{\omega}-2)}$$

$$= (\omega-\bar{\omega}-2) + 1$$

$$\therefore \quad a - z\bar{z}/a = (\omega+\bar{\omega}-1).$$

Hence

$$H' = \begin{bmatrix} \omega+\bar{\omega}-2 & 0 \\ 0 & \omega+\bar{\omega}-1 \end{bmatrix}.$$

Examine the unit circle:

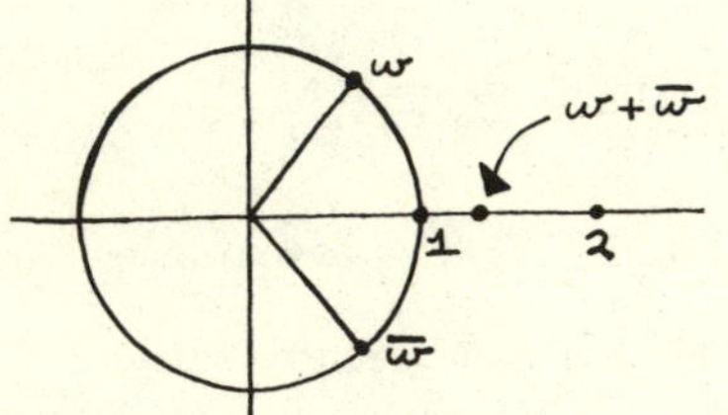

$$\omega \neq 1 \Rightarrow \omega+\bar{\omega}-2 < 0.$$

$$\therefore \quad \underline{\underline{\sigma_\omega = -1 + \mathrm{sgn}(\omega+\bar{\omega}-1)}}$$

Thus we need to know when $\omega+\bar{\omega} = 1$. $\left\{\begin{matrix} \omega = a+bi \\ \omega+\bar{\omega} = 1 \end{matrix}\right\} \Rightarrow a = 1/2$ $\Rightarrow b = \pm\sqrt{3}/2$. This is no surprise, since we know that the form goes singular at the roots of the Alexander polynomial. We can now draw a diagram to indicate the values of $\sigma_\omega(\quad) = \sigma_\omega(K)$ for different $\omega \in S^1$:

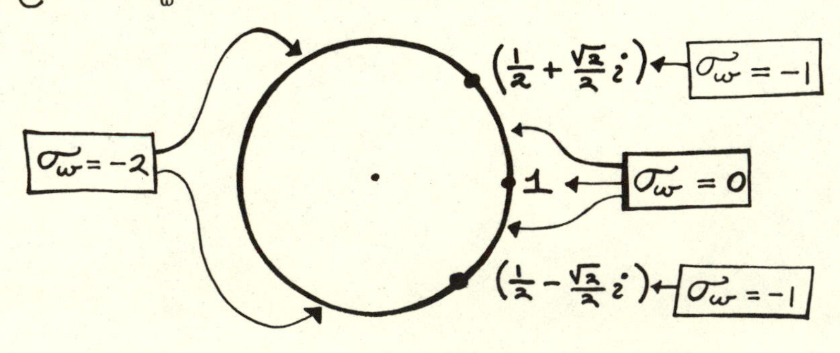

It is easy to use this diagram to calculate

$$\sigma(N_\ell(K)) \underset{\text{def.}}{\equiv} \sigma(K,\ell)$$

for $\ell = 1,2,3,4,\cdots$. Just remember (Theorem 12.6) that $\sigma(K,\ell) = \sum_{i=1}^{\ell-1} \sigma_{\omega^i}(K)$ where $\omega = e^{2\pi i/\ell}$, and plot these points on the diagram: Let $\sigma[\ell] = \sigma(K,\ell)$.

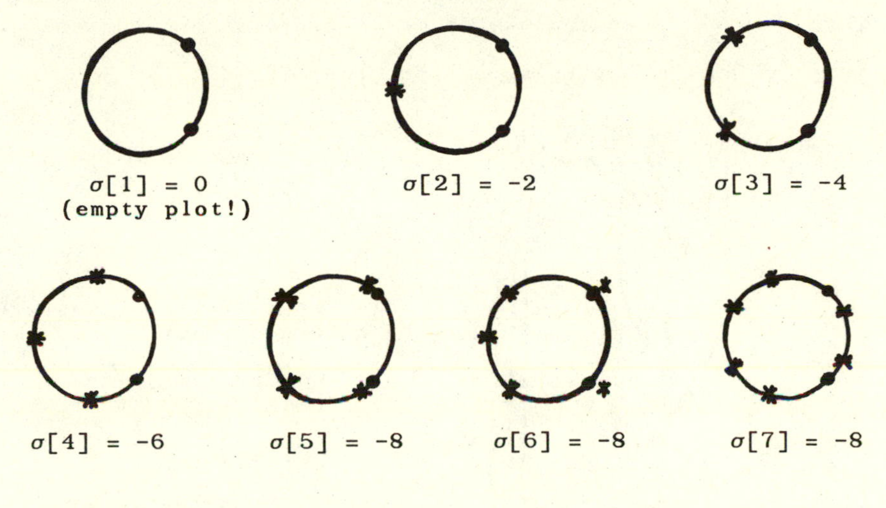

$\sigma[1] = 0$ (empty plot!) $\sigma[2] = -2$ $\sigma[3] = -4$

$\sigma[4] = -6$ $\sigma[5] = -8$ $\sigma[6] = -8$ $\sigma[7] = -8$

<u>Exercise</u>. $\sigma[\ell+6] = \sigma[\ell]-8 \quad \forall_\ell \geq 1$.

The last exercise can be generalized:

<u>Exercise 12.7</u>. Let $K \subset S^3$ be a knot such that the roots of the Alexander polynomial $\Delta_K(t)$ are contained in the dth roots of unity. Let $\sigma[\ell] = \sigma[\ell](K)$ denote the signature of the ℓ-fold branched cover of D^4, branched along a pushed-in spanning surface for K. Prove the

PERIODICITY THEOREM. *There exists a constant* c *such that* $\sigma[\ell+d] = c+\sigma[\ell]$ *for all* $\ell \geq 1$. (See [DK], [N].)

TORUS KNOTS

A torus knot of type (a,b), denoted K[a,b], has Seifert pairing (with respect to a Seifert surface for the usual drawing) $\theta = -\Lambda_a \otimes \Lambda_b$. There are many ways to see this. For the moment, take it as an exercise and contemplate the form of spanning surface for K[3,4]:

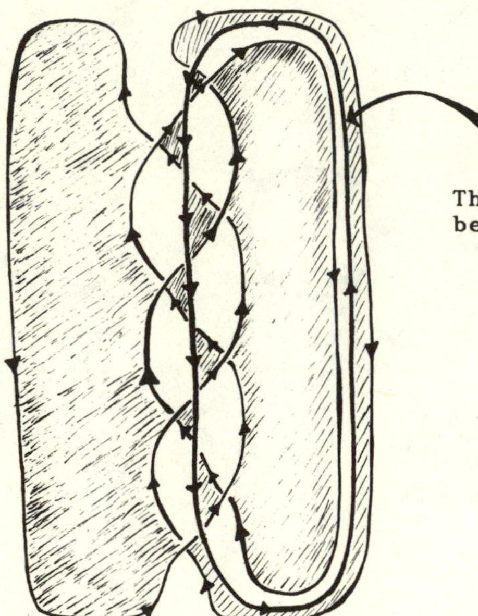

In three-dimensional space, the surface F[a,b] has an action of $(Z/aZ) \times (Z/bZ)$ corresponding exactly to our algebraic actions for Λ_a and Λ_b. Thus we have $x : x^a = 1$, $y : y^b = 1$ acting as isometries of θ. Consequently, we can apply the same algebraic technique

<u>that we used for cyclic branched covers and 4-manifolds to diagonalize</u>:

$$-\Lambda_a \otimes \Lambda_b \underset{\mathbb{C}}{\sim} -\Omega_a \otimes \Omega_b$$

and conclude that the signature of $K[a,b]$ is the signature of

$$(-\Omega_a \otimes \Omega_b) + (-\Omega_a \otimes \Omega_b)^* = M_{a,b}$$

where Ω_a is as shown below.

$$\Omega_a = \begin{bmatrix} 1-\omega & & & & \\ & 1-\omega^2 & & & \\ & & 1-\omega^3 & & \\ & & & \ddots & \\ & & & & 1-\omega^{a-1} \end{bmatrix}$$

with $\omega = e^{2\pi i/a}$. Our formalism shows that $\Lambda_a \underset{\mathbb{C}}{\sim} \Omega_a$ where $\underset{\mathbb{C}}{\sim}$ means congruence of matrices. that is, there is an invertible complex matrix P such that $P^*\Lambda_a P = \Omega_a$. This change of basis respects the isometry x so that we obtain $-\Omega_a \otimes \Omega_b$ as isomorphic to the Seifert form over the complex numbers.

The signature of the matrix $M_{a,b}$ is given by the formula

$$\sigma[a,b] = \sum_{\substack{1 \le i \le a-1 \\ 1 \le j \le b-1}} \operatorname{sgn} \operatorname{Re}[-(1-\omega^i)(1-\tau^j)]. \qquad (\tau = e^{2\pi i/b})$$

Here

$$\operatorname{sgn}(\alpha) = \begin{cases} +1 & \text{if } \alpha > 0 \\ -1 & \text{if } \alpha < 0 \\ 0 & \text{otherwise.} \end{cases}$$

To determine these signs more explicitly, note that

$$1-e^{i\theta} = (e^{-i\theta/2}-e^{i\theta/2})e^{i\theta/2}$$

$$= -2i \sin(\theta/2)e^{i\theta/2}$$

$$\therefore \quad -(1-\omega^{k})(1-\tau^{\ell}) = 4 \sin(\pi k/a)\sin(\pi\ell/b)\cdot e^{i\pi(k/a+\ell/b)}$$

$$\therefore \quad \operatorname{sgn}(\operatorname{Re}[-(1-\omega^{k})(1-\tau^{\ell})]) = \operatorname{sgn}(\operatorname{Re}(e^{i\pi(k/a+\ell/b)}))$$

$$= \operatorname{sgn}(\cos(i\pi(k/a+\ell/b))$$

$$= \begin{cases} +1 & \text{if } -\frac{1}{2} < k/a+\ell/b < +\frac{1}{2} \pmod 2 \\ -1 & \text{if } \frac{1}{2} < k/a+\ell/b < 3/2 \pmod 2 \end{cases}$$

$$\therefore \quad \epsilon[k,\ell] \underset{\text{def}}{\equiv} \begin{cases} +1 & \text{if } 0 < \frac{k}{a} + \frac{\ell}{b} + \frac{1}{2} < 1 \pmod 2 \\ -1 & \text{if } 1 < \frac{k}{a} + \frac{\ell}{b} + \frac{1}{2} < 2 \pmod 2 \end{cases}$$

We have the explicit formula:

$$\sigma(K[a,b]) = \sum_{\substack{1 \le k \le a-1 \\ 1 \le \ell \le b-1}} \epsilon[k,\ell]$$

for the signature of the torus knot $K[a,b]$.

For example, if $a = 2$, $b = 3$ then

$$k = 1, \quad \ell = 1,2 : \frac{1}{2} + \frac{1}{3} + \frac{1}{2} = 1 + \frac{1}{3} \Rightarrow \epsilon[1,1] = -1,$$

$$\frac{1}{2} + \frac{2}{3} + \frac{1}{2} = 1 + \frac{2}{3} \Rightarrow \epsilon[1,2] = -1.$$

Hence $\sigma(K[2,3]) = -2$, as we know.

Since these computations work just as well for links as for knots we can set up a table of signatures $\sigma[a,b]$:

a \ b	2	3	4	5	6	7	8	9	10	11	12	13
2	-1											
3	-2	-4										
4	-3	-6	-7									
5	-4	-8	-8	-12								
6	-5	-8	-11	-16	-17							
7	-6	-8	-14	-16	-18	-24						
8	-7	-10	-15	-20	-23	-30	-31					
9	-8	-12	-16	-24	-26	-32	-32	-40				
10	-9	-14	-19	-24	-29	-34	-39	-48	-49			
11	-10	-16	-22	-24	-34	-40	-42	-48	-50	-60		
12	-11	-16	-23	-28	-35	-42	-47	-54	-59	-70	-71	
13	-12	-16	-24	-32	-36	-48	-52	-56	-64	-72	-72	-84

(These signatures are brought to you courtesy of the Radio Shack Model 100 Portable Computer.)

You are strongly urged to try your hand at proving some of the patterns that leap to the eye. It is true that for fixed a, the signatures $\sigma[a,k] = f(k)$ have a quasi-periodicity to the effect: d = least common multiple $(a,2)$ then $f(k+d) = f(k)+c$ for a fixed constant c. This is actually a case of the periodicity theorem [Exercise 12.7].

But there are other patterns. Look at the signature of the links of type (a,a):

-1, -4, -7, -12, -17, -24, -31, -40, -49, -60, -71, -84
3 3 5 5 7 7 9 9 11 11 13

The successive differences indicate the pattern.

These facts imply that $\sigma[a,b] \neq 0$ for all a,b. Hence no torus knot is a slice knot.

XIII

SIGNATURE THEOREMS

We will return to knots and cyclic branched coverings in the next section. Here we prove general results about the signature of a manifold. (Unless otherwise specified, homology is taken with real coefficients.)

THEOREM 13.1 (Novikov Addition Theorem). *Let* M^{4n} *be a 4n-dimensional manifold that is obtained by gluing two manifolds along a common boundary. Then the signature of* M *is the sum of the signatures of these manifolds. Thus if* $M = Y_+ \cup Y_-$ *where* Y_+ *and* Y_- *are 4n-manifolds,* $X = Y_+ \cap Y_-$ *is a 4n-1 manifold* ($X = \partial Y_+$ and $X = \partial Y_-$), *then* $\sigma(M) = \sigma(Y_+) + \sigma(Y_-)$. *(Orientations compatible with this pasting.)* (See [AS].)

Proof: Use the Mayer-Victoris sequence to decompose $H_{2n}(M) \cong G_+ \oplus G_- \oplus A \oplus B$ where

$$G_+ = \text{Image}(H_{2n}(Y_+) \longrightarrow H_{2n}(M))$$

$$G_- = \text{Image}(H_{2n}(Y_-) \longrightarrow H_{2n}(M))$$

$$A = \{x \in H_{2n-1}(X) \mid i_+x = i_-x = 0\}$$

$$B = H_{2n}(X)/\{x \in H_{2n}(X) \mid j(x) = 0 \text{ in } H_{2n}(M)\}.$$

<u>Here</u>:

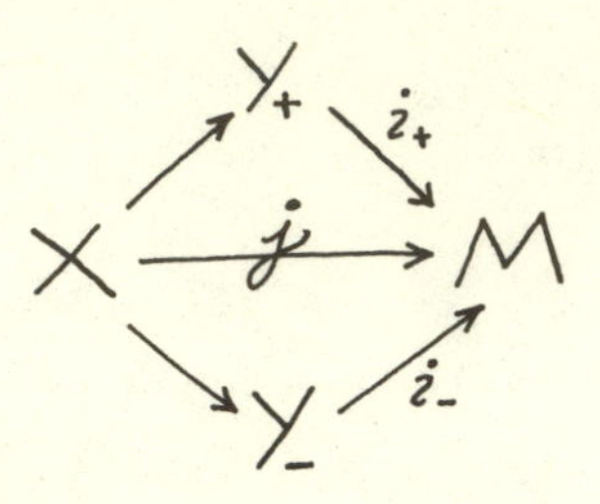

the diagram of inclusions

Thus

G_+ = 2n-cycles on Y_+.

G_- = 2n-cycles on Y_-.

A = cycles in X of dim(2n-1) bounding in Y_+ and Y_-.

B = 2n-cycles in X that live in M.

We leave this decomposition as an exercise, but note: In the Mayer-Vietoris Sequence for $M = Y_+ \cup Y_-$ we have

$$\begin{array}{ccccccc} \cdots \longrightarrow & H_{2n}(M) & \longrightarrow & H_{2n-1}(X) & \xrightarrow[f]{} & H_{2n-1}(Y_+) \oplus H_{2n-1}(Y_-) & \longrightarrow \cdots \\ & \downarrow & \nearrow & & & & \\ & A & & & & & \end{array}$$

since A = Kernel(f). Therefore we have a surjection $H_{2n}(M) \longrightarrow A \longrightarrow 0$. A becomes a direct summand of $H_{2n}(M)$ by lifting back: Given $[x] \in H_{2n-1}(X)$ with $i_+[x] = 0 = i_-[x]$ there are chains α_+, α_- on Y_+ and Y_- respectively such that $\partial\alpha_+ = x$, $\partial\alpha_- = x$. Thus $\alpha_+ - \alpha_-$ is a 2n-cycle on M. Let $[\hat{x}] = [\alpha_+ - \alpha_-]$. Then $\hat{A} \subset H_{2n}(M)$ as the actual direct summand.

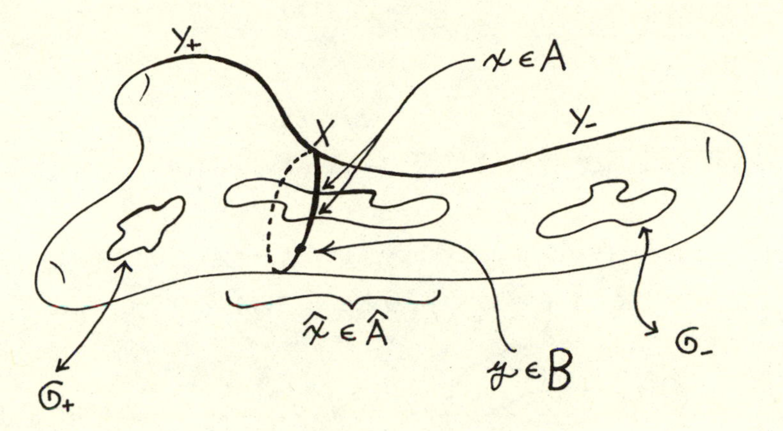

$$M = Y_+ \cup Y_-$$

Now observe the following basic fact about intersections of cycles:

$$\boxed{x \in A,\ y \in H_{2n}(X) \Rightarrow x \cdot y = \hat{x} \cdot j(y)}$$

Here the left $\cdot$ denotes the intersection pairing $H_{2n-1}(X) \times H_{2n}(X) \longrightarrow Z$ and the right $.$ denotes the pairing $H_{2n}(M) \times H_{2n}(M) \longrightarrow Z$. One dimension down, the picture is:

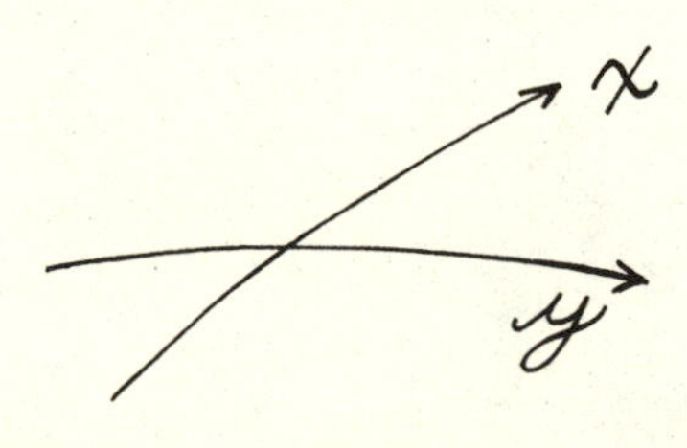

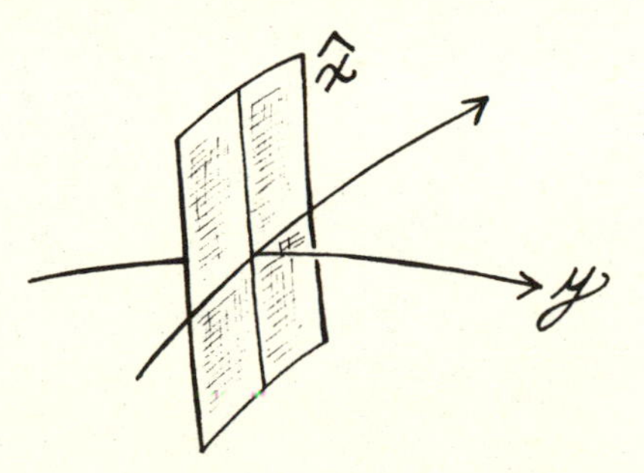

Claim: Given $y \in G_{2n}(X)$, then $j(y) = 0 \Longleftrightarrow x \cdot y = 0$ for all $x \in A_{2n-1}(X)$.

Proof: $\Longrightarrow$: $j(y) = 0$ given.

$x \cdot y = \hat{x} \cdot j(y) = 0$

$\Longleftarrow$: $j(y) \neq 0$ given. Then by Poincaré duality on M $\exists\, \alpha \in H_2(M)$ with $\alpha \cdot j(y) \neq 0$

$\Rightarrow \alpha = \hat{x}$ some $x \in H_{2n-1}(M)$

$\therefore\ 0 \neq \alpha \cdot j(y) = \hat{x} \cdot j(y) = x \cdot y$. ∎

We conclude, from this claim, that <u>A and B are Poincaré dual on M</u> (via duality of A and B on X). Therefore we can choose bases so that the intersection form on $H_{2n}(M)$ looks like:

$\bullet$	G_+	G_-	$\hat{A}$	B
G_+	*	0	*	0
G_-	0	*	*	0
$\hat{A}$	*	*	*	I
B	0	0	I	0

where I is an identity matrix. This implies that $\sigma(M) = \sigma(\cdot|G_+) + \sigma(\cdot|G_-)$. Hence $\sigma(M) = \sigma(Y_+) + \sigma(Y_-)$. ∎

THEOREM 13.2. *Let* M^{4n} *be a compact, oriented, 4n-dimensional manifold. Suppose that* M^{4n} *forms the boundary of a compact oriented manifold* N^{4n+1}. *Then* $\sigma(M^{4n}) = 0$.

Proof: Let $j : H_{2n}(M) \longrightarrow H_{2n}(N)$ denote the map induced by inclusion. And let $A = \{x \in H_{2n}(M) \mid j(x) = 0\}$ denote the kernel of j. Note that $x \in A$ implies that there exists $X \in H_{2n+1}(N,M)$ with $\partial X = x$. Choose a lifting $\alpha : A \longrightarrow H_{2n+1}(N,M)$ such that $\partial\alpha(x) = x$ for all $x \in A$. Then we have, for $x,y \in H_2(M)$, the formula $x \cdot y = j(x) \cdot \alpha(y)$ where the first intersection denotes intersection numbers in M, and the second denotes intersection numbers in N.

If $a,b \in A$ then $a \cdot b = j(a) \cdot \alpha(b) = 0$.

If $b \in H_{2n}(M)$ and $j(b) \neq 0$, then by Poincaré-Lefschetz duality we have an $X \in H_{2n+1}(N,M)$ with $j(b) \cdot X \neq 0$. Hence $b \cdot (\partial X) \neq 0$ and $(\partial X) \in A$.

Let $\{a_1, \cdots, a_r\}$ be a basis for A and $\{\hat{a}_1, \cdots, \hat{a}_r\} \subset H_{2n}(M)$ be dual (Poincaré dual) in the sense that $a_i \cdot \hat{a}_j = \delta_{ij}$. Since A is $\perp$ to A we know that $\{a_1, a_2, \cdots, a_r, \hat{a}_1, \cdots, \hat{a}_r\}$ is linearly independent.

Claim: This is a basis for $H_{2n}(M)$.

To see the claim, suppose $x \in H_{2n}(M)$ and let $\omega = \sum_i (x \cdot a_i) a_i + \sum_j (x \cdot a_j) a_j$. Then $j(x-\omega) \cdot H_{2n+1}(N,M) = (x-\omega) \cdot A = \{0\}$. Hence $j(x-\omega) = 0$ and therefore $(x-\omega) \in A$, whence $x = \omega$.

These remarks show that M^{4n} has intersection form $\begin{bmatrix} 0 & I \\ I & * \end{bmatrix}$ and hence $\sigma(M) = 0$. This completes the proof. ∎

Remark: It follows from the argument that if $M^{2n} = \partial N^{2n+1}$ then

$$\dim(\mathrm{Ker}(H_n(M) \longrightarrow H_n(N)) = \frac{1}{2} \dim H_n(M).$$

Remark: Two closed, compact, oriented manifolds M_1^{4n}, M_2^{4n} are said to be <u>cobordant</u> if there exists a compact, oriented manifold N^{4n+1} with $\partial N = M_1 \cup (-M_2)$ (disjoint union). Theorem 13.2 implies that $0 = \sigma(\partial N) = \sigma(M_1) - \sigma(M_2)$. Hence $\sigma(M_1) = \sigma(M_2)$. <u>Signature is a cobordism invariant</u>.

Remark: We can now prove that the signatures associated with cyclic branched covers of knots are independent of the choice of spanning surface in the four-ball.

PROPOSITION 13.3. *Let $K \subset S^3$ be an oriented knot or link. Let $F \subset D^4$ be any properly embedded surface which is oriented with boundary K. Let $N_a(F)$ denote the a-fold cyclic covering of D^4 branched along F. Then $\sigma(N_a(F))$ depends only on the knot or link $K \subset S^3$. By our previous work this means that $\sigma(N_a(F)) = \sigma_a(K)$.*

Proof: Let $F' \subset D'^4$ be another surface bounding K. Then $\mathcal{S} = F \cup -F' \subset D^4 \cup -D'^4 = S^4$ is a compact oriented surface embedded in S^4. We conclude that there exists a 3-manifold $\mathcal{F} \subset D^5$ bounding $\mathcal{S} \subset S^4$. Hence $N_a^5(\mathcal{F})$ is a 5-manifold with boundary $N_a^4(\mathcal{S}) = N_a^4(F) \cup -N_a^4(F')$. Hence $\sigma(N_a^4(\mathcal{S})) = 0$, and by Novikov, $\sigma(N_a^4(F)) = \sigma(N_a^4(F'))$. ■

Another fundamental property of the signature is the

PRODUCT THEOREM 13.4. *Let* M_1^{4n}, M_2^{4m} *be compact oriented manifolds, then* $\sigma(M_1 \times M_2) = \sigma(M_1)\sigma(M_2)$.

Proof: It suffices (by Kunneth Theorem) to prove that for the tensor product of bilinear forms on vector spaces V_1, V_2 over $\mathbb{R}$, $\sigma(V_1 \otimes V_2) = \sigma(V_1)\sigma(V_2)$. Let $\mathcal{A} = \{a_1, \cdots, a_k\}$ be a basis for V_1, $\mathcal{B} = \{b_1, \cdots, b_\ell\}$ be a basis for V_2. Let $\langle\ ,\ \rangle : V_i \times V_i \longrightarrow \mathbb{R}$ represent the forms. We may assume that they are individually diagonalized. Hence $\{a_i \otimes b_j\}$ is a diagonalizing basis for $V_1 \otimes V_2$. If $\sigma(V_1) = P_1 - N_1$ and $\sigma(V_2) = P_2 - N_2$ where P_1 = number of i such that $\langle a_i, a_i\rangle > 0$ and N_1 = number of i such that $\langle a_i, a_i\rangle < 0$. (Similarly for P_2 and N_2). Then

$$\begin{aligned}\sigma(V_1 \otimes V_2) &= (P_1P_2 + N_1N_2) - (P_1N_2 + P_2N_1) \\ &= (P_1 - N_1)(P_2 - N_2) \\ &= \sigma(V_1)\sigma(V_2). \quad ■\end{aligned}$$

Remark: If M^k is a compact oriented manifold and 4 does not divide $k = \dim(M)$, define the signature of M to be zero: $\sigma(M^k) = 0$ if $4|k$. Then $\sigma(M_1 \times M_2) = \sigma(M_1)\sigma(M_2)$ for manifolds of arbitrary dimension.

Example: Complex Projective Space. Complex projective space $\mathbb{CP}^2$ can be described in a number of ways. It is the set of complex lines in $\mathbb{C}^3$. Hence

$$\mathbb{CP}^2 = \{\langle z_0, z_1, z_2\rangle \mid (z_0, z_1, z_2) \in \mathbb{C}^3 - \{0,0,0\}\}.$$

Here, $\langle\ ,\ ,\ \rangle$ denotes homogeneous coordinates so that $\langle z_0, z_1, z_2\rangle = \langle \lambda z_0, \lambda z_1, \lambda z_2\rangle$ whenever $\lambda \neq 0$. By reformulating this version you can show that $\mathbb{CP}^2 = D^4 \cup_H S^2$ where $H : S^3 \longrightarrow S^2$ is the Hopf map and $\cup_H$ denotes the mapping cylinder on $\partial D^4 = S^3 \longrightarrow S^2$. The Hopf map is the map $S^3 \longrightarrow S^3/S^1$ where S^1 = unit circle in $\mathbb{C}$ acts on $S^3 = \{(z_0, z_1) \,\big|\, |z_0|^2 + |z_1|^2 = 1\}$ by $\lambda(z_0, z_1) = (\lambda z_0, \lambda z_1)$. From this, one sees that $H_2(\mathbb{CP}^2) \cong Z$ and the generator has self-intersection +1. Hence $\sigma(\mathbb{CP}^2) = +1$.

G-*SIGNATURE*

In studying branched covering spaces we have been looking at manifolds with a cyclic group action. We found that signatures decomposed into sums of signatures of eigenspaces. These patterns fit into a more general context. We will outline this context and discuss how to compute signatures of 4-dimensional manifolds admitting a cyclic action. This is called the G-signature theorem (for 4-manifolds). We shall use it in Chapter XVII to study slice knots.

Let $G = C_d$ be a cyclic group of order d. Suppose that G acts smoothly on N^{2n}, preserving orientation. N^{2n} is a compact, oriented manifold. Assume for now that n is even. Let $B(\ ,\ ) : H_n(N;\mathbb{R}) \times H_n(N;\mathbb{R}) \longrightarrow \mathbb{R}$ be the intersection form. (Symmetric since n is even.) Since the group acts on N, we have $B(gx,gy) = B(x,y)$ for all $g \in G$.

Thus we have the following algebraic situation: A symmetric form $B(\ ,\ ) : V \times V \longrightarrow \mathbb{R}$, V a vector space over $\mathbb{R}$. Assume, for simplicity, that $B(\ ,\ )$ is nondegenerate. For the cyclic action, we are given a linear transformation $g : V \longrightarrow V$ with $g^d = 1$ and $B(gx,gy) = B(x,y)$ for all $x,y \in V$.

LEMMA 13.5. *Let* V *be as above. Then there exist subspaces* $V^+, V^- \subset V$ *such that*

(1) $V = V^+ \oplus V^-$.

(2) B *is positive definite on* V^+.
B *is negative definite on* V^-.

(3) $g(V^+) = V^+$, $g(V^-) = V^-$.

Proof: By averaging the standard inner product we can

choose a positive definite inner product $\langle\ ,\ \rangle : V \times V \longrightarrow \mathbb{R}$ that is equivariant with respect to g. Define a linear transformation $A : V \longrightarrow V$ by $\langle Ax,y\rangle = B(x,y)$. Note that the adjoint A^* is defined by $\langle A^*x,y\rangle = \langle x,Ay\rangle$.

Claim: $Ag = gA$.

Proof:
$$\begin{aligned}\langle Agx,gy\rangle &= B(gx,gy)\\ &= B(x,y)\\ &= \langle Ax,y\rangle\\ &= \langle gAx,gy\rangle .\end{aligned}$$

Claim: $A = A^*$.

Proof:
$$\begin{aligned}\langle A^*x,y\rangle &= \langle x,Ay\rangle\\ &= \langle Ay,x\rangle\\ &= B(y,x)\\ &= B(x,y)\\ &= \langle Ax,y\rangle .\end{aligned}$$

Thus g acts on the eigenspaces of A. That is, if $Av = \lambda v$ then $Agv = gAv = g\lambda v = \lambda gv$. Since $A = A^*$, all eigenvalues of A are real. Let V^+ = the direct sum of the positive eigenspaces, and V^- = the direct sum of the negative eigenspaces.

DEFINITION. *With* V *and* g *as above, define* $\mathrm{Sign}(V,g) = \sigma(V,g) \equiv \mathrm{tr}(g|V^+) - \mathrm{tr}(g|V^-)$. *This is the* g-*signature of* (V,B) (tr *denotes trace*).

In order to see that the g-signature is well-defined, and also to make it more concrete it is convenient to reformulate it by using the complex numbers, just as we have done for cyclic branched covers. As we shall see, we have been computing (the equivalent of) g-signatures all along!

Given $B : V \times V \longrightarrow \mathbb{R}$, define $\overline{B} : \overline{V} \times \overline{V} \longrightarrow \mathbb{C}$ where $\overline{V} = V \otimes C$ via the formula $\overline{B}(x \otimes \alpha, y \otimes \beta) = \alpha\overline{\beta}B(x,y)$ for $\alpha,\beta \in \mathbb{C}$ and $x,y \in V$. Extend linearly to obtain a hermitian form $\overline{B}$. Then, as we have found for branched covers, one can decompose $\overline{V}$ into eigenspaces of $g : \overline{V} \longrightarrow \overline{V}$ ($g^d = 1$). The possible eigenvalues are $1,\omega,\omega^2,\cdots,\omega^{d-1}$ where $\omega \in e^{2\pi i/d}$. Thus

$$\overline{V} = \oplus_{m=0}^{d-1} \overline{V}(m)$$

where $\overline{V}(m)$ denotes the eigenspace corresponding to ω^m. This is an orthogonal decomposition with respect to $\overline{B}$.

COROLLARY 13.6. $\sigma(V,g) = \sum_{m=0}^{d-1} \omega^m \sigma(\overline{V}(m))$ *where* $\sigma(\overline{V}(m))$ *is the ordinary signature of* $\overline{B}|\overline{V}(m)$.

Proof: Immediate from the definitions.

We shall nearly always use the g-signatures in this concrete (albeit complex) form. It enjoys many of the properties of the ordinary signature (and, of course, agrees with it for $g = 1$).

Exercise 13.7. Let $G : V \longrightarrow V$, $g' : V' \longrightarrow V'$ be maps of order d, d' as above. Then $g \otimes g' : V \otimes V' \longrightarrow V \otimes V'$. Show that

$$\sigma(V \otimes V', g \otimes g') = \sigma(V,g)\sigma(V',g').$$

Thus g-signatures multiply when we take topological products.

EXTENSION TO SKEW FORMS

Now suppose that $B : V \times V \longrightarrow \mathbb{R}$ is skew-symmetric. That is, suppose that $B(x,y) = -B(y,x)$ for all $x,y \in V$. This is the case for intersection forms on surfaces, and generally for the intersection form on $H_n(M^{2n})$ when M^{2n} is a compact, closed, oriented manifold, and n is odd.

Given $g : V \longrightarrow V$, $g^d = 1$ we wish to define a g-signature for the skew-form B. It will also satisfy a product formula in relation to the ordinary signature. Consequently, we will be able to use it to calculate g-signatures of some four-manifolds by taking products of surfaces with symmetries.

Let $\overline{B} : \overline{V} \times \overline{V} \longrightarrow \mathbb{C}$ be the extension of the form B over the complex numbers $\mathbb{C}$. Let $\overline{V}_{\mathbb{R}}$ denote the underlying real vector space. [Let $\mathrm{Im}(a+ib) = ib$ denote the imaginary part of the complex number $a+ib$.] Then $i\mathrm{Im}(\overline{B}) : \overline{V}_{\mathbb{R}} \times \overline{V}_{\mathbb{R}} \longrightarrow \mathbb{R}$ is bilinear and symmetric.

DEFINITION. $\sigma(V,B) = \frac{1}{2}(\sigma(\overline{V}_{\mathbb{R}}, i\mathrm{Im}(\overline{B}))$ *This gives a definition of the signature for skew forms.*

Discussion: We are given that $B(x,y) = -B(y,x)$. Hence on conjugating the complex form, we have $\overline{\overline{B}(x,y)} = -\overline{B}(y,x)$. Since $\overline{a+bi} = -(a+bi)$ if and only if $a = 0$, we conclude that $\overline{B}(x,x)$ is purely imaginary. Hence we can rewrite $\overline{B} = iM$ where M is a hermitian form. Then $\sigma(V,B) = \sigma(-M)$ where the latter is the usual signature.

DEFINITION. *For g-signatures, take the eigenspace decomposition as before and define* $\sigma(V,g) = \sum_{m=0}^{d-1} \omega^m \sigma(\overline{V}(m))$ *where* $\sigma(\overline{V}(m))$ *denotes the skew-signature as defined above.*

(V,B), B symmetric $\Rightarrow \sigma(V,g) \in \mathbb{R}$
(V,B), B skew-symmetric $\Rightarrow \sigma(V,g) \in i\mathbb{R}$
(If $g = 1 \Rightarrow \sigma(V,1) = 0$.)

Remark: $z = a+ib \Rightarrow a = \frac{1}{2}(z+\overline{z})$

$$ib = \frac{1}{2}(z-\overline{z}).$$

THEOREM 13.8. *Let* M_1, M_2, N *be compact oriented manifolds.*

(1) *If* $N = M_1 \cup M_2$ *where* M_1 *and* M_2 *are matched along their boundaries and* $g : N \longrightarrow N$ *is a cyclic action restricting to the pieces* M_1 *and* M_2, *then* $\sigma(N,g) = \sigma(M_1,g)+\sigma(M_2,g)$.

(2) *If* $M = W$ *and a* g-*action on* M *extends to an action on* W, *then* $\sigma(M,g) = 0$.

(3) *If* g *acts on* M *and* M', *closed oriented manifolds, and we let* g *act on* $M \times M'$ *by* $g(x,y) = (gx,gy)$, *then* $\sigma(M\times M',g) = \sigma(M,g)\sigma(M',g)$.

Remark: This result has been stated for manifolds of arbitrary dimension. We define

$$\sigma(N^{2n},g) = \begin{cases} \underline{\text{signature}} & \text{if } n \text{ even} \\ \underline{\text{skew-signature}} & \text{if } n \text{ odd} \end{cases}$$

and

$$\sigma(N^{2n+1},g) = 0.$$

The proof of this theorem will be omitted. It is a straight generalization of the theorems we have already proved for $g = 1$.

Example 13.9: g-signature for cyclic action on a surface. (See[G2].)

Let m be an odd integer > 0, and let F_0 be a surface obtained by attaching m bands between two discs, arranged cyclically so that we have an action $g : F_0 \to F_0$. Here g = rotation through $2\pi/m$, as shown below:

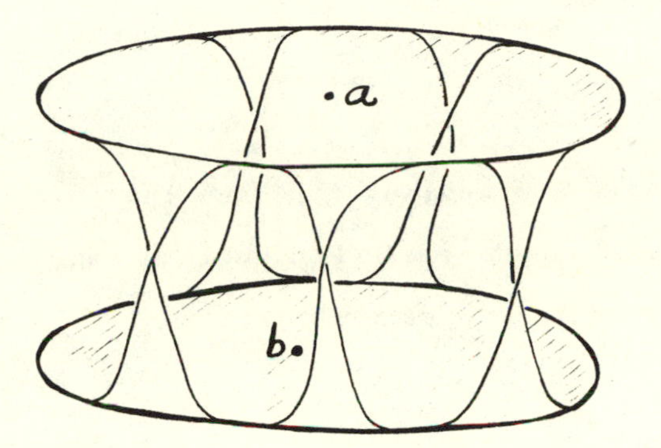

F_0 for $m = 5$

$g = 2\pi/5$ rotation about axis through a and b

<u>Exercise</u>. Show that for m odd as above, F_0 has a single boundary component.

Now let $F = F_0 \cup_\partial D^2$ be the surface obtained by capping off the boundary of F_0 with a disk. Then F is a closed surface with an action of $C_m = Z/mZ$ and <u>three</u> fixed points.

<u>We are interested in the relation between the g-signature and the form of the action near the fixed point set</u>.

Consequently, we make note of the amount of rotation induced at each fixed point. Now F has the fixed points $a, b \in F_0$ and $c \in D^2$ (c is the origin of D^2).

Claim: g^k for $\gcd(k,m) = 1$ (gcd denotes the greatest common divisor) induces the following rotations about fixed points of F:

point	rotation angle
a	$2\pi(k/m)$
b	$2\pi(k/m)$
c	$-2\pi\left[\left[\frac{m+1}{2}\right](k/m)\right]$

Proof: The first two parts are obvious. To see the rotation for c we divide the boundary of F_0 up into segments and look at the g-action. For example, let $m = 3$:

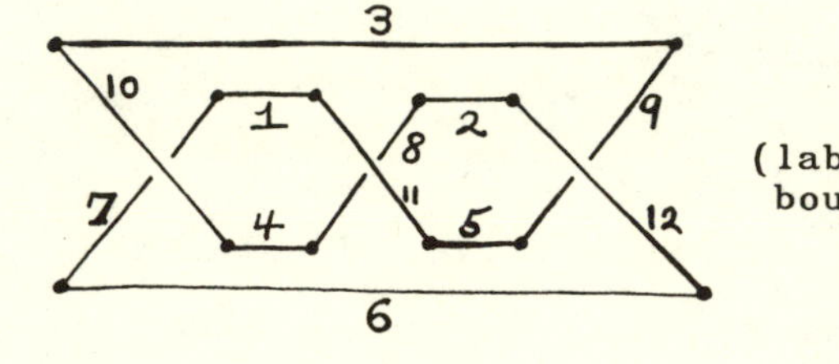

(labelling segments in boundary)

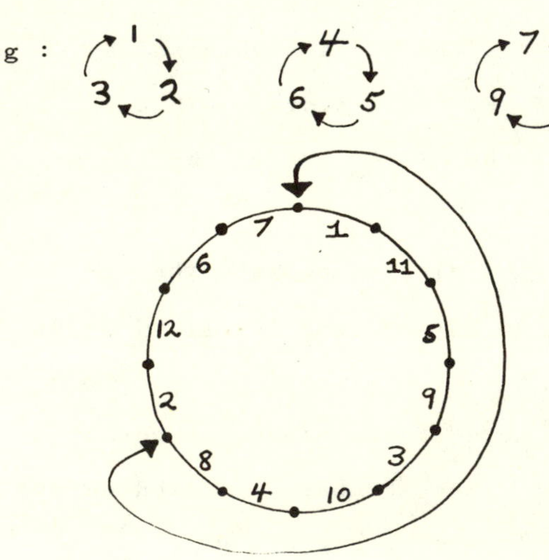

This arc indicates the induced rotation about c. It is 8/12 of a full rotation.

$$\frac{8}{12} = \frac{2}{3} = \frac{(3+1)/2}{3}$$

In general, g induces a rotation about c of $\left[\frac{m+1}{2m}\right]2\pi$.

We leave the remainder of this example as an exercise. Let x denote a cycle that passes through two consecutive bands. Show that $\{g^s x \mid 0 \le s \le m-1\}$ generates $H_1(F;Z)$ and that

$$g^s x \cdot g^t x = \begin{cases} -1 & s = t+1 \\ +1 & s = t-1 \\ 0 & \text{otherwise} \end{cases}.$$

Produce eigenvectors by averaging (just as we did for cyclic branched covers) and show that

$$\sigma(F,g^2) = -i\tan(\pi s(m-1)/2m).$$

Example 13.10. Let $F = F(\pi/3)$ be the surface obtained by identifying sides on the polygon, as shown below:

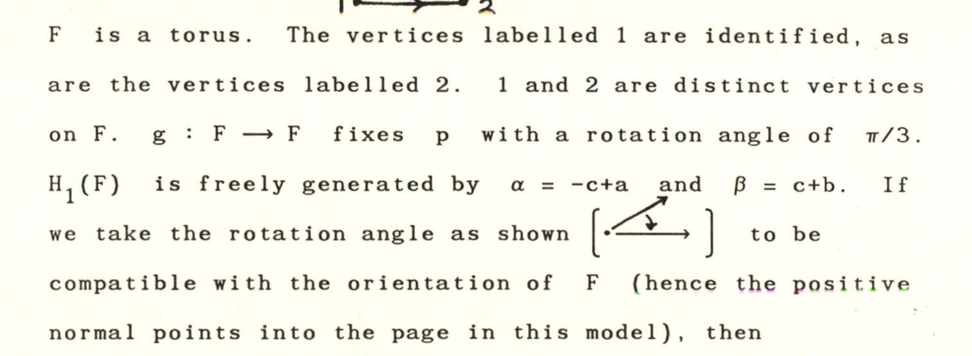

F is a torus. The vertices labelled 1 are identified, as are the vertices labelled 2. 1 and 2 are distinct vertices on F. $g : F \longrightarrow F$ fixes p with a rotation angle of $\pi/3$. $H_1(F)$ is freely generated by $\alpha = -c+a$ and $\beta = c+b$. If we take the rotation angle as shown [diagram] to be compatible with the orientation of F (hence the positive normal points into the page in this model), then

$\alpha \cdot \beta = +1$. We have

$$g\alpha = +a+b = +\alpha+\beta$$

$$g\beta = c+(-a) = -\alpha.$$

Thus the matrix T of g with respect to the basis $\{\alpha, \beta\}$ is $T = \begin{bmatrix} +1 & -1 \\ 1 & 0 \end{bmatrix}$. This has characteristic polynomial $f(\lambda) = \lambda^2 - \lambda + 1$, hence characteristic values

$$\underset{e^{i\pi/3}}{\underset{\|}{\frac{1}{2} + \frac{\sqrt{3}}{2}\,i}} \quad \text{and} \quad \underset{e^{-i\pi/3}}{\underset{\|}{\frac{1}{2} - \frac{\sqrt{3}}{2}\,i}}.$$

Let $\omega = e^{i\pi/3}$. We seek eigenvectors v such that $gv = \omega v$. Since $\omega^2 = \omega - 1$ we see that $v = \omega\alpha + \beta$ and $\overline{v} = \overline{\omega}\alpha + \beta$ are a basis of eigenvectors. (You can calculate v from T, of course.) Now

$$\begin{aligned} i(v \cdot v) &= i(\omega\alpha+\beta)\cdot(\omega\alpha+\beta) \\ &= i(\omega\alpha\cdot\beta + \beta\cdot\omega\alpha) \\ &= i(\omega - \overline{\omega}) \\ &= i\left[2i\,\frac{\sqrt{3}}{2}\right] \end{aligned}$$

$$\therefore \quad i(v \cdot v) = -\sqrt{3} < 0$$

$$\text{and} \quad i(\overline{v} \cdot \overline{v}) = +\sqrt{3} > 0.$$

Thus

$$\begin{aligned} \sigma(F, g) &= \omega(-1) + \overline{\omega}(+1) \\ &= -\omega + \overline{\omega} \\ &= -2i\sin(\pi/3). \end{aligned}$$

Now note:

$$\sin(\theta) = 2\sin(\theta/2)\cos(\theta/2)$$
$$= 2\sin(\theta/2)^2\cot(\theta/2).$$

Here

$$\theta = \pi/3,\ \sin(\theta/2) = \sin(30^\circ) = \frac{1}{2}$$
$$= \sin(\pi/3) = \frac{1}{2}\cot(\pi/6).$$
$$\therefore\quad \sigma(F,g) = -i\cot(\pi/6).$$

<u>Exercise 13.11</u>. Generalize this example to obtain an action of order 3 $(\theta = 2\pi/3)$.

<u>Hint:</u>

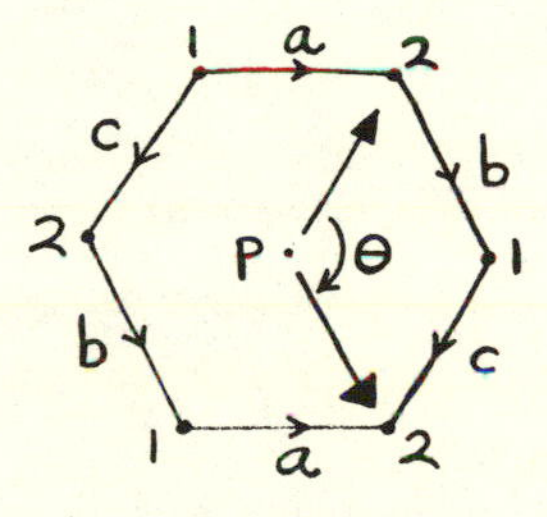

$\theta = 2\pi/3$

fixes 1 and 2 and p. Cut out disks around 1 and 2 and add a tube to obtain an action with one fixed point. Show $\sigma(F(2\pi/3),g) = -i\cot((2\pi/3)/2)$.

<u>Exercise 13.12</u>. Construct surfaces $F(\theta)$ for all $\theta = 2\pi/m$ and <u>one</u> fixed point. Show

$$\sigma(F(\theta),g) = -i\cot(\theta/2).$$

(This example follows a suggestion of José Montesinos.)

DEFINITION 13.13. *Let* M_1^n, M_2^n *be oriented, closed compact manifolds. Let* X *be any space. We say that continuous maps* $\alpha : M_1 \longrightarrow X$ *and* $\beta : M_2 \longrightarrow X$ *are* <u>*bordant*</u> $(\alpha \sim \beta)$ *if there exists a compact oriented manifold* N^{n+1} *and a map* $F : N^{n+1} \longrightarrow X$ *such that* $\partial N = M_1 \cup -M_2$ *(disjoint union), and* $\alpha = F|M_1$, $-\beta = F|-M_2$ $(-\beta = \beta$ *composed with an orientation reversing diffeomorphism of* $M_2)$. *We can write* $F\big|_{\partial N} = \alpha - \beta$.

Let $\Omega_n(X)$ denote the set of bordism classes of maps of closed oriented n-manifolds into X. (If $\alpha : M^n \longrightarrow X$ represents a bordism class, we allow M^n to have more than one component.)

Then $\Omega_n(X)$ has the structure of an abelian group: If $\alpha : M_1 \longrightarrow X$ and $\beta : M_2 \longrightarrow X$, then $\alpha+\beta : M_1 \cup M_2 \longrightarrow X$ where $\cup$ denotes disjoint union. Since $\partial(M\times[0,1]) = M \cup (-M)$ we see that $\alpha+(-\alpha) \sim 0$. The element 0 is the empty map (or, the map $S^n \longrightarrow * \in X$). Clearly $\Omega_0(X) \cong H_0(X;Z)$.

PROPOSITION 13.14. $\Omega_1(X) \cong H_1(X;Z)$.

Proof: We may assume that X is connected. It is easy to check that every element of $\Omega_1(X)$ is represented by a map $\alpha : S^1 \longrightarrow X$. Define $\phi : \pi_1(X,*) \longrightarrow \Omega_1(X)$ by the formula $\phi[\alpha] = \langle\alpha\rangle$ where [ ] denotes homotopy class, and $\langle\ \rangle$

denotes bordism class. Now verify that ϕ is onto, well-defined, and a homomorphism of groups. As a result, $\mathrm{Ker}\ \phi \supset [\pi_1,\pi_1]$ (= commutator subgroup of π_1) since $\Omega_1(X)$ is abelian. We now show that $[\pi_1,\pi_1] \supset \mathrm{Ker}\ \phi$: If $[\alpha] \in \mathrm{Ker}\ \phi$, then there exists a surface N with $\partial N = S^1$ and a map $F : N \longrightarrow X$ such that $F|\partial N = \alpha$. Let δ denote ∂N as an element of $\pi_1(N)$. Then we know that $\delta \in [\pi_1(N),\pi_1(N)]$. Hence $[\alpha] = F|\partial N[\delta] = F(\delta)$ $\in [\pi_1(X),\pi_1(X)]$. Hence $\mathrm{Ker}\ \phi \subset [\pi_1(X),\pi_1(X)]$. Hence $\mathrm{Ker}\ \phi = [\pi_1 X,\pi_1 X]$. Thus $\Omega_1(X) \cong \pi_1 X/[\pi_1 X,\pi_1 X] \cong H_1(X;Z)$.

PROPOSITION 13.15 [following [G2]].

(1) $\Omega_m(X) \longrightarrow H_n(X;Z)$ *is an isomorphism for* $n \leq 3$.

(2) *The sequence* $\Omega_4(*) \longrightarrow \Omega_4(X) \longrightarrow H_4(X;Z)$ *is exact.*

Proof: Let $f : K \longrightarrow X$ represent a class in $H_n(X)$, for some oriented n-cycle K. Assume, by induction, that K is a manifold away from its $(n-k)$ skeleton. Then the link L of an $(n-k)$-simplex σ of K will be an oriented $(k-1)$-manifold. Suppose $L = \partial M$ for some oriented k-manifold M. Since the joins $\sigma * L$ and $\sigma * M$ are contractible, $f|\sigma * L$ extends to a map $F : \sigma * M \longrightarrow X$, and $(f\times id) \cup F$ then defines a map from the oriented $(n+1)$ chain $(K\times I) \underset{(\sigma * L)\times\{1\}}{\cup} \sigma * M$ into X. This provides a homology between f and $f' : K' \longrightarrow X$, where

$K' = (K-\sigma * L) \underset{\partial\sigma * L}{\cup} \partial\sigma * M$, and $f' = f|(K-\sigma * L) \cup F|\partial\sigma * M$.

Doing this, if possible, for all $(n-k)$-simplexes, one obtains a representative of the original class in $H_n(X)$ by an n-cycle which is a manifold away from its $(n-k-1)$ skeleton.

Since $\Omega_1(*) = \Omega_2(*) = \Omega_3(*) = 0$, this shows that $\Omega_n(X) \longrightarrow H_n(X)$ is surjective for $n \leq 4$.

The same procedure applied to $(n+1)$-chains shows that $\Omega_n(X) \longrightarrow H_n(X)$ is injective for $n \leq 3$. ∎

The following result is true in all dimensions, but we prove it for the dimensions of our interest (2 and 4).

THEOREM 13.16. *Let* g *generate a free* $Z/mZ = C_m$ *action on a closed manifold* M *of dimension* 2 *or* 4. *Then* $\sigma(g,M) = 0$.

Proof: By 13.15, $\Omega_2(BZ_m) = 0$ (BZ_m denotes the classifying space for Z_m; hence $\Omega_2(BZ_m)$ corresponds to bordism classes of 2-manifolds with free Z_m-action). Hence every free Z_m-action on a closed 2-manifold is freely Z_m null-bordant. Hence it has $\sigma(M,g) = 0$.

In dimension 4, 13.15 implies that $\Omega_4 \longrightarrow \Omega_4(BZ_m)$ is onto. Hence every free Z_m-action on a closed 4-manifold M is freely Z_m-bordant to the multiplication action on $Z_m \times (M/Z_m)$. You can check that if g leaves no component of M invariant, then $\sigma(M,g) = 0$. Hence $\sigma(g, Z_m \times (M/Z_m)) = 0$.

This completes the proof. ■

COROLLARY 13.17. *Let* Z_m *act on* M^{2n} *and assume that the fixed point set is a smooth submanifold (perhaps a number of components of different dimensions). Then, for* $g \neq 1$, $\sigma(M,g)$ *depends only on the action of* Z_m *on a tubular neighborhood of the fixed point set.*

Remark: Strictly speaking, what we prove in this corollary, is that if M and M' have the same fixed-point set structure, then $\sigma(M,g) = \sigma(M',g)$.

We really want a result that states $\sigma(M,g) = \sum_c \#(c)$ where c runs over the components of the fixed point set, and $\#(c)$ denotes the specific contribution of this component. For surfaces, this comes about as follows: For each angle $\theta = 2\pi k/m$ there exists a surface $F(\theta)$ with one fixed point for a cyclic action g of order m such that g is rotation by $2\pi k/m$ about this point. For example:

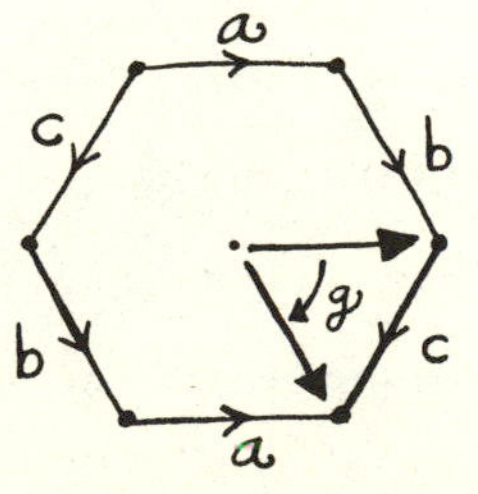

gives $F(2\pi/6)$. See Example 13.10 and Exercises 13.11 and

13.12. This implies that $\sigma(F(\theta),g) = -i\cot(\theta/2)$.

Let $B(\theta) = F(\theta)$ - Interior N where N is a disk neighborhood of the fixed point. Then $B(\theta)$ has a free action of type $2\pi/m$, and

$$\sigma(B(\theta),g) = \sigma(F(\theta),g) = \alpha(\theta) = -i\ \cot(\theta/2).$$

Suppose F is any surface and F has a g-action with fixed points of type $\theta_1,\cdots,\theta_k$. Form $\overline{F} = (F-\bigcup_i \text{Int } N(\theta_i)) \cup \bigcup_i -B(\theta)$. This is a closed manifold with a free action, and hence $\sigma(\overline{F},g) = 0$. Therefore, Novikov $\Rightarrow 0\quad \sigma(F-\bigcup_i \text{Int } N_i)-\sum_i \sigma(B(\theta_i))$. From this we conclude that

$$\sigma(F,g) = \sum_j \sigma(\theta_j) = -\sum_j i\cot(\theta_j/2).$$

<u>Hence the g-signature is a sum of specific contributions from the fixed points</u>.

Proof of Corollary 13.17: If $F \subset M^n$ is a component of the fixed-point set, then F has a tubular neighborhood $F \subset N \subset M$ so that $N \longrightarrow F$ is a ball-bundle with a Z_m-action (the cyclic group acts fiberwise on N). This is the import of the equivariant tubular neighborhood theorem [BR]. Thus we get a decomposition $M = (M\text{-Int}(N)) \cup N$ that is equivariant with respect to the group action.

Let $\mathcal{N}$ denote the union of such tubular neighborhoods over the components of the fixed point set. Write $M = \mathcal{M} \cup \mathcal{N}$ where $\partial\mathcal{M} = \partial\mathcal{N}$ and $G = Z_m$ acts on this decomposition. If $\mathcal{M}'$ is another manifold with $\partial\mathcal{M}' = \partial\mathcal{N}$ we can consider $M' = \mathcal{M}' \cup \mathcal{N}$. We wish to show that $\sigma(M',g) = \sigma(M,g)$.

Now $\mathcal{M} \cup (-\mathcal{M}')$ has a free G-action and it is a closed manifold. Therefore $\sigma(\mathcal{M} \cup (-\mathcal{M}'),g) = 0$. Hence $\sigma(\mathcal{M},g) = \sigma(\mathcal{M}',g)$. Now

$$\begin{aligned} \sigma(M',g) &= \sigma(\mathcal{M}' \cup \mathcal{N},g) \\ &= \sigma(\mathcal{M}',g) + \sigma(\mathcal{N},g) \\ &= \sigma(\mathcal{M},g) + \sigma(\mathcal{N},g) \\ \therefore \quad \sigma(M',g) &= \sigma(M,g). \end{aligned}$$

This completes the proof. ∎

While $\sigma(M,g)$ depends only on the g-action near the fixed-point set, this does not make it obvious how to calculate $\sigma(M,g)$! Some direct calculation is necessary.

Exercise 13.18. Go back to Example 13.10. We had fixed points of type $2\pi k/m$, $2\pi k/m$ and $-2\pi\left[\left[\frac{m+1}{2}\right](k/m)\right]$. We showed that $\sigma(F,g^k) = -i\tan(\pi k(m-1)/2m)$ by a direct calculation. This should equal the sum of the fixed point contributions. Therefore prove directly that

$$-i\cot(2\pi k/m) - i\cot(2\pi k/m) + i\cot\left[2\pi\left[\frac{m+1}{2}\right](k/m)\right]$$

$$\|$$

$$- i\tan(\pi k(m-1)/2m.$$

Exercise 13.19. In Exercises 13.11, 13.12 and in Example 13.8 we worked directly with closed surfaces having a cyclic action and a single fixed point. The purpose of this exercise is to show how these examples are related to fibered knots and their monodromy. Consider the trefoil knot K . This is the first (sic) example of a fibered knot. By this we mean that there is a smooth, locally trivial fibration $\phi : S^3 - K \longrightarrow S^1$ such that ϕ restricts to projection $K \times S^1 \longrightarrow S^1$ on some trivialization of the boundary of a tubular neighborhood of K. More concretely, this means that we can write S^3 as a union of surfaces F_t $(t \in S^1)$ such that $F_t \cap F'_t = K$ when $t \neq t'$. Each surface is smoothly embedded, and the family of surfaces varies continuously. In the case of the trefoil knot, this fiber structure is not hard to visualize. Write the knot on the surface of a torus and draw the Seifert spanning surface:

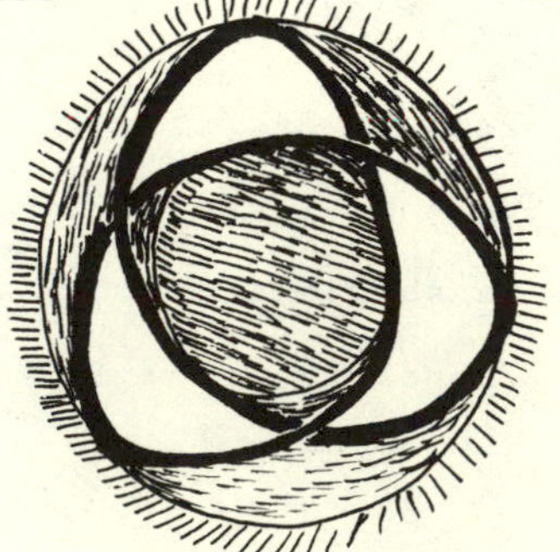

This surface F divides into three parts:

(1) an outer disk

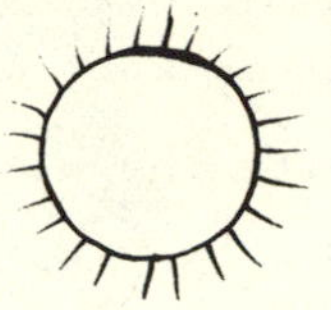

(including a point at infinity).

(2) an inner disk

(3) the twisted bands (interior to the torus).

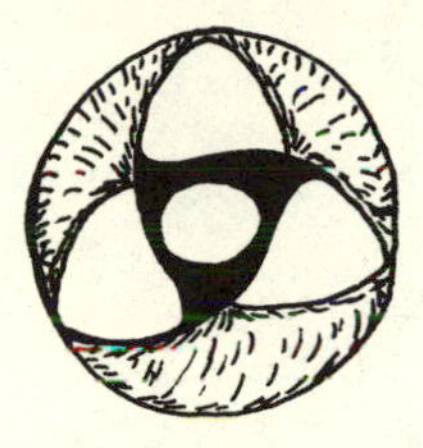

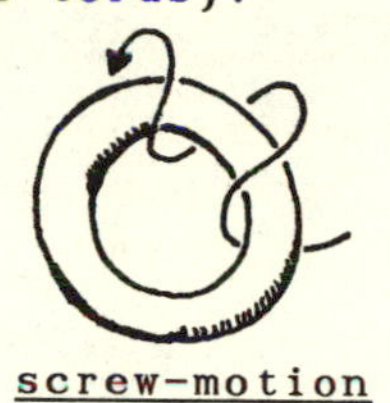

screw-motion

The bands have an obvious 3-fold symmetry and this can be realized by a time parametrized isotopy that is a "screw-turn" in the torus. As we perform this motion, the bands trace out a family of bands intersecting only along the knot. If we carry the disks along, then the inner disk must move up while the outer disk moves down, finally interchanging as we perform the whole movement.

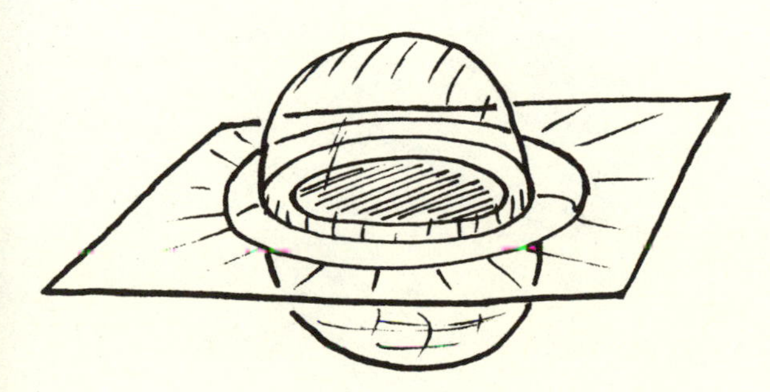

family of disks

Show: The final mapping $h : F \longrightarrow F$ that is a $\frac{1}{3}$ rotation of the bands, and an interchange of the disks, is fixed point free and slides the knot along itself by $2\pi/6$. Show that if $S = F \cup D^2$ with the action $h : S \longrightarrow S$ extended via a $2\pi/6$ rotation on the disk, then $h : S \longrightarrow S$ is identical to $g : F(2\pi/6) \longrightarrow F(2\pi/6)$ (Example 13.10).

Generalize: To fiber surfaces for other fibered knots. (Consider $(2,n)$ torus knots, for a start.) See [FR] for one discussion of the fibering for the figure eight knot.

XIV
G-SIGNATURE THEOREM FOR FOUR MANIFOLDS

The upshot of our work so far is the g-signature theorem for 2-manifolds:

$$\sigma(M^2, g) = \sum_p -i \cot(\theta_p/2)$$

where this sum extends over all isolated fixed points p. θ_p denotes the rotation angle at p measured in radians.

We now turn to the case of four-manifolds. We will only consider oriented fixed point sets and we make no classification of possible fixed point sets. Thus this is a quick trip through the g-signature theorem. The reader is referred to [G2] and [AS] for more details and other points of view.

We know that the g-signature is a sum of contributions from the fixed point sets. Suppose $p \in M^4$ is an isolated fixed point in the four-manifold M^4 with a neighborhood of the form $(B^2, \theta_1) \times (B^2, \theta_2)$ where $\theta_1 = 2\pi k/m$, $\theta_2 = 2\pi \ell/m$ are rotation angles for the individual disks. Call this an isolated fixed point of type $(B^4, \theta_1, \theta_2)$.

Now it is not hard to produce a closed surface $F(k)$ such that a tubular neighborhood of the fixed point set is isomorphic to $m(B^2; 2\pi k/m)$. For example, let $F(k)$ be an m-fold cyclic cover of S^2 branched along m points. Then $F(k) \times F(\ell)$ has a fixed point set of type $m^2(B^4; \theta_1, \theta_2)$ where $\theta_1 = 2\pi k/m$, $\theta_2 = 2\pi\ell/m$. Therefore the contribution from a single point of type $(B^4; \theta_1, \theta_2)$ is:

$$\begin{aligned}(1/m^2)\sigma(F(k) \times F(\ell), g) &= (1/m)\sigma(F(k), g)(1/m)\sigma(F(\ell), g) \\ &= (-i \cot(\theta_1/2))(-i \cot(\theta_2/2)) \\ &= -\cot(\theta_1/2)\cot(\theta_2/2).\end{aligned}$$

Hence, if M^4 has only isolated fixed points of type $(B^4; \theta_1, \theta_2)$, then

$$\sigma(M^4, g) = -\sum_{p} \cot(\theta_1/2)\cot(\theta_2/2).$$

Now let's suppose that M^4 has an oriented two-dimensional fixed point set F. F is a two-dimensional submanifold of M.

LEMMA 14.1. $F \cdot F = 0 \Rightarrow \sigma(M, g) = 0$. *(Here* $F \cdot F$ *denotes the self-intersection number of* F *in* M^4.)

Proof: $F \cdot F = 0$ implies that the normal bundle of F in M^4 is trivial. Hence we may choose a trivialization $N(F) \cong F \times D^2$. Choose a handlebody H with $F = \partial H$, and let $W = (M-F \times D^2) \cup_{\partial} (H \times S^1)$.

Let the cyclic group act on $H \times S^1$ via multiplication on S^1 by the appropriate root of unity, combined with the given action on the other part. Then the action on W is free. Hence for $g \neq 1$,

$$0 = \sigma(W,g) = \sigma(M\text{-Interior}(N(F)),g) + \sigma(H \times S^1,g).$$

But $\sigma(H \times S^1,g) = 0$. Hence

$$\sigma(M\text{-Interior}(N(F)),g) = 0$$

Thus $\sigma(M,g) = \sigma(F \times D^2,g) = 0$. ■

In order to go further, it is very useful to have a complex projective space in your pocket: Recall that $\mathbb{CP}^2 = \{\langle z_0,z_1,z_2\rangle\}$ (homogeneous coordinates). Let Z_m act on $\mathbb{CP}^2$ via

$$g\langle z_0,z_1,z_2\rangle = \langle \omega^r z_0,\omega^r z_1,z_2\rangle.$$

The fixed point set is

$$\underbrace{\{\langle z_0,z_1,0\rangle\}}_{\| \atop S^2} \cup \underbrace{\{\langle 0,0,1\rangle\}}_{\| \atop p}.$$

At p, the action is $(B^4;2\pi r/m,2\pi r/m)$. At S^2, the action is multiplication by ω^r on the normal bundle. Now g induces the identity map on $H^2(\mathbb{CP}^2)$, and hence $\sigma(\mathbb{CP}^2,g) = \sigma(\mathbb{CP}^2) = 1$.

Now suppose we have M^4 with cyclic action Z_m, and

the fixed point set is a surface F with action ω^r on the normal bundle. Let

$$W = (M^4,F) \#[-(F\cdot F)](\mathbb{CP}^2,S^2).$$

This means that we take equivariant connected sum of $-(F\cdot F)$ copies of $\mathbb{CP}^2$ along the fixed point sets. (-1 copy of $\mathbb{CP}^2$ is a copy of $-\mathbb{CP}^2$.) The fixed point set in W is then

$$-(F\cdot F)(B^4;2\pi r/m,2\pi r/m) + (F \#(-(F\cdot F)S^2).$$

Thus, in W, the 2-dimensional fixed point set has zero self-intersection. Consequently, the only contribution to the g-signature is from the fixed points that are isolated.

$$\therefore \quad \sigma(W,g) = -(F\cdot F)(-\cot^2(\pi r/m)).$$

But

$$\sigma(W,g) = \sigma(M^4,g) - (F\cdot F)\sigma(\mathbb{CP}^2,g)$$

$$\therefore \quad \sigma(W,g) = \sigma(M^4,g) - (F\cdot F)$$

$$\therefore \quad \sigma(M^4,g) = (F\cdot F) + (F\cdot F)\cot^2(\pi r/m)$$

$$\therefore \quad \boxed{\sigma(M^4,g) = (F\cdot F)\operatorname{cosec}^2(\pi r/m)}\ .$$

This exchange argument has determined for us the contribution from the 2-dimensional fixed point set.

If M^4 is a 4-manifold with cyclic action and isolated fixed points of type $(B^4;\theta_1,\theta_2)$, 2-dimensional oriented fixed surfaces with rotation angle ψ, then

$$\boxed{\sigma(M^4,g) = -\sum_{P} \cot(\theta_1/2)\cot(\theta_2/2) + \sum_{P}(F\cdot F)\operatorname{cosec}^2(\psi/2)} \quad .$$

This is the g-signature theorem for four-manifolds.

XV

SIGNATURE OF CYCLIC BRANCHED COVERINGS

(Chapters XV to XVII follow the paper [CG].) Let $\tilde{N} \xrightarrow{\pi} N$ be an m-fold cyclic branched covering of closed (compact, oriented) 4-manifolds, with branch set a surface $F \subset N$. Let $\tilde{F} = \pi^{-1}(F) \subset \tilde{N}$ denote the inverse image of the branch set.

<u>Exercise 15.1</u>. Show that $[\tilde{F}]^2 = \frac{1}{m}[F]^2$ where $[\tilde{F}]^2$ denotes the self-intersection of F in N, and $[F]^2$ denotes the self-intersection of F in N. *Hint:* let S be a closed surface, and let $L(S)$ = isomorphism classes of complex line bundles over S (hence D^2-bundles). Then $L(S) \cong [S, BS^1] = [S, \mathbb{CP}_\infty] = H^2(S;Z) \cong Z$. (Here $[\ ,\ ]$ denotes homotopy classes of maps.) Thus $L(S) \cong Z$. Since such a bundle trivializes over $S\text{-}D^2$ (D^2 a small disk in S), we write $S = S_+ \cup D^2_-$ and the bundle E comes from pasting $S_+ \times D^2$ along its boundary to $D^2_- \times D^2$. The pasting map $f : S^1 \times D^2 \longrightarrow S^1 \times D^2$ can be put in the form $f_n(\lambda, z) = (\lambda, \lambda^n z)$. Call this bundle E_n. Now $E_n \longrightarrow E_{nm}$. Show that in E_n, $[S]^2 = n$.

We want to compute the relationship between signatures

of $\tilde{N}$ and signatures of N. Let $\cdot$ denote the nonsingular hermitian intersection form on $H = H_2(\tilde{N};Z) \otimes \mathbb{C}$. Let $\tau : \tilde{N} \longrightarrow \tilde{N}$ and $\tau : H \longrightarrow H$ denote the automorphism induced by the covering translation: τ is assumed to be a rotation of $2\pi/m$ on the fibers of the normal bundle of $\tilde{F}$. Let $\omega = e^{2\pi i/m}$.

Let E_r = the ω^r-eigenspace of τ, $0 \leq r < m$. Thus $H = E_0 \oplus E_1 \oplus \cdots \oplus E_{m-1}$. Let $\epsilon_r(\tilde{N}) = \sigma(E_r)$.

PROPOSITION 15.2. $\epsilon_r(\tilde{N}) = \sigma(N) - 2[F]^2 r(m-r)/m^2$.

Proof: We know that

$$\sigma(\tilde{N},\tau^s) = \sum_{r=0}^{m-1} \omega^{rs}\epsilon_r(\tilde{N}).$$

<u>Exercise</u>: $\epsilon_0(\tilde{N}) = \sigma(N)$.

(*Hint:* Map $C_*(N) \xrightarrow{t} C_*(\tilde{N})$ via $t(x) = \sum_{i=0}^{m-1} \tau^i(\tilde{x})$. Use this transfer to show that

$$H_2(N) \cong \text{Invariant part of } H_2(\tilde{N}).)$$

Thus

$$\sigma(\tilde{N},\tau^s) - \sigma(N) = \sum_{r=1}^{m-1} \omega^{rs}\epsilon_r(\tilde{N}).$$

$$\therefore\ \epsilon_r(\tilde{N}) = \frac{1}{m} \sum_{s=1}^{m-1} (\omega^{-rs}-1)(\sigma(\tilde{N},\tau^s) - \sigma(N)).$$

$$\therefore \quad \epsilon_r(\tilde{N}) = \sigma(N) + \frac{1}{m} \sum_{s=1}^{m-1} (\omega^{-rs}-1)\sigma(\tilde{N},\tau^s).$$

But we know that

$$\sigma(\tilde{N},\tau^s) = [\tilde{F}]^2 \operatorname{cosec}^2(\pi s/m), \quad 0 < s < m$$

and by the first exercise $[\tilde{F}]^2 = [F]^2/m$.

$$\therefore \quad \epsilon_r(\tilde{N}) = \sigma(N) + \frac{[F]^2}{m^2} \sum_{s=1}^{m-1} (\omega^{-rs}-1)\operatorname{cosec}^2(\pi s/m).$$

Now

$$\sum_{s=1}^{m-1} (\omega^{-rs}-1)\operatorname{cosec}^2\left[\frac{\pi s}{m}\right] = -2 \sum_{s=1}^{m-1} \sin^2\left[\frac{\pi r s}{m}\right]\operatorname{cosec}^2\left[\frac{\pi s}{m}\right]$$

$$+i \sum_{s=1}^{m-1} \sin\left[\frac{2\pi r s}{m}\right]\operatorname{cosec}^2\left[\frac{\pi s}{m}\right].$$

Thus

$$\alpha = \sum_{s=1}^{m-1} (\omega^{-rs}-1)\operatorname{cosec}^2\left[\frac{\pi s}{m}\right]$$

$$= -2 \sum_{s=1}^{m-1} \sin^2\left[\frac{\pi r s}{m}\right]\operatorname{cosec}^2\left[\frac{\pi s}{m}\right]$$

(check that the imaginary part vanished).

$$\therefore \quad \alpha = -2 \sum_{s=1}^{m-1} \left[\frac{\xi^{rs}-\xi^{-rs}}{\xi^s-\xi^{-s}}\right]^2, \quad \boxed{\xi = e^{i\pi/m}}\ .$$

Now $\dfrac{X^r-Y^r}{X-Y} = X^{r-1} + X^{r-2}Y + \cdots + XY^{r-2} + Y^{r-1}$, hence

$$\alpha = -2 \sum_{s=1}^{m-1} (\xi^{s(r-1)} + \xi^{s(r-3)} + \cdots + \xi^{-s(r-1)})^2$$

$$\therefore \quad \alpha = -2 \sum_{s=1}^{m-1} P(\xi^s)$$

where $P(z) = (z^{r-1}+z^{r-3}+z^{r-5}+\cdots+z^{-(r-1)})^2$. Note that $P(z) = P(z^{-1})$, and that $\xi^{2m} = 1$. Thus

$$\sum_{s=1}^{m-1} P(\xi^s) = \frac{1}{2} \sum_{s=1}^{2m-1} P(\xi^s) - \frac{1}{2}(P(1)+P(-1))$$

$$= \frac{1}{2}\left[2m \sum_t [\text{coefficient of } z^{mt} \text{ in } P(z)]\right] - r^2$$

$$= mr - r^2.$$

The only coefficient contribution is from z^0. This occurs in r-ways since $P(z) = (\text{———})^2$ as above. Hence $\alpha = -2(m-r)r$.

Hence $\epsilon_r(\tilde{N}) = \sigma(N) + \frac{[F]^2}{m^2}(-2r(m-r))$. ■

PROPOSITION 15.3. $\sigma(\tilde{N}) = m\sigma(N) - [F]^2\left[\frac{m^2-1}{3m}\right]$.

Proof: $$\sigma(\tilde{N}) = \sum_{r=0}^{m-1} \epsilon_r(\tilde{N})$$

$$\therefore \quad \sigma(\tilde{N}) = m\sigma(N) - 2[F]^2 \sum_{r=1}^{m-1} \frac{r(m-r)}{m^2}.$$

Let

$$\Delta = \sum_{r=1}^{m-1} \frac{r(m-r)}{m^2} = \frac{1}{m} \sum_{r=1}^{m-1} r - \frac{1}{m^2} \sum_{r=1}^{m-1} r^2$$

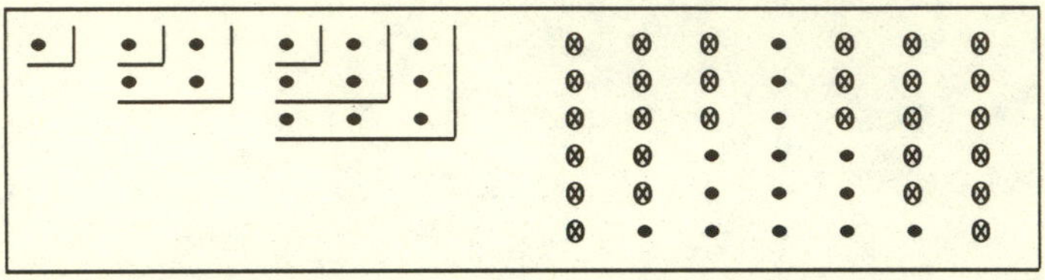

$$\Rightarrow 1^2 + 2^2 + \cdots + n^2 = \frac{1}{3} ((2n+1)(1+2+\cdots+n))$$

$$= \frac{1}{6} n(n+1)(2n+1).$$

$$\therefore \quad \Delta = \frac{1}{m} \left[\frac{(m-1)m}{2}\right] - \frac{1}{m^2} \left[\frac{1}{6}(m-1)(m)(2m-1)\right]$$

$$= \frac{3(m-1)m}{6m} - \frac{(m-1)(2m-1)}{6m}$$

$$\therefore \quad \Delta = (m^2-1)/6m.$$

XVI

AN INVARIANT FOR COVERINGS

Let M be a closed, oriented, 3-manifold and suppose there is a surjective homomorphism $\phi : H_1(M;Z) \to Z_m = Z/mZ$. Let $\tilde{M} \longrightarrow M$ be the corresponding covering space, with the generator of covering translations corresponding to $1 \in Z_m$.

Suppose that for some positive integer n there exists an mn-fold cyclic branched covering of 4-manifolds $\tilde{W} \longrightarrow W$ branched along $F \subset \text{Interior}(W)$ such that $\partial(\tilde{W} \longrightarrow W) = n(\tilde{M} \longrightarrow M)$, and such that the covering translation of $\tilde{W}$, inducing a $2\pi/m$ rotation in fibers normal to $\tilde{F}$, restricts on each component of $\partial\tilde{W}$ to the canonical covering transformation determined by ϕ.

For $0 < r < m$, define the <u>r-signature of (M,ϕ)</u> (also called the α-invariant; see [HNK]) by the formula:

$$\sigma_r(M,\phi) = \frac{1}{n}\left[\sigma(W) - \epsilon_r(\tilde{W}) - 2[F]^2\, \frac{r(m-r)}{m^2}\right].$$

(Recall that $\epsilon_r(\tilde{W})$ is the signature of the w^r-eigenspace $(w = e^{2\pi i/m})$ of $H_2(\tilde{W}) \otimes \mathbb{C}$.) Since, by Proposition 15.2, the right-hand side will vanish for W, a closed manifold, we conclude from the Novikov addition theorem that $\sigma_r(M,\phi)$ is well-defined, depending only upon (M,ϕ) and r.

In order to utilize the r-signature of a 3-manifold M it helps to have conditions under which this branched covering $\tilde{W} \longrightarrow W$ exists.

LEMMA 16.1. *Let* (M,ϕ) *be as above. Suppose* $M = \partial W$ *with* $H_1(W;Z_m) = 0$. *Then* $\tilde{M} \longrightarrow M$ *extends to a branched covering* (m-fold) $\tilde{W} \longrightarrow W$ *over a surface* $F \subset \text{Interior}(W)$ *such that the canonical covering translation of* $\tilde{M}$ *corresponds to rotation through* $2\pi/m$ *on each fiber of the normal bundle of* $\tilde{F} \subset \text{Interior}(\tilde{W})$.

Proof: $\phi \in \text{Hom}(H_1(M);Z_m) \cong H^1(M;Z_m)$. $H_1(W;Z_m) = 0$. Hence there exists a surface $F \subset \text{Interior}(W)$ such that the image in $H_2(W;Z_m)$ of $[F] \in H_2(W;Z)$ is Lefschetz dual to $\delta\phi \in H^2(W,M;Z_m) \cong \text{Hom}(H_2(W,M),Z_m)$. Hence $[F]\cdot x$ (modulo m) $= \delta\phi(x) = \phi(\partial x)$ for all $x \in H_2(W,M;Z)$.

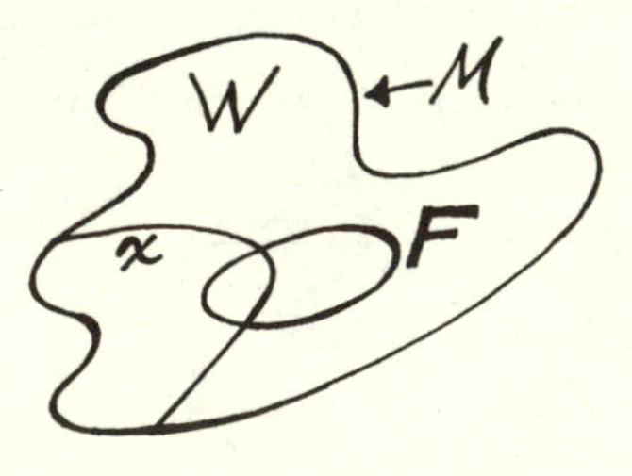

$x \cdot F = \phi(\partial x) \pmod m$

Let $\rho \in H^2(W,W-F;Z_m) \cong \text{Hom}(H_2(W,W-F),Z_m)$ be dual to the fundamental class in $H_2(F;Z_m)$.

$$\begin{array}{ccccccc} & & \psi & \longmapsto & \rho & & \\ 0 = H^1(W) & \longrightarrow & H^1(W-F) & \xrightarrow{\delta} & H^2(W,W-F) & \longrightarrow & H^2(W) \\ \downarrow & & \downarrow & & \downarrow & & \downarrow \\ 0 = H^1(W) & \longrightarrow & H^1(M) & \xrightarrow{\delta} & H^2(W,M) & \longrightarrow & H^2(W) \\ & & \phi & \longmapsto & \delta\phi & & \end{array}$$

Comparing the cohomology exact sequences of the pairs (W,M) and $(W,W-F)$ with Z_m coefficients, we see that $\rho = \delta\psi$ for some $\psi \in H^1(W-F;Z_m) \cong \mathrm{Hom}(H_1(W-F);Z_m)$ which extends ϕ.

Note also that since $[F]\cdot y \pmod m = \delta\psi(y) = \psi(\partial y)$ for all $y \in H_2(W,W-F)$, ψ evaluates to $1 \in Z_m$ on a meridian of F. Thus ψ defines the desired branched covering.

Remark: It is worth noting how the geometry of Alexander duality is working in this argument. Consider the picture one dimension down so that F is a curve in a 3-manifold:

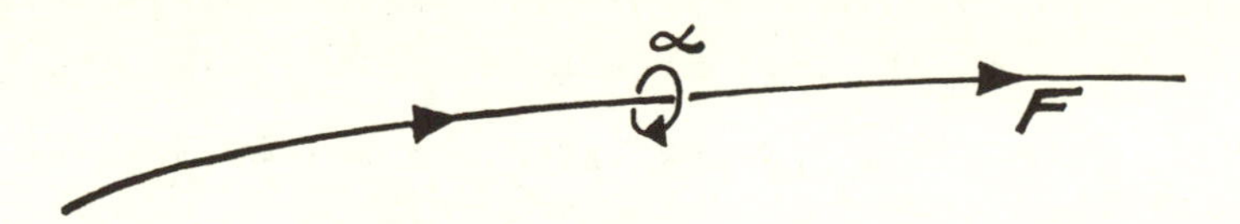

A cohomology element Φ that is dual to F will evaluate $\Phi(\alpha) = 1$ for the meridian α.

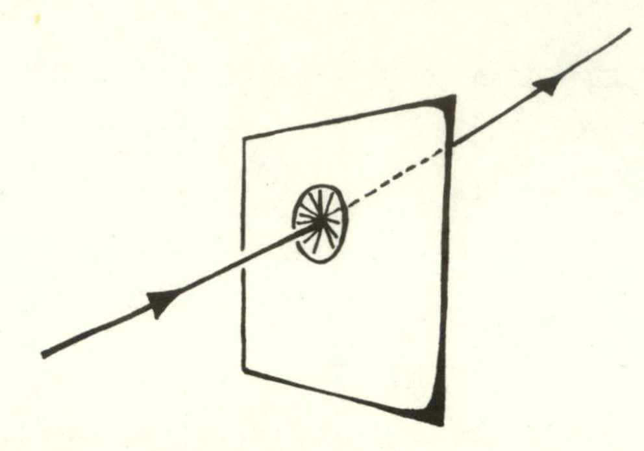

If $\alpha \subset S$ and S is a surface meeting F transversely at a point p, then $\Phi(\alpha) = 1$ corresponds geometrically to the intersection number of F and S at p.

This (pushed up a dimension) is the geometry of the correspondence of ψ and ϕ in the proof.

Remark: Another way to prove this lemma runs as follows: $H^1(X;Z_m) = [X,K(Z_m,1)]$ where $[\ ,\ ]$ denotes homotopy classes of maps and $K(Z_m,1)$ is an Eilenberg-Maclane space having fundamental group Z_m. Since in these applications, X is a finite-dimensional manifold we can replace $K(Z_m,1)$ by $S^{2n+1}/Z_m = \Delta_n(m)$ for n sufficiently large. Take the action of Z_m on S^{2n+1} to be $g(z_0,\cdots,z_n) = (\omega z_0,\cdots,\omega z_n)$ where $\omega = e^{2\pi i/m}$. Then $z_0 = 0$ gives an inclusion of $\Delta_{n-1}(m) \subset \Delta_n(m)$ as a submanifold (of codimension-two). Note that $\pi_1(\Delta_n - \Delta_{n-1}) \cong Z$ and that Δ_{n-1} has a normal disc-bundle, so that if we form the a-fold cyclic branched covering of Δ_n along Δ_{n-1}, then the fibers of this disk bundle will be mapped (from total space to base of the branched covering) as $z \longrightarrow z^a$.

Now, $\phi : H_1(M) \longrightarrow Z_m$ gives $\phi \in H^1(M;Z_m)$, hence $\phi : M \longrightarrow \Delta_n(m)$. That $H_1(W;Z_m) = 0$ means that ϕ extends to a mapping $\psi : W \longrightarrow \Delta_n(m)$. Make ψ transverse regular to $\Delta_{n-1}(m) \subset \Delta_n(m)$ so that $\psi^{-1}(\Delta_{n-1}(m)) = F \subset \text{Interior}(W)$ is a closed submanifold. The desired branched covering is then the pull-back of the m-fold branched covering $\tilde{\Delta}_n \longrightarrow \Delta_n$ of Δ_n along Δ_{n-1}.

$$\begin{array}{ccc} \tilde{W} & \longrightarrow & \tilde{\Delta}_n \\ \downarrow & & \downarrow \\ W & \xrightarrow{\psi} & \Delta_n . \end{array}$$

CALCULATING r-SIGNATURES FROM A SURGERY DESCRIPTION

A framed oriented link L with components $L_1, L_2, \cdots, L_n \subset S^3$ is a <u>surgery description</u> of (M,ϕ) if

(i) M is obtained by surgery on L (with this framing).

(ii) If $\bar{\mu}_i \in H_1(M)$ is the image of the class of a meridian μ_i of L_i, then $\phi(\bar{\mu}_i) = 1 \in Z_m$ for each $i = 1, 2, \cdots, n$.

Remark: What is meant here is that given a link component L_i, say

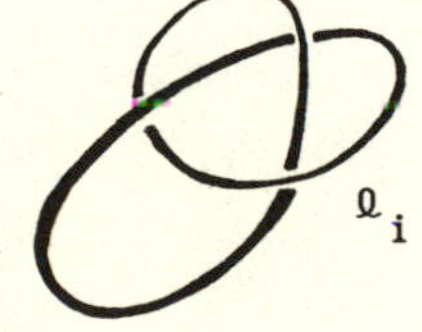

it has an associated integer ℓ_i (the framing). In the corresponding surgery, we cut out a tubular neighborhood of L_i, and sew back a solid torus so that its meridian is sewn to a curve running parallel to the component and linking it ℓ_i times.

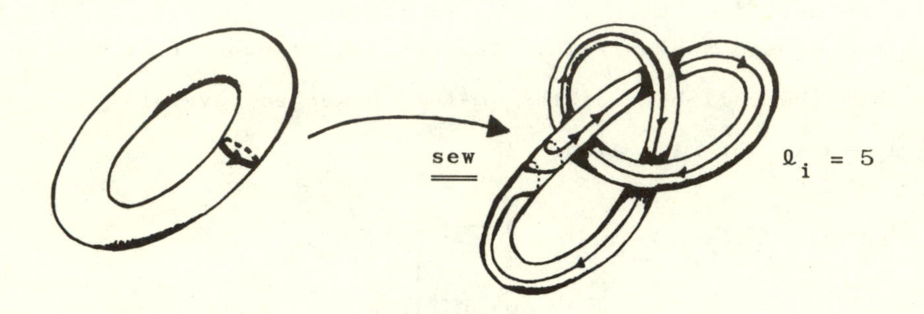

We wish to give a formula for $\sigma_r(M,\phi)$ in terms of this description.

Let $A = (a_{ij})$ be the matrix of linking numbers of the link components (plus framings on the diagonal):

$$a_{ij} = \ell k(L_i, L_j) \quad \text{if} \quad i \neq j.$$

$$a_{ii} = \ell_i, \quad \text{framing integer for} \quad L_i.$$

Let V be a (Seifert) spanning surface for L, and θ the corresponding Seifert matrix. Let $\omega = e^{2\pi i/m}$.

PROPOSITION 16.2. *For* $0 < r < m$,

$$\sigma_r(M,\phi) = \sigma(A) - \sigma((1-\omega^r)\theta + (1-\overline{\omega}^r)\theta') - 2\Big[\sum_{i,j} a_{ij}\Big]\frac{r(m-r)}{m^2}$$

Proof: Let W be the 4-manifold obtained by attaching n handles to the 4-ball B^4 along disjoint tubular neighborhoods of L according to the framing of L. Then W is simply connected, and $\partial W = M$.

We want a surface $F \subset \text{Interior}(W)$ such that

$$[F]\cdot x \pmod{m} = \delta\phi(x) = \phi(\partial x)$$

for all $x \in H_2(W,M)$.

Let $c_i \in H_2(W)$ be the class represented by the core of the ith 2-handle plus a cone on L_i.

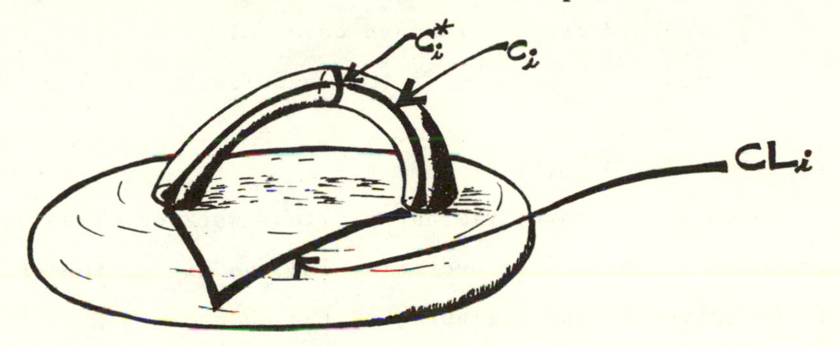

Then $H_2(W)$ is free abelian on $c_1,\cdots,c_n$. And $H_2(W,M)$ is free abelian on the duals $c_1^*,\cdots,c_n^*$ $(c_i\cdot c_j^* = \delta_{ij})$.

If $f = \sum_{i=1}^{n} c_i$, then $f\cdot c_i^* = 1$ for all i.

$\phi(\partial c_i^*) = \phi(\bar{\mu}_i) = 1 \in Z_m$. Therefore $f\cdot x \pmod{m} = \phi(\partial x)$ for all $x \in H_2(W,M)$.

To represent f, let V' be a push-in (to B^4) of the spanning surface V. Then $f = V' \cup$ (cores of the 2-handles) $\subset \text{Interior}(W)$ and represents f.

By the previous lemma, we now have $\tilde{W} \longrightarrow W$, an m-fold cyclic cover branched along F so that $\partial(\widetilde{W} \longrightarrow W) = \widetilde{M} \longrightarrow M$ (this being the cover defined by ϕ).

$$H_2(W) \text{ has intersection form } A.$$

$$\therefore \quad \sigma(W) = \sigma(A).$$

$$[F]^2 = (\sum c_i)^2 = \sum_{i,j} a_{ij}.$$

$\widetilde{W} = \widetilde{B} \cup \widetilde{H}$

$\widetilde{B}$ = m-fold cyclic branched cover of (B^4, V').

$\widetilde{H}$ = m-fold cyclic branched cover of

$(\cup(\text{2-handles}, \cup(\text{cores}))$.

Mayer-Vietoris $\Rightarrow \epsilon_r(\widetilde{W}) = \epsilon_r(\widetilde{B})$.

Hence we are done, via our previous work on eigenspace signatures for branched coverings, and the definition of $\sigma_r(M, \phi)$ given at the beginning of the chapter. ■

XVII

SLICE KNOTS

Let $K \subset S^3$ be a knot. Let q be a fixed prime, and let M_n be the q^n-fold branched cyclic cover of S^3, branched along K $(n = 1,2,3,\cdots)$. As we shall see, $H_*(M_n;Q) \cong H_*(S^3;Q)$.

Suppose we have an epimorphism $\phi : H_1(M) \longrightarrow Z_m$. Since the branched covering projection $M_n \xrightarrow{\pi_n} M_1$ induces a surjection on π_1, hence on H_1, the composition $\phi_n = \phi \circ \pi_n$ induces epimorphisms $\phi_n : H_1(M_n) \twoheadrightarrow Z_m$ for all n.

The main theorem of this section is due to Casson and Gordon [CG].

THEOREM 17.1. *Suppose* K *is a slice knot. Then there is a constant* c, *and a subgroup* G *of* $H_1(M_1)$ *with* $|G|^2 = |H_1(M_1)|$, *such that if* m *is a prime power and* $\phi : H_1(M_1) \longrightarrow Z_m$ *is an epimorphism satisfying* $\phi(G) = 0$, *then* $|\sigma_r(M_n,\phi_n)| < c$ *for all* n.

The proof is preceded by a few lemmas.

LEMMA 17.2. *Let* D *be a 2-disc in* B^4, *and let* V_n *be the* q^n*-fold branched cyclic cover of* (B^4,D), q *prime. Then* $\tilde{H}_*(V_n;Q) = 0$.

Proof: Let $\tilde{X}$ be the infinite cyclic cover of B^4-D. The short exact sequence

$$0 \longrightarrow C_*\tilde{X} \xrightarrow{t^m-1} C_*\tilde{X} \longrightarrow C_*\tilde{V}_n \longrightarrow 0.$$

(Here $m = q^n$, and $\tilde{V}_n$ the unbranched cover.) gives rise to a long exact sequence (following [M2])

$$\cdots \to \tilde{H}_i(\tilde{X};Z_q) \xrightarrow{t^{q^n}-1} \tilde{H}_i(\tilde{X};Z_q) \to \tilde{H}_i(V_n;Z_q) \to \tilde{H}_{i-1}(\tilde{X};Z_q) \to \cdots.$$

Note that V_n has replaced $\tilde{V}_n$. This would make a difference only at

$$\begin{array}{ccccccc} H_1(X;Z_q) & \longrightarrow & H_1(\tilde{V}_n;Z_q) & \longrightarrow & H_0(\tilde{X};Z_q) & \longrightarrow & 0 \\ & & \| & & \| & & \\ & & H_1(V_n;Z_q) \oplus Z_q & & Z_q & & \end{array}$$

Hence, by using reduced homology, we can obtain this replacement. The map t is, of course, the covering translation for the infinite cyclic cover.

Since $V_0 = B^4$, $t-1$ is an isomorphism on homology. Hence, with Z_q coefficients, $t^{q^n}-1 = (t-1)^{q^n}$ is also an isomorphism. Whence $\tilde{H}_*(V_n;Z_q) = 0$. Hence $\tilde{H}_*(V_n;Q) = 0$. ∎

Remark: The same technique shows that $H_*(M_n;Q) \cong H_*(S^3;Q)$ when M_n is the q^n-fold branched cyclic cover of S^3 along K.

LEMMA 17.3. *Let* V *be a* Q*-homology 4-ball. If the image of* $H_1(\partial V) \longrightarrow H_1(V)$ *has order* ℓ, *then* $H_1(\partial V)$ *has order* ℓ^2.

Proof: $H_2(\partial V) \cong H^1(\partial V) \cong \mathrm{Hom}(H_1(\partial V),Z) = 0$. Thus we have an exact sequence

$$0 \longrightarrow H_2(V) \longrightarrow H_2(V,\partial V) \longrightarrow H_1(\partial V) \longrightarrow H_1(V) \longrightarrow H_1(V,\partial V) \longrightarrow 0.$$

Duality and universal coefficients $\Rightarrow |H_2(V)| = |H_1(V,\partial V)|$, $|H_2(V,\partial V)| = |H_1(V)|$. This implies the result. ∎

LEMMA 17.4. *Let* X *be a connected complex. Suppose that* $\pi_1(X)$ *is finitely generated, and that* $H_1(X)$ *is finite and* $H_1(X;Z_p)$ *cyclic for some prime* p. *Let* $\tilde{X} \longrightarrow X$ *be a* p^a*-fold cyclic covering. Then* $H_1(\tilde{X};Q) = 0$.

Remarks: The hypotheses are necessary. For example, let $T : S^2 \times S^1 \longrightarrow S^2 \times S^1$ be the map given by the formula

$$T(x,\lambda) = (-x,\bar{\lambda})$$

where $x \longmapsto -x$ is the antipodal map, and $\lambda \longmapsto \bar{\lambda}$ is complex conjugation. Then $S^2 \times S^1/T = \mathbb{RP}^3 \# \mathbb{RP}^3$. Hence we have a 2-fold covering space

$$\begin{array}{c} S^2 \times S^1 \\ \downarrow \\ \mathbb{RP}^3 \# \mathbb{RP}^3 = B \end{array}$$

with $H_1(B) = Z_2 \oplus Z_2$ while $H_1(S^2 \times S^1) = Z$.

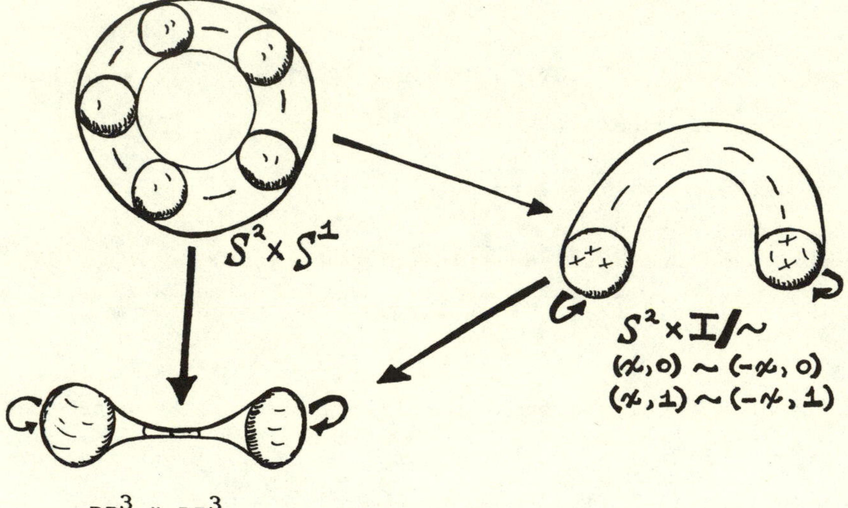

This example corresponds to the algebraic fact that the mapping from Z to $Z_2 * Z_2$ ($*$ denotes free product) defined below is an injection:

$$Z = (c| \ ) \longrightarrow (a,b \mid a^2 = b^2 = 1) = Z_2 * Z_2$$
$$c \longmapsto ab.$$

Similarly, we can make a geometric version for the map $Z \longrightarrow Z_2 * Z_3$ given by;

$$Z = (c|\ ) \longrightarrow (a,b \mid a^2 = 1, b^3 = 1) = Z_2 * Z_3$$
$$c \longmapsto ab.$$

Here $G/G' = Z_6$ when $G = Z_2 * Z_3$. Thus the hypothesis of Lemma 17.4 is again violated.

To prove Lemma 17.4 it suffices to verify the following group-theoretic proposition. (To which we apply the same numbering.)

PROPOSITION 17.4. *Let* G *be a finitely generated group, and let* G' *denote the commutator subgroup of* G. *Suppose that* G/G' *is finite, and that* $(G/G') \otimes Z_p$ *is cyclic for some prime* p. *Let* H *be a normal subgroup of* G *such that* $H \supset G'$ *and* H *has index* p^k *in* G *for some positive integer* k. *Then* H/H' *is finite.*

Proof of Proposition 17.4: Observe that it suffices to prove that $(H/H') \otimes Z_p$ is finite since (H/H') infinite would entail the infinitude of $(H/H') \otimes Z_p$. Therefore, by tensoring with Z_p, it suffices to prove the Goldschmidt Lemma [HS]:

GOLDSCHMIDT LEMMA. *Let* G *be a finitely presented group with* G/G' *cyclic. Let* H *be a subgroup normal in* G *such that* $G' \subset H \subset G$ and H has index $p^r = [G:H]$ *where* p *is a prime. Then* H/H' *is a finite group.*

Proof: By hypothesis, G/H' is finitely generated. (Note that H' is normal in G.) And

$$(G/H')' = G'/H',$$

$$(G/H')/(G/H')' = G/G' \text{ cyclic order } < \infty.$$

And $(G'/H') \subset (H/H') \subset (G/H')$

with $|G/H' : H/H'| = |G:H| = p^r.$

Thus we may assume, inductively, that $H' = 0$ and that H is a finitely generated abelian group.

Now let T be the torsion subgroup of H. Then T is normal in G and $(H/T) \subset (G/T)$ has index p^r. The quotient is abelian, and so $(H/T) \supset (G/T)'$. This shows that we are reduced to considering the case of H free abelian.

Thus we shall suppose that H is free abelian and $H \neq 0$.

LEMMA. *Let P be a finitely generated p-group with P/P' cyclic. Then P is cyclic.*

Proof of Lemma: Let Z denote the center of P. Then $Z \neq 0$ since a p-group has nontrivial center [BM]. Thus P/Z satisfies the hypothesis of the lemma, and has lower order. Therefore we may assume that P/Z is cyclic. This implies that $P = Z$, and hence P is cyclic. ■

To return to the proof of the Goldschmidt Lemma: Consider the diagram below.

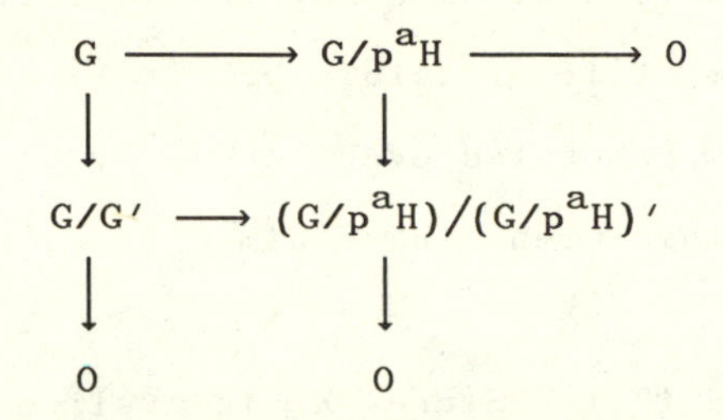

G/G' cyclic $\Rightarrow (G/p^aH)/(G/p^aH)'$ cyclic. Therefore, by the lemma, (G/p^aH) is cyclic. Hence $p^aH \supset G'$ for all a. Since H is free abelian, this implies that $G' = 0$. Hence G is finite. This is a contradiction.

This completes the proof of the Goldschmidt Lemma, and hence the proof of Proposition 17.4. ■

LEMMA 17.5. *Let* X *be a finite connected complex and* $\tilde{X} \longrightarrow X$ *a regular infinite cyclic covering. Let* F *be a field. If* $H_1(X;F) \cong F$, *then* $\dim H_1(\tilde{X};F)$ *is finite.*

Proof: Apply the long exact sequence associated with the exact sequence of chain complexes

$$1 \longrightarrow C_*\tilde{X} \xrightarrow{t-1} C_*\tilde{X} \longrightarrow C_*X \longrightarrow 1.$$

We have

$$\begin{array}{ccccccccc} H_1\tilde{X} & \xrightarrow{t-1} & H_1\tilde{X} & \longrightarrow & H_1\tilde{X} & \xrightarrow{\cong} & H_1\tilde{X} & \longrightarrow & 0 \\ & & & & \| & & \| & & \\ & & & & F & & F & & \end{array}$$

Therefore $H_1\tilde{X} \xrightarrow{(t-1)} H_1\tilde{X}$ is surjective. Hence as a $Z[F]$-module $H_1\tilde{X}$ cannot have any free-cyclic summand, since every element is divisible by $(t-1)$. Consequently, $H_1\tilde{X}$ is finitely generated over $Z[F]$ by elements of finite order. This means that $\dim H_1(\tilde{X};F)$ is finite. ∎

Proof of Theorem 17.1: Since K is a slice knot, $(S^3,K) = \partial(B^4,D)$ for some 2-disc $D \subset B^4$. Let V_n denote the q^n-fold branched cyclic cover of (B^4,D). Thus $\partial V_n = M_n$. By 17.2, $\tilde{H}_*(V_n;Q) = 0$. Let $i_0 : H_1(M_n) \longrightarrow H_1(V_n)$ be induced by inclusion (integral coefficients here). Let $G = \text{Kernel}(i_1)$. By 17.3, $|G|^2 = |H_1(M_1)|$.

This G will be the subgroup referred to in the statement of Theorem 17.1. We must now produce a constant c such that if m is a prime power, and $\phi : H_1(M_1) \longrightarrow Z_m$ is an epimorphism with $\phi(G) = 0$, then $|\sigma_r(M_n,\phi_n)| < c$ for all n. The exact value of the constant will appear at the end of this proof.

Suppose now that $m = p^a$, p prime. Assume that we are given an epimorphism $\phi : H_1(M_1) \longrightarrow Z_{p^a}$ with $\phi(G) = 0$. Since $\phi(G) = 0$, there is an epimorphism $\psi : H_1(M_1) \longrightarrow Z_{p^a}$ making the diagram below commute.

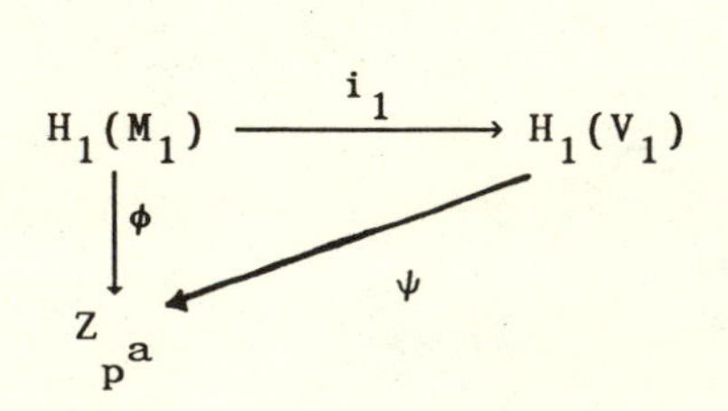

Composing ψ with the epimorphism $H_1(V_n) \longrightarrow H_1(V_1)$ induced by the branched covering projection gives a commutative diagram

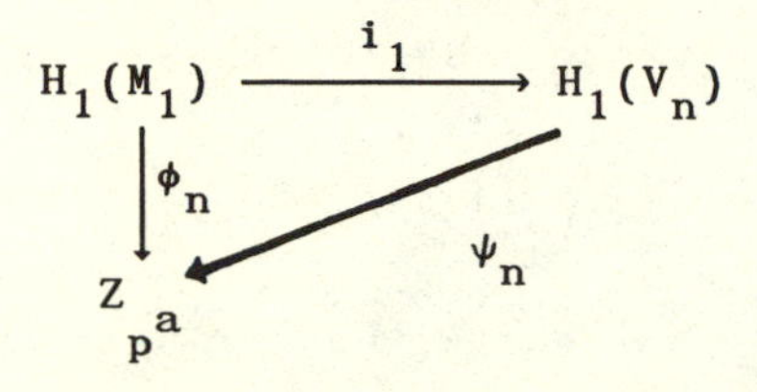

for all n.

Let $d_n = \dim H_1(V_n;Z_p)$. (Note that $H_1(M_n;Z_p)$ is cyclic by Lemma 17.4.) By doing appropriate surgery on (d_n-1) circles in the interior of V_n we may obtain W_n with $H_1(W_n;Z_p)$ cyclic, and a commutative diagram

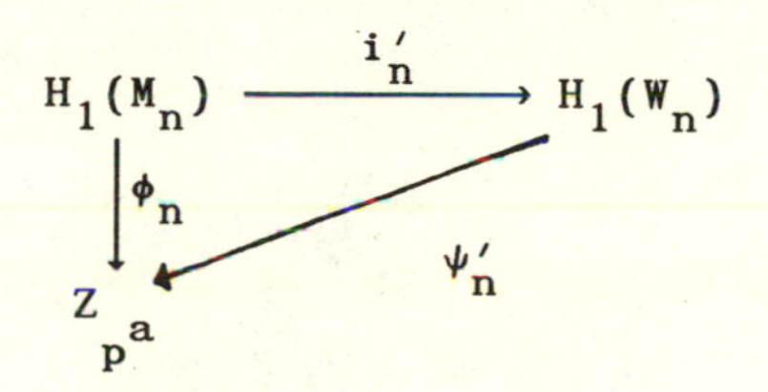

where i'_n is the inclusion, and ψ'_n is surjective. Let $\widetilde{W}_n \longrightarrow W_n$ be the p^a-fold cyclic covering induced by ψ'_n. Then $\partial(\widetilde{W}_n \longrightarrow W_n)$ consists of the p^a-fold cyclic covering $\widetilde{M}_n \longrightarrow M_n$ induced by ϕ_n.

Since $\widetilde{H}_*(V_n;Q) = 0$, the Euler characteristic $\chi(V_n) = 1$. Hence $\chi(W_n) = \chi(V_n) + 2(d_n-1) = 2d_n-1$, giving $\chi(\widetilde{W}_n) = p^a(2d_n-1)$. By 17.4, $H_1(\widetilde{W}_n;Q) \cong 0$. Therefore $H_3(\widetilde{W}_n;Q) \cong H_1(\widetilde{W}_n,\partial\widetilde{W}_n;Q)$ has dimension 0. [$\longrightarrow H_1(\widetilde{W}_n;Q) \longrightarrow H_1(\widetilde{W}_n,\partial\widetilde{W}_n;Q) \longrightarrow H_0(\partial\widetilde{W}_n;Q) \longrightarrow H_0(\widetilde{W}_n;Q)$ is the appropriate part of the exact sequence of the pair, and $H_0(\partial\tilde{W}_n;Q) \longrightarrow H_0(\tilde{W}_n;Q)$ is an isomorphism.]

Hence $\dim H_2(\tilde{W}_n;Q) = p^a(2d_{n-1}-1)-1$. Since signature is unaffected by surgery, $\sigma W_n = \sigma V_n = 0$. Hence

$$|\sigma_r(M_n,\phi_n)| \le p^a(2d_{n-1})-1.$$

Finally, let $\tilde{X}$ denote the infinite cyclic cover of $X = B^4-D$, and let $t : H_1(\tilde{X};Z_p) \longrightarrow H_1(\tilde{X};Z_p)$ be the automorphism induced by the canonical covering translation. Then

$$H_1(V_n;Z_p) \cong \text{Cokernel}(t^{q^n}-1).$$

[See the exact sequemce we used in 17.2.]

In particular, $d_n \le d = \dim H_1(\tilde{X};Z_p)$, which is finite by 4.5. Hence we may set $c = |G|(2d-1)-1$ to complete the proof. ■

Remark: This argument is a generalization of the remark that for any knot $K \subset S^3$, $|\sigma(K)| \le 2\bar{g}(K)$ where $\bar{g}(K)$ denotes the least genus among surfaces in B^4 that span K.

<u>Exercise</u>. Prove the inequality $|\sigma(K)| \le 2\bar{g}(K)$, and use it to show that a $(2,n)$ torus knot has $\bar{g} = (n-1)/2$.

XVIII

CALCULATING σ_r FOR GENERALIZED STEVEDORE'S KNOT

We consider the knots K_k $(k \in Z)$ as shown below:

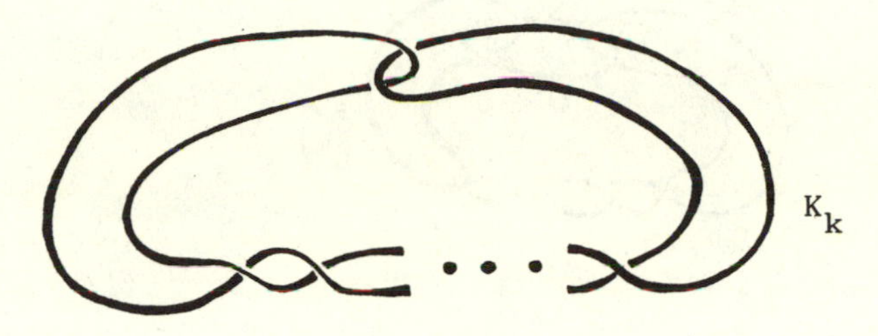

<u>k twists of 2π</u>

(Compare Chapter VIII, Example 8.4.)

K_k has a Seifert surface of genus 1 with corresponding Seifert matrix $\begin{bmatrix} -1 & 1 \\ 0 & k \end{bmatrix}$. Thus, as we showed in Example 8.4 of Chapter VIII, K_k is algebraically slice exactly when $4k+1 = \ell^2$ for some integer ℓ. The first two values give the unknot and the stevedore's knot which are indeed slice. However,

THEOREM 18.1 (Casson and Gordon). K_k *is slice only if* $k = 0,2$.

Proof: If K_k is slice, then it is algebraically slice. Hence, for some fixed k such that $4k+1 = \ell^2$, let M_n be the 2^n-fold branched cyclic cover of (S^3, K_k).

For any divisor m of ℓ we have epimorphisms $\phi : H_1(M_1) \cong Z_{\ell^2} \longrightarrow Z_m$ which necessarily satisfy $\phi(G) = 0$ where $G \subset H_1(M_1)$ has order ℓ. We compute $\sigma_r(M_n, \phi_n)$ (for suitable ϕ) by using the following surgery description:

By doing surgery with framing $+1$ on the unknotted curve J, we obtain S^3 so that the other curve, K, becomes K_k.

By an isotopy of S^3, the above figure becomes:

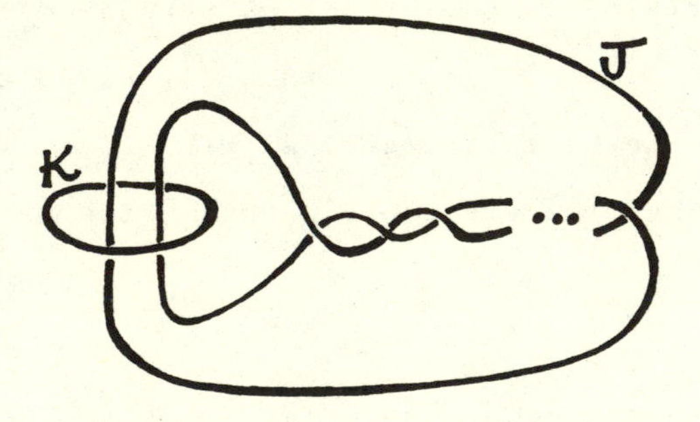

Then M_n, the 2^n-fold branched cyclic cover of (S^3, K_k), is obtained by surgery on the link L consisting of the 2^n lifts of J in the 2^n-fold branched cyclic cover of (S^3, K). The latter is just S^3, and L is illustrated below.

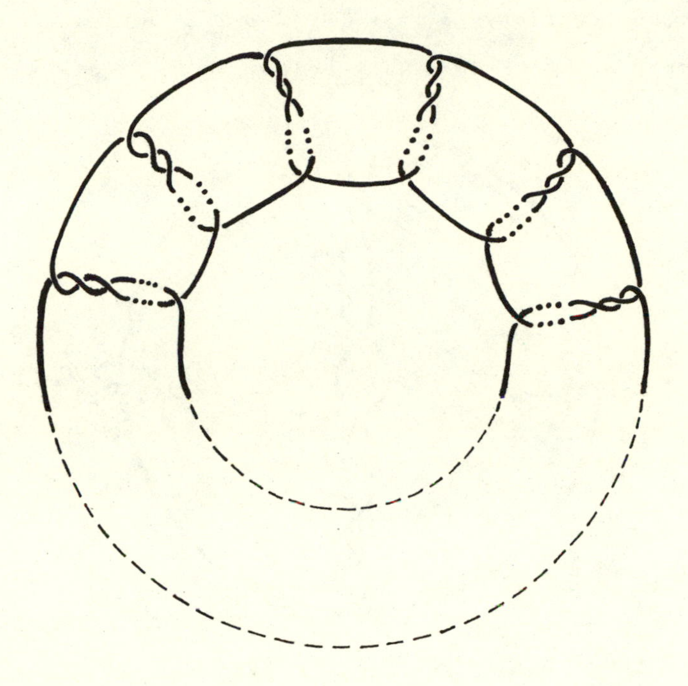

To determine the appropriate framing x of a component L_i of L, choose (temporarily) an equivariant orientation of L. Consider a 2-chain C_i whose boundary is a slightly pushed-off copy of L_i determined by the framing of L_i. This projects to a 2-chain C whose boundary is a similarly defined push-off of J. Consideration of the intersections of $\bigcup_i C_i$ and C with L and J respectively gives, for each i,

$$1 = \text{framing of } J = x + \sum_{j \neq i} \ell k(L_i, L_j)$$

$$\therefore \quad 1 = x - 2k$$

$$\therefore \quad x = 2k+1.$$

We must now consider $\phi_n : H_1(M_n) \longrightarrow Z_m$. M_1 is obtained by surgery on the framed link below:

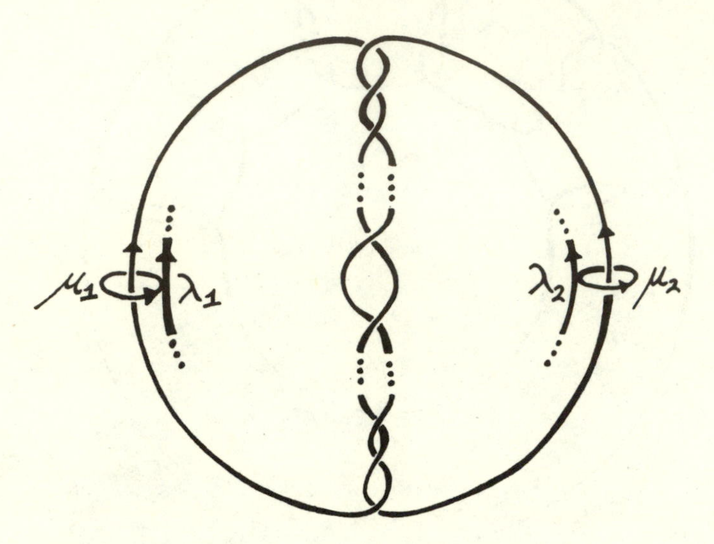

(note the nonequivariant orientation).

Let λ_1, μ_1 and λ_2, μ_2 be oriented longitude-meridian pairs for the two components, where λ_i is chosen to be null-homologous in the complement of the ith component.

The first homology of the complement of the link is free-abelian on μ_1, μ_2, and we have

$$\lambda_1 = 2k\mu_2, \quad \lambda_2 = 2k\mu_1.$$

Surgery has the effect of adding the relations

$$\lambda_1 + (2k+1)\mu_1 = 0, \quad \lambda_2 + (2k+1)\mu_2 = 0.$$

Thus if $\bar{\mu}_i$ denotes the image of μ_i in $H_1(M_1)$, we see that $H_1(M_1)$ is cyclic of order $4k+1 = \ell^2$, generated by $\bar{\mu}_1 = \bar{\mu}_2$. Hence $\phi : H_1(M_1) \longrightarrow Z_m$ can be chosen to satisfy $\phi(\bar{\mu}_1) = \phi(\bar{\mu}_2) = 1$.

More generally, we give the link L which yields M_n, the alternating orientation shown in the figure below.

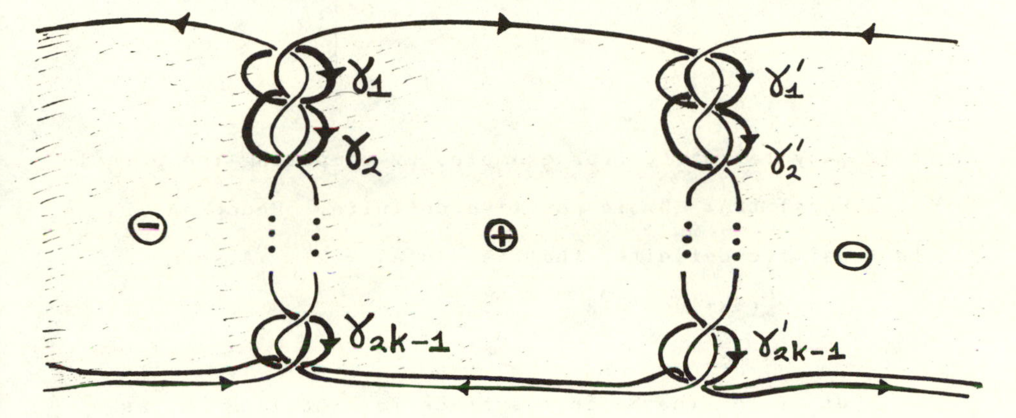

Recalling that ϕ_n is defined to be the composition $H_1(M_n) \longrightarrow H_1(M_1) \xrightarrow{\phi} Z_m$, it follows that we then have $\phi(\bar{\mu}_i) = 1$ for each i, where $\bar{\mu}_i \in H_1(M_n)$ corresponds to the meridian of the ith component L_i of L. Thus L is a surgery presentation for (M_n, ϕ_n) in the sense of Chapter XVI.

The linking matrix A of L is the $2^n \times 2^n$ matrix

$$\begin{bmatrix} 2k+1 & k & 0 & \cdots & 0 & k \\ k & 2k+1 & k & & & 0 \\ 0 & k & 2k+1 & & & \vdots \\ \vdots & & & & & k \\ k & 0 & \cdots & 0 & k & 2k+1 \end{bmatrix}.$$

Note that $A = kB+I$ where B is

$$\begin{bmatrix} 2 & 1 & 0 & \cdots & 0 & 1 \\ 1 & 2 & 1 & & & 0 \\ \vdots & & & \ddots & & \vdots \\ & & & & & 0 \\ 0 & & & & 2 & 1 \\ 1 & 0 & \cdots & 0 & 1 & 2 \end{bmatrix}.$$

It is easy to verify (for example, by computing the principal minors) that B is positive definite. Hence A is also positive definite, that is, $\sigma(A) = 2^n$. Also $\sum_{i,j} a_{ij} = 2^n(4k+1) = 2^n\ell^2$.

Let V be the Seifert surface for the link L as illustrated in the figure on p. 359. The 2^n $(2k-1)$-element sets of which $\{\gamma_1,\cdots,\gamma_{2k-1}\},\{\gamma_1',\cdots,\gamma_{2k-1}'\}$ is a typical pair, together with δ, determine a basis for $H_1(V)$.

In the corresponding Seifert matrix, S_n, $\gamma_1,\cdots,\gamma_{2k-1}$ contribute the $(2k-1)\times(2k-1)$ block

$$C = \begin{bmatrix} -1 & 1 & 0 & \cdots & 0 \\ 0 & -1 & 1 & & \vdots \\ & & -1 & & \\ \vdots & & & & 0 \\ & & & \ddots & 1 \\ 0 & \cdots & & & -1 \end{bmatrix}.$$

and $\gamma_1',\cdots,\gamma_{2k-1}'$ contribute the transpose C^T of C. Thus S_n is the $(2^n(2k-1)+1) \times (2^n(2k-1)+1)$ matrix

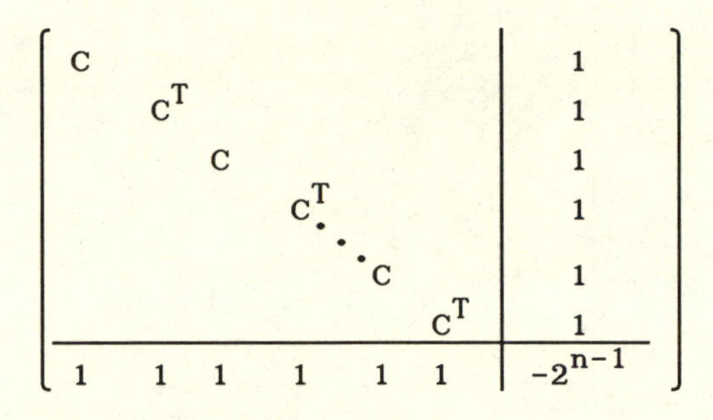

(In the last row (last column) there are 2^n equally spaced 1's separated by zeroes.) Let $\omega = e^{2\pi i/m}$. Write $S_{n,r}$ for the hermitian matrix $(1-\omega^{-r})S_n + (1-\omega^r)S_n^T$, and similarly for C. Then $S_{n,r}$ is

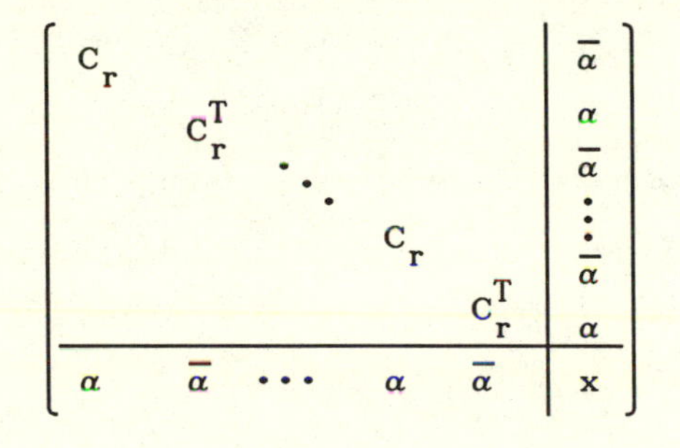

where
$$\alpha = 1-\omega^r$$
$$x = -2^{n-1}\,\alpha\bar{\alpha}.$$

Now choose P such that $PC_rP^* = D$ is diagonal, and let Q be the $(2^n(2k-1)+1) \times (2^n(2k+1)+1)$ matrix

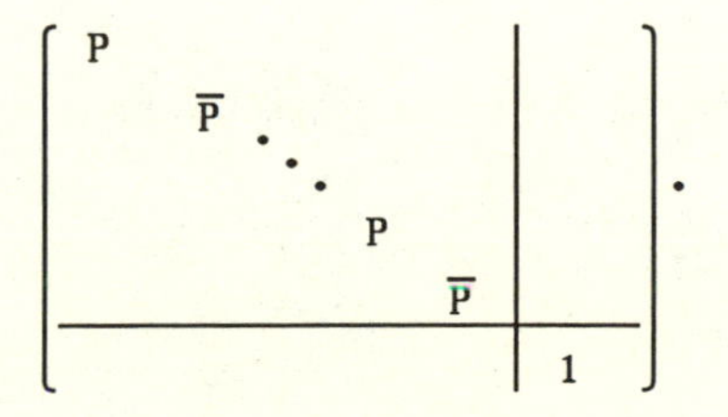.

Then $QS_{n,r}Q^*$ is

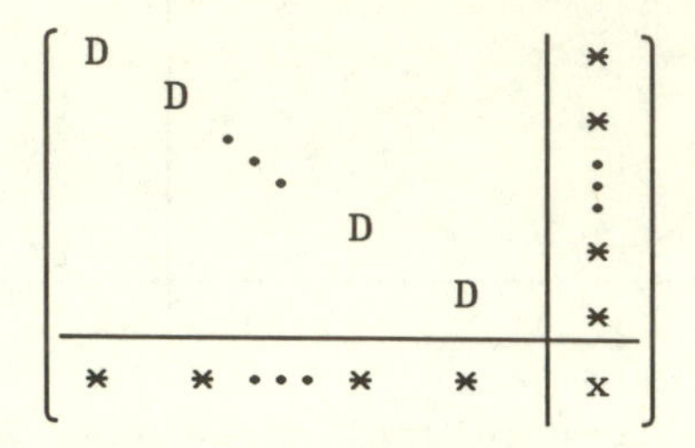

where the entries in the last row (and last column) are periodic with period $2(2k-1)$. We shall see later that C_r is nonsingular. Assuming this, use the diagonal entries of the D's in $QS_{n,r}Q^*$ to clear all the entries (except the last) from the last row and column. Because of the periodicity noted above, this process changes the entry x to $x+2^{n-1}y = 2^{n-1}(y-\alpha\bar{\alpha})$ for some y independent of n. Hence

$$\sigma(S_{n,r}) = 2^n\sigma(C_r) + \eta_r$$

where $|\eta_r| \le 1$ and η_r is independent of n.

By Proposition 16.2, for $0 < r < m$,

$$\sigma_r(M_n,\phi_n) = 2^n-2^n\sigma(C_r)-\eta_r-2^{n+1}r(m-r)(\ell/m)^2.$$

In particular, note the multiplicative relation

$$\sigma_r(M_n,\phi_n)+\eta_r = 2^n(\sigma_r(M_1,\phi)+\eta_r).$$

It follows from Theorem 17.1 that for K_k to be a slice we must have

$$\sigma_r(M_1,\phi)+\eta_r = 0$$

or, equivalently,

$$\sigma(C_r) = 1-2r(m-r)(\ell/m)^2,$$

for every prime power divisor m of ℓ, and every r, $0 < r < m$.

Since $C_{m-r} = \overline{C}_r$, and m is odd, there is no loss of generality in restricting to $0 < r \leq (m-1)/2$. Our theorem will follow easily from

LEMMA 18.2. *Suppose* m *is odd and* $0 < r \leq \frac{m-1}{2}$. *Then* $\sigma(C_r) = -2\left[\frac{2kr}{m}\right]-1$, *where* [ ] *denotes the greatest integer function.*

Proof: Let D_n be the $n\times n$ principal minor of C_r, $n = 1,\cdots,2k-1$. Then, writing $\alpha = 1-\omega^r$ as before, and expanding D_n by (say) the first row, we obtain the difference equation

$$D_n = -(\alpha+\overline{\alpha})D_{n-1}-\alpha\overline{\alpha}D_{n-2}, \quad n = 2,\cdots,2k-1.$$

Since the roots of the corresponding characteristic equation $x^2+(\alpha+\overline{\alpha})x+\alpha\overline{\alpha} = 0$ are $-\alpha$, $-\overline{\alpha}$, the general solution of this difference equation is $(-1)^n(A\alpha^n+B\overline{\alpha}^n)$, where A and B are arbitrary constants. Our initial values $D_0 = 1$, $D_1 = -(\alpha+\overline{\alpha})$ give $A = \alpha/(\alpha-\overline{\alpha})$, $B = -\overline{\alpha}/(\alpha-\overline{\alpha})$, hence $D_n = (-1)^n\left[\frac{\alpha^{n+1}-\alpha^{-n+1}}{\alpha-\overline{\alpha}}\right]$. Write $\alpha = \rho e^{i\theta}$, $\rho > 0$. Then $D_n = (-\rho)^n \frac{\sin(n+1)\theta}{\sin\theta}$. Also, since $\tan\theta = \frac{-\sin 2\pi r/m}{1-\cos(2\pi r/m)} = -\cot(\pi r/m)$ and since $-\pi/2 < \theta < 0$, we have $\theta = \pi r/m-\pi/2$. In particular, we see that $D_{2k+1} = \det C_r \neq 0$, a fact we used earlier. We also see that there are no two consecutive zeros among the D_n, so

$$\sigma(C_r) = (\text{number of permanences of sign}) - (\text{number of changes of sign})$$

in the sequence $D_0,\cdots,D_{2k-1}$ (where the 0's may be assigned either sign). Thus $\sigma(C_r) = 2c-(2k-1)$, where c = number of changes of sign of

$$\sin(n\theta) = \sin(n(\pi r/m-\pi/2)), \quad n = 1,\cdots,2k.$$

Write $r = (m-s)/2$, $1 \le s \le m-2$, s odd. Then

$$\sin(n(\pi r/m-\pi/2)) = -\sin(\pi ns/2m).$$

Hence c = number of changes of sign of $\sin(\pi ns/2m)$, $n = 1,\cdots,2k$,

$$c = \left[\frac{2ks}{m}\right] = \left[k-\frac{2kr}{m}\right] = k-1-\left[\frac{2kr}{m}\right].$$

Therefore, $\sigma(C_r) = 2\left[k-1-\left[\frac{2kr}{m}\right]\right]-(2k-1) = -2\left[\frac{2kr}{m}\right]-1$ as stated. ∎

Returning to the proof of Theorem 18.1, recall that

K_k slice implies $2r(m-r)(\ell/m)^2-1+\sigma(C_r) = 0$ for every prime-power divisor m of $\ell = 4k+1$, and every r, $0 < r < m$. By the lemma, this is equivalent to the condition that for every r, $0 < r \leq \frac{m-1}{2}$,

$$r(m-r)(\ell/m)^2-\left[\frac{2kr}{m}\right]-1 = 0.$$

Replacing $\left[\frac{2kr}{m}\right]$ by

$$\frac{(\ell^2-1)}{2m} = \frac{2kr}{m} > \left[\frac{2kr}{m}\right],$$

it follows that we must have

$$r(m-r)(\ell/m)^2- \frac{(\ell^2-1)r}{2m} -1 < 0.$$

Multiplying by $(2/r)$, we obtain

$$(m-2r)(\ell/m)^2 + 1/m - 2/r < 0,$$

and hence, since $(m/\ell) \leq 1$,

$$m + 1/m < 2(r + 1/r).$$

But putting $r = (m-1)/2$, the value which maximizes $r+1/r$, gives $m^2-4m-1 < 0$, which is clearly violated by (odd) $m > 3$. Moreover, if $\ell > 3$, then ℓ has a prime-power divisor $m > 3$. Hence K_k can be slice only if $\ell = 1,3$, that is, $k = 0,2$. ∎

XIX
SINGULARITIES, KNOTS AND BRIESKORN VARIETIES

A good reference for this section is Milnor's book [M3], *Singular Points of Complex Hypersurfaces*; also the original papers of Pham [PH] and Brieskorn [BK] and the notes by Hirzebruch and Zagier [HZ]. There is a large and continuing literature on this topic. Our intent here is to give a survey of examples and constructions. As we shall see, the subject of the topology of algebraic singularities is intimately related to knot theory and to the structure of branched covering spaces. In the case of the Brieskorn manifolds these ideas come together, so that the link of a Brieskorn singularity may be described completely in terms of knots and branched coverings (Example 19.12 of this chapter). In this sense many constructions of high-dimensional topology, including exotic spheres, may be seen as implicit in, or as generated from the deep three-dimensional knot-work of Alexander and Seifert. Since this early topological work owed much of its impetus to the desire to understand the topology of algebraic varieties, it is fitting that we end our tale of knots and manifolds in this realm.

Let $f(z_0, z_1, z_n) = f$ be a polynomial in $(n+1)$ complex variables. We define the <u>variety of</u> f by $V(f) = \{z \in \mathbb{C}^{n+1} | f(z) = 0\}$. The variety of f is its locus of zeroes.

When $f(0) = 0$ we define the link of f by the equation $L(f) = V(f) \cap S_\epsilon^{n+1}$ where S_ϵ^{2n+1} is a sphere about $0 \in \mathbb{C}^{n+1}$ of radius $\epsilon > 0$. Usually ϵ is chosen very small so that the topology of $L(f)$ and its embedding in S_ϵ^{2n+1} reflects the nature of the variety $V(f)$ at 0. In the most general case the link $L(f)$ will depend upon the choice of ϵ. However, under special conditions (such as an isolated singularity—see below) $L(f)$ will be independent of ϵ for sufficiently small ϵ.

A point $z \in V(f)$ is said to be a singularity of f if $\nabla_f(z)$ vanishes, where $\nabla_f = (\partial f/\partial z_0, \partial f/\partial z_2, \cdots, \partial f/\partial z_n)$ denotes the complex gradient (<u>not</u> the Conway polynomial!). A singularity is <u>isolated</u> if it has a neighborhood in $\mathbb{C}^{n+1}$ containing no other singularities of f. The polynomials $z_0^{a_0} + z_1^{a_1} + \cdots + z_n^{a_n}$, $(a_0, a_1, \cdots, a_n)$ an $(n+1)$-tuple of positive integers greater than or equal to 2, form a collection having isolated singularities at the origin. They will be referred to as <u>Brieskorn polynomials</u>. It is these polynomials that will occupy our attention in this chapter. The Brieskorn polynomials were first studied by Pham [PH] in relation to problems in particle physics.

Pham's calculations generalized earlier key calculations of Lefschetz [LF] for the behavior of $z_0^2 + z_1^2 + \cdots + z_n^2$. Brieskorn utilized Pham's calculations and recognized that the links of these polynomials comprised an extensive class of manifolds, providing, in particular, realizations of many exotic spheres.

DEFINITION 19.1. *Let* $\Sigma(a_0, \cdots, a_n) = L(z_0^{a_0} + \cdots + z_n^{a_n})$ *denote the link of the Brieskorn singularity defined by* $z_0^{a_0} + z_1^{a_1} + \cdots + z_n^{a_n} = 0$.

PROPOSITION 19.2. $\Sigma(a_0, a_1) \subset S^3$ *is a torus link of type* (a_0, a_1).

Proof: By definition, $\Sigma(a_0, a_1)$ is the set of points in $\mathbb{C}^2$ satisfying the equations $z_0^{a_0} + z_1^{a_1} = 0$, $|z_1|^2 + |z_2|^2 = 1$. (We use a sphere of radius one for this demonstration. That the link is independent of the radius is easy to verify for Brieskorn manifolds.) Let $z_0 = re^{i\theta}$, $z_1 = se^{i\phi}$. If r and s are real numbers satisfying $r^{a_0} + s^{a_1} = 0$, $r^2 + s^2 = 1$, then we can obtain further complex solutions via the condition

$$r^{a_0} e^{ia_0\theta} = -s^{a_1} e^{ia_1\phi} \Rightarrow e^{ia_0\theta} = e^{ia_1\phi}.$$

This defines a torus link of type (a_0, a_1) on the torus

parametrized by $(re^{i\theta}, se^{i\phi})$. That the whole link $\Sigma(a_0, a_1)$ arises in this form is left as an exercise for the reader.

Our next result shows how the higher-dimensional Brieskorn manifolds are cyclic branched coverings along lower-dimensional Brieskorn manifolds. Before proving this fact, we set up some useful notation:

Let $\Sigma = \Sigma(a) = \Sigma(a_0, a_1, \cdots, a_n)$ denote the Brieskorn manifold obtained as the link of the singularity $z_0^{a_0} + z_1^{a_1} + \cdots + z_n^{a_n}$. Here $\underline{a}$ is an abbreviation for the (n+1)-tuple $(a_0, \cdots, a_1)$. Let

$$\Sigma_k = \Sigma_k(a) = \Sigma(a_0, a_1, \cdots, a_n, k).$$

Thus $\Sigma(a) \subset S^{2n+1}$ while $\Sigma_k(a) \subset S^{2n+3}$.

PROPOSITION 19.3. *There is a map* $\pi : \Sigma_k \longrightarrow S^{2n+1}$ *exhibiting* Σ_k *as a* k-*fold cyclic branched cover of* S^{2n+1}, *with branch set* Σ.

Remark: It follows from this proposition that all of the Brieskorn manifolds are obtained by forming certain towers of branched coverings in the pattern

$$\begin{array}{ccc} & & \vdots \\ & & \downarrow \\ & \Sigma(a_0,a_1,a_2) \subset S^5 & \\ & \downarrow & \\ \Sigma(a_0,a_1) \subset S^3 & & \end{array}$$

So far, each embedding $\Sigma(a_0,\cdots,a_1) \subset S^{2n+1}$ gives rise to a branched covering manifold, which, by dint of our elementary algebraic geometery, is _itself_ embedded in a sphere so that the construction can continue. In fact, there are topological constructions for the embeddings as well. We will discuss these constructions shortly. Constructions of this type are motivated by the geometry and topology of algebraic singularities.

While we are on the subject of the relation of knot theory and singularities, it is worth remarking that any knot can be regarded as the link of a "singularity" although this is not necessarily algebraic: Given $K \subset S^n$ we have the cone on K, $CK \subset D^{n+1}$. The cone is a topological space with a singularity at the cone point ($CK = \{r\overline{x} \in D^{n+1} | \overline{x} \in K \subset S^n,\ 0 \leq r \leq 1\}$). The cone point is, by definition, the origin in D^{n+1}. This apparently simple remark is the key to amalgamating constructions in knot theory with properties of algebraic singularities.

It also is helpful to sketch immersions into $\mathbb{R}^3$ to see the geometry of the singularity. View the following figure.

An Immersion of CK in $\mathbb{R}^3$

Proof of 19.3: Parametrize $\mathbb{C}^{n+2} = \mathbb{C}^{n+1} \times \mathbb{C}$ as $\{(z_0, z_1, \cdots, z_n, x) = (z, x) \mid z_i \in \mathbb{C},\ x \in \mathbb{C}\}$. Let $f(z) = z_0^{a_0} + z_1^{a_1} + \cdots + z_n^{a_n}$. Then let $F : \mathbb{C}^{n+2} \longrightarrow \mathbb{C}$ be the polynomial $F(z,x) = f(z)+x^k$. Let $V(F)$ denote the variety of F. Thus

$$V(F) = \{(z,x) \in \mathbb{C}^{n+2} \mid f(z)+x^k = 0\}.$$

Define $p : V(F) \longrightarrow \mathbb{C}^{n+1}$ by the formula $p(z,x) = z$. The mapping p exhibits $V(F)$ as a branched covering of $\mathbb{C}^{n+1}$ with branching set $V(f) \subset \mathbb{C}^{n+1}$. We wish to modify this mapping to obtain $\pi : \Sigma_k \longrightarrow S^{2n+1}$.

First consider the restriction of p to Σ_k: $\Sigma_k = \{(z,x) \mid f(z)+x^k = 0,\ |z|^2+|x|^2 = 1\}$, $p : \Sigma_k \longrightarrow \mathcal{S} = p(\Sigma_k) \subset S^{2n+1}$. Since $p(z,x) = z$ we see that for

$\Sigma = \{(z,0)\} \subset \Sigma_k$, $p(z,0) = z$ and $p(\Sigma) = \Sigma \subset \mathbb{C}^{n+1}$. Thus $\Sigma \subset \mathcal{S}$, and Σ_k is a k-fold branched covering of $\mathcal{S}$ with branch locus Σ. It remains to show that $\mathcal{S}$ is ambient isotopic to $S^{2n+1} \subset \mathbb{C}^{n+1}$.

To this end, define an operation of the nonnegative real numbers, $\mathbb{R}^+$, on $\mathbb{C}^{n+1}$ via

$$\rho * z = \left[\rho^{1/a_0} z_0, \rho^{1/a_1} z_1, \cdots, \rho^{1/a_n} z_n\right]$$

for $\rho \in \mathbb{R}^+$, $z \in \mathbb{C}^{n+1}$. Note that $f(\rho * z) = \rho f(z)$.

Note also that $0 \notin \mathcal{S}$ since if $p(z,x) = 0$ then $z = 0$, whence $f(0)+x^k = 0$ whence $x^k = 0$, hence $x = 0$. But $(0,0) \notin \Sigma_k$ and $\mathcal{S} = p(\Sigma_k)$. Therefore, define $E : \mathcal{S} \longrightarrow S^{2n+1} = \{z \in \mathbb{C}^{n+1} \mid |z| = 1\}$ by the formula $E(z) = \rho * z$ for that unique $\rho > 0$ such that $|\rho * z| = 1$. We leave it as an exercise to show that $E : \mathcal{S} \longrightarrow S^{2n+1}$ is a diffeomorphism. Thus we have the diagram

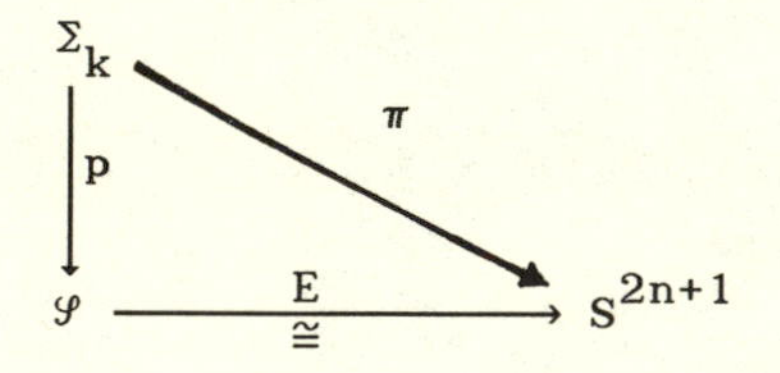

and we define $\pi = E \circ p$. Since Σ is invariant under E, we have shown that Σ_k is a k-fold branched covering of S^{2n+1} along $\Sigma \subset S^{2n+1}$. That it is a cyclic branched cover is also left as an exercise. This completes the proof.

Remark: Proposition 19.3 can be considerably generalized by replacing the directly constructed map $E : \mathcal{S} \longrightarrow S^{2n+1}$ by maps obtained through integrating vector fields. See [DK] and [KN]. In [DK] we show that the link $L(f(z)+x^k)$ is always a cyclic branched cover whenever $f(z)$ has an isolated singularity at the origin.

Example 19.4: Proposition 19.3 tells us that $\Sigma(2,2,2)$ is the 2-fold cyclic cover of S^3 branched along $\Sigma(2,2) \subset S^3$. The latter is the $(2,2)$ torus link, also known as the <u>Hopf Link</u> $\Lambda \subset S^3$.

Λ

Here we have said "the" branched cover, by which we mean the branched cover that corresponds to the representation

$$\begin{aligned} \pi_1(S^3-\Lambda) \cong Z \oplus Z &\longrightarrow Z \longrightarrow Z_k \\ (1,0) &\longmapsto 1 \longmapsto 1 \\ (0,1) &\longmapsto 1 \longmapsto 1 \end{aligned}$$

where $(1,0)$ and $(0,1)$ correspond to meridinal generators oriented positively around respective link components. The link is presented with linking number 1. We leave it as an exercise to show that this representation corresponds to the k-fold cyclic branched covering $\Sigma(2,2,k) \longrightarrow S^3$.

Returning to $\Sigma(2,2,2)$, it is amusing to reformulate this in two related ways:

1. $\Sigma(2,2,2) \cong T$ where T denotes the tangent circle bundle to the two-sphere S^2.

2. $\Sigma(2,2,2) \cong \mathbb{RP}^3$ where $\mathbb{RP}^3$ denotes real projective 3-space.

And also

3. $\mathbb{RP}^3 \cong SO(3)$ the group of orthogonal, orientation preserving linear transformations of $\mathbb{R}^3$.

Thus T, $\mathbb{RP}^3$, $SO(3)$ and $\Sigma(2,2,2)$ are all versions of the same space.

1. We use the algebraic geometry to see that $\Sigma(2,2,2) = T$ as follows:

$\Sigma(2,2,2) = \{(z_0, z_1, z_2) \mid z_0^2+z_1^2+z_n^2 = 0, |z_1|^2+|z_2|^2+|z_3|^2 = 1\}$.

Let $z_i = X_i + \sqrt{-1}\, Y_i$ for $i = 0,1,2$. Then $z_i^2 = (X_i^2 - Y_i^2) + 2\sqrt{-1}\, X_i Y_i$. Hence, letting $\overline{X} = (X_0, X_1, X_2)$, $\overline{Y} = (Y_0, Y_1, Y_2)$ and $\overline{X}\cdot\overline{Y} = X_0Y_0+X_1Y_1+X_2Y_2$, $\|\overline{X}\|^2 = X_0^2+X_1^2+X_2^2$, we have

$$z_0^2 + z_1^2 + z_2^2 = (\|X\|^2 - \|\overline{Y}\|^2) + 2\sqrt{-1}\,(\overline{X}\cdot\overline{Y}).$$

Thus,

$$\begin{aligned}\Sigma(2,2,2) &= \{(\overline{X},\overline{Y}) \in \mathbb{R}^3\times\mathbb{R}^3 \mid \|\overline{X}\|^2 = \|\overline{Y}\|^2, \overline{X}\cdot\overline{Y} = 0, \\ &\qquad \|\overline{X}\|^2+\|\overline{Y}\|^2 = 1\} \\ &= \{(\overline{X},\overline{Y}) \in \mathbb{R}^3\times\mathbb{R}^3 \mid \|\overline{X}\| = \tfrac{1}{2} = \|\overline{Y}\|,\ \overline{X}\cdot\overline{Y} = 0\}.\end{aligned}$$

This is precisely the set of pairs of points on S^2 $(\overline{X}, \|\overline{X}\| = \frac{1}{2})$ coupled with tangent vectors $\overline{Y}$ $(\overline{X}\cdot\overline{Y} = 0)$ of fixed length. Thus $\Sigma(2,2,2) \cong$ the tangent circle bundle to S^2.

2. To see that $\Sigma(2,2,2) \cong \mathbb{RP}^3$ it will suffice, by 19.3, to show that $\mathbb{RP}^3$ is the 2-fold branched covering of S^3 branched along the Hopf Link. To this end, let D^3 denote the unit 3-ball: $D^3 = \{\overline{X} \in \mathbb{R}^3 \mid \|\overline{X}\| \le 1\}$. Let $\tau : D^3 \longrightarrow D^3$ denote an 180° rotation about an axis of D^3 (straight line through the center).

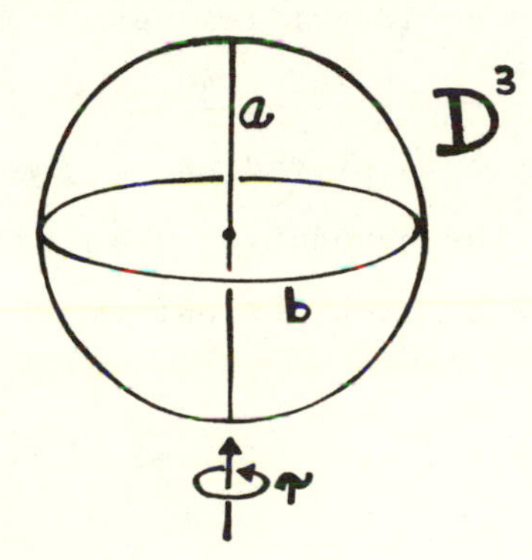

Let $\underline{a}$ denote this axis and $\underline{b}$ denote an equatorial circle on the boundary of D^3.

Now $\mathbb{RP}^3 = D^3/\sim$ where $\overline{x} \sim \overline{x}'$ if and only if $\|\overline{x}\| = \|\overline{x}'\| = 1$ and $\overline{x}' = -\overline{x}$. That is, $\mathbb{RP}^3$ is the 3-ball with antipodal boundary points identified. Since τ preserves antipodal pairs we obtain $\overline{\tau} : \mathbb{RP}^3 \longrightarrow \mathbb{RP}^3$, a map of order two that fixes pointwise $\widetilde{\Lambda} = \overline{a} \cup \overline{b}$ where $\overline{a}$ and $\overline{b}$ are the images of the axis $\underline{a}$ and equator $\underline{b}$ in $\mathbb{RP}^3$.

Note that both $\bar{a}$ and $\bar{b}$ are embedded circles in $\mathbb{RP}^3$. We leave it as an exercise to show that $\mathbb{RP}^3/(p \sim \bar{\tau}p)$ is the three sphere S^3 and that $\bar{a} \cup \bar{b}$ projects to the Hopf link in S^3. This completes the proof that $\Sigma(2,2,2) \cong \mathbb{RP}^3$.

3. One way to see that $\mathbb{RP}^3 \cong SO(3)$ is to prove directly that $SO(3)$ is homeomorphic to $D^3/\sim$. To see this, represent elements of $SO(3)$ by pairs $[\theta,\bar{v}]$ where $0 \leq \theta \leq \pi$ and $\bar{v}$ is a unit vector in $\mathbb{R}^3$. Then $[\theta,\bar{v}]$ represents a rotation of θ about the axis $\bar{v}$ (using the right-hand rule). Note that $[\pi,\bar{v}] = [-\pi,\bar{v}]$, and that otherwise there are no identifications. Then $SO(3) \longrightarrow D_{\pi}/\sim$ via $[\theta,\bar{v}] \longmapsto [\theta\bar{v}]$ shows that $SO(3)$ is homeomorphic to the ball of radius π, modulo antipodal identifications on the boundary.

This completes our tour of points of view on $\Sigma(2,2,2)$.

Example 19.5: Propositions 19.2 and 19.3 taken together show that $\Sigma(a,b,c)$ is

(a) The a-fold branched cover along $\Sigma(b,c)$.

(b) The b-fold branched cover along $\Sigma(a,c)$.

(c) The c-fold branched cover along $\Sigma(a,b)$.

Thus these three spaces are diffeomorphic.

Example 19.6: The Dodecahedral Space. The purpose of this example is to give proof that $\Sigma(2,3,5) = L(Z_0^2+Z_1^3+Z_2^5)$ is the dodecahedral space $\mathfrak{D}$. $\mathfrak{D}$ is a compact orientable three-dimensional manifold whose fundamental group $\hat{G} = \pi_1(\mathfrak{D})$ is the <u>binary dodecahedral group</u>. That is, $\hat{G}$ is a subgroup of $SU(2)$ (which double covers the rotation group $SO(3)$). Let $\pi : SU(2) \longrightarrow SO(3)$ be this double covering. Then $\hat{G} = \pi^{-1}(G)$ where $G \subset SO(3)$ is the dodecahedral subgroup of $SO(3)$. That is, G is the group of rotational symmetries of an icosahedron or a dodecahedral (they are dual) in Euclidean three-dimensional space.

The dodecahedral space is an important example in topology. Its history goes all the way back to Poincaré. In fact, it is the first counterexample to a precursor to the Poincaré conjecture. The precursor would state that a three-manifold M with $H_1(M) = \{0\}$ is the 3-sphere. Dodecahedral space $\mathfrak{D}$ has a <u>perfect</u> but nontrivial fundamental group. Thus $\pi_1(\mathfrak{D}) \neq \{1\}$, but $H_1(\mathfrak{D}) = \{0\}$. Recall that the <u>Poincaré conjecture</u> in dimension three states: <u>A compact connected three-manifold</u> M <u>with</u> $\pi_1(M) = \{1\}$ <u>is homeomorphic to the three-sphere</u> S^3. It remains unproved to this day.

The textbook [ST] by Seifert and Threlfall contains an excellent account of the combinatorial topology of $\mathfrak{D}$. We shall show that $\Sigma(2,3,5) \cong S^3/\hat{G}$ with a natural covering space action of $\hat{G}$ on S^3. See also the book [DV] by DuVal, and the papers [M1] by Milnor and [OW] by Orlik and Wagreich.

First recall the definition of the Lie group $SU(2)$. As a space, $SU(2)$ is diffeomorphic to S^3. In fact, it can be defined as the group of unit-length quaternions. We give the definition in terms of complex valued 2×2 matrices:

$$SU(2) = \left\{\left\{\begin{bmatrix} z & w \\ -\overline{w} & \overline{z}\end{bmatrix}\right\} \;\middle|\; z,w \in \mathbb{C},\ |z|^2+|w|^2 = 1\right\}$$

$$\therefore \quad SU(2) = \left\{A \;\middle|\; \begin{matrix} A \ \text{ is a } 2\times 2 \text{ complex matrix, and} \\ AA^* = I, \quad \mathrm{Det}(A) = 1 \end{matrix}\right\}.$$

Here A^* denotes the conjugate transpose, and I denotes the identity matrix.

Since $S^3 = \{(z,w) \in \mathbb{C}\times\mathbb{C} \mid |z|^2+|w|^2 = 1\}$, it is manifest that $SU(2) \cong S^3$.

Now let $\mathbb{C}^+$ denote the one-point compactification of $\mathbb{C}$. Thus $\mathbb{C}^+ \cong S^2$. We may also describe S^2 as $\mathbb{CP}^1$ via homogeneous coordinates:

$$S^2 \cong \mathbb{CP}^1 = \{\langle z,w\rangle \mid (z,w) \in \mathbb{C}^2-\{0\}\}.$$

Here $\langle z,w\rangle$ denotes the equivalence class of (z,w) where $(z,w) \sim (\lambda z,\lambda w)$ for any nonzero complex number λ. $\mathbb{CP}^1$ is the set of complex lines through the origin in $\mathbb{C}^2$.

Let $\mathcal{L}$ denote the set of <u>linear</u> <u>fractional</u> <u>transformations</u> <u>of</u> C^+. Thus

$$\mathcal{L} = \{T : \mathbb{C}^+ \longrightarrow \mathbb{C}^+ \mid T(\alpha) = (z\alpha+w)/(-\overline{w}\alpha+\overline{z})\}.$$

Here $\begin{bmatrix} z & w \\ -\overline{w} & \overline{z}\end{bmatrix}$ is an element of $SU(2)$. The linear fractional transformations derive from the natural action of $SU(2)$ on $\mathbb{CP}^1$:

> $\begin{bmatrix} a & b \\ c & d \end{bmatrix} \in SU(2)$, $\langle x,y \rangle \in \mathbb{CP}^1 \Longrightarrow$
>
> $\begin{bmatrix} a & b \\ c & d \end{bmatrix}\langle x,y \rangle = \langle ax+by, cx+dy \rangle$. The latter is equal to $(ax+by)/(cx+dy)$ in $\mathbb{C}^+$ where $1/0 = \infty =$ the extra point in the one-point compactification.

Note that $\alpha \in \mathbb{C}^+$ corresponds to $\langle \alpha, 1 \rangle \in \mathbb{CP}^1$ for $\alpha \neq \infty$, and that $\infty \in \mathbb{C}^+$ corresponds to $\langle 1,0 \rangle \in \mathbb{CP}^1$.

Since A and $-A$ in $SU(2)$ give rise to the same element of $\mathcal{L}$, it is easy to see that the map $\pi : SU(2) \longrightarrow \mathcal{L}$ is 2 to 1 and onto. In fact, $\mathcal{L}$ is isomorphic with $SO(3)$. The isomorphism can be made explicit through a specific choice of stereographic projection $St : S^2 \longrightarrow \mathbb{C}^+$ where

$$S^2 = \{(x_1,x_2,x_3) \in \mathbb{R}^3 \mid \|x\| = 1\}.$$

If $G \subset \mathcal{L}$ is a subgroup, let $\hat{G} \subset SU(2)$ denote $\pi^{-1}(G)$.

With these preliminaries completed, we now turn to the action of $SU(2)$ on the ring $R = \mathbb{C}[X,Y]$ of polynomials over $\mathbb{C}$ in two variables. It is through this action that we shall prove that $\Sigma(2,3,5) \cong S^3/\hat{G}$. $SU(2)$ acts on R as follows: Let $\sigma = \begin{bmatrix} a & b \\ -\bar{b} & \bar{a} \end{bmatrix} \in SU(2)$, and let $F(X,Y) \in R$. Then F^σ will denote the result of applying σ to F. F^σ is defined by the formula:

$$F^{\sigma}(X,Y) = F(aX+bY,-\bar{b}X+\bar{a}Y).$$

Given a finite subgroup $\hat{G} \subset SU(2)$, we seek $R^{\hat{G}} = \{F \in R \mid F^{\sigma} = F \ \forall\sigma \in \hat{G}\}$, the ring of polynomials invariant under the action of $\hat{G}$. We shall see that for the binary dodecahedral group $\hat{G}$ there are three generators for R: H_1, H_2, H_3 satisfying the relation $H_1^2+H_2^3+H_3^5 = 0$. Thus $R^{\hat{G}} = \mathbb{C}[Z_0,Z_1,Z_2]/(Z_0^2+Z_1^3+Z_2^5)$ and from this it will follow that $\Sigma(2,3,5) \cong S^3/\hat{G}$. The details follow as below.

First we look at the action of $SU(2)$ on R. Note that if F is a homogeneous polynomial, then so is F^{σ}. (The polynomial F is homogeneous if all single terms have the same total degree $d = i+j$.) Since any polynomial is a sum of homogeneous polynomials, it suffices to determine which homogeneous polynomials are invariant under $\hat{G}$.

Now observe that if $F \in R$ is a homogeneous polynomial, then $F = \prod_{i=1}^{k}(a_iX+b_iY)$ where $a_i,b_i \in \mathbb{C}$. Let this correspond to the following "polynomial" with "roots" in $\mathbb{C}^+ = \mathbb{CP}^1$:

$$F \ \underline{\text{corresponds to}} \ f = \prod_{i=1}^{k}(z-\langle a_i,b_i\rangle).$$

Call f the <u>formal polynomial</u> corresponding to the homogeneous polynomial F. Let $\mathcal{H}$ denote the collection of these formal polynomials, and note that $\mathcal{H}$ is in one-to-one correspondence with the set

$$(\text{homogeneous polynomials in } R)/\sim$$

where $F \sim \lambda F$ for any nonzero complex number λ.

G acts on $\mathcal{H}$ by: Given $g \in G$, let $\sigma \in \hat{G} \subset SU(2)$ be an element projecting to g. Then $f^g = f^\sigma$ where f^σ is the formal polynomial corresponding to F^σ (F corresponds to f). More specifically: If $F = \Pi(a_iX+b_iY)$ and $\sigma = \begin{bmatrix} a & b \\ -\bar{b} & \bar{a} \end{bmatrix}$, then

$$F^\sigma(X,Y) = \Pi(a_i(aX+bY) + b_i(-\bar{b}X+\bar{a}Y))$$

$$\therefore\ F^\sigma(X,Y) = \Pi((aa_i-\bar{b}b_i)X + (a_ib+b_i\bar{a})Y)$$

Consequently, F^σ corresponds to f^σ where

$$f^\sigma = \Pi(z-\langle ba_i+a\bar{b}_i,\ -aa_i+\bar{b}b_i\rangle)$$

$$= \Pi\left[z-\begin{bmatrix} a & -b \\ \bar{b} & a \end{bmatrix}\langle b_i,-a_i\rangle\right]$$

$$\therefore\ f^\sigma = \Pi(z-\sigma^{-1}\langle b_i,-a_i\rangle)$$

$$(f = \Pi(z-\langle b_i,-a_i\rangle))$$

Conclusion: f^σ is obtained from f by transforming the "roots" of f via the inverse of the linear fractional transformation corresponding to σ.

Here is a summary of what we have done so far:

1. If F is homogeneous and invariant under $\hat{G}$, then the corresponding formal polynomial $f \in \mathcal{H}$

is invariant under G.

2. If $f \in \mathscr{H}$ is invariant under G and $F \in R$ corresponds to f, then for any $\sigma \in \hat{G}$, $F^{\sigma} = \lambda F$ for some nonzero complex number λ (depending upon σ). Since we assume G finite, this implies that λ _is a root of unity whose order divides the order of_ σ.

The Moral: In order to study $\hat{G}$-invariant polynomials in R, first study G-invariant formal polynomials in $\mathscr{H}$. _The latter correspond (via the roots) to collections of points in_ S^2 _(or in_ $\mathbb{C}^+$_) that are invariant under the action of_ G.

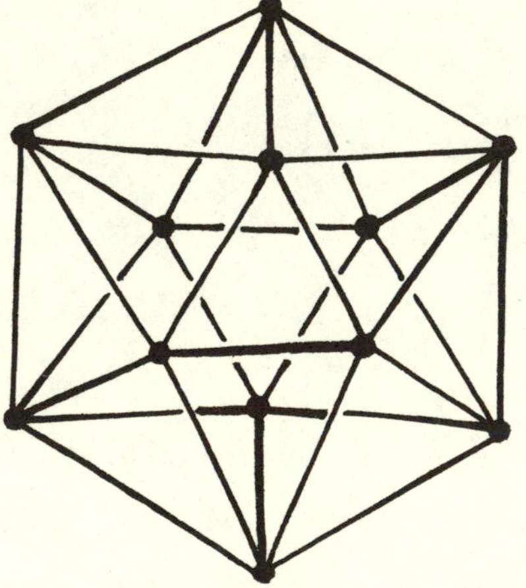

#vertices = V = 12

#edges = E = 20

#faces = FA = 30

The Icosahedron

Let G be the symmetry group of the icosahedron. Then (view the figure above) the icosahedron has $V = 12$

vertices, $E = 20$ edges, and $FA = 30$ faces. Let $\mathscr{V}$ denote the set of vertices, $\mathscr{E}$ the set of midpoints of edges, and $\mathscr{F}$ the set of midpoints of faces of the icosahedron. Then $\mathscr{V}$, $\mathscr{E}$ and $\mathscr{F}$ are G-invariant subsets (of S^2 via radial outward projection from the rectilinear icosahedral form). Any other G invariant subset noncongruent to $\mathscr{V}$, $\mathscr{E}$, or $\mathscr{F}$ will be a full orbit of 60 points.

Let f_1 denote the polynomial in $\mathscr{H}$ whose roots are the set $\mathscr{V}$, f_2 the polynomial with roots $\mathscr{E}$, and f_3 the polynomial with roots $\mathscr{F}$. Let F_1, F_2, F_3 be any three corresponding polynomials in R.

Claim: F_1, F_2 <u>and</u> F_3 <u>are each</u> G <u>invariant.</u>

Proof of Claim: We prove the claim for F_1 and leave the rest for the reader. Since the roots of f_1 are the twelve vertices, it is possible that $\sigma \in G$ may multiply factors of F by a 10th root of unity λ. However, a look at the geometry of the situation shows that 10 roots must then be permuted among themselves by σ and two left fixed. This means that F is multiplied by λ^{10}, hence it is left invariant. Similar considerations hold for the other divisors of 120.

We now make the following

Claim: If F is any homogeneous polynomial in R^G of

degree 60, then $F = K_1F_1^5 + K_2F_2^3$ for some constants $K_1, K_2 \in \mathbb{C}$.

Proof: Let V correspond to the (formal) polynomial f. Then we may choose $p \in S^2 - \mathcal{V} \cup \mathcal{F} \cup \mathcal{E}$, a point in the complement of the special invariant sets, and constants K_1, K_2 such that

$$f(p) = K_1f_1^5(p) + K_2f_2^3(p).$$

Hence (by invariance)

$$(f - K_1f_1^5 - K_2f_2^3)(\sigma p) = 0$$

for all <u>60</u> points $\{\sigma p | \sigma \in G\}$. Therefore $f = K_1f_1^5 + K_2f_2^3$. Thus $F = K_1F_1^5 + K_2F_2^3$ at least up to a constant. This is sufficient to prove the claim.

As a result of this claim, we may choose constants K_1, K_2, K_3 such that if $H_1 = K_1F_1$, $H_2 = K_2F_2$, $H_3 = K_3F_3$, then

$$\underline{H_1^5 + H_2^3 + H_3^2 = 0}$$

Furthermore, we have shown that R^G is generated by H_1 and H_2.

THEOREM 19.7. *Let* $\mathcal{S} = \mathbb{C}[A,B,C]/(A^5+B^3+C^2)$ *be the quotient ring of the ring of polynomials in three variables (with complex coefficients) by the relation* $A^5+B^3+C^2$.

Define a map $\psi : \mathcal{S} \longrightarrow R^G$ *by extending* $\psi(A) = H_1$, $\psi(B) = H_2$, $\psi(C) = H_3$. *Then* ψ *is an isomorphism of rings.*

Proof: ψ is onto, and we know that $\dim_{\mathbb{C}} R^G = 2$ (since there is no relation between H_1 and H_2). Therefore the ideal $(A^5+B^3+C^2)$ must in fact be the kernel of ψ. Otherwise the dimensions would not compare. ∎

(Compare with [KL].)

Now let $V = V(Z_1^5+Z_2^3+Z_3^2) \subset \mathbb{C}^3$ be the Brieskorn Variety (5,3,2). And define $\phi : \mathbb{C}^2 \longrightarrow V$ by the map

$$\phi(\alpha) = (H_1(\alpha), H_2(\alpha), H_3(\alpha)).$$

PROPOSITION 19.8.

(1) ϕ *is surjective.*

(2) *If* $v \in V$, *then* $\phi^{-1}(v)$ *is an orbit under the action of* G *on* $\mathbb{C}^2$.

(3) $V \cong \mathbb{C}^3/G$.

(4) $\Sigma(5,3,2) \cong S^3/G$.

Proof: Using

$$\begin{array}{ccc} \psi : \mathbb{C}[A,B,C]/(A^5+B^3+C^2) & \longrightarrow & \mathbb{C}[X,Y]^G \\ \| & & \| \\ \mathcal{S} & & R^G \end{array}$$

and the inclusion $R^G \overset{i}{\subset} R$, it suffices to prove that the induced map on ring spectra $\text{Spec } R \longrightarrow \text{Spec } \mathcal{S}$ is surjective (for 1)). (See [SH] for algebraic geometery background.) However, it is easy to see that R is a finitely generated integral ring extension of $R^{\hat{G}}$. Hence $\mathrm{Spec}(R) \longrightarrow \mathrm{Spec}(R^{\hat{G}})$ is finite to one. Therefore $2 = \dim_{\mathbb{C}} R = \dim_{\mathbb{C}} R^{\hat{G}}$. Since the dimension of $\mathcal{S}$ is also 2, and $\psi : \mathcal{S} \longrightarrow R^{\hat{G}}$ is an isomorphism, we see that

$$\mathrm{Spec}\ R \twoheadrightarrow \mathrm{Spec}\ R^{\hat{G}} \xrightarrow{\cong} \mathrm{Spec}\ \mathcal{S}$$

so that $\mathrm{Spec}\ R \longrightarrow \mathrm{Spec}\ \mathcal{S}$ is surjective. This translates via the Nullstellensatz [SH] to the statement that $\phi : \mathbb{C} \longrightarrow V$ is surjective.

For the second part it is necessary to show that $\phi(\alpha) = \phi(\alpha') \Longrightarrow \alpha' = \hat{g}\alpha$ for some $\hat{g} \in \hat{G}$. Since we may assume that α,α' are not zero, let $\overline{\alpha},\overline{\alpha}'$ denote the corresponding elements of $S^2 = \mathbb{CP}^1$. Similarly, let g be the element of $SO(3)$ corresponding to $\hat{g}$. Then from $\phi : \mathbb{C}^2 \longrightarrow V$ we obtain $\overline{\phi} : \mathbb{CP}^1 \longrightarrow (V-\{0\})/\mathbb{C}^*$ ($\mathbb{C}^*$ = the nonzero complex numbers). Now $\overline{\phi}(\overline{\alpha}) = \overline{\phi}(\overline{\alpha}')$ implies that all nonzero formal polynomials in $\mathbb{C}[z]^G$ take the same values on $\overline{\alpha}$ and $\overline{\alpha}'$. Let $f(z) = \prod_{g\in G}(z-g\overline{\alpha})$. Then $f(\overline{\alpha}) = 0$ and hence $f(\overline{\alpha}') = 0$. Thus $\overline{\alpha}' = g\overline{\alpha}$ for some $g \in G$. Transferring to $\hat{G}$ we conclude that $\lambda\hat{g}\alpha = \alpha'$ for some $\lambda \in \mathbb{C}^*$. Hence $\phi(\lambda\alpha) = \phi(\alpha)$ for some $\lambda \in \mathbb{C}^*$.

Thus we are reduced to showing that $\phi(\lambda\alpha) = \phi(\alpha)$ implies that $\lambda\alpha = \hat{h}\alpha$ for some $\hat{h} \in G$. Now $\phi(\lambda\alpha) = \phi(\alpha)$

means that

$$H_1(\alpha) = H_1(\lambda\alpha) = \lambda^{30}H_1(\alpha)$$

$$H_2(\alpha) = H_2(\lambda\alpha) = \lambda^{20}H_2(\alpha)$$

$$H_3(\alpha) = H_3(\lambda\alpha) = \lambda^{12}H_3(\alpha).$$

Consider the various cases:

Case 1: $H_1(\alpha)$, $H_2(\alpha)$ and $H_3(\alpha)$ all nonzero. Then $\lambda^{30} = \lambda^{20} = \lambda^{12} = 1$. Hence $\lambda^2 = 1$. Thus $\lambda = \pm 1$. Since -1 is an element of $\hat{G}$, we conclude that $\phi(\alpha) = (-\hat{g})\alpha$, as desired.

Case 2: If $H_1(\alpha) = 0$, while $H_2(\alpha)$ and $H_3(\alpha)$ are both nonzero, then $\lambda^{10} = \lambda^{12} = 1$, hence $\lambda^4 = 1$. However, $H_1(\alpha) = 0$ implies that α is a midpoint of an edge of the icosahedron. There is an order two symmetry g $(g^2 = 1)$ that rotates by 180° about an axis passing through the midpoints of opposite edges. Therefore $\hat{g}^2 = -1$ and $\hat{g}$ has order four. (That is, $\hat{g}$ exists.) Consequently, we can realize the fourth root of unity with $\hat{g} \in G$ as desired.

The other cases follow by similar geometry. This proves part (2). Part (3) follows from parts (1) and (2). Finally, to see part (3) use the same argument as in the proof of 19.3 to slide points onto the standard sphere. This completes the proof. ■

Note that it follows from our discussion that the dodecahedral space is obtained as the 2-fold branched covering M of S^3 with branch set a $(3,5)$ torus knot. It is a good exercise to show that $\pi_1(M) \cong \hat{G}$, and a more challenging exercise to show directly that $M_2(K_{3,5})$ and $S^3/\hat{G}$ are homeomorphic (even diffeomorphic) manifolds!

Exercise. To prove that $\hat{G}$ is perfect:

(i) Let G be the symmetry group of the icosahedron, $G \subset SO(3)$. Show that G is isomorphic to A_5, the group of even permutations on five letters. (HINT: Represent the five letters a,b,c,d,e as collections of four faces such that no two faces in any collection have edges or vertices in common.)

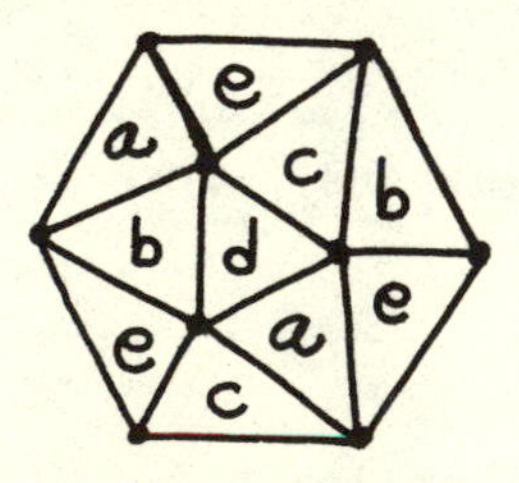

(ii) Show that A_5 is perfect. (Show that every element of A_5 is a product of 3-cycles, and that every 3-cycle is a commutator.)

(iii) Show that if S^3 = unit quaternions and if $u, v \in S^3$ such that $u^2 = v^2 = -1$ so that u and v are unit vectors in $S^3 \subset \mathbb{R}^3 = \{ai+bj+ck\}$ $(a,b,c \in \mathbb{R})$ with $u \perp v$ ($\perp$ denotes euclidean perpendicularity), then $uvu^{-1}v^{-1} = -1$. Thus -1 is a commutator in S^3. Show that -1 is a commutator in G. (*Hint:* This corresponds to finding two 180° rotations of the icosahedron having perpendicular axes.)

A few comments about the quaternions are germane to this last exercise. We regard

$$\mathbb{R}^4 = \{t+ai+bi+ck \mid t,a,b,c \in \mathbb{R}\}.$$

Quaternionic multiplication on $\mathbb{R}^4$ is generated by the identities $i^2 = j^2 = k^2 = ijk = -1$. (Plus associativity and distributivity.) The <u>pure quaternions</u> $\mathbb{R}^3 = \{ai+bj+ck\}$ constitute euclidean three-space, and the unit sphere $S^2 = \{ai+bj+ck = u \mid a^2+b^2+c^2 = 1\}$ has the property that $u \in S^3$ if and only if $u^2 = -1$. Thus any quaternion $g \in S^3$ can be written as $g = e^{u\theta} = \cos(\theta) + u\sin(\theta)$ where $0 \leq \theta \leq 2\pi$ and $u \in S^2$. We define $\pi : S^3 \longrightarrow SO(3)$ by the map $\pi(f)(v) = gv\overline{g}$ where $\overline{g} = e^{-u\theta}$. It is not hard to see that $\pi(g)$ <u>is a rotation about the axis</u> u <u>by the angle</u> 2θ. This is the quaternionic version of the double covering of $SO(3)$ by $SU(2)$.

Example 19.9. The Milnor Fibration: In [M3], Milnor proves the following theorem.

FIBRATION THEOREM. *Let* $f : \mathbb{C}^{n+1} \longrightarrow \mathbb{C}$ *be a complex polynomial mapping with an isolated singularity at the origin. Let* $K = V(f) \cap S_\epsilon^{2n+1}$ *denote the link of the singularity. Then* $\phi : S_\epsilon^{2n+1} - K \longrightarrow S^1$ *is a smooth fibration, where the mapping* ϕ *is defined by* $\phi(z) = f(z)/|f(z)| = \arg(f(z))$.

Thus, links of isolated singularities have fibered complements. At this stage it is worth generalizing the term <u>knot</u> to denote <u>any codimension two smooth submanifold of a sphere</u>. Thus Milnor's theorem is that links of isolated singularities are fibered knots.

In particular, the fibration theorem states that the map

$$S^3 - \Sigma(a,b) \xrightarrow{\phi} S^1$$

$$\phi(z_0, z_1) = \arg(z_0^a + z_1^b)$$

gives the fiber structure for the (a,b) torus knot (or link if $\gcd(a,b) > 1$). Recall that we have explained the geometry of a fiber structure for $S^3 - \Sigma(a,b)$ in Exercise 13.17.

In this example we see how the Fibration Theorem works in the case of the Brieskorn varieties. The reader is referred to Milnor's book for the full theorem.

Let $f(z) = z_0^{a_0} + z_1^{a_1} + \cdots + z_n^{a_n}$ and view this polynomial as a mapping $f : \mathbb{C}^{n+1} \longrightarrow \mathbb{C}$. Let $\mathbb{C}^*$ denote $\mathbb{C}-\{0\}$ and let $W = f^{-1}(\mathbb{C}^*) \subset \mathbb{C}^{n+1}$. Our first assertion is

LEMMA 19.10. $W \xrightarrow{f} \mathbb{C}^*$ *is a smooth fiber bundle.*

Proof: In order to prove this lemma we must examine how to locally trivialize the mapping. Recall that we have defined $\rho * z = (\rho^{1/a_0} z_0, \rho^{1/a_1} z_1, \cdots, \rho^{1/a_n} z_n)$ for positive real numbers ρ. Note that $f(\rho * z) = \rho f(z)$. For $0 \leq \theta \leq 2\pi$ we define $h_\theta(z)$ by the formula

$$h_\theta(z) = (\omega_0^\theta z_0, \omega_1^\theta z_1, \cdots, \omega_n^\theta z_n)$$

where $\omega_k = e^{i/a_k}$. We see that $\rho * h_\theta : f^{-1}(z) \to f^{-1}(\rho e^{i\theta} z)$. Thus these maps can be used to produce the local trivialization.

We now restrict the bundle of Lemma 19.10 to produce another bundle that is relevant to fibering the complement of $\Sigma(a_0, a_1, \cdots, a_n)$. Let $E_{\delta,\epsilon}$ denote the set defined below:

$$E_{\delta,\epsilon} = \{z \in \mathbb{C}^{n+1} \mid |f(z)| = \delta,\ |z| \leq \epsilon\}.$$

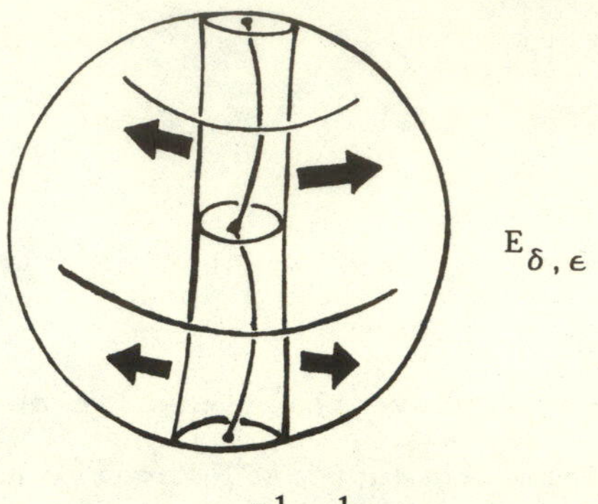

$$E_{\delta,\epsilon} = f^{-1}(S^1_\delta) \cap D_\epsilon$$

If we choose $0 < \delta << \epsilon$, then it is easy to see that $f : E_{\delta,\epsilon} \longrightarrow S^1_\delta$, $z \longmapsto f(z)$ is a C^∞-fiber bundle. Since $E_{\delta,\epsilon}$ sits inside the ball D^{2n+2}_ϵ of radius ϵ we see that its boundary is the boundary of a tubular neighborhood of $\Sigma = \Sigma(a_0, a_1, \cdots, a_n) \subset S^{2n+1}_\epsilon$. In fact, $E_{\delta,\epsilon}$ deforms to give a fiber structure on the complement of this tubular neighborhood. The deformation involves expanding points of $E_{\delta,\epsilon}$ via $z \longrightarrow \rho * z$ for ρ such that $|\rho * z| = \epsilon$. This creates a fiber structure $S^{2n+1}_\epsilon - N(\Sigma) \xrightarrow{\phi} S^1$ via $\phi(z) = \arg(f(z))$. Here $\partial N(\Sigma) = E_{\delta,\epsilon} \cap S^{2n+1}_\epsilon$. This gives most of Milnor's theorem for this special case. He manages to fiber the full complement by more careful analysis.

It is worth understanding the geometery of this fibration in more detail. We may take $\epsilon = 1$ and note that the fibers of $\phi : S^{2n+1} - N(\Sigma) \longrightarrow S^1$ are deformation retracts of $\{z \in \mathbb{C}^{n+1} | f(z) = 1\} = F$.

Now $F = \{z \in \mathbb{C}^{n+1} \mid z_0^{a_0} + z_1^{a_1} + \cdots + z_n^{a_n} = 1\}$, and F has as deformation retract a wedge of $(a_0-1)(a_1-1)\cdots(a_n-1)$ spheres of dimension n. This occurs as follows:

(1) $F \supset \{\bar{r} \in \mathbb{R}^{n+1} | r_0^{a_0} + r_1^{a_1} + \cdots + r_n^{a_n} = 1,\ r_i \geq 0\} = \mathfrak{R}$.

(2) F is invariant under multiplication of the ith coordinate by an a_i^{th} root of unity.

(3) Thus $F \supset \{\bar{r} \in \mathfrak{R}\} \bullet (\Omega a_0 \times \Omega a_1 \times \cdots \times \Omega a_n)$ where Ωa_i = group of a_i^{th} roots of unity. That is, $F \supset \{(r_0\omega_0, \cdots, r_n\omega_n) \mid \bar{r} \in \mathfrak{R},\ \omega_i \in \Omega a_i\} = \mathcal{S}$.

It is good exercise to show

(a) this last set $\mathcal{S}$ is a deformation retract of F.

(b) $\mathcal{S} \cong \Omega a_0 * \Omega a_1 * \cdots * \Omega a_n$ (where $*$ denotes join)

$$\cong \underbrace{(S^0 \vee \cdots \vee S^0)}_{(a_0-1)} \vee \cdots \vee \underbrace{(S^0 \vee \cdots \vee S^0)}_{(a_n-1)}$$

$$\cong \bigvee_{(a_0-1)(a_1-1)\cdots(a_n-1)} S^n .$$

See [BK] for more details.

Here is a visualization for the two-variable case: The fiber is $F : z_0^{a_0} + z_1^{a_1} = 1$. Define $\pi : F \longrightarrow \mathbb{C}$ by $\pi(z_0, z_1) = z_1$. Then π is a branched covering of the complex plane branched along the a_1^{st} roots of unity. We

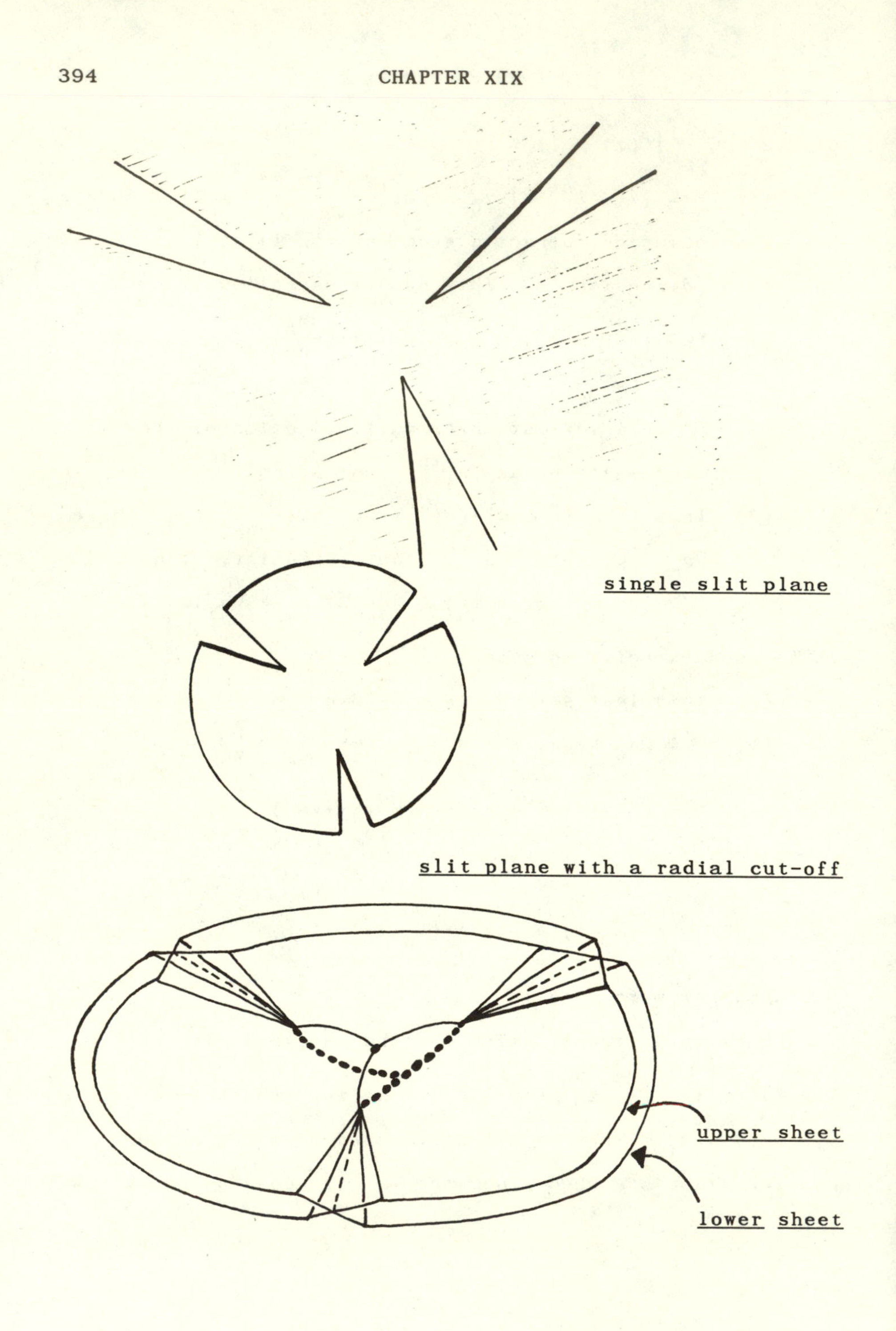
single slit plane
slit plane with a radial cut-off
upper sheet
lower sheet

can see F by creating a cut-and-paste picture of the branched covering. This is obtained by slitting the complex plane along rays emanating from each root of unity. For example, take $F : z_0^2 + z_1^3 = 1$. This surface construction is illustrated in the figure on p. 394. Note that $\Omega_2 * \Omega_3$ appears as the form

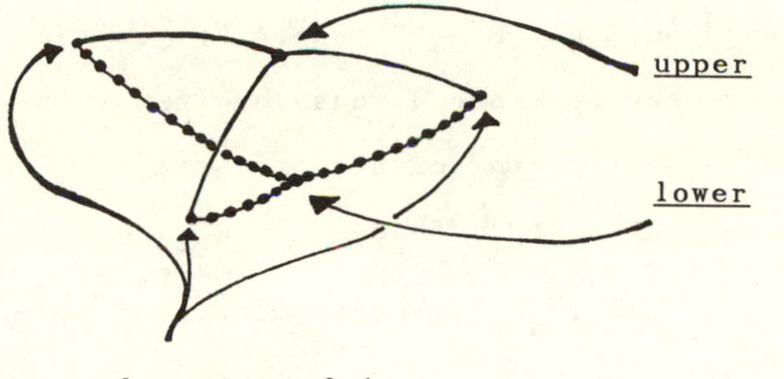

Note also how a projection of the (2,3) torus knot appears in the boundary of this representation:

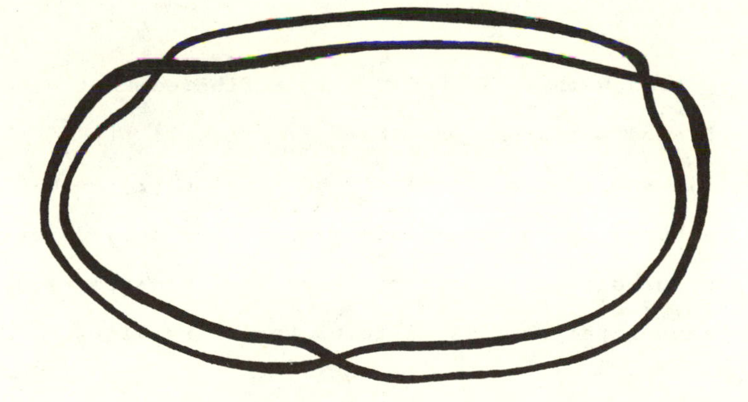

These same patterns hold true for the more general case of

$z_0^{a_0} + z_1^{a_1} = 1$.

If we replace $z_0^{a_0} + z_1^{a_1} = 1$ by $z_0^{a_0} + z_1^{a_1} = \delta$ then, as δ approaches 0, the $\Omega a_0 * \Omega a_1$ part shrinks to a point, until at 0, F has degenerated to the cone on the (a_0, a_1) knot.

The structure of the fibration $\phi : S^{2n-1} - N(\Sigma) \longrightarrow S^1$ is given by the monodromy $h : F \longrightarrow F$ where F is the fiber. It is easy to see from our discussion that this monodromy consists in multiplying each coordinate by the corresponding root of unity. Thus

$h(z_0, z_1, \cdots, z_n) = (\omega_0 z_0, \cdots, \omega_n z_n)$ where $\omega_i = e^{2\pi_i/a_i}$.

This means that $S^{2n+1} - N(\Sigma)$ is diffeomorphic to $F \times I/(h(x),0) \sim (x,1)$. From this description it is possible to compute many things—including the Alexander polynomial of $\Sigma \subset S^{2n+1}$.

Exercise. Show that if $K \subset S^3$ is a fibered knot with fiber F and monodromy $h : F \longrightarrow F$, then $\Delta_K(t) \doteq \mathrm{Det}(H - tI)$ where H = the matrix of $h_* : H_1(F) \longrightarrow H_1(F)$ for some basis of this homology group. Use this description and our discussion of Brieskorn manifolds to recompute the Alexander polynomials of torus knots and links.

Example 19.11: The Empty Knot. The simplest Brieskorn polynomial is $f(z_0) = z_0^{a_0}$. Here $f : S^1 \longrightarrow S^1$, $z_0 \mapsto z_0^{a_0}$ is the Milnor fibration. The "knot" $\Sigma(a_0)$ is the empty set! Nevertheless, this knot has a fiber, and it has a Seifert pairing with respect to this fiber. We calculate the pairing.

Let $F(a_0)$ be the fiber of the Milnor fibration for this empty knot. Then

$$F(a_0) = f^{-1}(1)$$
$$= \{\omega \in S^1 \mid \omega^{a_0} = 1\}$$

$\therefore$ $F(a_0) = \Omega a_0$, the set of $a_0{}^{th}$ roots of unity.

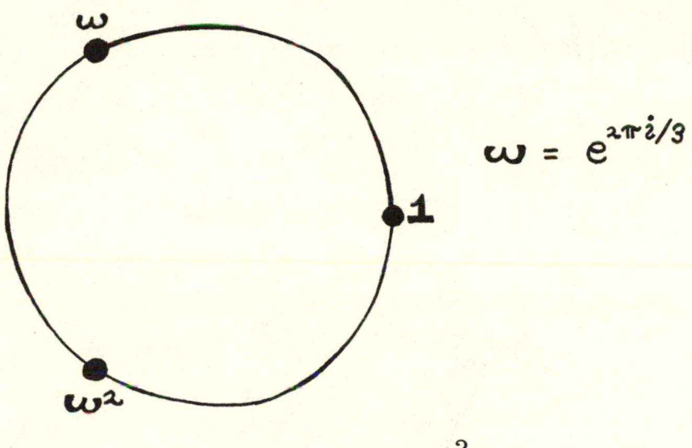

$F(3) = \Omega_3 = \{1, \omega, \omega^2\}$

The Milnor Fiber for the Empty Knot of Degree Three

By letting $\omega = e^{2\pi i/a_0}$ denote an $a_0{}^{th}$ root of unity, we have that $\Omega a_0 = \{1, \omega, \omega^2, \cdots, \omega^{a_0 - 1}\}$ represents the "spanning surface" for the empty knot of degree a_0. The Seifert pairing is defined in reduced homology:

$$\theta a_0 : \tilde{H}_0(\Omega a_0) \times \tilde{H}_0(\Omega a_0) \longrightarrow Z.$$

We may take the push-off in the normal direction to Ωa_0 to be generated by a small counter-clockwise rotation of S^1. Let x^* denote the result of so pushing a chain x.

Note that the generators for $\tilde{H}_0(\Omega a_0)$ are $\langle 1-\omega\rangle$, $\langle \omega-\omega^2\rangle$, $\langle \omega^2-\omega^3\rangle,\cdots,\langle \omega^{a_0-2}-\omega^{a_0-1}\rangle$. These form a basis. If we let $e_0 = \langle 1-\omega\rangle$, $e_1 = \langle \omega-\omega^2\rangle$ and generally $e_k = \langle \omega^k-\omega^{k+1}\rangle$ where k is taken modulo a_0, then $\tilde{H}_1(\Omega a_0)$ has basis $\{e_0,e_1,\cdots,e_{a_0-2}\}$. Also $e_k = \omega^k e_0$ in the sense of the multiplicative action of the roots of unity on $\tilde{H}_0$.

The Seifert pairing is defined by the formula $\theta(a,b) = \ell k(a^*,b)$. Here we see that

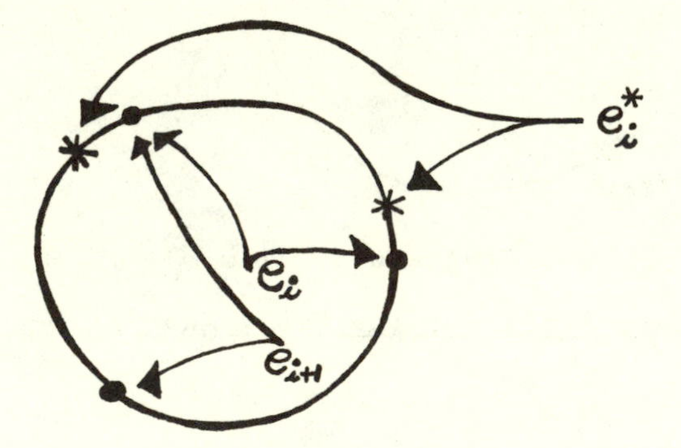

$$\begin{cases} \theta(e_i,e_i) = \ell k(e_i^*,e_i) = +1 \\ \theta(e_i,e_{i+1}) = \ell k(e_i^*,e_{i+1}) = -1 \end{cases}$$

and otherwise $\theta(e_i, e_j) = 0.$

This means that <u>for the empty knot of degree</u> a <u>the Seifert pairing has matrix</u>

$$\Lambda_a = \begin{bmatrix} 1 & -1 & & & \\ & 1 & -1 & & \\ & & 1 & -1 & \\ & & & \ddots & 1 \\ & & & & 1 \end{bmatrix}$$

<u>with respect to the basis</u> $\{e_0, e_1, \cdots, e_{a-2}\} = \mathcal{B}$.

We have already encountered this matrix in Chapter 12 where the intersection on the a-fold cyclic branched covering of D^4 along a pushed-in Seifert surface for a knot K has the form $\theta \otimes \Lambda_a + \theta^T \otimes \Lambda_a^T$ (T denotes transpose), and θ is the given Seifert pairing for $K \subset S^3$. This connection with the formalism of the empty knot is not spurious. It is in fact, the first instance of a unified arena of constructions which happen both in studying singularities and in studying knot theory. Most of the rest of this chapter will be devoted to an outline of these constructions, which we have elsewhere called the <u>cyclic suspension</u> ([N]) and the <u>knot product</u> ([KN],[K6]).

Example 19.12: The Cyclic Suspension. In this example we explain how the empty knot of degree a is related to the a-fold cyclic branched covering. Note that any knot (codimension two embedding in a sphere) can be regarded as the link of a singularity that is not necessarily algebraic. First view the figure on p. 371. Here we have indicated a singularity associated with a knot $K \subset S^3$ obtained by forming the cone $CK \subset D^4$. The cone is illustrated as a projection into three-dimensional space.

If we really want to think of $CK \subset D^4$ as analogous to an algebraic singularity, then there should be a function $f : D^4 \longrightarrow D^2$ such that $f^{-1}(0) = CK$. This is analogous to a mapping from $\mathbb{C}^n \longrightarrow \mathbb{C}$ in the complex case. Such a mapping can always be obtained. The construction is as follows: Let $K \subset S^n$ be a smooth codimension two submanifold with trivial normal bundle $N(K) \cong K \times D^2 \longrightarrow S^n$. Then an obstruction theory argument shows that there is a mapping $\alpha : S^n\text{-IntN}(K) \longrightarrow S^1$ that is smooth and that restricts to $pr : K \times S^1 \longrightarrow S^1$, the projection on the boundary of the tubular neighborhood. The mapping α represents a generator of $H^1(S^n\text{-}K)$ when K is connected, and an oriented sum of generators in the case of a link. In the case of a knot in the three-sphere, this mapping may be visualized by first constructing a spanning surface for K, then splitting S^3 along the spanning surface, then writing a Morse function to $[0,1]$ from the split manifold. In any case, α may be chosen smooth, so that $\alpha^{-1}(p)$ is a smooth spanning surface for K, for a dense set of $p \in S^1$ (via the Morse lemma [M]). When K is a fibered knot, α is a smooth fibration. By its construction, α may be extended to $\overline{\alpha} : S^n \longrightarrow D^2$ by taking its union with the projection $K \times D^2 \longrightarrow D^2$. Finally, let $f : D^{n+1} \longrightarrow D^2$ be the <u>cone on</u> $\overline{\alpha}$. That is, $f(ru) = r\overline{\alpha}(u)$ where $u \in S^n$ and $0 \leq r \leq 1$. We shall call $f : D^{n+1} \longrightarrow D^2$ a <u>generator</u> of the knot $K \subset S^n$. See [KN] for discussion of the details of construction and uniqueness of generators.

Note that it follows from the discussion that a generator $f : D^{n+1} \longrightarrow D^2$ for a fibered knot, <u>itself</u> gives rise to a fibration $f : D^{n+1} - CK \longrightarrow D^2 - \{0\}$. This is exactly analogous to the fibration $\mathbb{C}^n - V(f) \longrightarrow \mathbb{C}^*$ discussed in Lemma 19.10 for the Brieskorn polynomials. In general, if $f : D^{n+1} \longrightarrow D^2$ is a generator then $f^{-1}(p)$ will generically be a codimension two submanifold of D^{n+1} with boundary ambient isotopic to K.

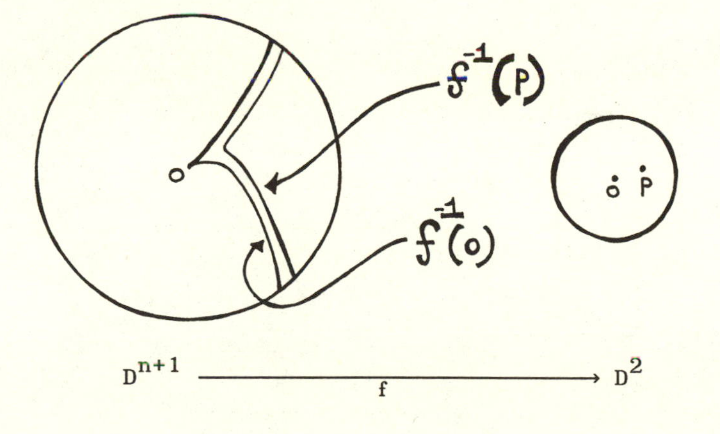

We now observe how to use a generator $f : D^2 \longrightarrow D^2$ to construct cyclic branched covers:

(i). Let $\underline{a} : D^2 \longrightarrow D^2$ denote the mapping $\underline{a}(z) = z^a$ for a fixed positive integer a. Note that $\underline{a}$ is a generator of the empty knot of degree a. [In this context, the cone over the empty set is a single point.]

(ii) Given any generator $f : D^{n+1} \longrightarrow D^2$ let f^* denote a slight displacement of f that is obtained by composing f with a diffeomorphism $\epsilon : D^2 \longrightarrow D^2$ that moves 0 to a point p with $f^{-1}(p)$ a smooth submanifold. We require that $\epsilon|\partial D^2$ is the identity.

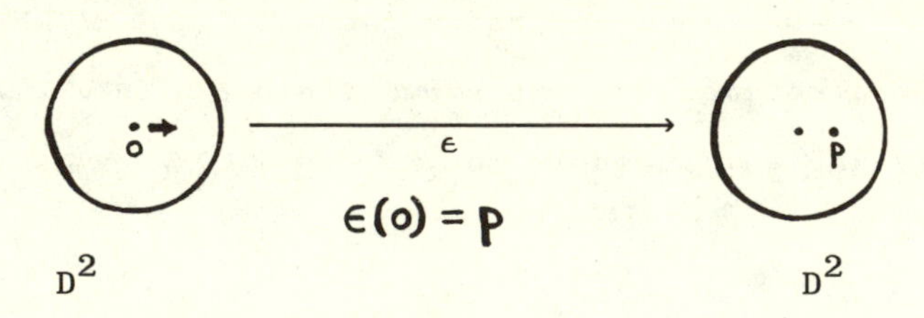

Thus $f^{*-1}(0) \subset D^{n+1}$ is a smooth spanning manifold for $K \subset \partial D^{n+1}$.

(iii) Form the <u>pull-back diagram</u>

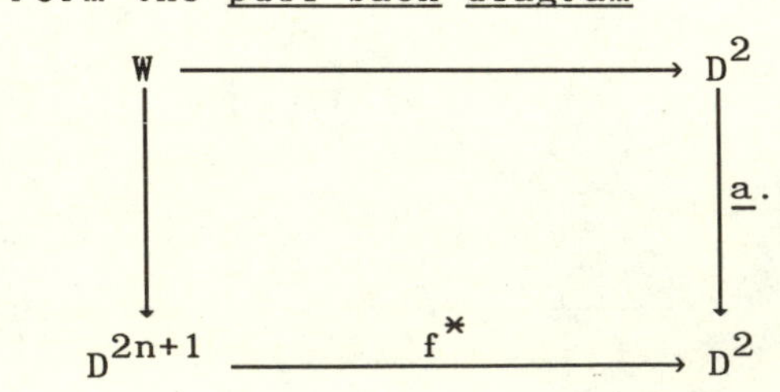

Then $W = \{(x,z) \in D^{n+1} \times D^2 \mid f^*(x) = z^a\}$ is (manifestly) the _a-fold cyclic branched covering of_ D^{n+1} _branched along_ $f^{*-1}(0) - F$.

This topological construction for the branched covering makes W the precise analog of a variety $\epsilon = z^a - f(x)$ (where $f(x)+\epsilon$ corresponds to $f^*(x)$). Again see [KN] for the precise comparison theorems. By creating branched coverings in this fashion, we get much more than just a relation with the case of algebraic varieties. We also get the embedding $W \subset D^{n+1} \times D^2$ and hence an embedding $\partial W \subset \partial(D^{n+1} \times D^2) \cong S^{n+2}$. Now ∂W _is the a-fold cyclic branched cover of_ S^n _with branch set_ K. Let $M_a(K) \longrightarrow S^n$ denote this branched cover. Then we have proved the

THEOREM [KN]. *Let* $K \subset S^n$ *be a knot (codimension two embedding) and let* $M_a(K) \longrightarrow S^n$ *denote the a-fold cyclic covering of* S^n *with branch set* K. *Then there is a natural embedding of* $M_a(K)$ *in* S^{n+2}. *Thus we have a tower of embeddings and branched coverings:*

$$
\begin{array}{ccccc}
 & & & & \cdots \\
 & & & & \downarrow \\
 & & M_{a'}(M_a(K)) & \longrightarrow & S^{n+4} \\
 & & \downarrow & & \\
M_a(K) & \longrightarrow & S^{n+2} & & \\
\downarrow & & & & \\
K \subset S^n & & & &
\end{array}
$$

These embeddings are the topological analogs of the embeddings already discussed for algebraic varieties. That is, the embedding $\text{Link}(f(x)+z^a) \subset S^{n+2}$ is obtained as ambient isotopic to $M_a(\text{Link}(f(x)) \subset S^{n+2}$ where $\text{Link}(f(x)) \subset S^n$.

In the particular case of the Brieskorn manifolds, everything actually begins with the empty knots! For consider the diagram

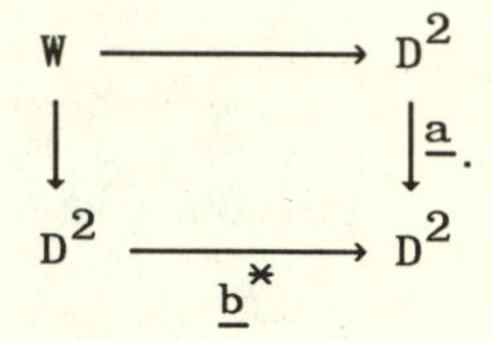

This describes the construction of a surface $W \subset D^2 \times D^2$ whose boundary $\partial W \subset S^3$ is the (a,b) torus knot (link).

If we let $K \otimes [a] \subset S^{n+2}$ denote the knot obtained from $K \subset S^n$ by embedding the branched covering along K into S^{n+2}, then we have

$$[a_0] \otimes [a_1] \subset S^3 \quad \text{torus knot (link)}$$

and, generally, $[a_0] \otimes [a_1] \otimes \cdots \otimes [a_n] \subset S^{2n+1}$ is ambient isotopic to the Brieskorn manifold

$$\Sigma(a_0, a_1, \cdots, a_n) \subset S^{2n+1}.$$

This result is not just formal. For by analyzing the construction of the cyclic suspension $K \otimes [a] \subset S^{n+2}$ more closely one can conclude information about the Seifert pairing for a surface $F_a \subset S^{n+2}$ with $\partial F_a = K \otimes [a]$. The result is

THEOREM [KN]. *Let* $F \subset S^n$ *be a spanning manifold for* $K \subset S^n$ *with Seifert pairing* $\theta : H_*(F) \times H_*(F) \longrightarrow Z$. *Then* $K \otimes [a] \subset S^{n+2}$ *has a spanning manifold* F_a *with Seifert pairing* $\theta \otimes \Lambda_a$ *where* Λ_a *is the Seifert pairing for the empty knot of degree* a.

We indicate briefly in the next example how this result is proved. Please note that this explains how Λ_a appears in the formula for the intersection form on the branched covering $N_a(F)$ of Chapter 12. In these terms

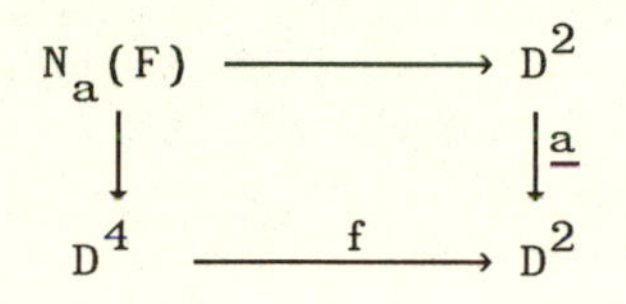

where f is a generator for $K \subset S^3$. We have $M_a(K) = \partial N_a(F) \subset \partial(D^4 \times D^2) = S^5$ with Seifert pairing $\theta \otimes \Lambda_a$ for θ a Seifert pairing for $K \subset S^3$. We did not yet indicate that $N_a(F)$ <u>itself</u> embeds in S^5 with this Seifert pairing $\theta \otimes \Lambda_a$. Nevertheless, this is the case and the proof is a generalization of our argument that pushed Brieskorn fibers $\Sigma z_i^{a_i} = \epsilon$ to Milnor fibers in the sphere. The upshot is an embedding $N_a(F) \subset S^5$ with Seifert form $\theta \otimes \Lambda_a$. Hence it has intersection form $\theta \otimes \Lambda_a + (\theta \otimes \Lambda_a)^T$ (since the intersection form is the sum of the Seifert form and its transpose in this dimension). This intersection form was the result of direct calculation in Chapter 12.

As a specific example, consider the dodecahedral space $\Sigma(2,3,5)$. According to the above results, $\Sigma(2,3,5)$ bounds a manifold $N(2,3,5) \subset S^5$ and $N(2,3,5)$ has intersection form $\pm(\theta+\theta^T)$ where θ is the Seifert form for a $(3,5)$ torus knot. Reference to the table after Exercise 12.7 then shows that Sign $N(2,3,5) = \pm 8$. Thus these results about Seifert forms and embeddings lead to various signature calculations. An exactly analogous calculation shows that $\Sigma(3,5,2,2,2) = \Sigma$ also bounds a manifold of signature ± 8. This leads [M3], [BK], to the identification of $\Sigma(3,5,2,2,2)$ as an exotic sphere. Thus, the Milnor sphere is three cyclic suspensions of a $(3,5)$ torus knot. It is obtained by classical branched covering constructions.

Example 19.13: Products of Knots. The cyclic suspension generalizes to a product construction that corresponds to the link of the sum of two singularities. This is obtained by replacing (in the cyclic suspension) the empty knot generator $\underline{a} : D^2 \longrightarrow D^2$ by <u>any generator</u> $\lambda : D^{m+1} \longrightarrow D^2$ <u>for a fibered knot</u> $L \subset S^m$. The pull-back diagram then becomes:

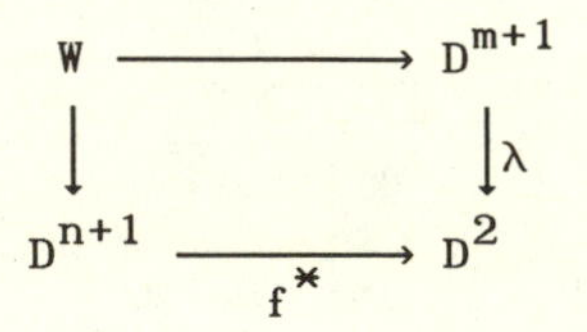

$$\begin{array}{ccc} W & \longrightarrow & D^{m+1} \\ \downarrow & & \downarrow \lambda \\ D^{n+1} & \xrightarrow[f^*]{} & D^2 \end{array}$$

We define $K \otimes L = \partial W \subset \partial(D^{n+1} \times D^{m+1}) \cong S^{n+m+1}$. Thus, given a knot $K \subset S^n$ and a fibered knot $L \subset S^m$ (fibered is needed to make the construction well-defined) there is a new composite or product knot $K \otimes L \subset S^{n+m+1}$.

The construction is built to be a straightened version of the link of the sum of two singularities. In fact it is true that if $f(x)$ and $g(y)$ are polynomial singularities with separate sets of variables x and y, then

$$\text{Link}(f+g) \cong \text{Link}(f) \otimes \text{Link}(g)$$

where $\cong$ denotes ambient isotopy of the corresponding knots.

In terms of the construction, the generators $f : D^{n+1} \longrightarrow D^2$ and $\lambda : D^{m+1} \longrightarrow D^2$ give rise to a new generator $\phi : D^{n+1} \times D^{m+1} \longrightarrow D^2$ that is essentially the difference map $f(x)-\lambda(y)$. One can show that a nonsingular fiber of ϕ has the homotopy type of the join of non-singular fibers of f and λ individually. Furthermore, this join structure is preserved via a deformation of the large nonsingular fiber into the join sphere $S^{n+m+1} = \partial(D^{n+1} \times D^{m+1})$. From this one shows that $K \otimes L$ bounds a manifold $\mathcal{F} \subset S^{n+m+1}$ with Seifert form the <u>tensor product</u> $\theta_K \otimes \theta_L$ of respective Seifert forms for K and L individually.

The product construction has useful corollaries. We conclude by mentioning just one. Let Λ :  $\subset S^3$ denote the Hopf Link. Then $K \subset S^n \longrightarrow K \otimes \Lambda \subset S^{n+4}$ takes spherical knots to spherical knots, and it generates the isomorphism of Levine knot concordance groups $C_n \xrightarrow{\cong} C_{n+4}$. See [KN], [L1].

A good deal more can be said about the knot product. Other geometric interpretations are available, and there are connections with spinning and twist-spinning as well.

Example 19.14: The 8-fold Periodicity of $\Sigma(k,2,2,2,\cdots,2)$ *(an odd number of 2's).* Let $\sum_k^{4n+1}$ denote the Brieskorn manifold $\Sigma(k,2,2,2,\cdots,2)$ with $(2n+1)$ 2's (other than k). $\sum_k^{4n+1}$ bounds a handle-body whose structure is analogous (see [E], [K7] for details) to the spanning surface for a $(2,k)$ torus link. Furthermore, the operation of band exchange

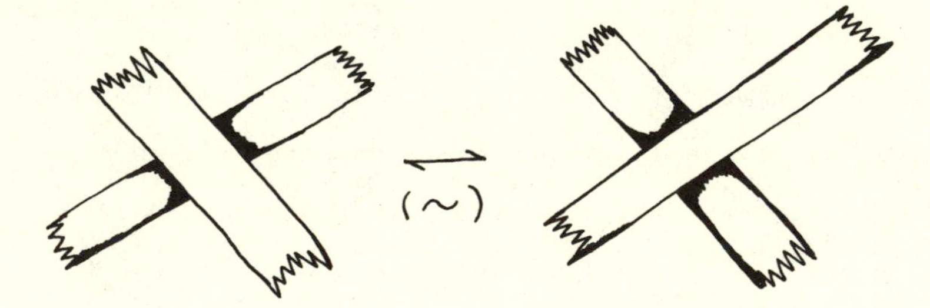

results in a diffeomorphism (via handle-sliding) of this handle-body and hence a diffeomorphism of its boundary. As a result, we obtain an 8-fold periodicity in the list of manifolds $\sum_k^{4n+1}$, $k = 2,3,4,\cdots$. The periodicity follows from a corresponding periodicity in the band-exchange classes of the corresponding spanning surfaces. This is a good example of how low dimensional knot theory can influence the properties of high dimensional manifolds. The (2,k) torus links have spanning surfaces in the pattern:

$(2,2) = K_2$

$(2,3) = K_3$

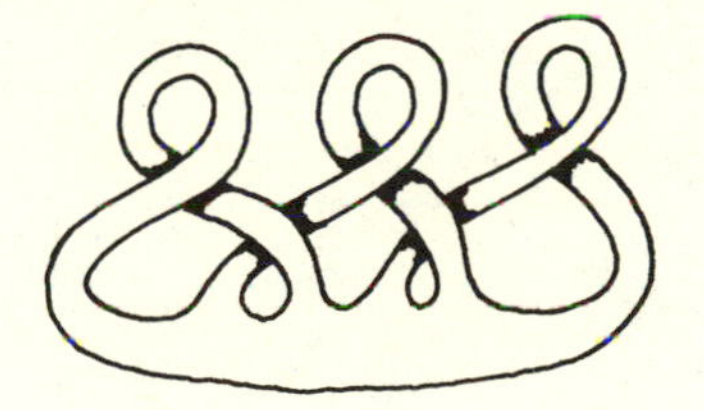

$(2,4) = K_4$

$(2,5) = K_5$

Topological script will be used to notate the periodicity. (See Chapter VI, Sections 6.3 and 6.13 of these notes.)

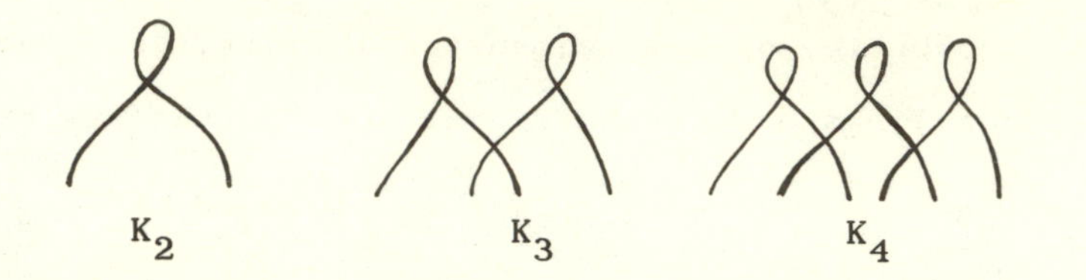

are the corresponding script representations.

Now we note that we use

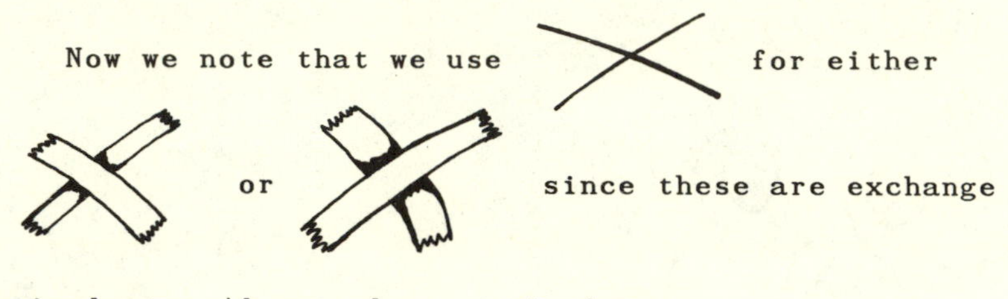

for either ... or ... since these are exchange equivalent. Also we have equivalences

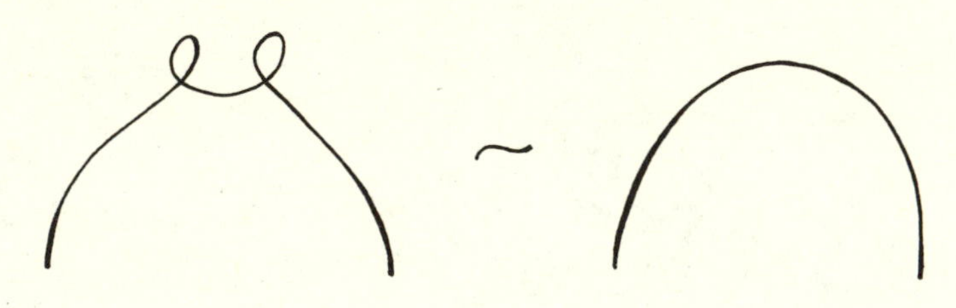

and

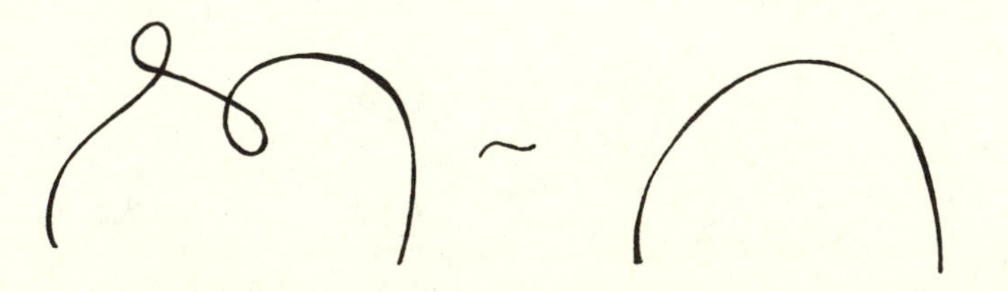

since these can be accomplished by ambient isotopy and exchange on the corresponding bands. Thus

Let [figure] be denoted by Λ_0 and [figure] by $\Lambda_1 = K_2$.

Thus $\underline{K_4 \sim K_3 \# \Lambda_0}$.

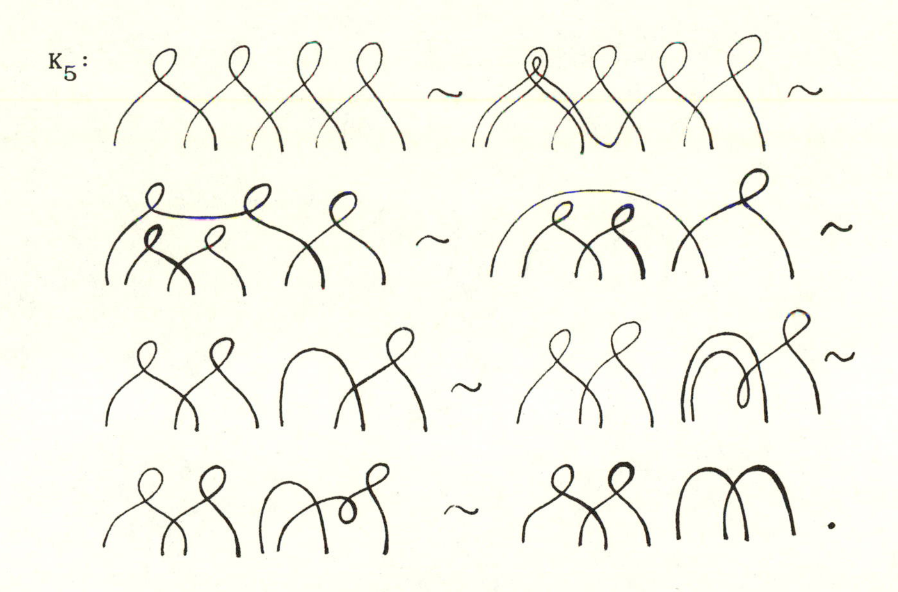

Let H_0 denote 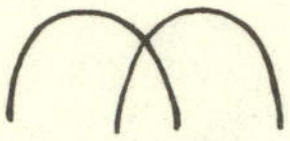 . Since H_0 represents the trivial knot: $X \# H_0 \sim X$.

Thus $\underline{K_5 \sim K_3 \# H_0 \sim K_3}$.

K_6:

$\sim$

$\sim$

$\sim$

.

$\therefore$ $\underline{K_6 \sim K_5 \# \Lambda_1}$.

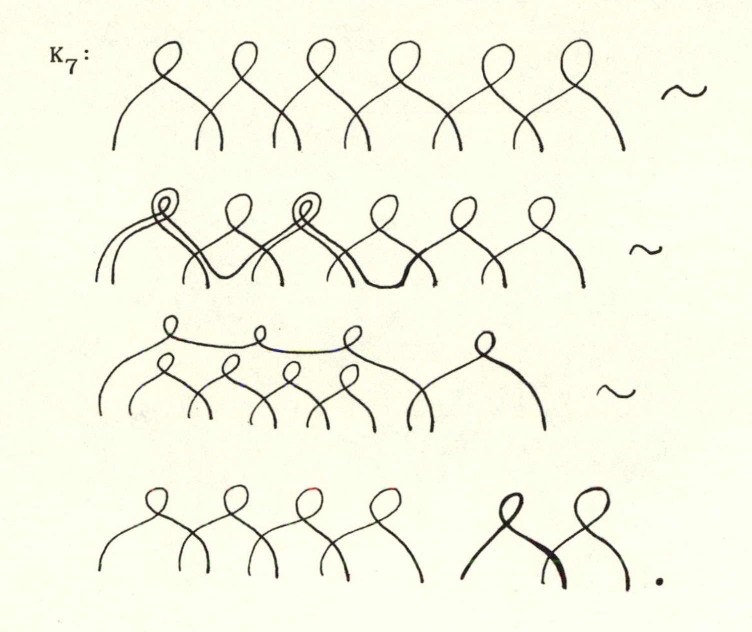

Thus $\underline{K_7 \sim K_5 \# K_3}$.

Continuing in this pattern, we find that:

$K_2 \sim \Lambda_1$	$K_3 \sim K_3$
$K_4 \sim K_3 \# \Lambda_0$	$K_5 \sim K_3 \# H_0 \sim K_3$
$K_6 \sim K_5 \# \Lambda_1$	$K_7 \sim K_5 \# K_3$
$K_8 \sim K_7 \# \Lambda_0$	$K_9 \sim K_7 \# H_0 \sim K_7$
$K_{10} \sim K_9 \# \Lambda_1$	

Thus $K_{10} \sim K_7 \# \Lambda_1 \sim K_5 \# K_3 \# \Lambda_1 \sim K_3 \# K_3 \# \Lambda_1$. But we know (Section 6.3 of Chapter 6) that $K_3 \# K_3 \sim H_0 \# H_0$ $\sim$ (blank). Hence $K_{10} \sim \Lambda_1 \sim K_2$.

This begins the <u>8-fold</u> <u>periodicity</u>: $K_{\ell+8} \sim K_\ell$. The basic list is

$$K_2 \sim K_2$$
$$K_3 \sim K_3$$
$$K_4 \sim K_3 \# \Lambda_0$$
$$K_5 \sim K_5$$
$$K_6 \sim K_3 \# \Lambda_1 \sim H_0 \# \Lambda_1 \sim \Lambda_1$$
$$K_7 \sim K_3 \# K_3 \sim H_0$$
$$K_8 \sim \Lambda_0$$
$$K_9 \sim H_0$$

with $K_{\ell+8} \sim K_\ell$.

To go into the precise details of the relationship between the corresponding manifolds and these links would take us too far afield. However, the list of manifolds is as follows:

K_2 — T^{4n+1} = tangent sphere bundle to S^{2n+1}
K_3 — Σ^{4n+1} = Kervaire sphere
K_4 — $\Sigma^{4n+1} \# S^{2n+1} \times S^{2n}$
K_5 — Σ^{4n+1}
K_6 — T^{4n+1}
K_7 — S^{4n+1}
K_8 — $S^{2n+1} \times S^{2n}$
K_9 — S^{4n+1} (see [K7], [DK]).

The Kervaire sphere is exotic in many dimensions (for example, Σ^9 is exotic). Under these circumstances the exoticity is detected by the Arf invariant, which in this context corresponds to the Arf invariant of the corresponding (2,k) torus knot. The connected sum of two Kervaire spheres is diffeomorphic to the standard sphere S^{4n+1}. The handle-sliding geometry of this diffeomorphism is depicted via topological script in the equivalence $K_3 \# K_3 \sim H_0 \# H_0$.

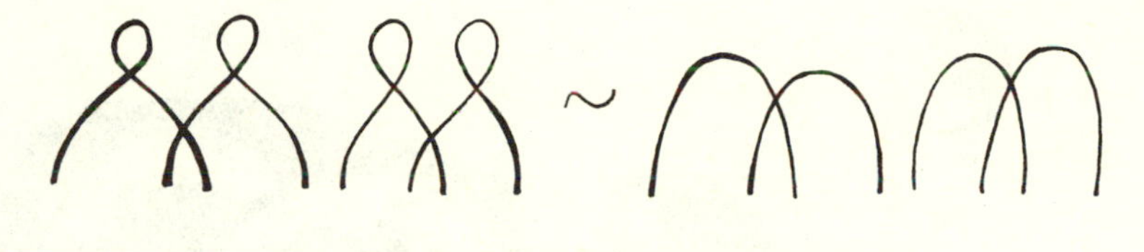

$$\Sigma^9 \# \Sigma^9 \cong S^9$$

Epilogue: This final chapter has been a sketch of relationships between knot theory and manifolds in geometric topology. We have hardly touched on the beginnings of many topics such as the work of Thurston, or the Kirby Calculus and its application to 4-manifolds. The subject of knots and algebraic varieties could expand to another book. Therefore it is time for this writing to stop. I hope these pages have given the reader a taste for the surprising variety, fascination and mathematical pleasure that is the theory of knots.

> "Existence, by nothing bred,
> Breeds everything.
> Parent of the universe,
> It smooths rough edges,
> Unties hard knots,
> Tempers the sharp sun,
> Lays blowing dust,
> Its image in the well spring never fails.
> But how was it conceived? - this
> Image
> Of no other sire."

[From *The Way of Life* by Lao Tsu, translated by Witter Bynner; Capricorn Books, 1944.]

APPENDIX

GENERALIZED POLYNOMIALS AND A STATE MODEL FOR THE JONES POLYNOMIAL

The purpose of this appendix is to present some developments about generalized polynomials and the Jones polynomial.

We shall first describe a two-variable invariant of knots and links, distinct from the generalized polynomial discussed in Chapter VI (Sections 18, 19, 20).

This polynomial specializes to an invariant called the <u>bracket polynomial</u> that has a particularly nice definition as a sum over states of the planar diagram. The bracket is our elementary construction for the Jones polynomial, and it allows the proof of results that appear to be unavailable from other points of view.

In particular, one can show the topological invariance of the number of crossings in a reduced alternating projection ([7],[12], [16] and [18]). This settles a conjecture that goes back to the creation of the knot tables by Tait, Kirkman and Little over a century ago.

Related work by Kunio Murasugi and by Morwen Thistlethwaite settles old conjectures and brings forth new information about alternating knots and links ([16], [18], [19]).

THE L AND F POLYNOMIALS

Our first topic is a two-variable Laurent polynomial invariant of regular isotopy (denoted L) and its normalized companion invariant of ambient isotopy (denoted F). The axioms for L are as follows:

1. To each unoriented knot or link L there is associated a Laurent polynomial $L_K(a,z)$ in the variables a and z. Regularly isotopic (see Chapter VI, Section 19) links receive identical polynomials.

2. If the links H, H′, K, K′ differ at the site of a single crossing as indicated below, then

$$L_H + L_{H'} = z(L_K + L_{K'}).$$

3. The L-polynomial behaves on curls according to the formulas below.

$$L = aL$$
$$L = a^{-1}L$$

4. The value of L on an unknotted circle is one.

$$L_O = 1.$$

See [9] for a detailed exposition of this invariant.

The associated oriented invariant of ambient isotopy is obtained, just as in our discussion of the generalized polynomial, by multiplying L by the variable α raised to the negative of the writhe to form a polynomial denoted F_K. Thus

$$F_K = a^{-w(K)}L_K.$$

The F-polynomial behaves similarly to the generalized polynomial (called G in our notes—Chapter VI). In particular, F applied to the mirror image of a link K, is obtained by replacing a by (a^{-1}) in the polynomial for K. F distinguishes many knots and links from their mirror images. It appears that F is better at this mirror game than the polynomial G (the generalized Jones polynomial). In calculations of knots up to twelve crossings, F distinguishes more knots from their mirror images than G. Specifically, F and G both fail on the chiral knot 9_{42} (Reidemeister numbering). The first knot shown chiral by F and not by G is 10-130 (Thistlethwaite numbering). There is (so far!) no failure by F that is not also a failure by G. (These remarks are based on the calculations of Morwen Thistlethwaite, and our own computations in collaboration with Ivan Handler of Chicago, Illinois, and Marco Ilic at the Universita di Bologna, Bologna, Italy.)

Here are the first few L-polymonials, and the corresponding values of F:

1. $L_{\text{[diagram]}} + L_{\text{[diagram]}} = z(L_{\text{[diagram]}} + L_{\text{[diagram]}})$

$\Rightarrow a + a^{-1} = z(L_{\text{[diagram]}} + 1)$

$\therefore \quad L_{\text{[diagram]}} = \left[\frac{a+a^{-1}}{z} - 1\right].$

2. $L_{\text{[diagram]}} + L_{\text{[diagram]}} = z(L_{\text{[diagram]}} + L_{\text{[diagram]}}) = z(a^{-1}+a).$

$\therefore \quad L_{\text{[diagram]}} = (z-z^{-1})(a^{-1}+a)+1.$

3. $L_{\text{[diagram]}} + L_{\text{[diagram]}} = z\left[L_{\text{[diagram]}} + L_{\text{[diagram]}}\right]$

$\Rightarrow L_{\text{[diagram]}} = \left[\frac{a+a^{-1}}{z} - 1\right]L_{\text{[diagram]}}.$

Letting $\delta = \left[\frac{\alpha+\alpha^{-1}}{z}\right]-1$, and $\sqcup$ denote disjoint union, we have

$$L_{O \sqcup K} = \delta L_K$$

$$F_{O \sqcup K} = \delta F_K.$$

4.  T ... Λ

$L_T + a = z(a^{-2} + L_\Lambda)$

$L_T = -(2a+a^{-1}) + z(a^{-2}+1) + z^2(a+a^{-1})$

$F_T = a^{-3}L_T$

$F_T = -(2a^{-2}+a^{-4}) + z(a^{-5}+a^{-3}) + z^2(a^{-2}+a^{-4}).$

(5) 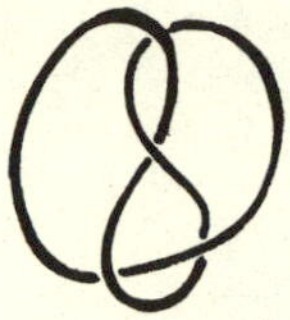 E $\quad w(E) = 0 \Longrightarrow L_E = F_E.$

$$L_E = (-a^{-2}-a^2-1)-z(a+a^{-1})+z^2(a^2+2+a^{-2})+z^3(a+a^{-1}).$$

The L-polynomial generalizes the Q-polynomial of Ho [5] and (independently) Brandt, Lickorish and Millet [4]. To obtain the Q-polynomial from L, set a equal to 1.

Remarkably, the F-polynomial generalizes the original Jones polynomial. We shall explain this relationship by first constructing a special case of L (the bracket polynomial - ⟨K⟩). (See [14] for a different proof of this fact.) Following this appendix is a table of values of the L-polynomial for all knots of up to nine crossings.

THE BRACKET POLYNOMIAL

The bracket polynomial is a special case of the L-polynomial that is easily seen to be well-defined. (Recall that the main subtlety in the theory of the generalized polynomial resides in its intricate inductive definition. The situation is the same for L.) To this end, ⟨K⟩ will satisfy the following identities:

$$\left\{\begin{array}{ll} 1. & \langle \times \rangle = A\langle \asymp \rangle + B\langle\,)(\, \rangle \\ & \langle \times \rangle = B\langle \asymp \rangle + A\langle\,)(\, \rangle . \\ 2. & \langle O \sqcup K \rangle = d\langle K \rangle . \\ 3. & \langle O \rangle = 1 . \end{array}\right\}$$

Here O denotes the unknot; $\sqcup$ denotes disjoint union; A, B, d are commuting algebraic variables whose relationship is not yet determined.

Thus property 1 states a straightforward recursion that will hold for the bracket polynomials of three *unoriented* links differing at the site of one crossing. Property 2 says that the presence of an extra unknotted circle anywhere in the diagram multiplies the bracket by d. Property 3 normalizes the bracket on the unknotted circle.

In order to see that the bracket polynomial is well-defined we bring forth the notion of a *chromatic state* of an (unoriented) link diagram.

Given any link diagram K with N crossings, there are 2^N associated diagrams without crossings, each obtained from K by splicing out each of its crossings in one of the two possible ways. We can indicate the type of splitting for a given crossing by placing *markers* on the universe as shown below.

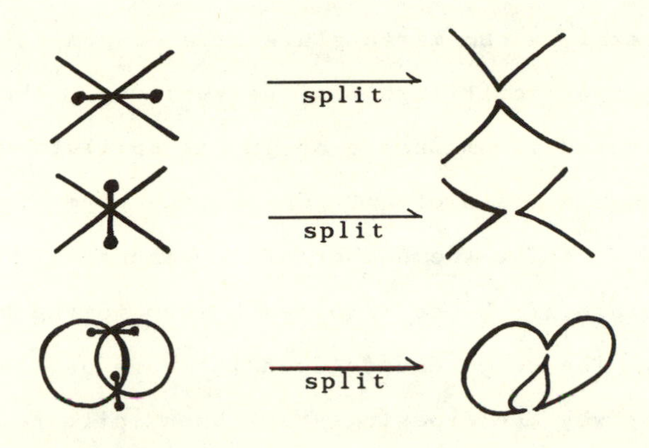

Each marker pierces a crossing and touches the two regions that are amalgamated by the split. See Figure A1 for the states of the trefoil diagram.

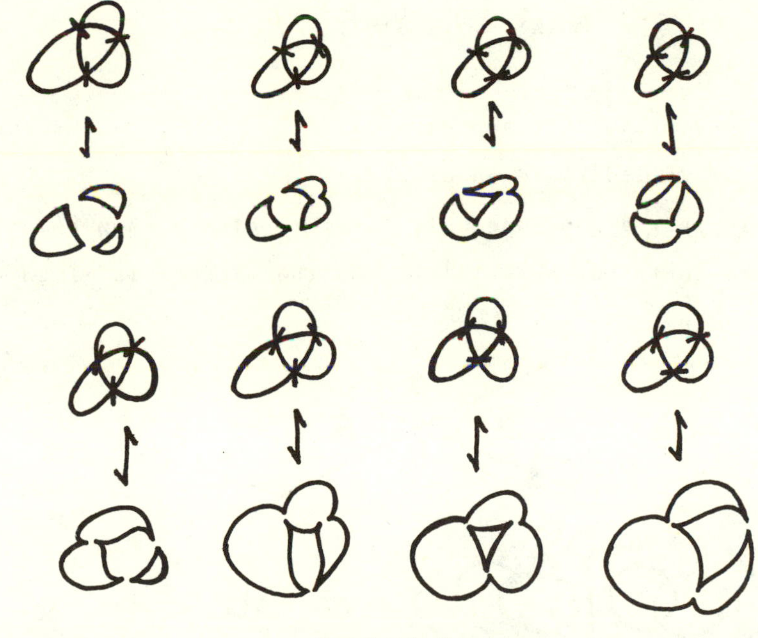

Trefoil States

FIGURE A1

In general, a chromatic state of a diagram is one of these choices of splittings for the vertices. The state may be indicated via markers, or in its split form.

If S is a state of K, define the <u>norm</u> of S, denoted $|S|$, to be the number of components of S.

Each state of K contributes a term to the bracket polynomial. The term depends upon the norm of the state and upon the way the crossings have been split to form S. To obtain this part of the contribution, let $\langle K|S\rangle$ denote the product of all variables (A or B) touched by state markers for S. See Figure A1. Then

$$\langle K\rangle = \sum_{S} \langle K|S\rangle d^{|S|-1}.$$

This gives an explicit representation for $\langle K\rangle$, and shows that it is well-defined on diagrams.

Note that a good mnemonic for the relationship between the state-contributions $\langle K|S\rangle$ and the markers is given by the labelling

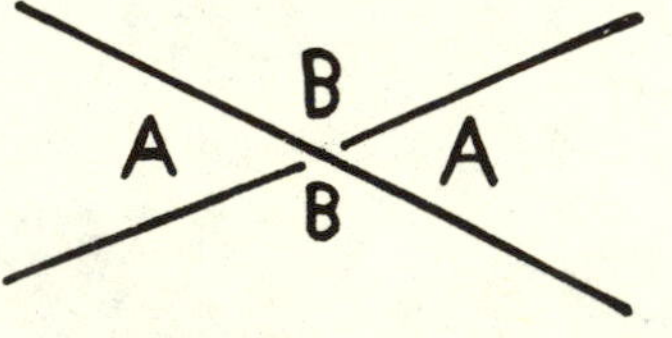

Thus

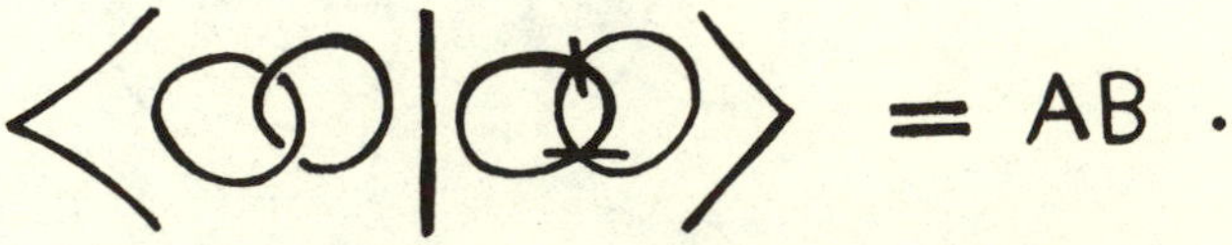

$= AB$.

This also brings us back to the question asked in Section 15 of Chapter VI: What labellings are appropriate for link diagrams? There we explained the labellings that give rise to the Conway polynomial as a state summation. Here a simpler labelling scheme gives the bracket polynomial.

We have called these chromatic states because the bracket is a direct analogue of the chromatic polynomial (see Chapter VI, Sections 6.14, 6.16 of these notes). For a detailing of this relationship see [8], [10], [11]. By translating a planar graph into an alternating knot diagram and choosing A and B correctly, the scheme for calculating the bracket actually <u>is</u> the chromatic polynomial.

Turning to topology: By choosing $B = A^{-1}$ and $d = -A^2-A^{-2}$, $\langle K \rangle$ becomes an invariant of regular isotopy and a special case of the L-polynomial: The sum of the identities

$$\langle \diagup\!\!\!\diagdown \rangle = A\langle \asymp \rangle + A^{-1}\langle\,)(\,\rangle$$

$$\langle \diagdown\!\!\!\diagup \rangle = A^{-1}\langle \asymp \rangle + A\langle\,)(\,\rangle$$

gives $$\langle \diagup\!\!\!\diagdown \rangle + \langle \diagdown\!\!\!\diagup \rangle = z(\langle \asymp \rangle + \langle\,)(\,\rangle)$$

for $z = A+A^{-1}$. And for this choice of B and d one finds that the L-axioms are satisfied with $a = -A^3$.

Letting f_K denote the associated invariant of

ambient isotopy for oriented links K, we have

$$f_K = (-A^3)^{-w(K)} \langle K \rangle .$$

In fact f_K is a version of the original Jones polynomial! Specifically,

$$V_K(t) = f_K(t^{-1/4}).$$

To see this, note that by multiplying the following expansions by the reciprocals of A and B respectively, and then subtracting, we obtain an exchange identity for the bracket:

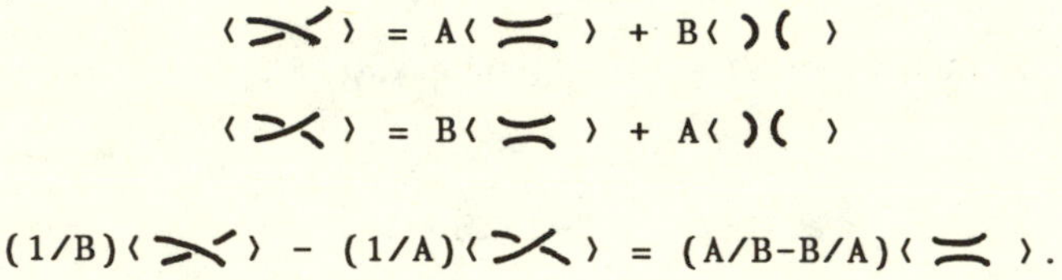

$$\langle \times \rangle = A\langle \asymp \rangle + B\langle)(\rangle$$

$$\langle \times \rangle = B\langle \asymp \rangle + A\langle)(\rangle$$

$$(1/B)\langle \times \rangle - (1/A)\langle \times \rangle = (A/B-B/A)\langle \asymp \rangle .$$

Normalizing this equation to an identity about f, we obtain

$$-A^4 f_{\times} + A^{-4} f_{\times} = (A^2-A^{-2}) f_{\rightrightarrows} .$$

With the indicated substitutions, this becomes the (defining) identity for the Jones polynomial. (See Section 18 of Chapter VI.) $[B = A^{-1},\ a = -A^3,\ A = t^{-1/4}]$.

APPLYING THE BRACKET

Here we indicate a few steps in applying the bracket polynomial to alternating knots and links. A key result is our determination of the maximal and minimal degrees in the variable A for ⟨K⟩ when K is an alternating diagram [7].

To understand this result it is useful to observe the following fact about the checkerboard shading of a connected alternating diagram: Call a shaded crossing of <u>type A</u> if the two shaded regions correspond to the label A in the bracket polynomial. (See Figure A2.) Otherwise call the shaded crossing type B.

Then it is apparent from Figure A2 that <u>all the shaded crossings in a connected alternating diagram have the same type</u>. Thus we shall assume that the diagram is shaded so that all crossings have type A. The shaded regions will be referred to as the black regions, and the unshaded regions as the white regions.

By choosing the state S so that all the state markers fall on the shaded parts (hence of type A), we see (as in Figure A2) that the number of components in this state is equal to the number of white regions in the shading. Consequently, the maximal degree contribution of S to the bracket is the degree in A of <u>V+2(W-1)</u> where V is the number of vertices (crossings) in the diagram, and W denotes the number of white regions. (Recall that each state contributes a power of A multiplied by a power of d, and that the power of d is one less than the number of components in the state. Note that d is the negative

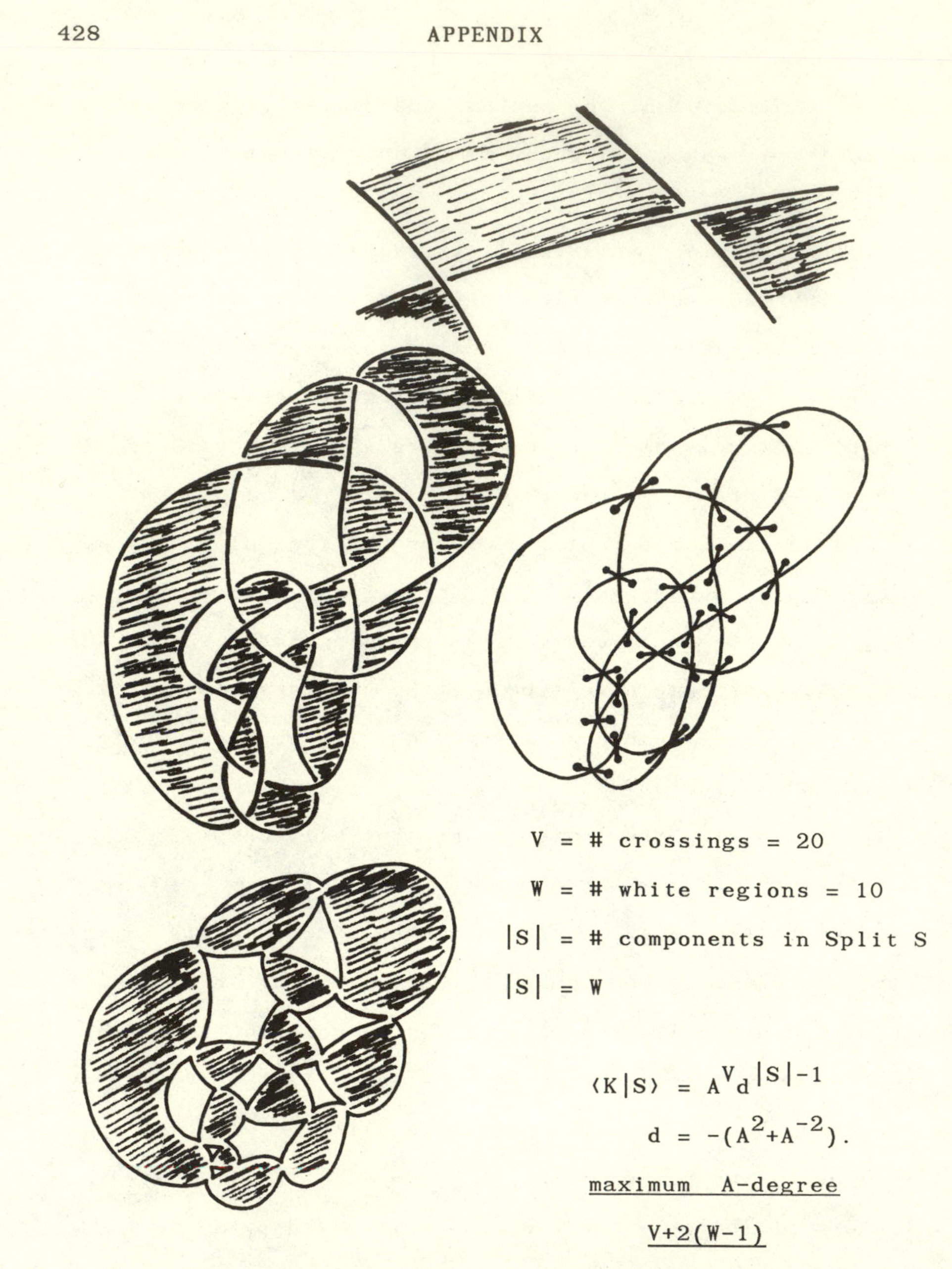

Maximal State Contribution for Alternating Projection

FIGURE A2

sum of the square of A and the square of its inverse.)

It then turns out to be easy to verify that for a reduced diagram (one that has no isthmus—see below) this state S actually contributes the maximal degree to the bracket polynomial.

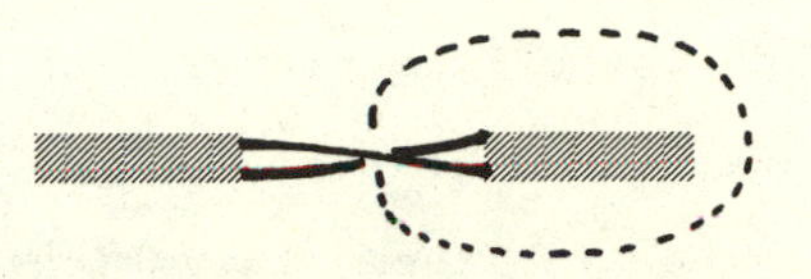

an isthmus

PROPOSITION [7]. *Let* K *be a connected, reduced alternating link diagram. Let* K *be shaded so that all the crossings are of type* A. *Let* K *have* V *crossings,* W *white regions, and* B *black regions. Then* ⟨K⟩ *has maximal* A-*degree* <u>max</u> *and minimal* A-*degree* <u>min</u>, *where*

$$\max = V+2(W-1)$$

$$\min = -V-2(B-1).$$

This proposition has a number of remarkable consequences. For the first, consider the difference diff = max - min. This is a topological invariant of K (since ⟨K⟩ is determined up to multiplication by a power of A.) But

$$\begin{aligned} \text{diff} &= (V+2(W-1))-(-V-2(B-1)) \\ &= 2V+2(W+B)-4 \\ &= 2V+2(V+2)-4 \\ &= 4V. \end{aligned}$$

This calculation implies that the number of crossings in the diagram of a reduced alternating projection is an invariant of its topological type. (Compare [7], [16]. [18].) The same techniques show that V is the minimal number of crossings for any projection of the link K.

Another application of the proposition concerns chirality. With the same hypotheses we see that when the diagram K has writhe w, then the ambient isotopy invariant polynomial f_K has maximal and minimal degrees $\max f$ and $\min f$ where

$$\max f = -3w + \max = -3w + V + 2(W-1)$$

$$\min f = -3w + \min = -3w - V - 2(B-1).$$

If K is ambient isotopic to its mirror image (we say achiral), then it follows that $-\min f = \max f$. Using the formulas above, this implies that

$$3w(K) = W-B.$$

This is a very strong restriction that allows one to easily prove that many alternating knots and links are chiral. Note that for writhe zero, achirality demands that the number of black regions equals the number of white regions.

Let $B(K)$ denote the graph obtained from K by placing one vertex in each black region, and joining two vertices by an edge whenever there is a crossing in K

incident to both their regions. This is the graph of the black regions. Similarly, let $W(K)$ be the graph of the white regions. We conjecture that a prime, reduced alternating diagram of writhe zero is achiral if and only if the graphs $B(K)$ and $W(K)$ are isomorphic.

REPRESENTATIONS AND BRAIDS

The bracket also provides an entry into the representation theory associated with the Jones polynomial. In order to see this we define a diagram monoid based on the patterns shown in Figure A3. Here diagrams with free ends are multiplied as braids, while multiplication by the closed loop δ denotes disjoint union. As the figure shows, the resulting monoid has relations

$$\left\{\begin{array}{l} h_i^2 = \delta h_i \\ h_i h_{i+1} h_i = h_i \\ h_{i+1} h_i h_{i+1} = h_{i+1} \\ h_i h_j = h_j h_i, \quad |i-j| \geq 2 \end{array}\right\}$$

(Relations for D_n, $i = 1,2,\cdots,n-1$.)

The original Jones polynomial [6] was defined via a representation into an abstract algebra satisfying these multiplicative relations. (Let $e_i = h_i/\delta$, $\tau = 1/\delta$. Then $e_i^2 = e_i$, $e_i e_{i+1} e_i = \tau e_i$, etc.) The diagram monoid forms the beginning of a direct geometric connection between these algebras and the theory of braids. In our terms the

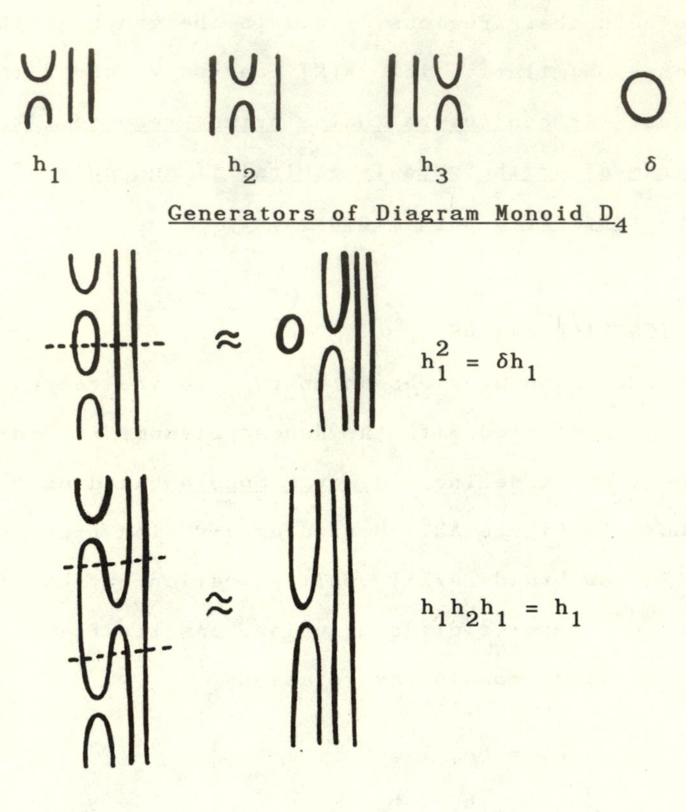

Examples of Diagram Algebra Relations

$h_1 h_3 = h_3 h_1$

FIGURE A3

relation $\langle \asymp \rangle = A\langle \asymp \rangle + B\langle)(\rangle$ becomes the pattern for a representation of the braid group (see [1]) into the algebra with the above relations and $\delta = -(A^2+B^2)$, $B = 1/A$. If σ_i is the ith braid generator then the representation is given by the formula

$$\rho(\sigma_i) = Ah_i + B1 \quad .$$

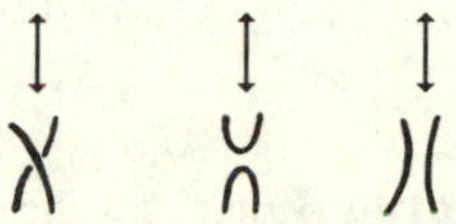

For a more complete exposition of this direction, see [4]. For a discussion of the braid monoid (mixing h_i's and braid generators) see [3], [8], [9], [11], [20].

MORE ABOUT TAIT CONJECTURES

The L-polynomial and the bracket yield even more information about alternating knots. First of all, Thistlethwaite [19] has observed the following result:

THEOREM (Thistlethwaite). *Let* K *be a connected, reduced alternating diagram. Then the highest degree term in* z *in* L_K *has the form* $k(a+a^{-1})z^{n-1}$ *where* n *is the number of crossings in the diagram* K, *and* k *is a positive integer.*

This theorem has the immediate corollary that for K reduced, alternating w(K) is an ambient isotopy invariant of K. The writhe is a topological invariant for reduced alternating links. (To prove the corollary, just note that $F_K = \alpha^{-w(K)}L_K$ is an ambient isotopy invariant of K. To prove the theorem, work inductively with the identity $L_{\times} + L_{\times'} = z(L_{\asymp} + L_{)(})$ using the fact that the two splices $\asymp$, $)($ of an alternating diagram are both alternating, and that if $\times$ is reduced then one of $\asymp$ $)($ is also reduced.)

The first alternating knot to have a coefficient $k \neq 1$ in $k(\alpha+\alpha^{-1})z^{n-1}$ is the knot 8_{16} of the Reidemeister tables.

There is a curious history to this problem about the writhe. The writhe is known not to be a topological invariant for nonalternating diagrams. One example of this is the so-called Perko pair:

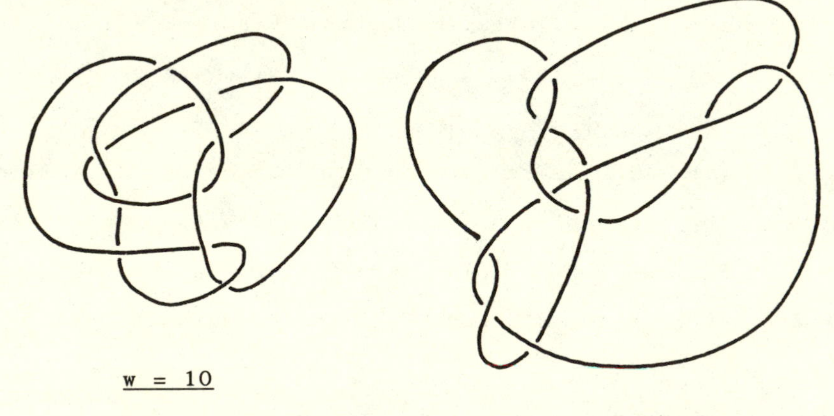

These two diagrams are ambient isotopic, with different

writhes. They sat distinguished on Little's knot tables until Perko discovered in the 1970's that they are in fact ambient isotopic.

Nevertheless, we now know that $w(K)$ is an invariant for reduced alternators, and hence that <u>for K reduced, alternating and $w(K) \neq 0$, K is necessarily chiral</u>.

Now let's return to the formula K achiral $\Rightarrow$ $3w(K) = W-B$ that we proved earlier in this section (under the hypothesis reduced and alternating). In fact, what we did was look at the extremal degrees of f_K. For <u>any</u> K reduced, alternating the <u>sum</u> $S(K)$ of these extremal degrees is an ambient isotopy invariant of K. And we see that this sum is given by the formula: $S(K) = -6w(K)+2(W-B)$. Thus $W-B$, the difference between black and white region counts, is also an invariant for K reduced alternating. (Murasugi [15] found a very clever proof of this and the invariance of the writhe by observing that the <u>signature</u> of K reduced and alternating is given by the formula $\sigma(K) = \frac{1}{2}(W-B) - \frac{1}{2}(w(K))$.)

Thus we see that a number of structural features (number of crossings, the writhe, the black/white defect) must be the same in any reduced alternating projection of a given knot. These facts lend credence to the <u>Tait conjecture</u> that states <u>that two reduced alternating projections are ambient isotopic if and only if one can be obtained from the other by flyping</u>.

A <u>flype</u> is a 180-degree rotation of a tangle in the form

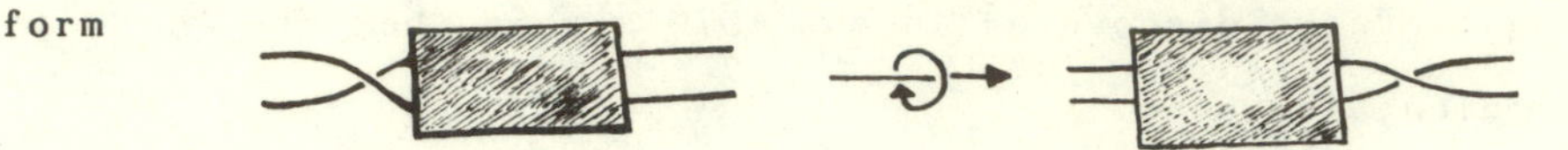

This preserves alternating structure. (Depending on whether your diagram is in the plane or on the two-sphere, you may need a slightly more technical statement. We leave this to you.) It is very instructive to experiment with examples in relation to this conjecture.

GRAPH THEORY AND STATISTICAL MECHANICS

The bracket polynomial is a direct relative of the chromatic polynomial in graph theory. To see the relationship, consider the <u>dichromatic polynomial</u> $Z_G(q,v)$ defined by the recursion:

1. $Z_{\text{—•—•—}} = Z_{\text{—•} \; \text{•—}} + vZ_{\text{—•—}}$.
2. $Z_{\bullet \; H} = qZ_H$.

(Compare with Chapter VI, Section 16.) For $v = -1$, we have $C_G(q) = Z_G(v,-1)$ is the number of proper vertex colorings of the graph G.

Example: $Z_{\bullet\!\!-\!\!\bullet} = Z_{\bullet \; \bullet} + vZ_{\bullet} = q^2+vq$.

Associate a universe $M(G)$ to each planar graph G by the method indicated below:

Note that G is then the graph associated with the checkerboard shading of M(G):

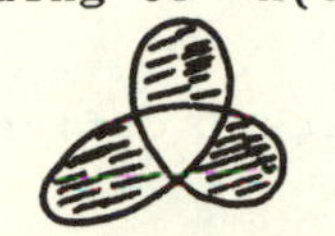 

Thus we can translate the dichromatic polynomial into a recursion about shaded universes:

1′. 

2′. $Z_{\mathscr{C}H} = qZ_H$ if $\mathscr{C}$ is a shaded component without crossings that is disjoint from H.

(e.g., Z $= q^2$).

Example: Z = Z + vZ .

∴ Z = (q+v)Z .

Example: Z = (q+v)Z = (q+v)q.

We can continue this translation and write the dichromatic polynomial in bracket form. To do this, let K(G) be the link diagram obtained from M(G) by placing A-type crossings at each crossing. Thus

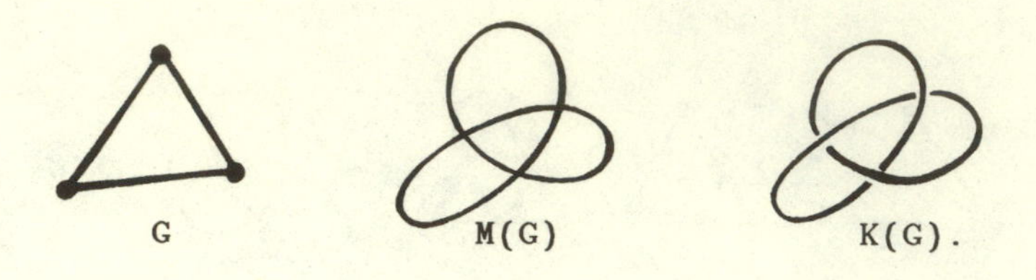

Then K(G) is an alternating link associated with the planar graph G.

Let $\{K\}$ denote the special bracket defined by:

1. $\{\asymp\!\!\!\!\times\} = (q^{-\frac{1}{2}}v)\{\asymp\} + \{\supset\subset\}$.

2. $\{O \sqcup K\} = q^{\frac{1}{2}}\{K\}$

$$\{O\} = q^{\frac{1}{2}}.$$

Then we have [10], [11].

PROPOSITION. *Let G be a planar graph with N vertices. Then the dichromatic polynomial for G can be expressed as a bracket for K(G) by the formula*

$$Z_G(q,v) = q^{N/2}\{K(G)\}$$

Example: •——• (G) ∞ (K(G)) $N = \lambda$

$$\{\infty\} = q^{-\frac{1}{2}}v\{\infty\} + \{OO\}$$

$$= q^{-\frac{1}{2}}v(q^{\frac{1}{2}}) + (q^{\frac{1}{2}})^2$$

$$\{\infty\} = v + q$$

$$q^{N/2}\{\infty\} = q(v+q).$$

Another Formula

Now here is another formula for the dichromatic polynomial: Suppose that we wish to color G with q colors from the discrete set $\{1,2,\cdots,q\}$ (so q is a positive integer). Let G have vertices labelled $i,j,\cdots$ and let $\langle i,j\rangle$ denote an edge in G from i to j. Let $\sigma = \{\sigma_i\}$ denote <u>any</u> choice of colors for the vertices of G with σ_i the value at the vertex i. <u>Then</u>

$$Z_G(q,v) = \sum_{\sigma} \prod_{\langle i,j\rangle} (1+v\delta(\sigma_i,\sigma_j))$$

where δ denotes the Kronecker delta; $\delta(x,y) = 1$ if $x = y$ and $\delta(x,y) = 0$ if $x \neq y$. It is easy to see that this formula is correct by verifying the recursion relation.

Statistical Mechanics

We can now explain a relationship with statistical mechanics. In statistical mechanics, one considers a system with states σ of energy $E(\sigma)$, and a <u>partition function</u> $Z = \sum_{\sigma} e^{-\frac{1}{kT}E(\sigma)}$ where k is a constant (Boltzman's constant), and T denotes temperature. The partition function can be used to obtain many properties of the system. In particular, one wants to know how it behaves for very large systems. A particular model (<u>Potts Model</u> [1]) that is studied consists in a planar graph G

with states σ consisting of assignments of q "colors" (spins, e.g.) at each vertex of the lattice G. The energy for a state σ is taken to be

$$E(\sigma) = \sum_{\langle i,j \rangle} \delta(\sigma_i, \delta_j),$$

Thus local coincidences of spins contribute to the energy. ($\langle i,j \rangle$ is an edge in G as explained above.) We then have:

$$Z = \sum_{\sigma} e^{-\frac{1}{kT}E(\sigma)}$$

$$= \sum_{\sigma} e^{-\frac{1}{kT} \sum_{\langle i,j \rangle} \delta(\sigma_i, \sigma_j)}$$

$$= \sum_{\sigma} \prod_{\langle i,j \rangle} \left[e^{-\frac{1}{kT}\delta(\sigma_i, \sigma_j)} \right]$$

$$\therefore \quad Z = \sum_{\sigma} \prod_{\langle i,j \rangle} (1 + v\delta(\sigma_i, \sigma_j))$$

$$\text{where} \quad v = e^{-\frac{1}{kT}} - 1.$$

Thus _the partition function for the Potts model on a planar graph G is a dichromatic polynomial_. Hence if Z_G is the Potts model partition function on a planar graph with N vertices, then

$$Z_G = q^{N/2}\{K(G)\}$$

for q the number of Potts states, and $v = e^{-\frac{1}{kT}} - 1$.

One significance of this translation is that $\{K(G)\}$ can be computed through the algebra of the diagram monoid. This same algebraic structure occurs in statistical physics as the Temperley-Lieb algebra (see [1]). Vaughan Jones was the first to point out the algebraic relation between his polynomial and the Potts model [6]. Here we have given a direct diagrammatic translation.

REFERENCES
(Appendix)

1. R.J. Baxter. *Exactly Solved Models in Statistical Mechanics*. Academic Press (1982).

2. J.S. Birman. *Braids, Links, and Mapping Class Groups*. Annals of Mathematics Studies. Princeton University Press (1974).

3. J.S. Birman, H. Wenzel. Braids, link polynomials and a new algebra (preprint 1986).

4. R.D.Brandt, W.B.R. Lickorish, K.C. Millett. A polynomial invariant for unoriented knots and links. Invent. Math. 84, 503-573 (1986).

5. C.F. Ho. A new polynomial invariant for knots and links—preliminary report. AMS Abstracts, Vol. 6, #4, Issue 39(1985), 300.

6. V.F.R. Jones. A polynomial invariant for links via von Neumann algebras. Bull. Amer. Math. Soc. 12(1985), 103-112,

7. L.H. Kauffman. State models and the Jones polynomial. (To appear in Topology.)

8. L.H. Kauffman. New invariants in knot theory (preprint 1986).

9. L.H. Kauffman. An invariant of regular isotopy announcement 1985, preprint 1986).

10. L.H. Kauffman. Statistical Mechanics and the Jones polynomial (preprint 1986).

11. L.H. Kauffman. *Sign and Space, Knots and Physics* (book in preparation).

12. M. Kidwell. On the degree of the Brandt-Lickorish-Millett polynomial of a link (preprint 1986).

13. T.P. Kirkman. The enumeration, description and construction of knots with fewer than 10 crossings. Trans. Royal Soc. Edinburgh 32(1865), 281-309.

14. W.B.R. Lickorish. A relationship between link polynomials. Math. Proc. Camb. Phil. Soc. (to appear).

15. C.N. Little. Nonalternate ± knots. Trans. Royal Soc. Edinburgh (35) (1889), 663-664.

16. K. Murasugi. Jones polynomials and classical conjectures in knot theory I and II (preprints 1986).

17. P.B. Tait. On Knots I, II, III. Scientific Papers Vol. I, Cambridge University Press, London, 1898, 273-347.

18. M. Thistlethwaite. A spanning tree expansion of the Jones polynomial (to appear in Topology).

19. M. Thistlethwaite. Kauffman's polynomial and alternating links (preprint 1986).

20. D. Yetter. Markov Algebras (preprint 1986).

KNOT TABLES AND THE L-POLYNOMIAL

These tables list the knots up through nine crossings, as given in the classic book, *Knotentheorie* [R], by Reidemeister. In Reidemeister's tables, no distinction is made between an alternating diagram and its mirror image. Our pictures choose one of the mirror pairs in each case. The drawings each have the sign of the crossings indicated, and the writhe of the given diagram (see the Appendix) of that diagram. The L-polynomial is an invariant of regular isotopy and hence is diagram dependent. Certain patterns are easier to see using it. The corresponding 2-variable ambient isotopy invariant F_K is given by the formula $F_K = a^{-w(K)}L_K$ where $w(K)$ is the writhe of the diagram K. Hence this is immediate from the table.

Up through eight crossings, I have written out the polynomials in the usual high school algebra form. After nine crossings I use a code that goes as follows:

$$(a+a^{-1})Z^2 + (1+a^{-2})Z^1 + (-2a-a^{-1})Z^0$$

$$\Updownarrow$$

$$(\overline{1}|\ 1,\ \overline{1}|\ -1\,|2)(\overline{1}|\ 0,\ \overline{1}|\ -2\,|1)(\overline{-2}|\ 1,\ \overline{-1}|\ -1\,|0).$$

Thus,

$$\overline{n}|\ m \longleftrightarrow na^{m}$$

$$\overline{n}|\ m,\ \overline{r}|\ s \longleftrightarrow na^{m} + ra^{s}$$

$$(\overline{\lambda}|\ n) \longleftrightarrow \lambda Z^{n}$$

$$(T)(U) \longleftrightarrow T + U.$$

The individual parts of this code correspond directly to the polynomial, and there should be no difficulty in reading it. A last example: here is the L-polynomial for the knot 9_1 juxtaposed with its code:

$$\begin{aligned}
& (a+a^{-1})Z^{8} \\
+ \ & (1+a^{-2})Z^{7} \\
+ \ & (-8a-7a^{-1}+a^{-3})Z^{6} \\
+ \ & (-6-5a^{-2}+a^{-4})Z^{5} \\
+ \ & (21a+16a^{-1}-4a^{-3}+a^{-5})Z^{4} \\
+ \ & (10+6a^{-2}-3a^{-4}+a^{-6})Z^{3} \\
+ \ & (-20a-14a^{-1}+3a^{-3}-2a^{-5}+a^{-7})Z^{2} \\
+ \ & (-4-a^{-2}+a^{-4}-a^{-6}+a^{-8})Z^{1} \\
+ \ & (5a+4a^{-1})Z^{0}
\end{aligned}$$

$$\Updownarrow$$

$$(\overline{1}|\ 1,\overline{1}|\ {-1}\,|8)(\overline{1}|\ 0,\overline{1}|\ {-2}\,|7)(\overline{-8}|\ 1,\overline{-7}|\ {-1},\overline{1}|\ {-3}\,|6)$$

$$(\overline{-6}|\ 0,\overline{-5}|\ {-2},\overline{1}|\ {-4}\,|5)(\overline{21}|\ 1,\overline{16}|\ {-1},\overline{-4}|\ {-3},\overline{1}|\ {-5})\,|4)$$

$$(\overline{10}|\ 0,\overline{6}|\ {-2},\overline{-3}|\ {-4},\overline{1}|\ {-6}\,|3)(-\overline{20}|\ 1,\overline{-14}|\ {-1},\overline{3}|\ {-3},\overline{-2}|\ {-5},\overline{+1}|\ {-7}\,|2)$$

$$(\overline{-4}|\ 0,\overline{-1}|\ {-2},\overline{1}|\ {-4},-\overline{1}|\ {-6},\overline{1}|\ {-8}\,|1)(\overline{5}|\ 1,\overline{4}|\ {-1}\,|0).$$

3_1:

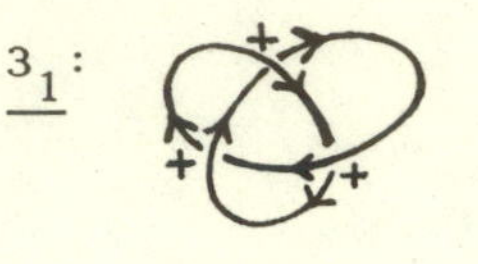

w = 3

$$L = (a+a^{-1})Z^2 + (1+a^{-2})Z^1 + (-2a-a^{-1})Z^0$$

4_1:

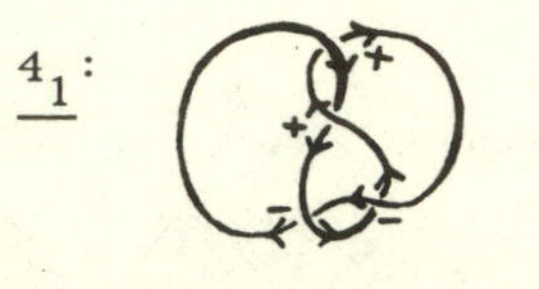

w = 0

$$L = (a+a^{-1})Z^3 + (a^2+2+a^{-2})Z^2 + (-a-a^{-1})Z^1 + (-a^2-1-a^{-2})Z^0$$

5_1:

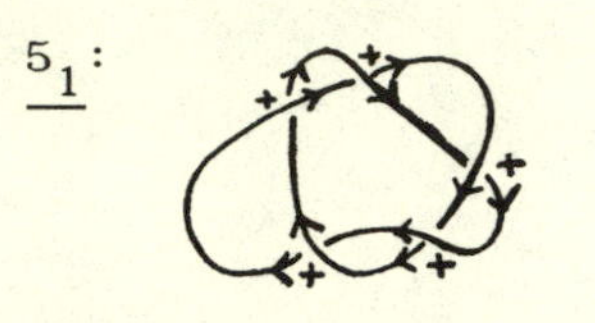

w = 5

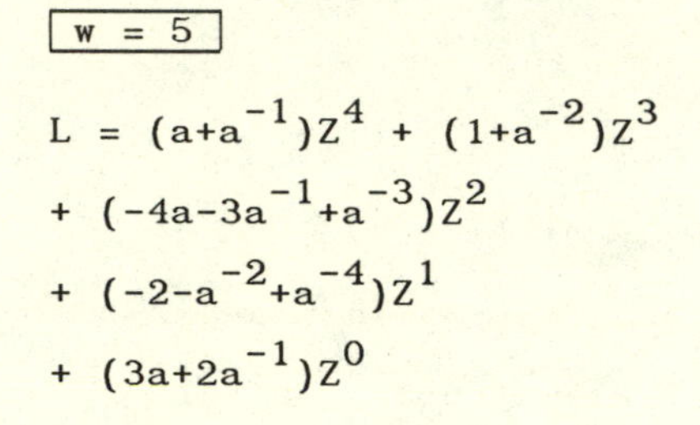

$$L = (a+a^{-1})Z^4 + (1+a^{-2})Z^3 + (-4a-3a^{-1}+a^{-3})Z^2 + (-2-a^{-2}+a^{-4})Z^1 + (3a+2a^{-1})Z^0$$

5_2:

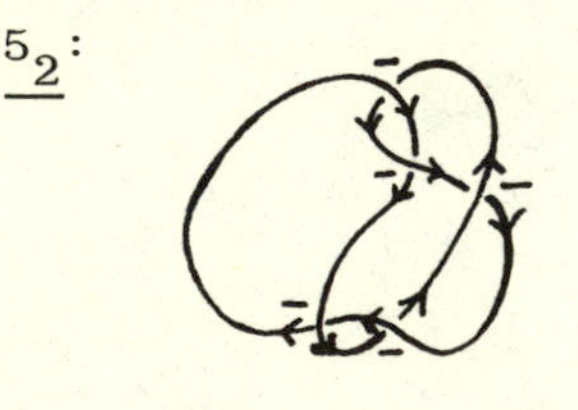

w = -5

$$L = (a+a^{-1})Z^4 + (a^2+2+a^{-2})Z^3 + (-2a-a^{-1}+a^{-3})Z^2 + (-2a^2-2)Z^1 + (a+a^{-1}-a^{-3})Z^0$$

6_1:

w = -2

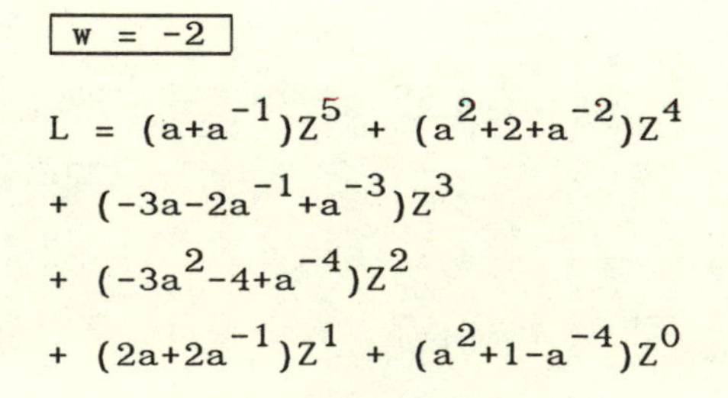

$$L = (a+a^{-1})Z^5 + (a^2+2+a^{-2})Z^4 + (-3a-2a^{-1}+a^{-3})Z^3 + (-3a^2-4+a^{-4})Z^2 + (2a+2a^{-1})Z^1 + (a^2+1-a^{-4})Z^0$$

$\underline{6_2}$:

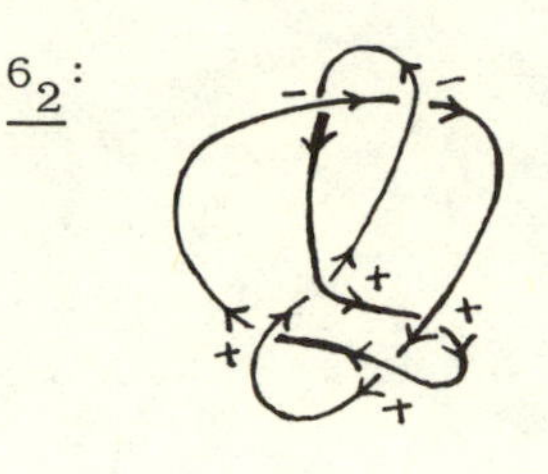

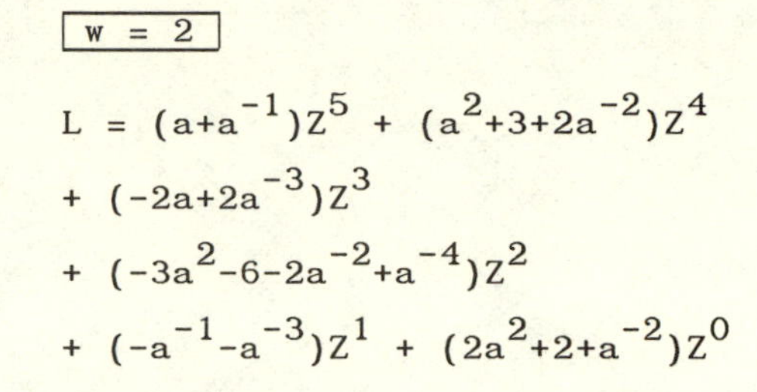

$\boxed{w = 2}$

$$L = (a+a^{-1})Z^5 + (a^2+3+2a^{-2})Z^4$$
$$+ (-2a+2a^{-3})Z^3$$
$$+ (-3a^2-6-2a^{-2}+a^{-4})Z^2$$
$$+ (-a^{-1}-a^{-3})Z^1 + (2a^2+2+a^{-2})Z^0$$

$\underline{6_3}$:

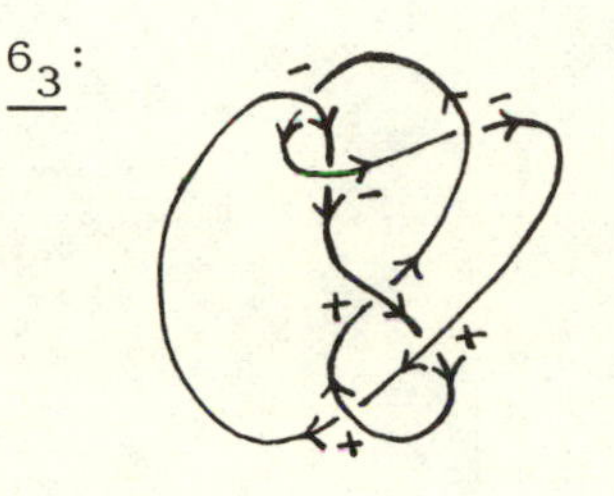

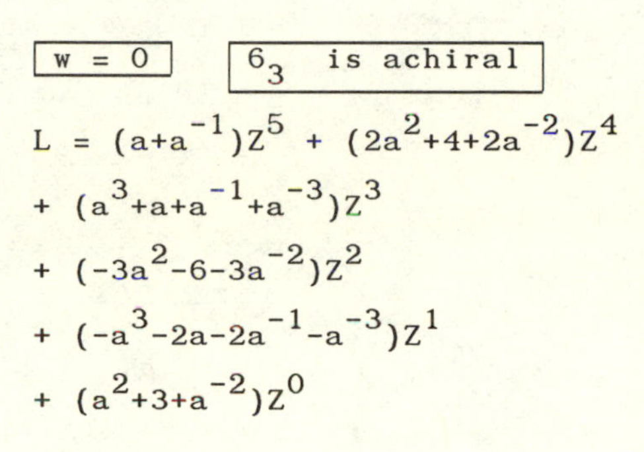

$\boxed{w = 0}$ $\boxed{6_3 \text{ is achiral}}$

$$L = (a+a^{-1})Z^5 + (2a^2+4+2a^{-2})Z^4$$
$$+ (a^3+a+a^{-1}+a^{-3})Z^3$$
$$+ (-3a^2-6-3a^{-2})Z^2$$
$$+ (-a^3-2a-2a^{-1}-a^{-3})Z^1$$
$$+ (a^2+3+a^{-2})Z^0$$

$\underline{7_1}$:

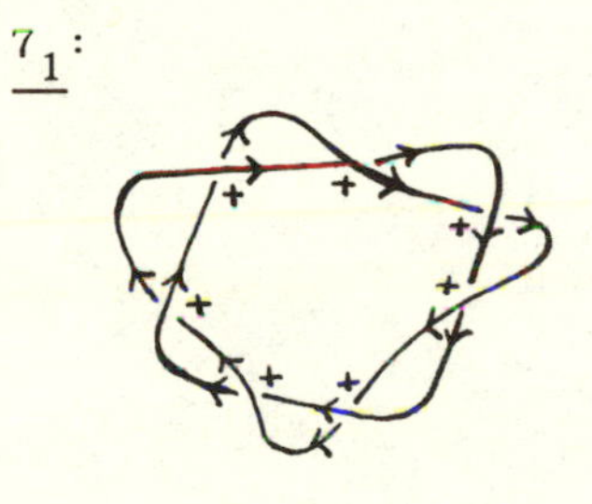

$\boxed{w = 7}$

$$L = (a+a^{-1})Z^6 + (1+a^{-2})Z^5$$
$$+ (-6a-5a^{-1}+a^{-3})Z^4$$
$$+ (-4-3a^{-2}+a^{-4})Z^3$$
$$+ (10a+7a^{-1}-2a^{-3}+a^{-5})Z^2$$
$$+ (3+a^{-2}-a^{-4}+a^{-6})Z^1$$
$$+ (-4a-3a^{-1})Z^0$$

$\underline{7_2}$:

$\boxed{w = -7}$

$$L = (a+a^{-1})Z^6 + (a^2+2+a^{-2})Z^5$$
$$+ (-4a-3a^{-1}+a^{-3})Z^4$$
$$+ (-4a^2-6-a^{-2}+a^{-4})Z^3$$
$$+ (4a+3a^{-1}+a^{-5})Z^2$$
$$+ (3a^2+3)Z^1 + (-a-a^{-1}-a^{-5})Z^0$$

$\underline{7_3}$:

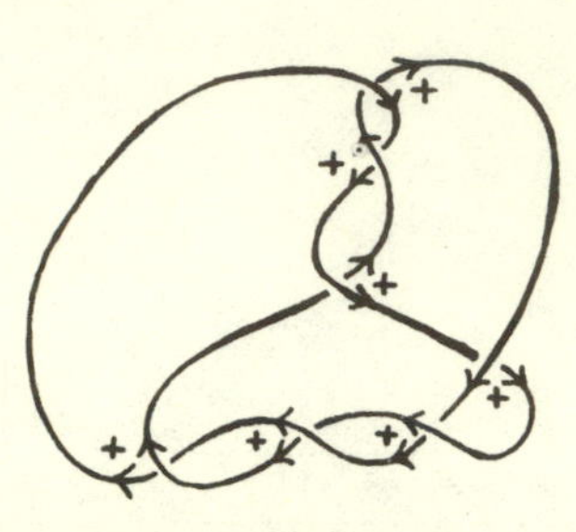

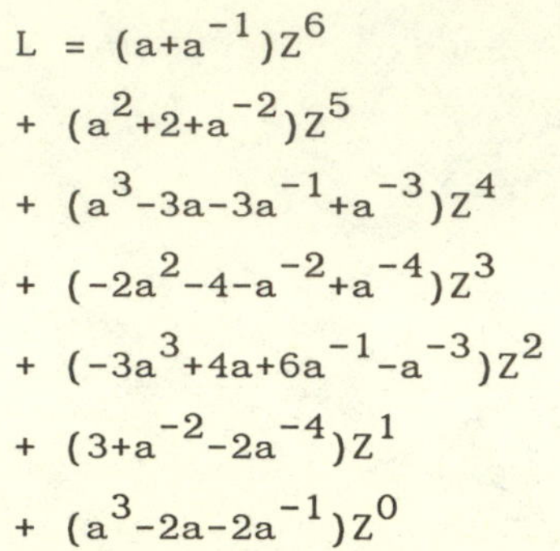

$\boxed{w = +7}$

$$L = (a+a^{-1})Z^6$$
$$+ (a^2+2+a^{-2})Z^5$$
$$+ (a^3-3a-3a^{-1}+a^{-3})Z^4$$
$$+ (-2a^2-4-a^{-2}+a^{-4})Z^3$$
$$+ (-3a^3+4a+6a^{-1}-a^{-3})Z^2$$
$$+ (3+a^{-2}-2a^{-4})Z^1$$
$$+ (a^3-2a-2a^{-1})Z^0$$

$\underline{7_4}$:

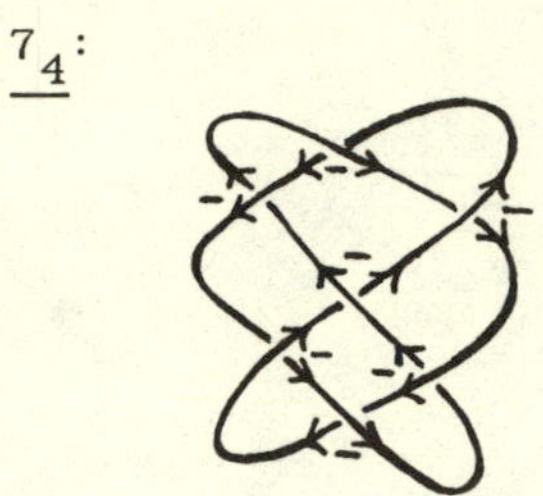

$\boxed{w = -7}$

$$L = (a+a^{-1})Z^6$$
$$+ (a^2+3+2a^{-2})Z^5$$
$$+ (-3a+3a^{-3})Z^4$$
$$+ (-4a^2-8-2a^{-2}+2a^{-4})Z^3$$
$$+ (2a-3a^{-1}-4a^{-3}+a^{-5})Z^2$$
$$+ (4a^2+4)Z^1$$
$$+ (-a+2a^{-3})Z^0$$

$\underline{7_5}$:

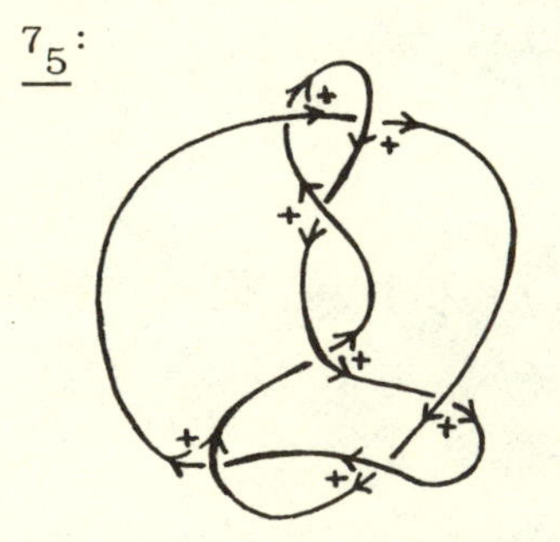

$\boxed{w = +7}$

$$L = (a+a^{-1})Z^6$$
$$+ (a^2+3+2a^{-2})Z^5$$
$$+ (a^3-a+2a^{-3})Z^4$$
$$+ (-a^2-4-2a^{-2}+a^{-4})Z^3$$
$$+ (-3a^3+a^{-1}-2a^{-3})Z^2$$
$$+ (-a^2+1+a^{-2}-a^{-4})Z^1$$
$$+ (2a^3-a^{-1})Z^0$$

$\underline{7_6}$:

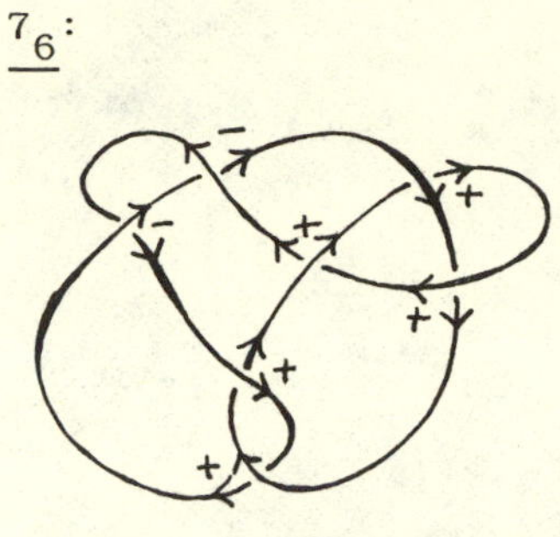

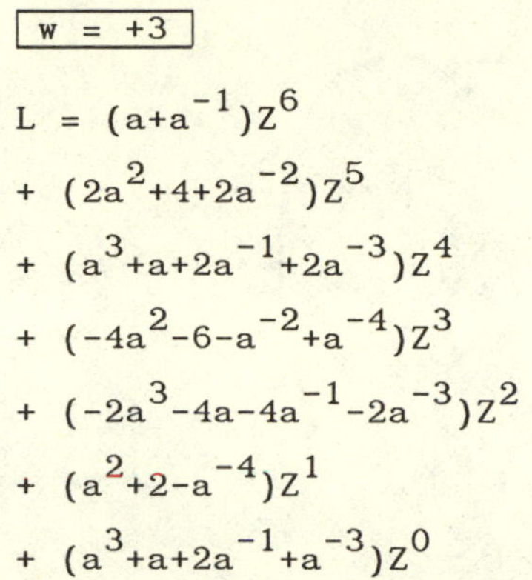

$\boxed{w = +3}$

$$L = (a+a^{-1})z^6$$
$$+ (2a^2+4+2a^{-2})z^5$$
$$+ (a^3+a+2a^{-1}+2a^{-3})z^4$$
$$+ (-4a^2-6-a^{-2}+a^{-4})z^3$$
$$+ (-2a^3-4a-4a^{-1}-2a^{-3})z^2$$
$$+ (a^2+2-a^{-4})z^1$$
$$+ (a^3+a+2a^{-1}+a^{-3})z^0$$

$\underline{7_7}$:

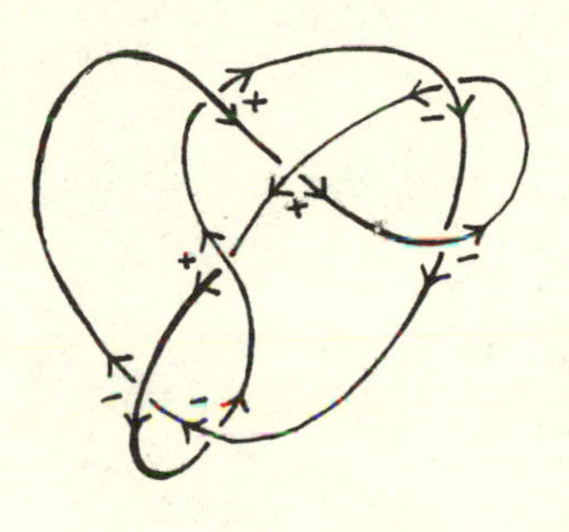

$\boxed{w = -1}$

$$L = (a+a^{-1})z^6$$
$$+ (2a^2+5+3a^{-2})z^5$$
$$+ (a^3+2a+4a^{-1}+3a^{-3})z^4$$
$$+ (-4a^2-8-3a^{-2}+a^{-4})z^3$$
$$+ (-2a^3-6a-7a^{-1}-3a^{-3})z^2$$
$$+ (2a^2+3+a^{-2})z^1$$
$$+ (a^3+2a+2a^{-1})z^0$$

$\underline{8_1}$:

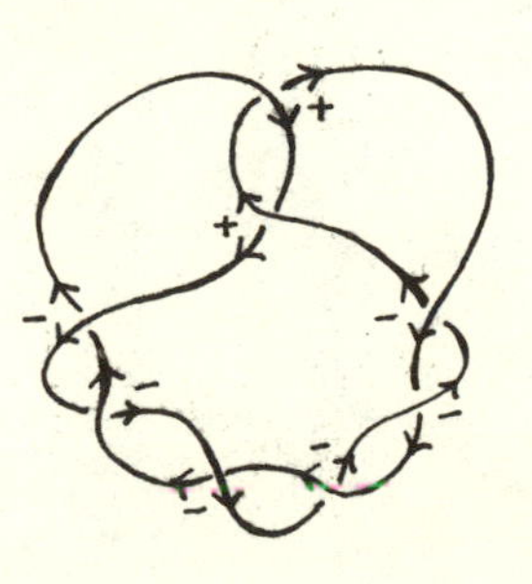

$\boxed{w = -4}$

$$L = (a+a^{-1})z^7 + (a^2+2+a^{-2})z^6$$
$$+ (-5a-4a^{-1}+a^{-3})z^5$$
$$+ (-5a^2-8-2a^{-2}+a^{-4})z^4$$
$$+ (7a+5a^{-1}-a^{-3}+a^{-5})z^3$$
$$+ (6a^2+7+a^{-6})z^2$$
$$+ (-3a-3a^{-1})z^1$$
$$+ (-a^2-1-a^{-6})z^0$$

$\underline{8_2}$:

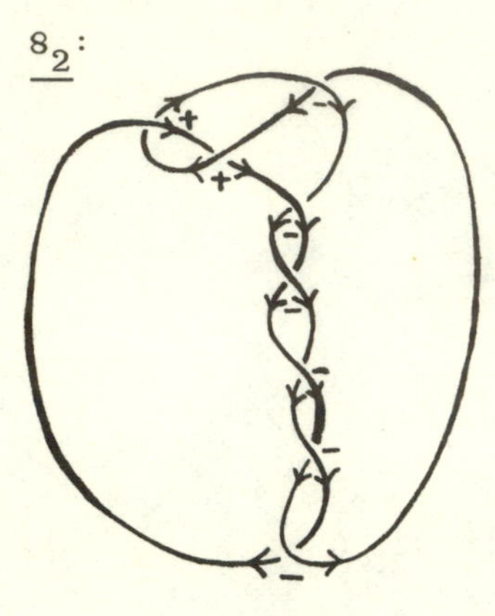

w = -4

$$L = (a+a^{-1})Z^7 + (2a^2+3+a^{-2})Z^6$$
$$+ (2a^3-2a-4a^{-1})Z^5$$
$$+ (2a^4-5a^2-12-5a^{-2})Z^4$$
$$+ (2a^5-2a^3-a+3a^{-1})Z^3$$
$$+ (a^6-a^4+3a^2+12+7a^{-2})Z^2$$
$$+ (-a^5-a^3+a+a^{-1})Z^1$$
$$+ (-a^2-3-3a^{-2})Z^0$$

$\underline{8_3}$:

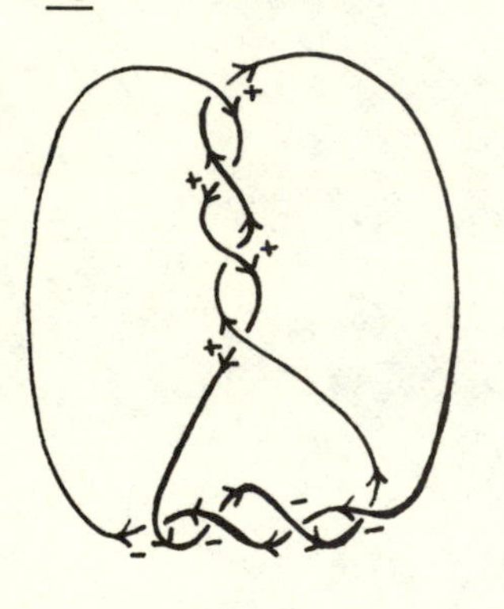

w = 0

8_3 is achiral

$$L = (a+a^{-1})Z^7 + (a^2+2+a^{-2})Z^6$$
$$+ (a^3-4a-4a^{-1}+a^{-3})Z^5$$
$$+ (a^4-2a^2-6-2a^{-2}+a^{-4})Z^4$$
$$+ (-2a^3+8a+8a^{-1}-2a^{-3})Z^3$$
$$+ (-3a^4+a^2+8+a^{-2}-3a^{-4})Z^2$$
$$+ (-4a-4a^{-1})Z^1$$
$$+ (a^4-1+a^{-4})Z^0$$

$\underline{8_4}$:

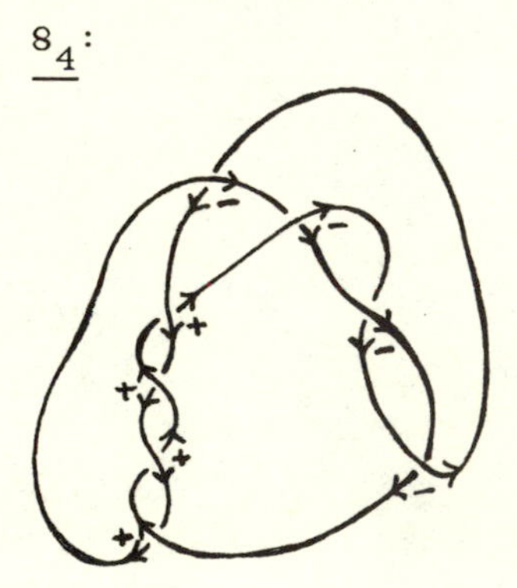

w = 0

$$L = (a+a^{-1})Z^7 + (2a^2+3+a^{-2})Z^6$$
$$+ (3a^3-a-4a^{-1})Z^5$$
$$+ (3a^4-3a^2-11-5a^{-2})Z^4$$
$$+ (2a^5-5a^3-3a+4a^{-1})Z^3$$
$$+ (a^6-3a^4-a^2+10+7a^{-2})Z^2$$
$$+ (2a^3+a-a^{-1})Z^1$$
$$+ (a^4-2-2a^{-2})Z^0$$

$\underline{8_5}$:

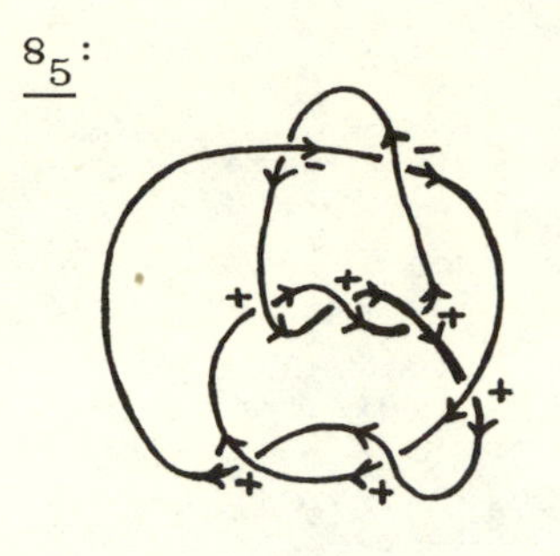

w = 4

$$L = (a+a^{-1})Z^7 + (a^2+4+3a^{-2})Z^6$$
$$+ (-3a+a^{-1}+4a^{-3})Z^5$$
$$+ (-5a^2-15-7a^{-2}+3a^{-4})Z^4$$
$$+ (-10a^{-1}-8a^{-3}+2a^{-5})Z^3$$
$$+ (8a^2+15+4a^{-2}-2a^{-4}+a^{-6})Z^2$$
$$+ (3a+7a^{-1}+4a^{-3})Z^1$$
$$+ (-4a^2-5-2a^{-2})Z^0$$

$\underline{8_6}$:

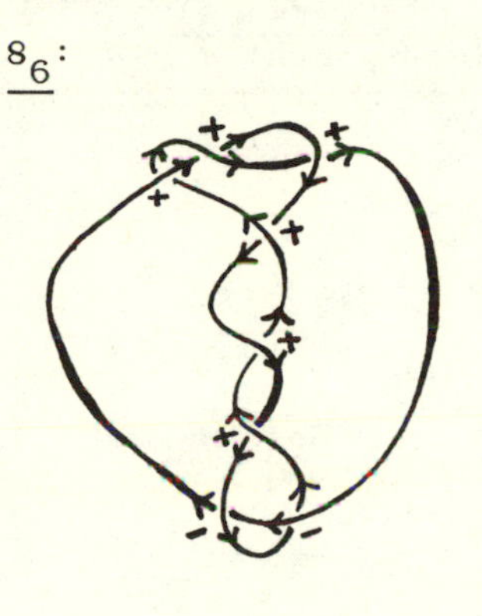

w = 4

$$L = (a+a^{-1})Z^7 + (a^2+3+2a^{-2})Z^6$$
$$+ (a^3-2a-a^{-1}+2a^{-3})Z^5$$
$$+ (a^4-6-4a^{-2}+a^{-4})Z^4$$
$$+ (-a^3+5a+2a^{-1}-4a^{-3})Z^3$$
$$+ (-3a^4-2a^2+6+3a^{-2}-2a^{-4})Z^2$$
$$+ (-a^3-3a-a^{-1}+a^{-3})Z^1$$
$$+ (2a^4+a^2-1-a^{-2})Z^0$$

$\underline{8_7}$:

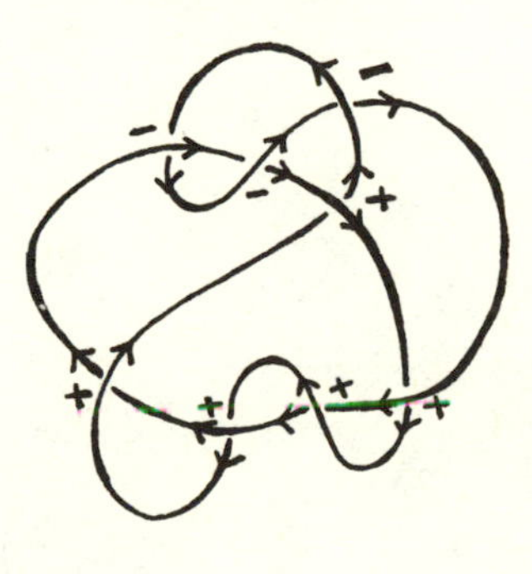

w = 2

$$L = (a+a^{-1})Z^7 + (2a^2+4+2a^{-2})Z^6$$
$$+ (a^3-a+2a^{-3})Z^5$$
$$+ (-7a^2-12-3a^{-2}+2a^{-4})Z^4$$
$$+ (-3a^3-3a-2a^{-1}-a^{-3}+a^{-5})Z^3$$
$$+ (6a^2+12+4a^{-2}-2a^{-4})Z^2$$
$$+ (a^3+2a+2a^{-1}-a^{-5})Z^1$$
$$+ (-a^2-4-2a^{-2})Z^0$$

$\underline{8_8}$:

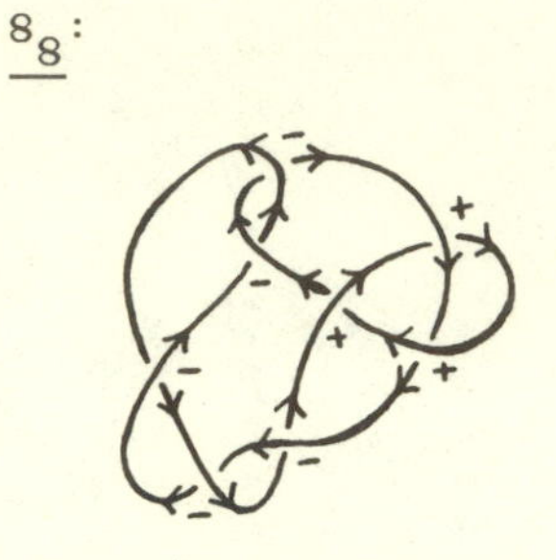

w = -2

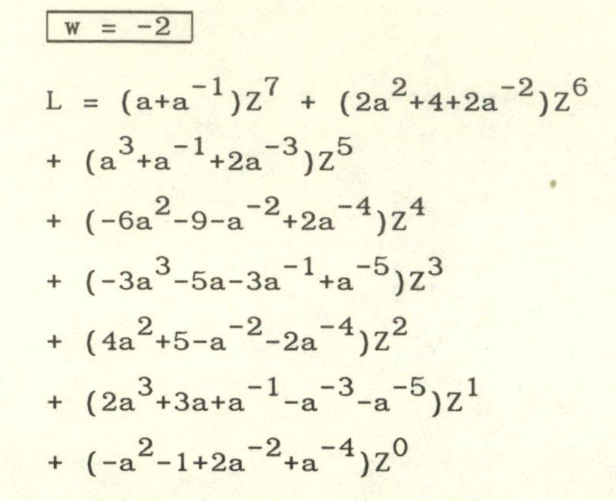

$$L = (a+a^{-1})Z^7 + (2a^2+4+2a^{-2})Z^6$$
$$+ (a^3+a^{-1}+2a^{-3})Z^5$$
$$+ (-6a^2-9-a^{-2}+2a^{-4})Z^4$$
$$+ (-3a^3-5a-3a^{-1}+a^{-5})Z^3$$
$$+ (4a^2+5-a^{-2}-2a^{-4})Z^2$$
$$+ (2a^3+3a+a^{-1}-a^{-3}-a^{-5})Z^1$$
$$+ (-a^2-1+2a^{-2}+a^{-4})Z^0$$

$\underline{8_9}$:

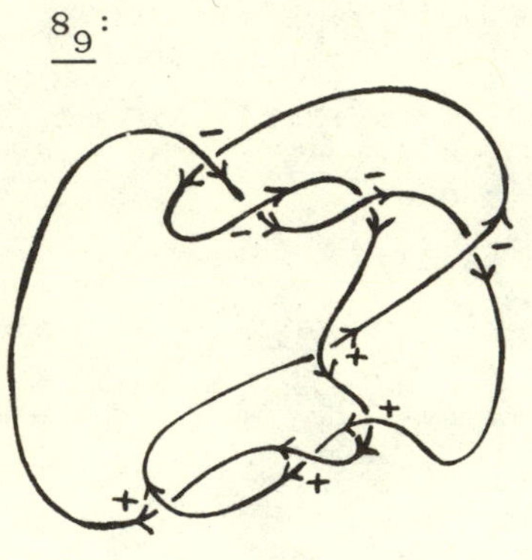

w = 0 | 8_9 is achiral

$$L = (a+a^{-1})Z^7 + (2a^2+4+2a^{-2})Z^6$$
$$+ (2a^3+2a^{-3})Z^5$$
$$+ (a^4-4a^2-10-4a^{-2}+a^{-4})Z^4$$
$$+ (-4a^3-a-a^{-1}-4a^{-3})Z^3$$
$$+ (-2a^4+4a^2+12+4a^{-2}-2a^{-4})Z^2$$
$$+ (a^3+a+a^{-1}+a^{-3})Z^1$$
$$+ (-2a^2-3-2a^{-2})Z^0$$

$\underline{8_{10}}$:

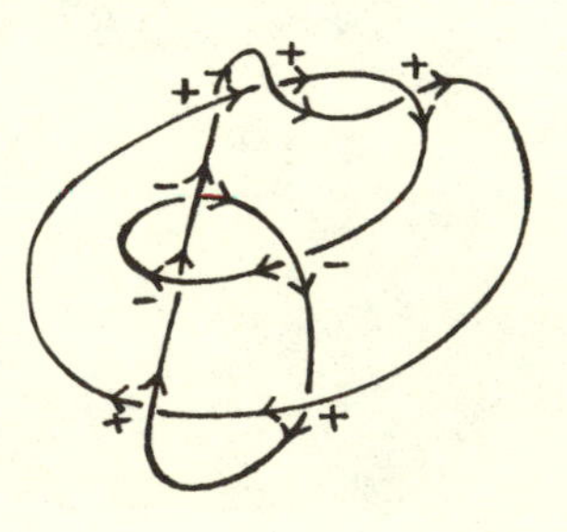

w = 2

$$L = (a+a^{-1})Z^7 + (2a^2+5+3a^{-2})Z^6$$
$$+ (a^3+a+3a^{-1}+3a^{-3})Z^5$$
$$+ (-6a^2-13-5a^{-2}+2a^{-4})Z^4$$
$$+ (-3a^3-8a-9a^{-1}-3a^{-3}+a^{-5})Z^3$$
$$+ (5a^2+12+6a^{-2}-a^{-4})Z^2$$
$$+ (2a^3+5a+6a^{-1}+2a^{-3}-a^{-5})Z^1$$
$$+ (-2a^2-6-3a^{-2})Z^0$$

$\underline{8_{11}}$:

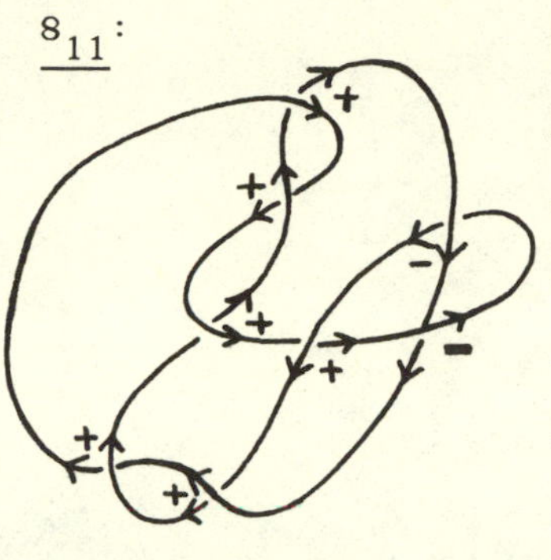

w = 4

$L = (a+a^{-1})Z^7 + (2a^2+4+2a^{-2})Z^6$
$+ (2a^3+a+a^{-1}+2a^{-3})Z^5$
$+ (a^4-2a^2-7-3a^{-2}+a^{-4})Z^4$
$+ (-3a^3-2a-3a^{-1}-4a^{-3})Z^3$
$+ (-2a^4+6+2a^{-2}-2a^{-4})Z^2$
$+ (a+3a^{-1}+2a^{-3})Z^1$
$+ (a^4-a^2-2-a^{-2})Z^0$

$\underline{8_{12}}$:

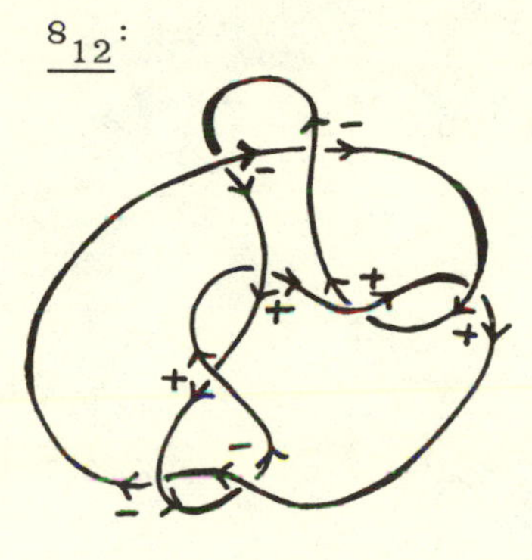

w = 0

8_{12} is achiral

$L = (a+a^{-1})Z^7 + (2a^2+4+2a^{-2})Z^6$
$+ (2a^3+2a+2a^{-1}+2a^{-3})Z^5$
$+ (a^4-a^2-4-a^{-2}+a^{-4})Z^4$
$+ (-3a^3-3a-3a^{-1}-3a^{-3})Z^3$
$+ (-2a^4-2a^2-2a^{-2}-2a^{-4})Z^2$
$+ (a^3+a^{-3})Z^1$
$+ (a^4+a^2+1+a^{-2}+a^{-4})Z^0$

$\underline{8_{13}}$:

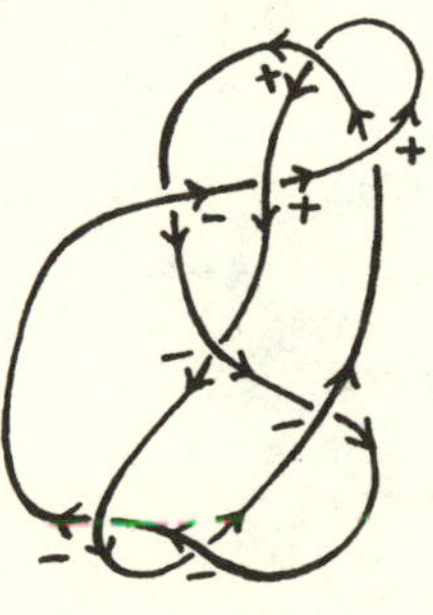

w = 2

$L = (a+a^{-1})Z^7 + (2a^2+5+3a^{-2})Z^6$
$+ (a^3+a+4a^{-1}+4a^{-3})Z^5$
$+ (-6a^2-11-2a^{-2}+3a^{-4})Z^4$
$+ (-3a^3-7a-9a^{-1}-4a^{-3}+a^{-5})Z^3$
$+ (5a^2+7-2a^{-4})Z^2$
$+ (2a^3+4a+3a^{-1}+a^{-3})Z^1$
$+ (-a^2-2)Z^0$

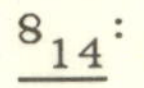

$\underline{8_{14}}$:

w = -4

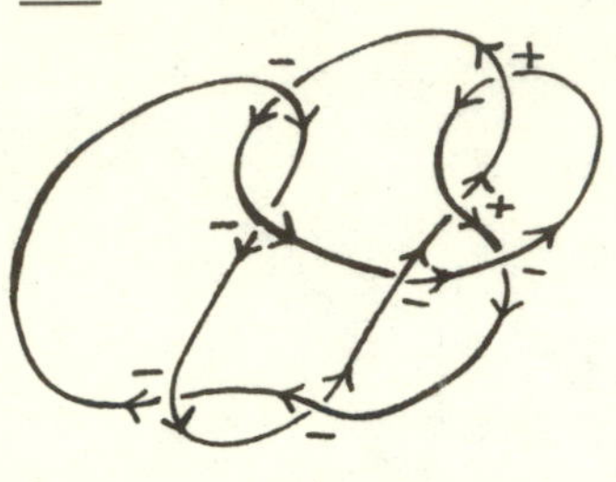

$$L = (a+a^{-1})Z^7 + (3a^2+5+2a^{-2})Z^6 + (3a^3+4a+3a^{-1}+2a^{-3})Z^5 + (a^4-4a^2-7-a^{-2}+a^{-4})Z^4 + (-5a^3-8a-6a^{-1}-3a^{-3})Z^3 + (-a^4+a^2+3-a^{-2}-2a^{-4})Z^2 + (a^3+3a+3a^{-1}+a^{-3})Z^1 + (a^{-4})Z^0$$

$\underline{8_{15}}$:

w = 8

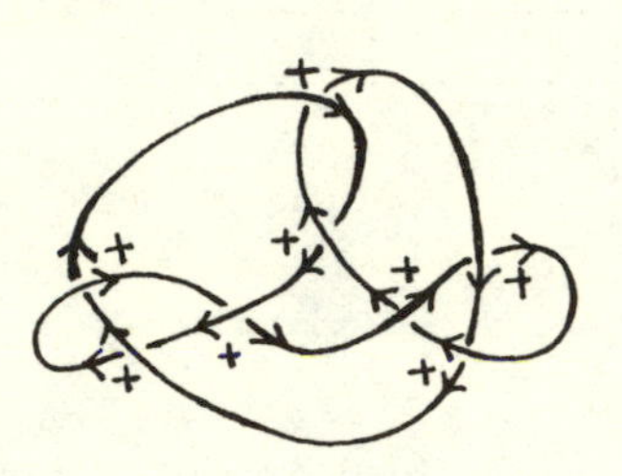

$$L = (a+a^{-1})Z^7 + (3a^2+6+3a^{-2})Z^6 + (2a^3+5a+6a^{-1}+3a^{-3})Z^5 + (a^4-5a^2-10-3a^{-2}+a^{-4})Z^4 + (-2a^3-11a-14a^{-1}-5a^{-3})Z^3 + (-2a^4+5a^2+8-a^{-4})Z^2 + (6a+8a^{-1}+2a^{-3})Z^1 + (a^4-3a^2-4-a^{-2})Z^0$$

$\underline{8_{16}}$:

w = 2

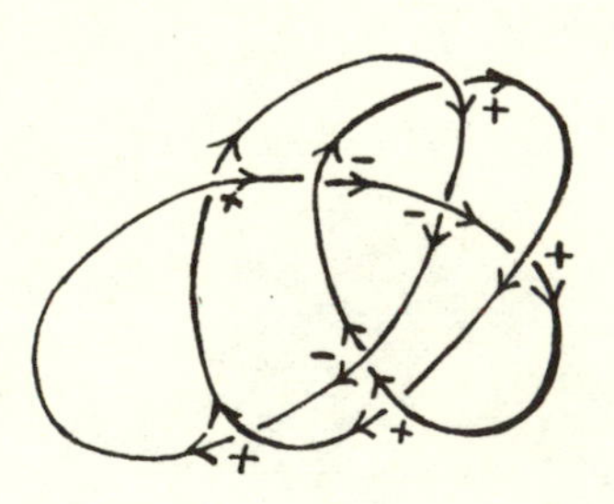

$$L = (2a+2a^{-1})Z^7 + (3a^2+8+5a^{-2})Z^6 + (a^3-a+3a^{-1}+5a^{-3})Z^5 + (-8a^2-18-7a^{-2}+3a^{-4})Z^4 + (-2a^3-6a-10a^{-1}-5a^{-3}+a^{-5})Z^3 + (5a^2+10+4a^{-2}-a^{-4})Z^2 + (a^3+3a+4a^{-1}+2a^{-3})Z^1 + (-2-a^{-2})Z^0$$

$\underline{8_{17}}$:

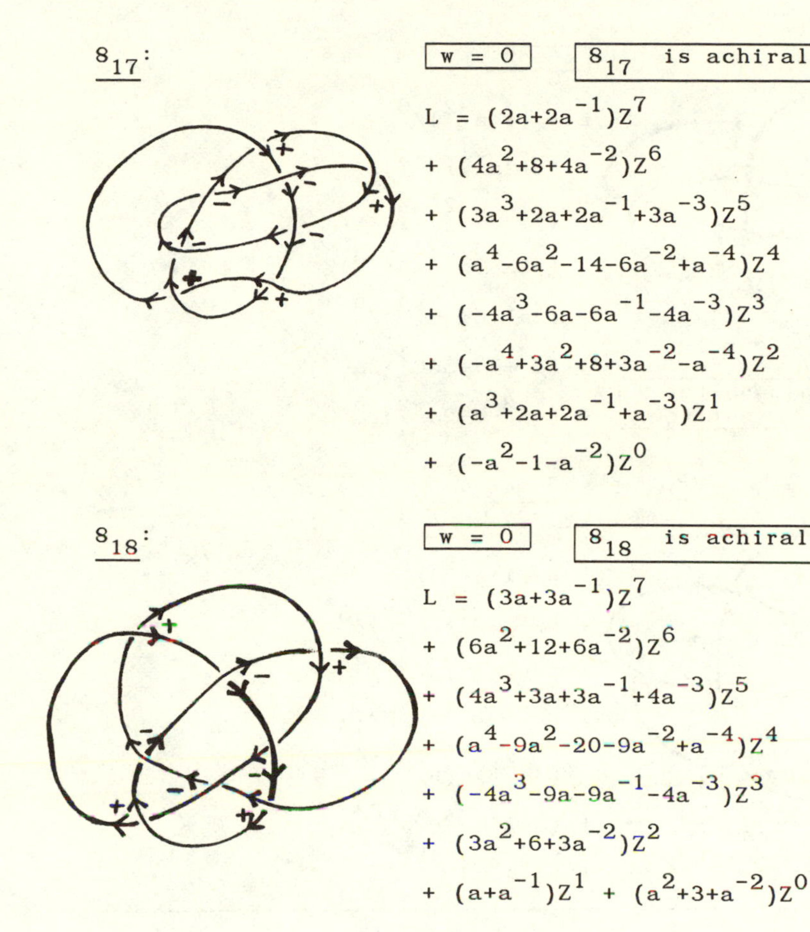

$w = 0$

8_{17} is achiral

$$L = (2a+2a^{-1})Z^7$$
$$+ (4a^2+8+4a^{-2})Z^6$$
$$+ (3a^3+2a+2a^{-1}+3a^{-3})Z^5$$
$$+ (a^4-6a^2-14-6a^{-2}+a^{-4})Z^4$$
$$+ (-4a^3-6a-6a^{-1}-4a^{-3})Z^3$$
$$+ (-a^4+3a^2+8+3a^{-2}-a^{-4})Z^2$$
$$+ (a^3+2a+2a^{-1}+a^{-3})Z^1$$
$$+ (-a^2-1-a^{-2})Z^0$$

$\underline{8_{18}}$:

$w = 0$

8_{18} is achiral

$$L = (3a+3a^{-1})Z^7$$
$$+ (6a^2+12+6a^{-2})Z^6$$
$$+ (4a^3+3a+3a^{-1}+4a^{-3})Z^5$$
$$+ (a^4-9a^2-20-9a^{-2}+a^{-4})Z^4$$
$$+ (-4a^3-9a-9a^{-1}-4a^{-3})Z^3$$
$$+ (3a^2+6+3a^{-2})Z^2$$
$$+ (a+a^{-1})Z^1 + (a^2+3+a^{-2})Z^0$$

$\underline{8_{19}}$:

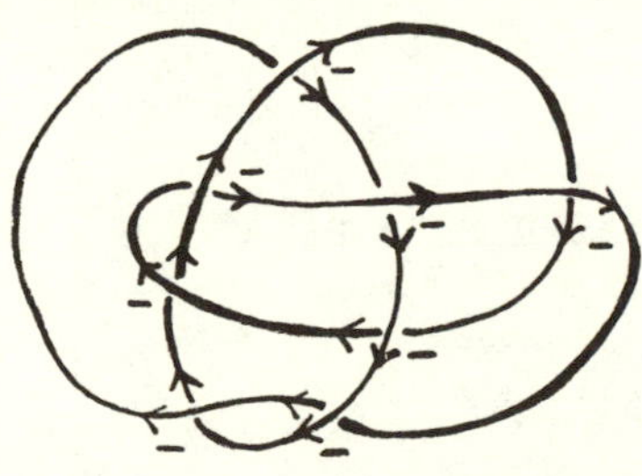

$w = -8$

8_{19} is the first nonalternating knot.

$$L = (1+a^{-2})Z^6$$
$$+ (a+a^{-1})Z^5$$
$$+ (-6-6a^{-2})Z^4$$
$$+ (-5a-5a^{-1})Z^3$$
$$+ (10+10a^{-2})Z^2$$
$$+ (5a+5a^{-1})Z^1 + (-a^2-5-5a^{-2})Z^0$$

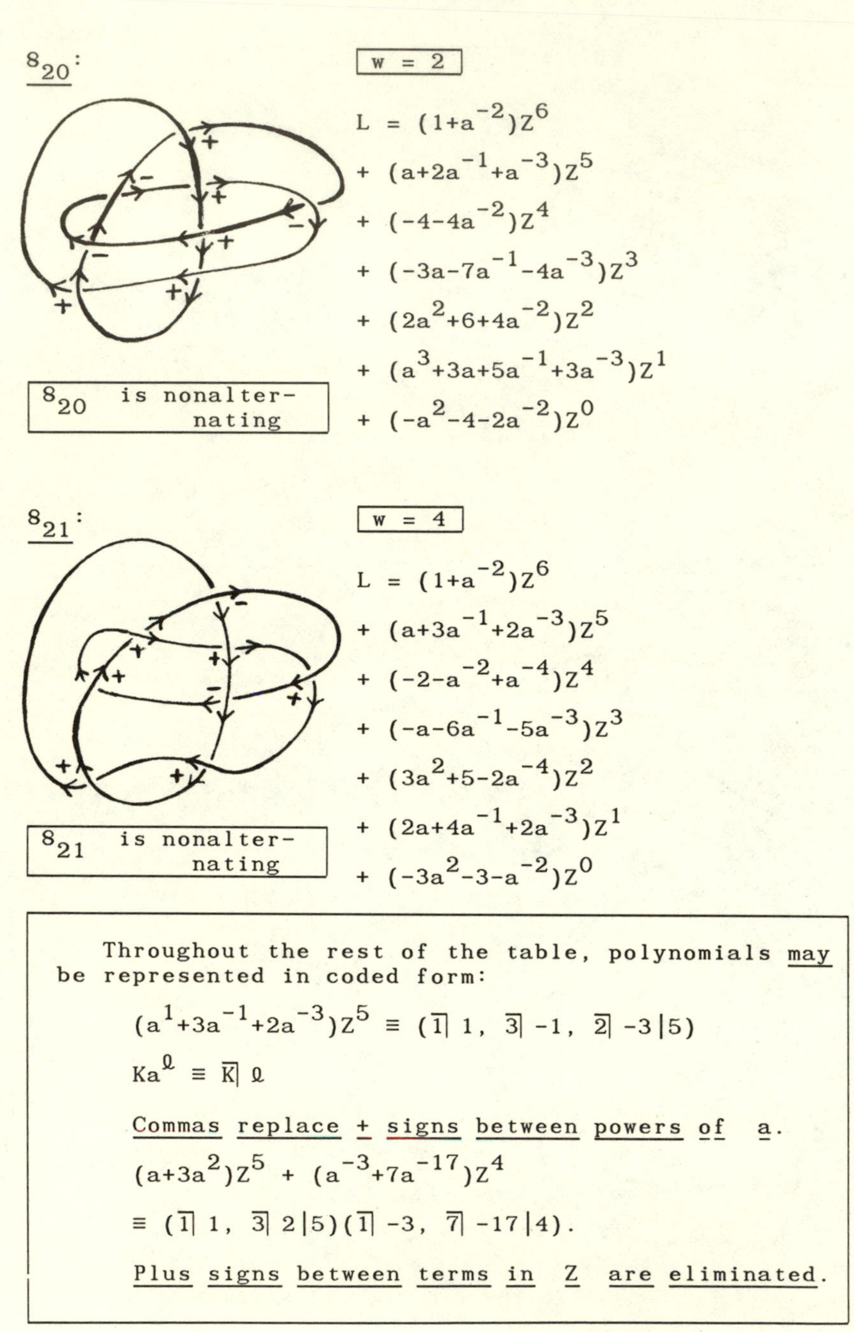

$\underline{8_{20}}$:

$\boxed{w = 2}$

$$L = (1+a^{-2})Z^6$$
$$+ (a+2a^{-1}+a^{-3})Z^5$$
$$+ (-4-4a^{-2})Z^4$$
$$+ (-3a-7a^{-1}-4a^{-3})Z^3$$
$$+ (2a^2+6+4a^{-2})Z^2$$
$$+ (a^3+3a+5a^{-1}+3a^{-3})Z^1$$
$$+ (-a^2-4-2a^{-2})Z^0$$

8_{20} is nonalternating

$\underline{8_{21}}$:

$\boxed{w = 4}$

$$L = (1+a^{-2})Z^6$$
$$+ (a+3a^{-1}+2a^{-3})Z^5$$
$$+ (-2-a^{-2}+a^{-4})Z^4$$
$$+ (-a-6a^{-1}-5a^{-3})Z^3$$
$$+ (3a^2+5-2a^{-4})Z^2$$
$$+ (2a+4a^{-1}+2a^{-3})Z^1$$
$$+ (-3a^2-3-a^{-2})Z^0$$

8_{21} is nonalternating

Throughout the rest of the table, polynomials <u>may</u> be represented in coded form:

$$(a^1+3a^{-1}+2a^{-3})Z^5 \equiv (\overline{1}|\ 1,\ \overline{3}|\ -1,\ \overline{2}|\ -3\,|5)$$

$$Ka^{\ell} \equiv \overline{K}|\ \ell$$

<u>Commas replace + signs between powers of a.</u>

$$(a+3a^2)Z^5 + (a^{-3}+7a^{-17})Z^4$$

$$\equiv (\overline{1}|\ 1,\ \overline{3}|\ 2|5)(\overline{1}|\ -3,\ \overline{7}|\ -17|4).$$

<u>Plus signs between terms in Z are eliminated.</u>

$\underline{9_1}$:

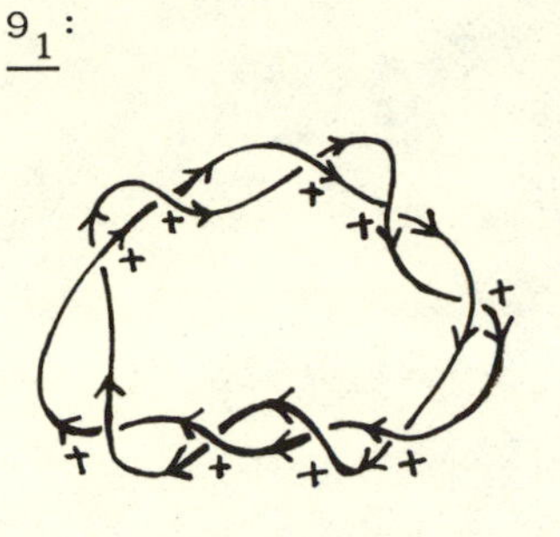

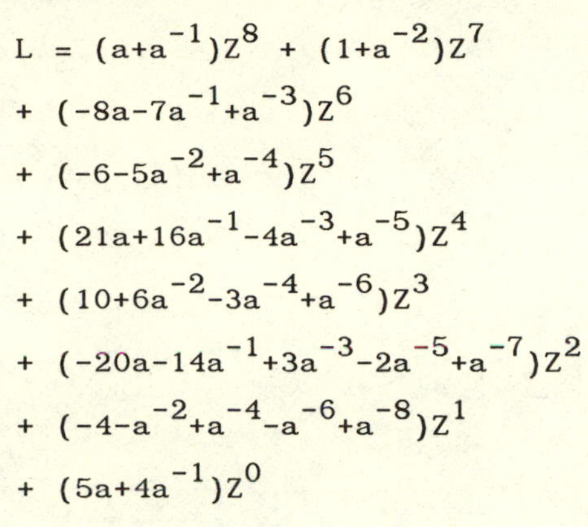

w = 9

$$L = (a+a^{-1})Z^8 + (1+a^{-2})Z^7$$
$$+ (-8a-7a^{-1}+a^{-3})Z^6$$
$$+ (-6-5a^{-2}+a^{-4})Z^5$$
$$+ (21a+16a^{-1}-4a^{-3}+a^{-5})Z^4$$
$$+ (10+6a^{-2}-3a^{-4}+a^{-6})Z^3$$
$$+ (-20a-14a^{-1}+3a^{-3}-2a^{-5}+a^{-7})Z^2$$
$$+ (-4-a^{-2}+a^{-4}-a^{-6}+a^{-8})Z^1$$
$$+ (5a+4a^{-1})Z^0$$

$\underline{9_2}$:

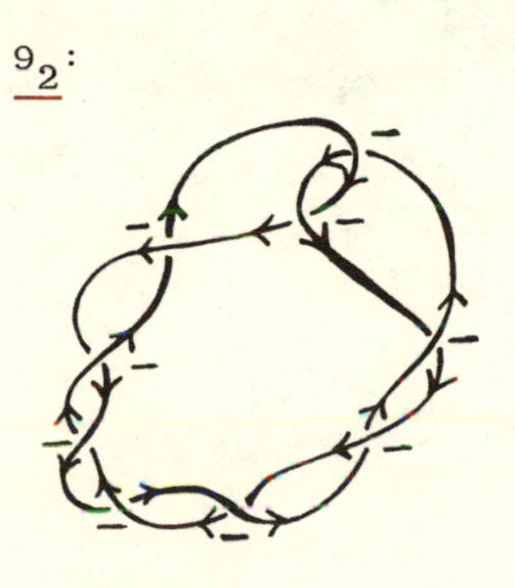

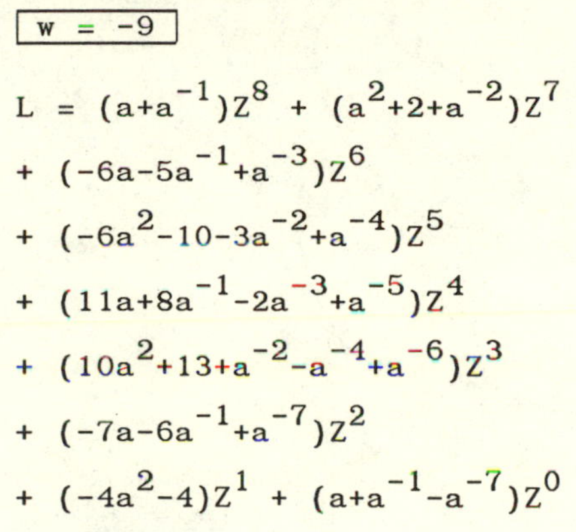

w = -9

$$L = (a+a^{-1})Z^8 + (a^2+2+a^{-2})Z^7$$
$$+ (-6a-5a^{-1}+a^{-3})Z^6$$
$$+ (-6a^2-10-3a^{-2}+a^{-4})Z^5$$
$$+ (11a+8a^{-1}-2a^{-3}+a^{-5})Z^4$$
$$+ (10a^2+13+a^{-2}-a^{-4}+a^{-6})Z^3$$
$$+ (-7a-6a^{-1}+a^{-7})Z^2$$
$$+ (-4a^2-4)Z^1 + (a+a^{-1}-a^{-7})Z^0$$

$\underline{9_3}$:

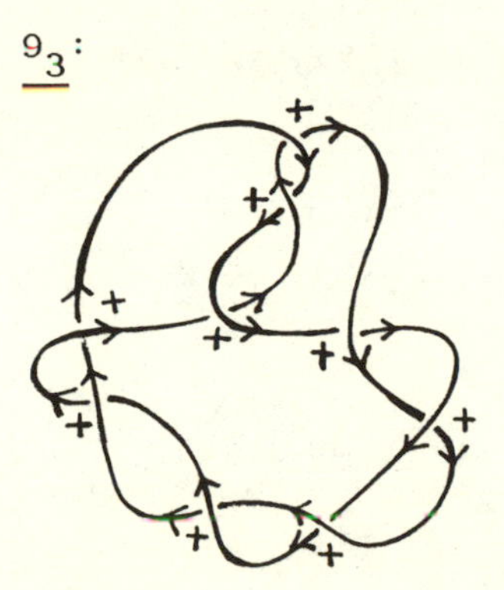

w = 9

$$L = (a+a^{-1})Z^8 + (a^2+2+a^{-2})Z^7$$
$$+ (a^3-5a-5a^{-1}+a^{-3})Z^6$$
$$+ (-4a^2-8-3a^{-2}+a^{-4})Z^5$$
$$+ (-5a^3+9a+11a^{-1}-2a^{-3}+a^{-5})Z^4$$
$$+ (3a^2+9+4a^{-2}-a^{-4}+a^{-6})Z^3$$
$$+ (6a^3-9a-11a^{-1}+3a^{-3}-a^{-5})Z^2$$
$$+ (-4-a^{-2}+a^{-4}-2a^{-6})Z^1$$
$$+ (-a^3+3a+3a^{-1})Z^0$$

$\underline{9_4}$:

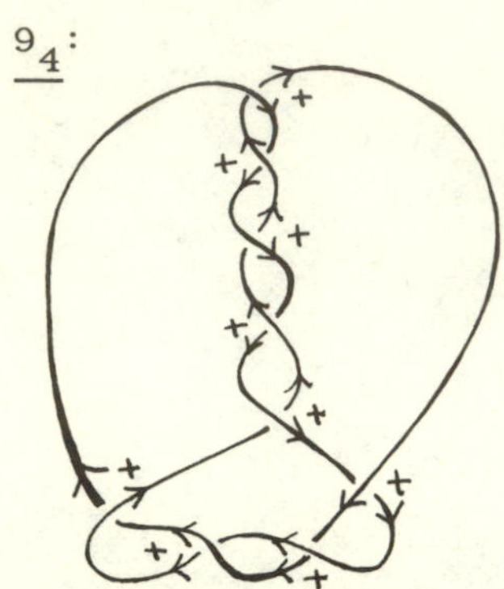

$\boxed{w = 9}$

$$L = (a+a^{-1})Z^{8} + (a^{2}+2+a^{-2})Z^{7}$$
$$+ (a^{3}-5a-5a^{-1}+a^{-3})Z^{6}$$
$$+ (a^{4}-3a^{2}-8-3a^{-2}+a^{-4})Z^{5}$$
$$+ (a^{5}-2a^{3}+11a+11a^{-1}-3a^{-3})Z^{4}$$
$$+ (-2a^{4}+4a^{2}+12+2a^{-2}-4a^{-4})Z^{3}$$
$$+ (-3a^{5}+a^{3}-7a-10a^{-1}+a^{-3})Z^{2}$$
$$+ (-4-a^{-2}+3a^{-4})Z^{1} + (a^{5}+2a+2a^{-1})Z^{0}$$

$\underline{9_5}$:

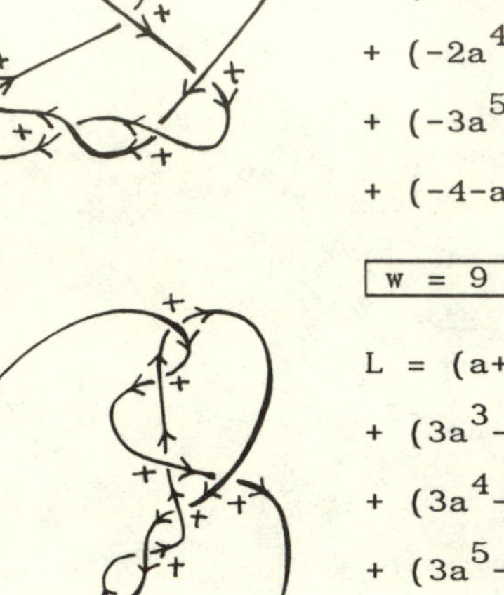

$\boxed{w = 9}$

$$L = (a+a^{-1})Z^{8} + (2a^{2}+3+a^{-2})Z^{7}$$
$$+ (3a^{3}-2a-5a^{-1})Z^{6}$$
$$+ (3a^{4}-5a^{2}-14-6a^{-2})Z^{5}$$
$$+ (3a^{5}-7a^{3}-3a+7a^{-1})Z^{4}$$
$$+ (2a^{6}-4a^{4}+a^{2}+18+11a^{-2})Z^{3}$$
$$+ (a^{7}-3a^{5}+3a^{3}+4a-3a^{-1})Z^{2}$$
$$+ (-6-6a^{-2})Z^{1} + (a^{5}-a^{3}+a^{-1})Z^{0}$$

$\underline{9_6}$:

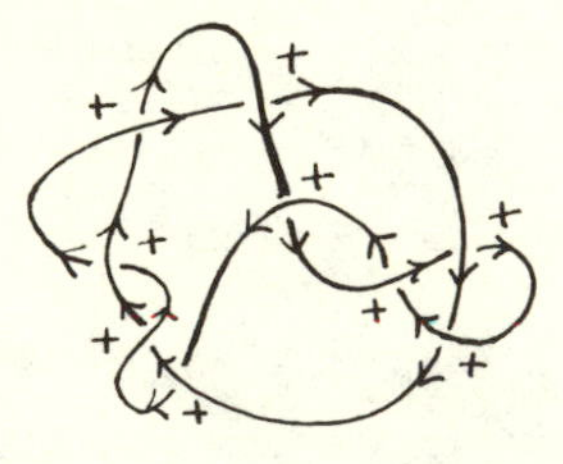

$\boxed{w = 9}$

$$L = (a+a^{-1})Z^{8} + (a^{2}+3+2a^{-2})Z^{7}$$
$$+ (a^{3}-3a-2a^{-1}+2a^{-3})Z^{6}$$
$$+ (-3a^{2}-10-5a^{-2}+2a^{-4})Z^{5}$$
$$+ (-5a^{3}+a+2a^{-1}-2a^{-3}+2a^{-5})Z^{4}$$
$$+ (8+6a^{-2}-a^{-4}+a^{-6})Z^{3}$$
$$+ (7a^{3}+a-3a^{-1}+a^{-3}-2a^{-5})Z^{2}$$
$$+ (2a^{2}-1-2a^{-2}-a^{-6})Z^{1}$$
$$+ (-3a^{3}-a+a^{-1})Z^{0}$$

$\underline{9_7}$:

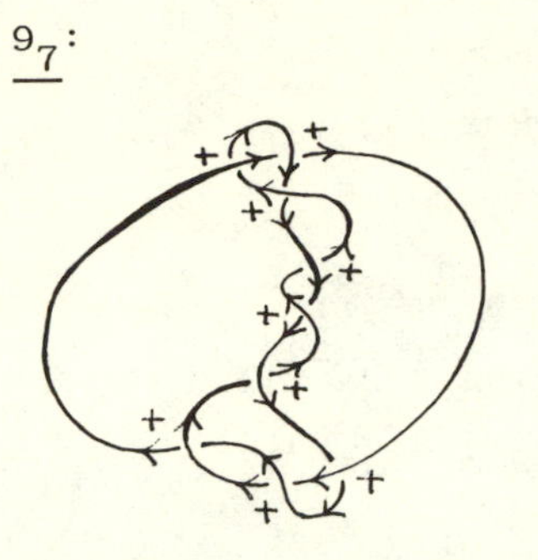

w = 9

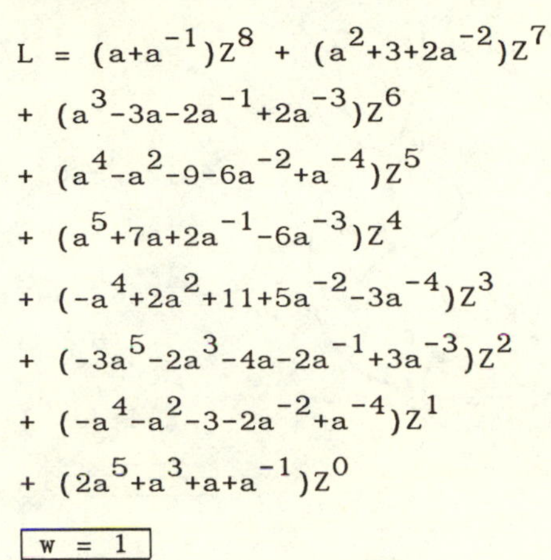

$$L = (a+a^{-1})Z^8 + (a^2+3+2a^{-2})Z^7$$
$$+ (a^3-3a-2a^{-1}+2a^{-3})Z^6$$
$$+ (a^4-a^2-9-6a^{-2}+a^{-4})Z^5$$
$$+ (a^5+7a+2a^{-1}-6a^{-3})Z^4$$
$$+ (-a^4+2a^2+11+5a^{-2}-3a^{-4})Z^3$$
$$+ (-3a^5-2a^3-4a-2a^{-1}+3a^{-3})Z^2$$
$$+ (-a^4-a^2-3-2a^{-2}+a^{-4})Z^1$$
$$+ (2a^5+a^3+a+a^{-1})Z^0$$

$\underline{9_8}$:

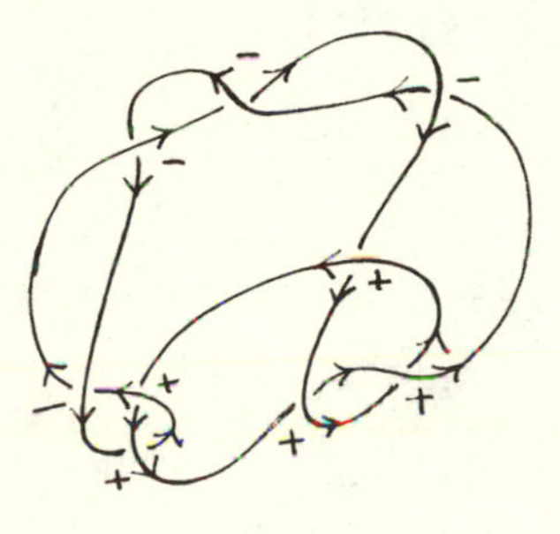

w = 1

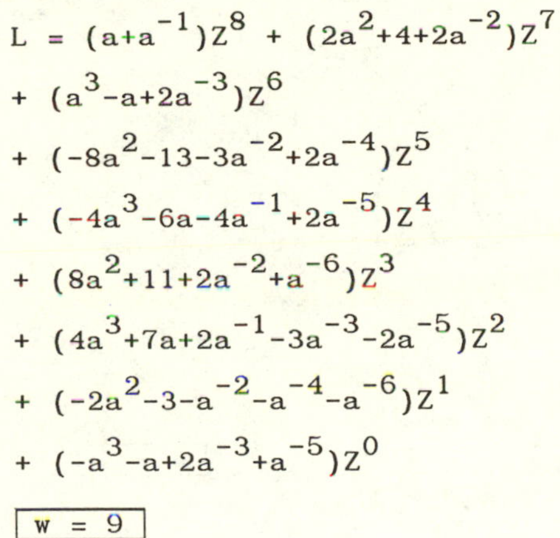

$$L = (a+a^{-1})Z^8 + (2a^2+4+2a^{-2})Z^7$$
$$+ (a^3-a+2a^{-3})Z^6$$
$$+ (-8a^2-13-3a^{-2}+2a^{-4})Z^5$$
$$+ (-4a^3-6a-4a^{-1}+2a^{-5})Z^4$$
$$+ (8a^2+11+2a^{-2}+a^{-6})Z^3$$
$$+ (4a^3+7a+2a^{-1}-3a^{-3}-2a^{-5})Z^2$$
$$+ (-2a^2-3-a^{-2}-a^{-4}-a^{-6})Z^1$$
$$+ (-a^3-a+2a^{-3}+a^{-5})Z^0$$

$\underline{9_9}$:

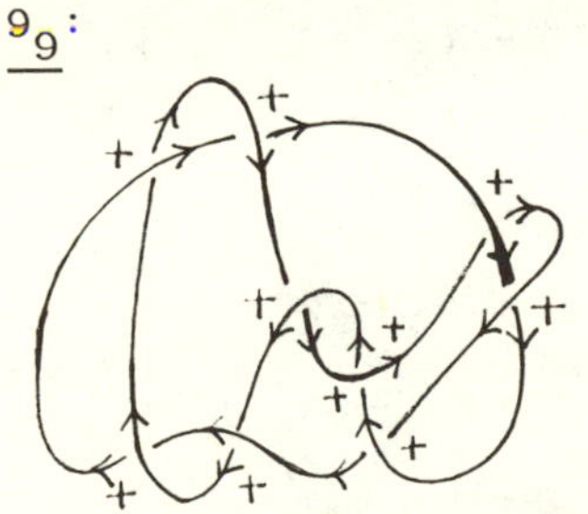

w = 9

$$L = (a+a^{-1})Z^8 + (a^2+3+2a^{-2})Z^7$$
$$+ (a^3-3a-a^{-1}+3a^{-3})Z^6$$
$$+ (-3a^2-8-2a^{-2}+3a^{-4})Z^5$$
$$+ (-5a^3+3a+2a^{-1}-4a^{-3}+2a^{-5})Z^4$$
$$+ (a^2+5-3a^{-4}+a^{-6})Z^3$$
$$+ (7a^3-3a-6a^{-1}+3a^{-3}-a^{-5})Z^2$$
$$+ (a^2-2+2a^{-4}-a^{-6})Z^1$$
$$+ (-2a^3+a+2a^{-1})Z^0$$

$\underline{9_{10}}$:

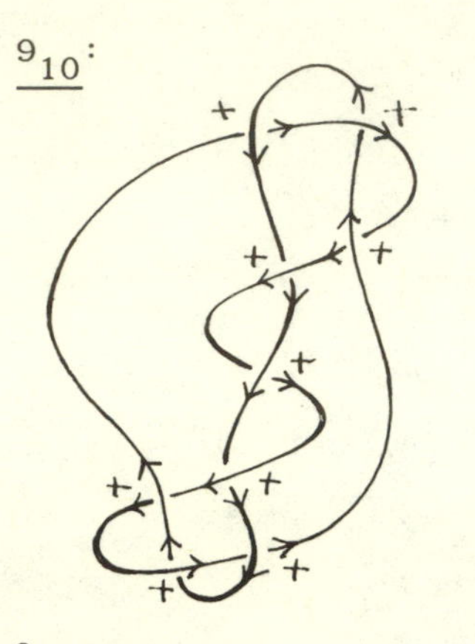

w = 9

$L = (a+a^{-1})Z^8 + (2a^2+3+a^{-2})Z^7$
$+ (3a^3-a-3a^{-1}+a^{-3})Z^6$
$+ (2a^4-3a^2-7-a^{-2}+a^{-4})Z^5$
$+ (a^5-7a^3+3a+9a^{-1}-2a^{-3})Z^4$
$+ (-3a^4+3a^2+9-a^{-2}-4a^{-4})Z^3$
$+ (-2a^5+7a^3-2a-11a^{-1})Z^2$
$+ (-4+4a^{-4})Z^1 + (-2a^3+a+2a^{-1})Z^0$

$\underline{9_{11}}$:

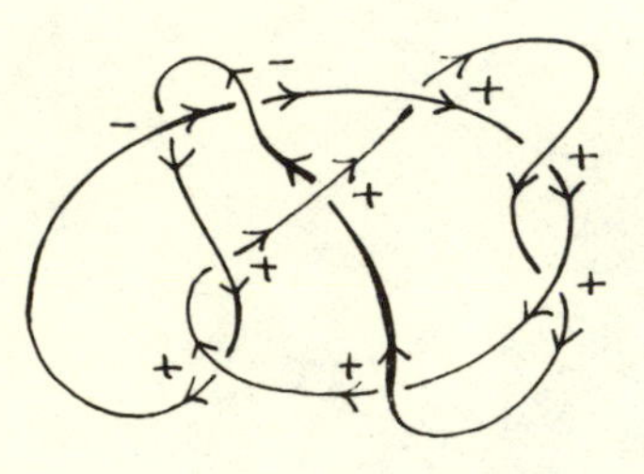

w = 5

$L = (a+a^{-1})Z^8 + (2a^2+4+2a^{-2})Z^7$
$+ (a^3-a+a^{-1}+3a^{-3})Z^6$
$+ (-8a^2-12-a^{-2}+3a^{-4})Z^5$
$+ (-4a^3-5a-7a^{-1}-4a^{-3}+2a^{-5})Z^4$
$+ (8a^2+9-3a^{-2}-3a^{-4}+a^{-6})Z^3$
$+ (4a^3+5a+6a^{-1}+4a^{-3}-a^{-5})Z^2$
$+ (-a^2-2+2a^{-2}+2a^{-4}-a^{-6})Z^1$
$+ (-a^3-a-3a^{-1}-2a^{-3})Z^0$

$\underline{9_{12}}$:

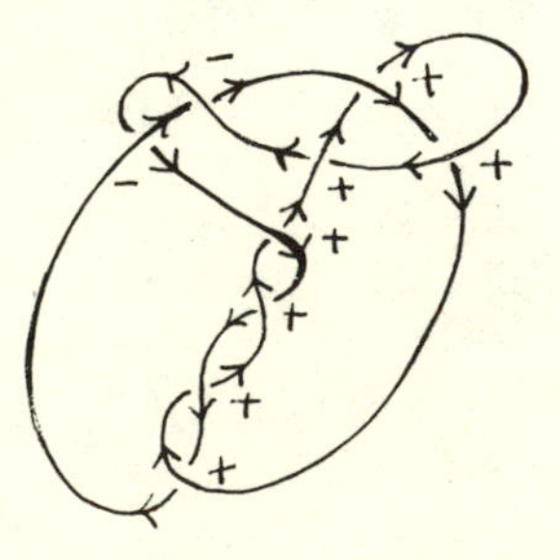

w = 5

$L = (a+a^{-1})Z^8 + (2a^2+4+2a^{-2})Z^7$
$+ (2a^3+2a^{-3})Z^6$
$+ (2a^4-3a^2-11-5a^{-2}+a^{-4})Z^5$
$+ (a^5-a^3-a-5a^{-1}-6a^{-3})Z^4$
$+ (-3a^4+4a^2+13+3a^{-2}-3a^{-4})Z^3$
$+ (-2a^5-2a^3+3a+7a^{-1}+4a^{-3})Z^2$
$+ (-2a^2-4-a^{-2}+a^{-4})Z^1$
$+ (a^5-a-2a^{-1}-a^{-3})Z^0$

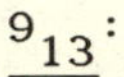

9_{13}:

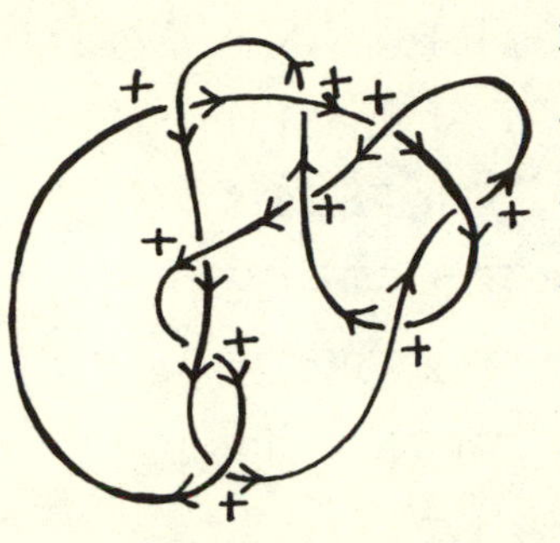

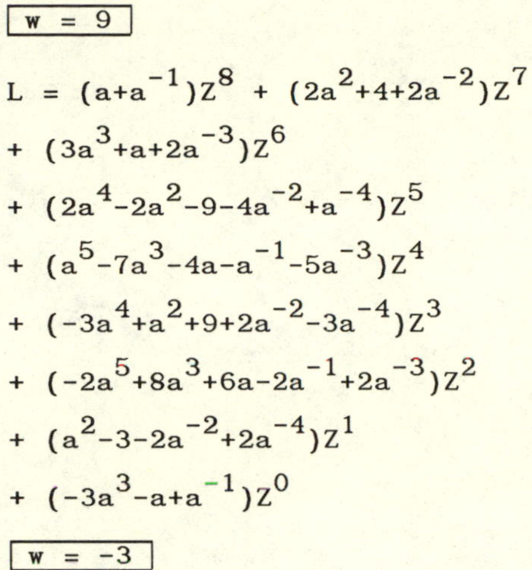

w = 9

$$L = (a+a^{-1})Z^8 + (2a^2+4+2a^{-2})Z^7$$
$$+ (3a^3+a+2a^{-3})Z^6$$
$$+ (2a^4-2a^2-9-4a^{-2}+a^{-4})Z^5$$
$$+ (a^5-7a^3-4a-a^{-1}-5a^{-3})Z^4$$
$$+ (-3a^4+a^2+9+2a^{-2}-3a^{-4})Z^3$$
$$+ (-2a^5+8a^3+6a-2a^{-1}+2a^{-3})Z^2$$
$$+ (a^2-3-2a^{-2}+2a^{-4})Z^1$$
$$+ (-3a^3-a+a^{-1})Z^0$$

9_{14}:

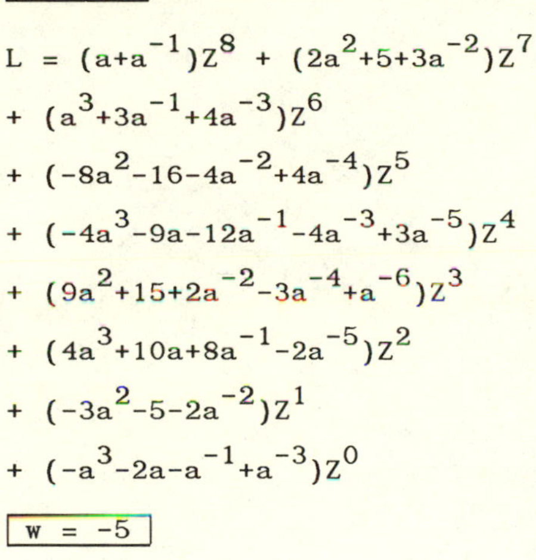

w = -3

$$L = (a+a^{-1})Z^8 + (2a^2+5+3a^{-2})Z^7$$
$$+ (a^3+3a^{-1}+4a^{-3})Z^6$$
$$+ (-8a^2-16-4a^{-2}+4a^{-4})Z^5$$
$$+ (-4a^3-9a-12a^{-1}-4a^{-3}+3a^{-5})Z^4$$
$$+ (9a^2+15+2a^{-2}-3a^{-4}+a^{-6})Z^3$$
$$+ (4a^3+10a+8a^{-1}-2a^{-5})Z^2$$
$$+ (-3a^2-5-2a^{-2})Z^1$$
$$+ (-a^3-2a-a^{-1}+a^{-3})Z^0$$

9_{15}:

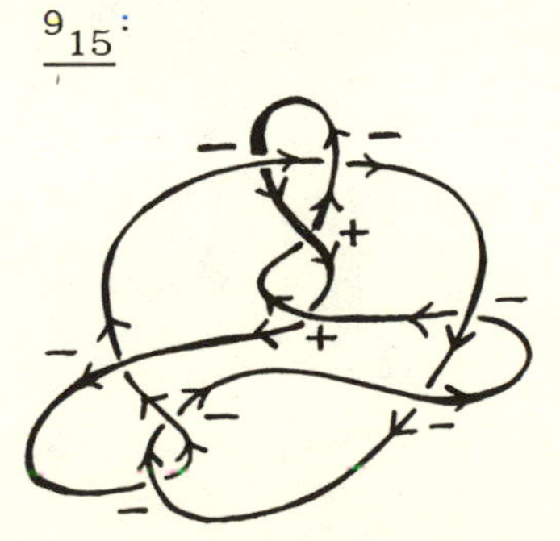

w = -5

$$L = (a+a^{-1})Z^8 + (2a^2+4+2a^{-2})Z^7$$
$$+ (2a^3+a+a^{-1}+2a^{-3})Z^6$$
$$+ (a^4-3a^2-7-a^{-2}+2a^{-4})Z^5$$
$$+ (-5a^3-4a+a^{-5})Z^4$$
$$+ (-3a^4-a^2+5-3a^{-4})Z^3$$
$$+ (3a^3+2a-2a^{-1}-3a^{-3}-2a^{-5})Z^2$$
$$+ (2a^4+a^2-1+a^{-2}+a^{-4})Z^1$$
$$+ (-a^3-a+a^{-1}+a^{-3}+a^{-5})Z^0$$

$\underline{9_{16}}$:

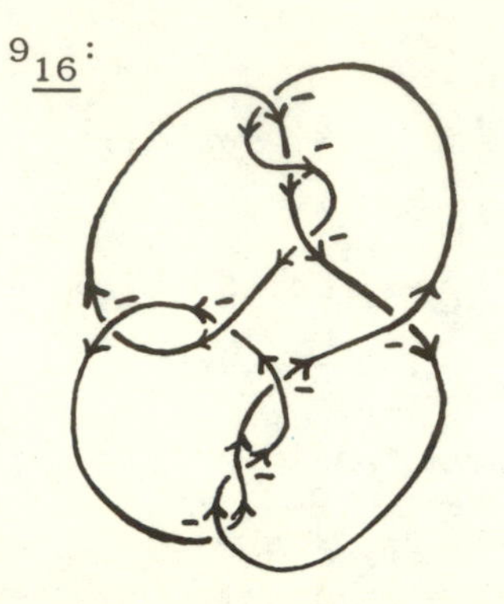

$\boxed{w = -9}$

$L = (\overline{1|}\ 1, \overline{1|}\ -1\,|\,8)(\overline{3|}\ 2, \overline{4|}\ 0, \overline{1|}\ -2\,|\,7)$

$(\overline{5|}\ 3, \overline{3|}\ 1, \overline{-1|}\ -1, \overline{1|}\ -3\,|\,6)$

$(\overline{5|}\ 4, \overline{-1|}\ 2, \overline{-8|}\ 0, \overline{-2|}\ -2\,|\,5)$

$(\overline{3|}\ 5, \overline{-6|}\ 3, \overline{-8|}\ 1, \overline{-4|}\ -1, \overline{-5|}\ -3\,|\,4)$

$(\overline{1|}\ 6, \overline{-5|}\ 4, \overline{-5|}\ 2, \overline{-1|}\ 0, \overline{-2|}\ -2\,|\,3)$

$(\overline{-1|}\ 5, \overline{2|}\ 3, \overline{1|}\ 1, \overline{6|}\ -1, \overline{8|}\ -3\,|\,2)$

$(\overline{2|}\ 4, \overline{2|}\ 2, \overline{4|}\ 0, \overline{4|}\ -2\,|\,1)$

$(\overline{-3|}\ -1, \overline{-4|}\ -3\,|\,0)$

$\underline{9_{17}}$:

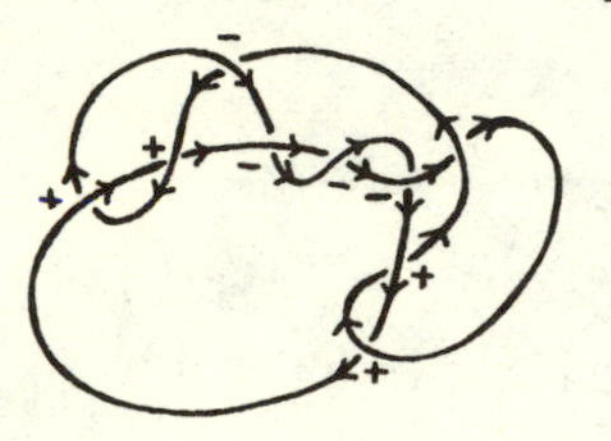

$\boxed{w = -1}$

$L = (\overline{1|}\ 1, \overline{1|}\ -1\,|\,8)(\overline{3|}\ 2, \overline{5|}\ 0, \overline{2|}\ -2\,|\,7)$

$(\overline{4|}\ 3, \overline{4|}\ 1, \overline{1|}\ -1, \overline{1|}\ -3\,|\,6)$

$(\overline{4|}\ 4, \overline{-2|}\ 2, \overline{-13|}\ 0, \overline{-7|}\ -2\,|\,5)$

$(\overline{3|}\ 5, \overline{-3|}\ 3, \overline{-14|}\ 1, \overline{-12|}\ -1, \overline{-4|}\ -3\,|\,4)$

$(\overline{1|}\ 6, \overline{-3|}\ 4, \overline{-4|}\ 2, \overline{6|}\ 0, \overline{6|}\ -2\,|\,3)$

$(\overline{-2|}\ 5, \overline{-1|}\ 3, \overline{9|}\ 1, \overline{13|}\ -1, \overline{5|}\ -3\,|\,2)$

$(\overline{1|}\ 4, \overline{3|}\ 2, \overline{1|}\ 0, \overline{-1|}\ -2\,|\,1)$

$(\overline{-2|}\ 1, \overline{-3|}\ -1, \overline{-2|}\ -3\,|\,0)$

$\underline{9_{18}}$:

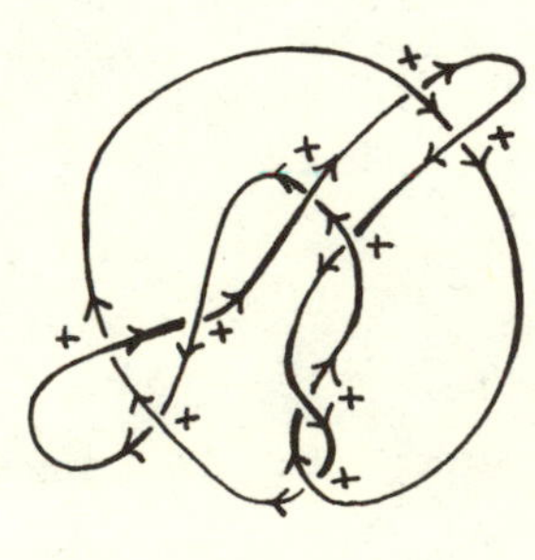

$\boxed{w = 9}$

$L = (\overline{1|}\ 1, \overline{1|}\ -1\,|\,8)(\overline{2|}\ 2, \overline{4|}\ 0, \overline{2|}\ -2\,|\,7)$

$(\overline{3|}\ 3, \overline{2|}\ 1, \overline{1|}\ -1, \overline{2|}\ -3\,|\,6)$

$(\overline{2|}\ 4, \overline{1|}\ 2, \overline{-5|}\ 0, \overline{-3|}\ -2, \overline{1|}\ -4\,|\,5)$

$(\overline{1|}\ 5, \overline{-4|}\ 3, \overline{-2|}\ 1, \overline{-2|}\ -1, \overline{-5|}\ -3\,|\,4)$

$(\overline{-2|}\ 4, \overline{-4|}\ 2, \overline{1|}\ 0, \overline{-3|}\ -4\,|\,3)$

$(\overline{-2|}\ 5, \overline{3|}\ 3, \overline{-2|}\ -1, \overline{3|}\ -3\,|\,2)$

$(\overline{2|}\ 2, \overline{2|}\ -4\,|\,1)(\overline{1|}\ 5, \overline{-1|}\ 3, \overline{1|}\ -1\,|\,0)$

9_{19}:

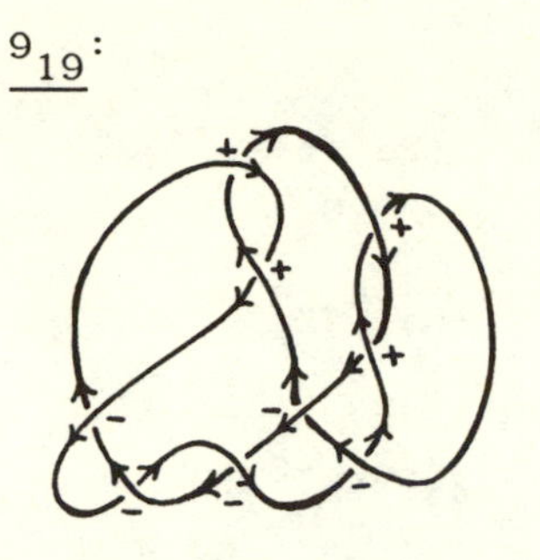

w = -1

$$L = (\overline{1|}\ 1,\overline{1|}\ -1\,|\,8)(\overline{3|}\ 2,\overline{5|}\ 0,\overline{2|}\ -2\,|\,7)$$
$$(\overline{3|}\ 3,\overline{3|}\ 1,\overline{2|}\ -1,\overline{2|}\ -3\,|\,6)$$
$$(\overline{1|}\ 4,\overline{-7|}\ 2,\overline{-11|}\ 0,\overline{-1|}\ -2,\overline{2|}\ -4\,|\,5)$$
$$(\overline{-8|}\ 3,\overline{-11|}\ 1,\overline{-4|}\ -1,\overline{1|}\ -5\,|\,4)$$
$$(\overline{-2|}\ 4,\overline{4|}\ 2,\overline{10|}\ 0,\overline{1|}\ -2,\overline{-3|}\ -4\,|\,3)$$
$$(\overline{4|}\ 3,\overline{8|}\ 1,\overline{3|}\ -1,\overline{-3|}\ -3,\overline{-2|}\ -5\,|\,2)$$
$$(\overline{-1|}\ 2,\overline{-3|}\ 0,\overline{-1|}\ -2,\overline{1|}\ -4\,|\,1)$$
$$(\overline{-1|}\ 1,\overline{1|}\ -3,\overline{1|}\ -5\,|\,0)$$

9_{20}:

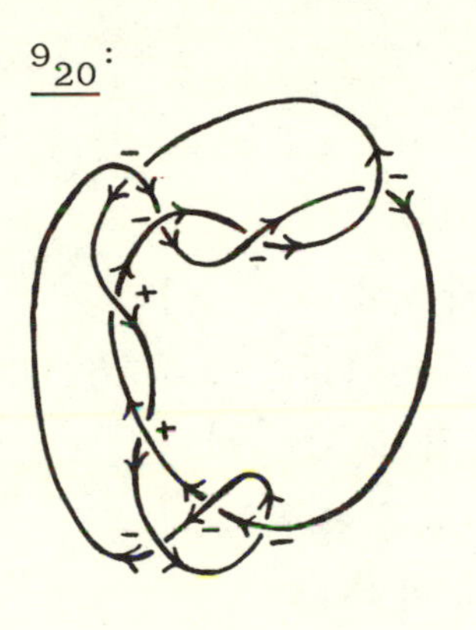

w = -5

$$L = (a+a^{-1})Z^8 + (3a^2+5+2a^{-2})Z^7$$
$$+ (5a^3+5a+a^{-1}+a^{-3})Z^6$$
$$+ (5a^4-12-7a^{-2})Z^5$$
$$+ (3a^5-6a^3-16a-11a^{-1}-4a^{-3})Z^4$$
$$+ (a^6-5a^4-7a^2+5+6a^{-2})Z^3$$
$$+ (-a^5+3a^3+10a+11a^{-1}+5a^{-3})Z^2$$
$$+ (2a^4+2a^2)Z^1+(-a^3-2a-2a^{-1}-2a^{-3})Z^0$$

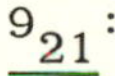

9_{21}:

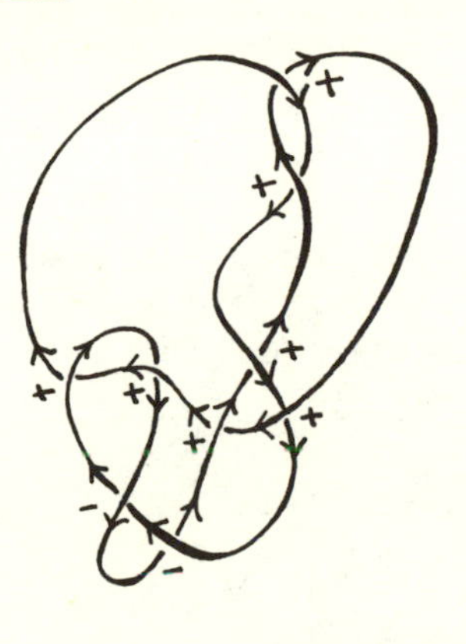

w = 5

$$L = (a+a^{-1})Z^8 + (3a^2+5+2a^{-2})Z^7$$
$$+ (4a^3+4a+2a^{-1}+2a^{-3})Z^6$$
$$+ (3a^4-3a^2-10-3a^{-2}+a^{-4})Z^5$$
$$+ (a^5-6a^3-9a-7a^{-1}-5a^{-3})Z^4$$
$$+ (-4a^4+2a^2+9-3a^{-4})Z^3$$
$$+ (-a^5+3a^3+6a+5a^{-1}+3a^{-3})Z^2$$
$$+ (-a^2-3+2a^{-4})Z^1$$
$$+ (-a^3-a^{-1}-a^{-3})Z^0$$

$\underline{9_{22}}$:

$\boxed{w = 1}$

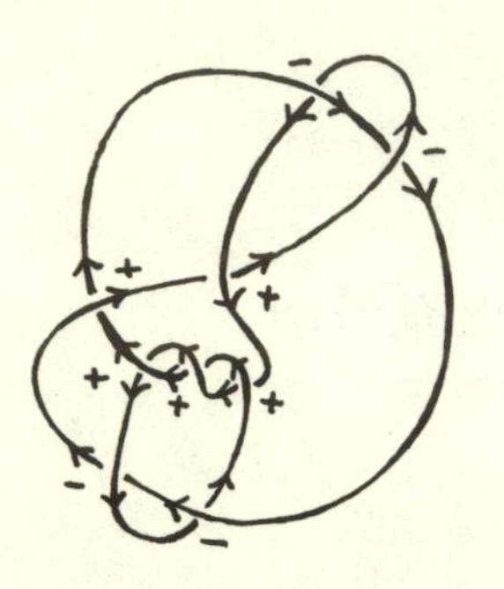

$$L = (a+a^{-1})Z^8 + (2a^2+6+4a^{-2})Z^7$$
$$+ (a^3+2a+7a^{-1}+6a^{-3})Z^6$$
$$+ (-7a^2-16-4a^{-2}+5a^{-4})Z^5$$
$$+ (-4a^3-15a-23a^{-1}-9a^{-3}+3a^{-5})Z^4$$
$$+ (7a^2+10-2a^{-2}-4a^{-4}+a^{-6})Z^3$$
$$+ (5a^3+16a+17a^{-1}+5a^{-3}-a^{-5})Z^2$$
$$+ (-2a^2-2+a^{-2}+a^{-4})Z^1$$
$$+ (-2a^3-4a-4a^{-1}-a^{-3})Z^0$$

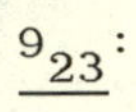

$\underline{9_{23}}$:

$\boxed{w = -9}$

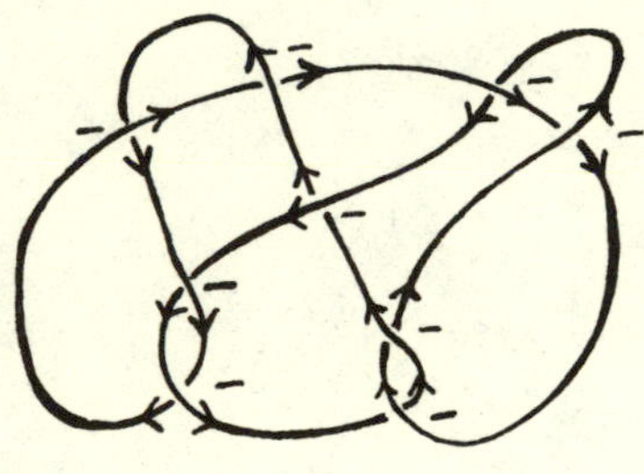

$$L = (\overline{1}|\;1, \overline{1}|\;-1\,|8)(\overline{3}|\;2, \overline{5}|\;0, \overline{2}|\;-2\,|7)$$
$$(\overline{3}|\;3, \overline{4}|\;1, \overline{4}|\;-1, \overline{3}|\;-3\,|6)$$
$$(\overline{1}|\;4, \overline{-5}|\;2, \overline{-6}|\;0, \overline{2}|\;-2, \overline{2}|\;-4\,|5)$$
$$(\overline{-7}|\;3, \overline{-10}|\;1, \overline{-8}|\;-1, \overline{-4}|\;-3, \overline{1}|\;-5\,|4)$$
$$(\overline{-2}|\;4, \overline{-2}|\;0, \overline{-6}|\;-2, \overline{-2}|\;-4\,|3)$$
$$(\overline{3}|\;3, \overline{3}|\;1, \overline{6}|\;-1, \overline{4}|\;-3, \overline{-2}|\;-5\,|2)$$
$$(\overline{1}|\;4, \overline{1}|\;2, \overline{4}|\;0, \overline{4}|\;-2\,|1)$$
$$(\overline{-2}|\;-1, \overline{-2}|\;-3, \overline{1}|\;-5\,|0)$$

$\underline{9_{24}}$:

$\boxed{w = 1}$

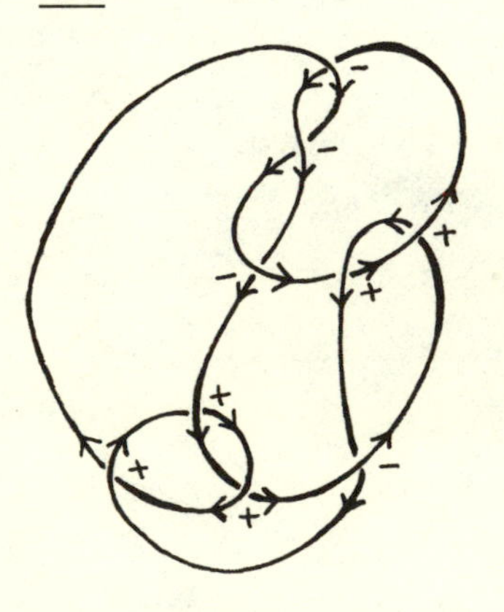

$$L = (a+a^{-1})Z^8 + (3a^2+5+2a^{-2})Z^7$$
$$+ (4a^3+5a+3a^{-1}+2a^{-3})Z^6$$
$$+ (3a^4-a^2-7-2a^{-2}+a^{-4})Z^5$$
$$+ (a^5-5a^3-11a-10a^{-1}-5a^{-3})Z^4$$
$$+ (-4a^4-3a^2+1-3a^{-2}-3a^{-4})Z^3$$
$$+ (-a^5+2a^3+9a+10a^{-1}+4a^{-3})Z^2$$
$$+ (a^4+2a^2+2+3a^{-2}+2a^{-4})Z^1$$
$$+ (-a^3-3a-5a^{-1}-2a^{-3})Z^0$$

$\underline{9_{25}}$:

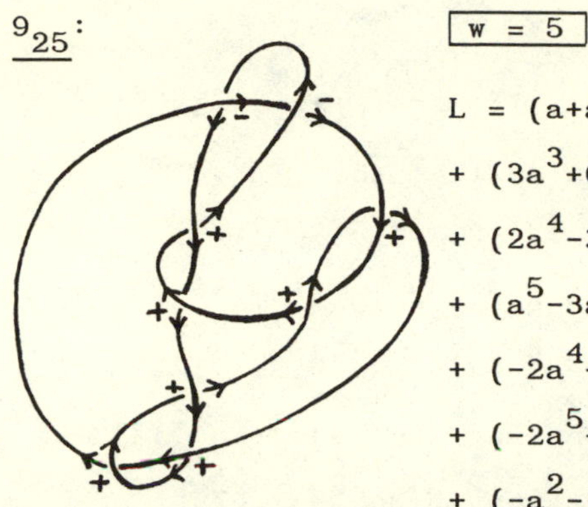

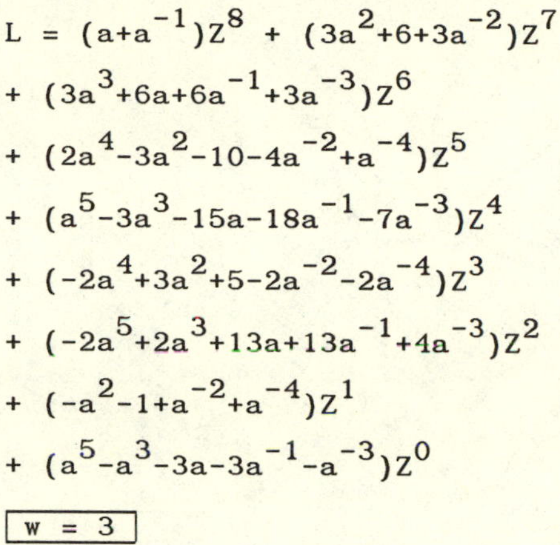

w = 5

$$L = (a+a^{-1})z^8 + (3a^2+6+3a^{-2})z^7$$
$$+ (3a^3+6a+6a^{-1}+3a^{-3})z^6$$
$$+ (2a^4-3a^2-10-4a^{-2}+a^{-4})z^5$$
$$+ (a^5-3a^3-15a-18a^{-1}-7a^{-3})z^4$$
$$+ (-2a^4+3a^2+5-2a^{-2}-2a^{-4})z^3$$
$$+ (-2a^5+2a^3+13a+13a^{-1}+4a^{-3})z^2$$
$$+ (-a^2-1+a^{-2}+a^{-4})z^1$$
$$+ (a^5-a^3-3a-3a^{-1}-a^{-3})z^0$$

$\underline{9_{26}}$:

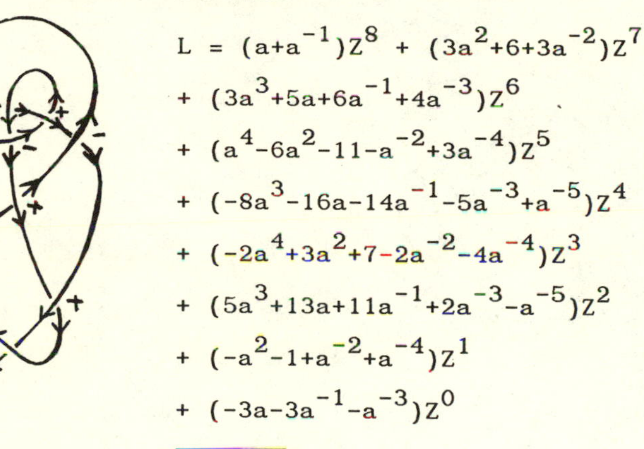

w = 3

$$L = (a+a^{-1})z^8 + (3a^2+6+3a^{-2})z^7$$
$$+ (3a^3+5a+6a^{-1}+4a^{-3})z^6$$
$$+ (a^4-6a^2-11-a^{-2}+3a^{-4})z^5$$
$$+ (-8a^3-16a-14a^{-1}-5a^{-3}+a^{-5})z^4$$
$$+ (-2a^4+3a^2+7-2a^{-2}-4a^{-4})z^3$$
$$+ (5a^3+13a+11a^{-1}+2a^{-3}-a^{-5})z^2$$
$$+ (-a^2-1+a^{-2}+a^{-4})z^1$$
$$+ (-3a-3a^{-1}-a^{-3})z^0$$

$\underline{9_{27}}$:

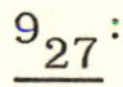

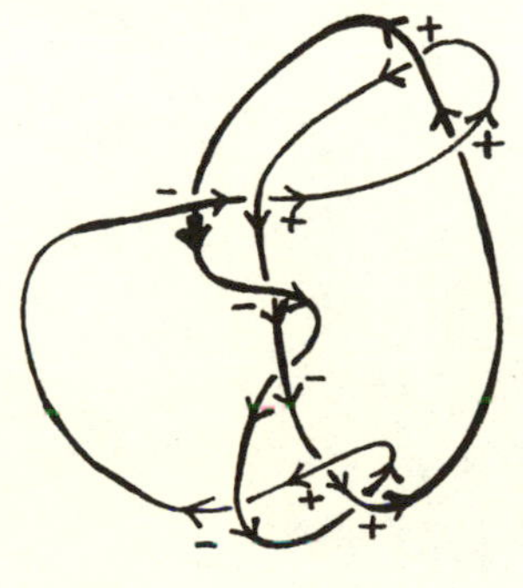

w = 1

$$L = (a+a^{-1})z^8 + (3a^2+6+3a^{-2})z^7$$
$$+ (4a^3+7a+6a^{-1}+3a^{-3})z^6$$
$$+ (3a^4-8-4a^{-2}+a^{-4})z^5$$
$$+ (a^5-5a^3-16a-17a^{-1}-7a^{-3})z^4$$
$$+ (-4a^4-4a^2-2a^{-2}-2a^{-4})z^3$$
$$+ (-a^5+3a^3+12a+12a^{-1}+4a^{-3})z^2$$
$$+ (a^4+2a^2+2+2a^{-2}+a^{-4})z^1$$
$$+ (-a^3-2a-3a^{-1}-a^{-3})z^0$$

$\underline{9_{28}}$:

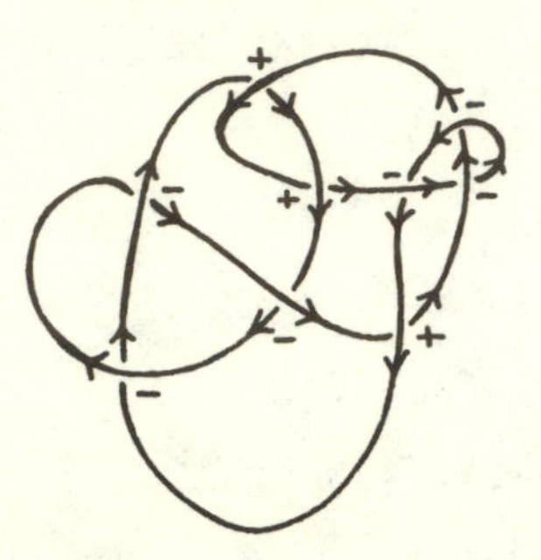

w = -3

$$L = (a+a^{-1})Z^8 + (3a^2+6+3a^{-2})Z^7$$
$$+ (4a^3+8a+7a^{-1}+3a^{-3})Z^6$$
$$+ (3a^4+2a^2-5-3a^{-2}+a^{-4})Z^5$$
$$+ (a^5-4a^3-17a-19a^{-1}-7a^{-3})Z^4$$
$$+ (-4a^4-9a^2-7-4a^{-2}-2a^{-4})Z^3$$
$$+ (-a^5+2a^3+12a+14a^{-1}+5a^{-3})Z^2$$
$$+ (2a^4+6a^2+6+3a^{-2}+a^{-4})Z^1$$
$$+ (-a^3-4a-5a^{-1}-a^{-3})Z^0$$

$\underline{9_{29}}$:

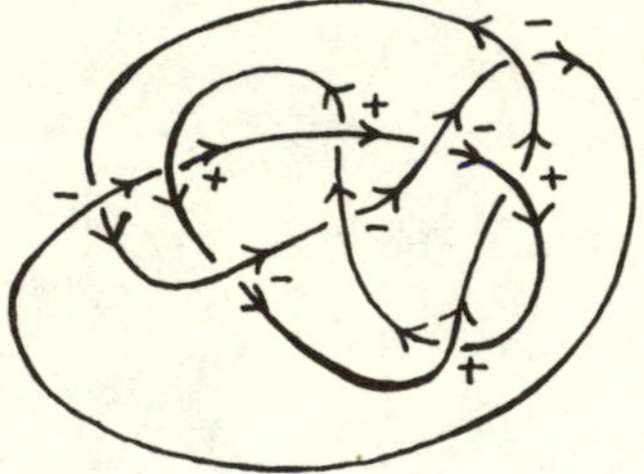

w = -1

$$L = (2a+2a^{-1})Z^8 + (6a^2+9+3a^{-2})Z^7$$
$$+ (8a^3+6a-a^{-1}+a^{-3})Z^6$$
$$+ (6a^4-8a^2-24-10a^{-2})Z^5$$
$$+ (3a^5-13a^3-24a-11a^{-1}-3a^{-3})Z^4$$
$$+ (a^6-5a^4-a^2+14+9a^{-2})Z^3$$
$$+ (8a^3+17a+12a^{-1}+3a^{-3})Z^2$$
$$+ (2a^4+2a^2-1-a^{-2})Z^1$$
$$+ (-2a^3-5a-3a^{-1}-a^{-3})Z^0$$

$\underline{9_{30}}$:

w = +1

$$L = (a+a^{-1})Z^8 + (4a^2+7+3a^{-2})Z^7$$
$$+ (5a^3+10a+8a^{-1}+3a^{-3})Z^6$$
$$+ (3a^4-2a^2-9-3a^{-2}+a^{-4})Z^5$$
$$+ (a^5-7a^3-23a-22a^{-1}-7a^{-3})Z^4$$
$$+ (-3a^4-2a^2-3a^{-2}-2a^{-4})Z^3$$
$$+ (-a^5+5a^3+17a+16a^{-1}+5a^{-3})Z^2$$
$$+ (a^4+a^2+1+2a^{-2}+a^{-4})Z^1$$
$$+ (-2a^3-4a-4a^{-1}-a^{-3})Z^0$$

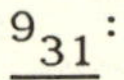

$\underline{9_{31}}$:

$\boxed{w = +3}$

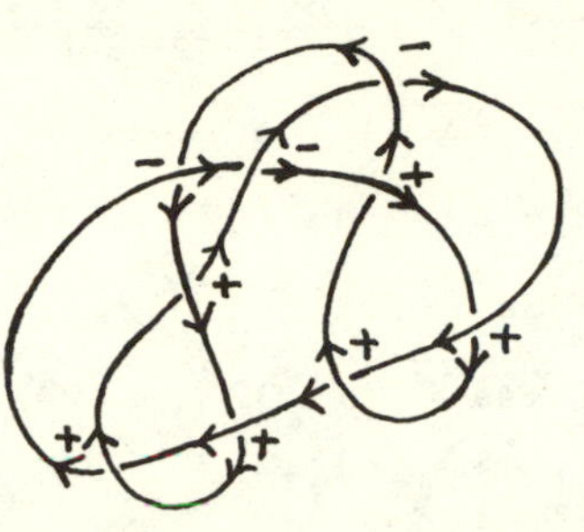

$$L = (a+a^{-1})z^8 + (3a^2+7+4a^{-2})z^7$$
$$+ (3a^3+8a+11a^{-1}+6a^{-3})z^6$$
$$+ (a^4-3a^2-7+a^{-2}+4a^{-4})z^5$$
$$+ (-7a^3-21a-23a^{-1}-8a^{-3}+a^{-5})z^4$$
$$+ (-2a^4-3a^2-5-8a^{-2}-4a^{-4})z^3$$
$$+ (5a^3+15a+13a^{-1}+3a^{-3})z^2$$
$$+ (a^4+3a^2+5+3a^{-2})z^1$$
$$+ (-a^3-4a-2a^{-1})z^0$$

$\underline{9_{32}}$:

$\boxed{w = 3}$

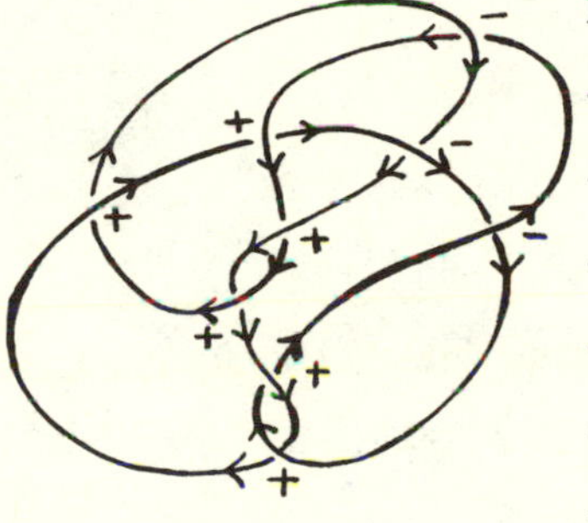

$$L = (2a+2a^{-1})z^8 + (5a^2+10+5a^{-2})z^7$$
$$+ (4a^3+6a+7a^{-1}+5a^{-3})z^6$$
$$+ (a^4-9a^2-18-5a^{-2}+3a^{-4})z^5$$
$$+ (-8a^3-19a-18a^{-1}-6a^{-3}+a^{-5})z^4$$
$$+ (-a^4+3a^2+9+2a^{-2}-3a^{-4})z^3$$
$$+ (3a^3+10a+12a^{-1}+4a^{-3}-a^{-5})z^2$$
$$+ (-a^2-2+a^{-4})z^1$$
$$+ (a^3-a-2a^{-1}-a^{-3})z^0$$

$\underline{9_{33}}$:

$\boxed{w = 1}$

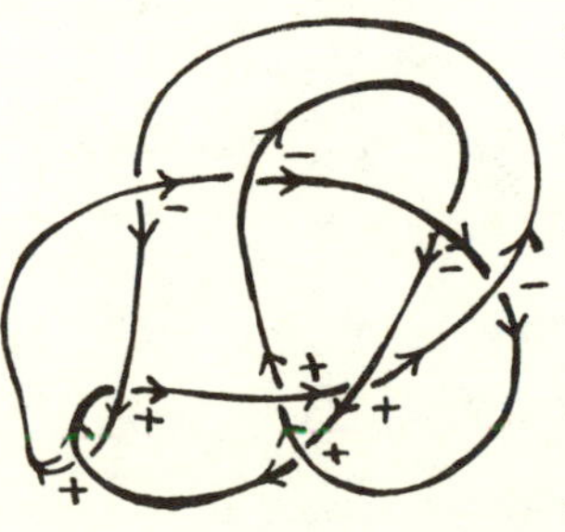

$$L = (2a+2a^{-1})z^8 + (6a^2+10+4a^{-2})z^7$$
$$+ (7a^3+9a+5a^{-1}+3a^{-3})z^6$$
$$+ (4a^4-5a^2-16-6a^{-2}+a^{-4})z^5$$
$$+ (a^5-9a^3-20a-16a^{-1}-6a^{-3})z^4$$
$$+ (-3a^4-a^2+5+a^{-2}-2a^{-4})z^3$$
$$+ (3a^3+9a+10a^{-1}+4a^{-3})z^2$$
$$+ (a^{-2}+a^{-4})z^1 + (-2a^{-1}-a^{-3})z^0$$

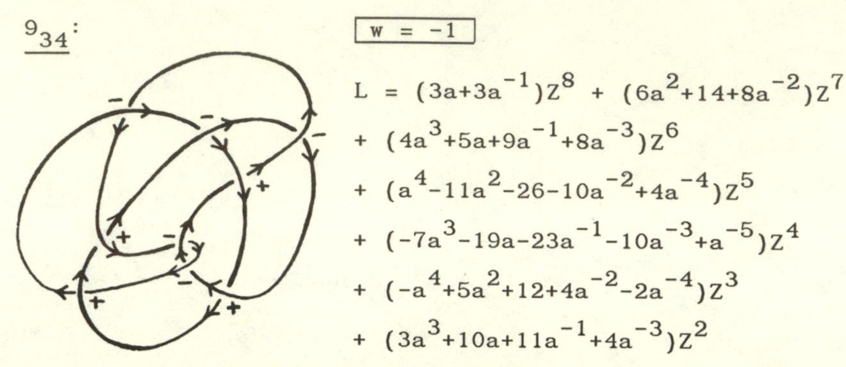

$\underline{9_{34}}$:

$\boxed{w = -1}$

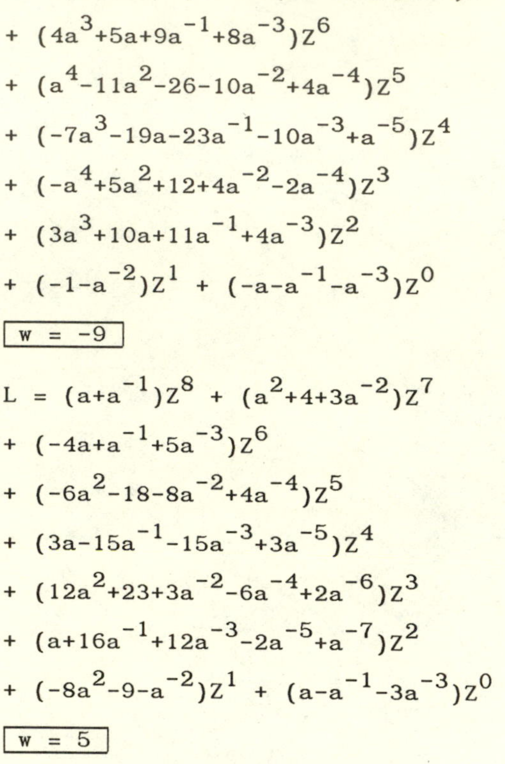

$$L = (3a+3a^{-1})Z^8 + (6a^2+14+8a^{-2})Z^7$$
$$+ (4a^3+5a+9a^{-1}+8a^{-3})Z^6$$
$$+ (a^4-11a^2-26-10a^{-2}+4a^{-4})Z^5$$
$$+ (-7a^3-19a-23a^{-1}-10a^{-3}+a^{-5})Z^4$$
$$+ (-a^4+5a^2+12+4a^{-2}-2a^{-4})Z^3$$
$$+ (3a^3+10a+11a^{-1}+4a^{-3})Z^2$$
$$+ (-1-a^{-2})Z^1 + (-a-a^{-1}-a^{-3})Z^0$$

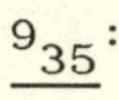

$\underline{9_{35}}$:

$\boxed{w = -9}$

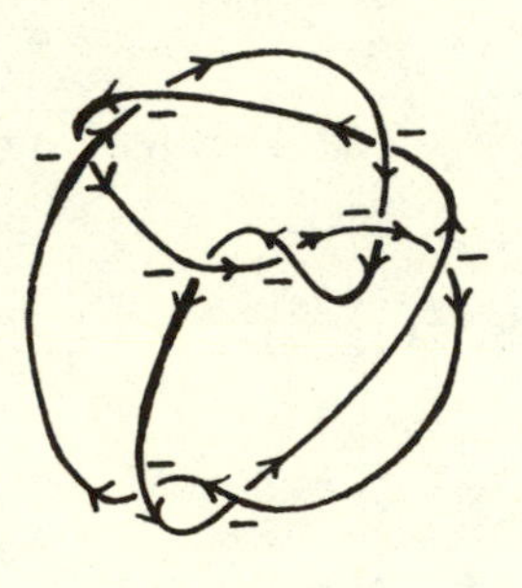

$$L = (a+a^{-1})Z^8 + (a^2+4+3a^{-2})Z^7$$
$$+ (-4a+a^{-1}+5a^{-3})Z^6$$
$$+ (-6a^2-18-8a^{-2}+4a^{-4})Z^5$$
$$+ (3a-15a^{-1}-15a^{-3}+3a^{-5})Z^4$$
$$+ (12a^2+23+3a^{-2}-6a^{-4}+2a^{-6})Z^3$$
$$+ (a+16a^{-1}+12a^{-3}-2a^{-5}+a^{-7})Z^2$$
$$+ (-8a^2-9-a^{-2})Z^1 + (a-a^{-1}-3a^{-3})Z^0$$

$\underline{9_{36}}$:

$\boxed{w = 5}$

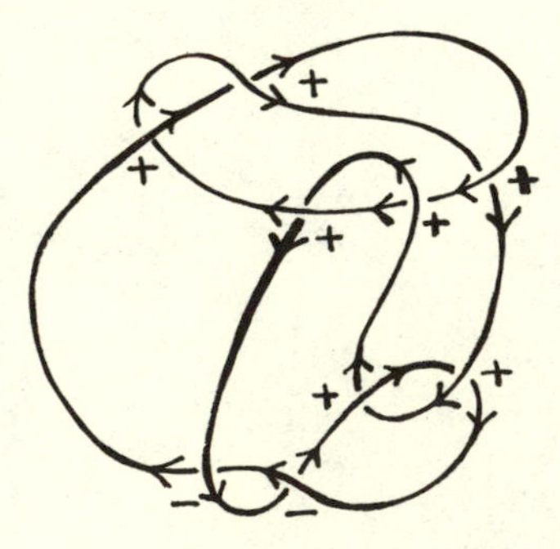

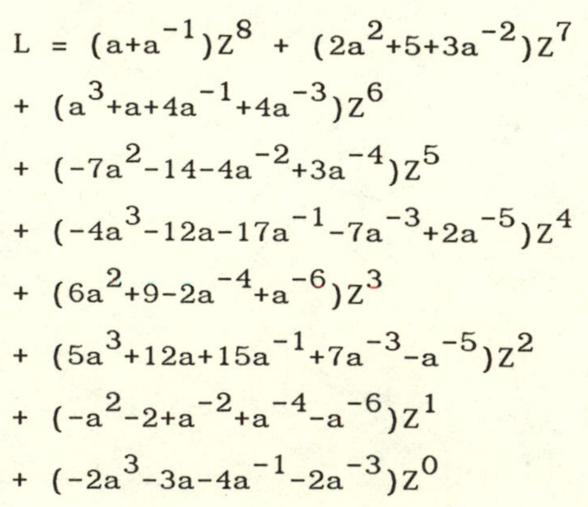

$$L = (a+a^{-1})Z^8 + (2a^2+5+3a^{-2})Z^7$$
$$+ (a^3+a+4a^{-1}+4a^{-3})Z^6$$
$$+ (-7a^2-14-4a^{-2}+3a^{-4})Z^5$$
$$+ (-4a^3-12a-17a^{-1}-7a^{-3}+2a^{-5})Z^4$$
$$+ (6a^2+9-2a^{-4}+a^{-6})Z^3$$
$$+ (5a^3+12a+15a^{-1}+7a^{-3}-a^{-5})Z^2$$
$$+ (-a^2-2+a^{-2}+a^{-4}-a^{-6})Z^1$$
$$+ (-2a^3-3a-4a^{-1}-2a^{-3})Z^0$$

9_{37}:

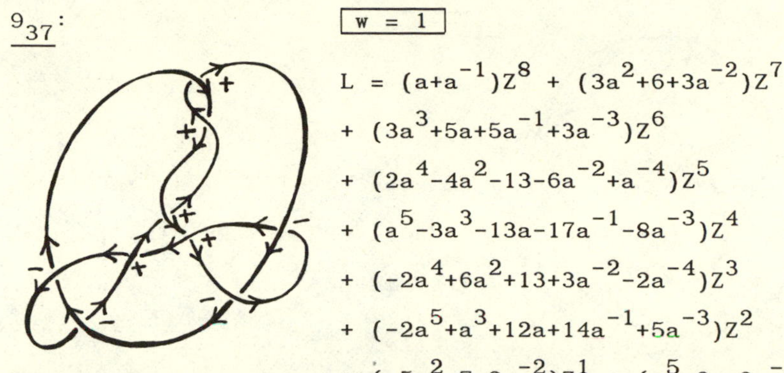

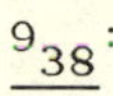

w = 1

$$L = (a+a^{-1})Z^8 + (3a^2+6+3a^{-2})Z^7$$
$$+ (3a^3+5a+5a^{-1}+3a^{-3})Z^6$$
$$+ (2a^4-4a^2-13-6a^{-2}+a^{-4})Z^5$$
$$+ (a^5-3a^3-13a-17a^{-1}-8a^{-3})Z^4$$
$$+ (-2a^4+6a^2+13+3a^{-2}-2a^{-4})Z^3$$
$$+ (-2a^5+a^3+12a+14a^{-1}+5a^{-3})Z^2$$
$$+ (-5a^2-7-2a^{-2})Z^1 + (a^5-2a-2a^{-1})Z^0$$

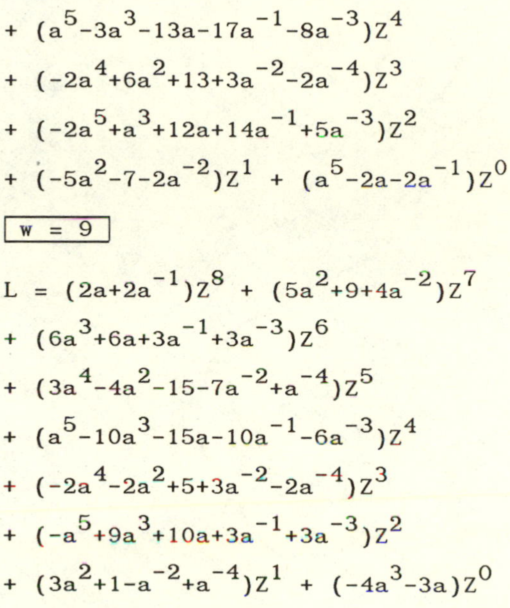

9_{38}:

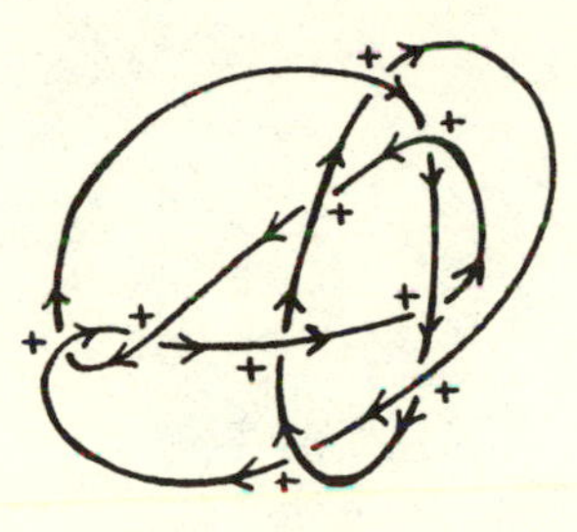

w = 9

$$L = (2a+2a^{-1})Z^8 + (5a^2+9+4a^{-2})Z^7$$
$$+ (6a^3+6a+3a^{-1}+3a^{-3})Z^6$$
$$+ (3a^4-4a^2-15-7a^{-2}+a^{-4})Z^5$$
$$+ (a^5-10a^3-15a-10a^{-1}-6a^{-3})Z^4$$
$$+ (-2a^4-2a^2+5+3a^{-2}-2a^{-4})Z^3$$
$$+ (-a^5+9a^3+10a+3a^{-1}+3a^{-3})Z^2$$
$$+ (3a^2+1-a^{-2}+a^{-4})Z^1 + (-4a^3-3a)Z^0$$

9_{39}:

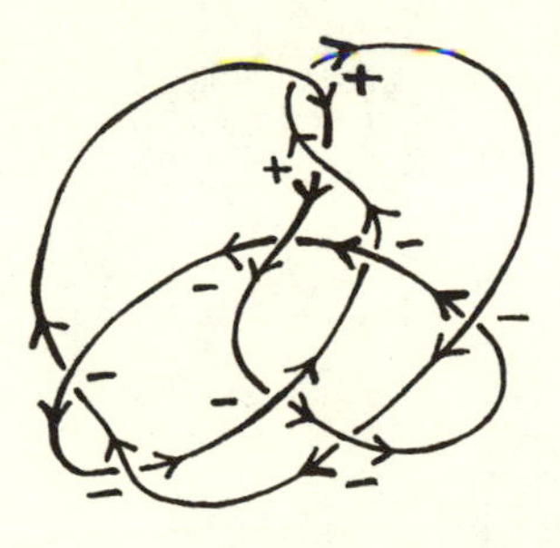

w = -5

$$L = (2a+2a^{-1})Z^8 + (4a^2+9+5a^{-2})Z^7$$
$$+ (3a^3+3a+5a^{-1}+5a^{-3})Z^6$$
$$+ (a^4-7a^2-18-7a^{-2}+3a^{-4})Z^5$$
$$+ (-6a^3-13a-15a^{-1}-7a^{-3}+a^{-5})Z^4$$
$$+ (-2a^4+2a^2+12+5a^{-2}-3a^{-4})Z^3$$
$$+ (3a^3+9a+12a^{-1}+5a^{-3}-a^{-5})Z^2$$
$$+ (a^4-a^2-3-a^{-2})Z^1$$
$$+ (-a^3-2a-2a^{-1}-2a^{-3})Z^0$$

$\underline{9_{40}}$:

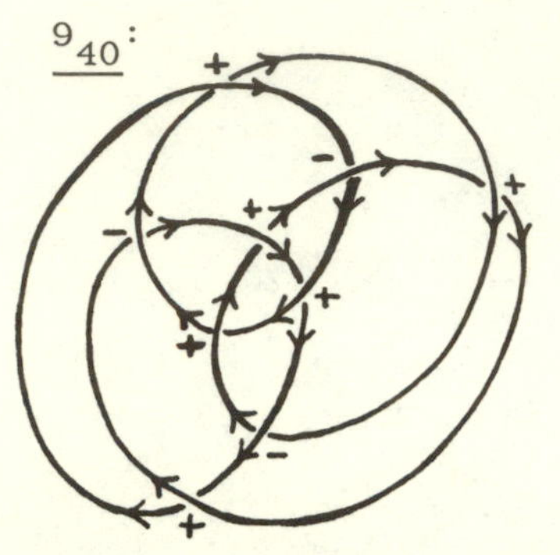

w = 3

$$L = (4a+4a^{-1})Z^8 + (8a^2+17+9a^{-2})Z^7$$
$$+ (5a^3+4a+7a^{-1}+8a^{-3})Z^6$$
$$+ (a^4-15a^2-32-12a^{-2}+4a^{-4})Z^5$$
$$+ (-7a^3-17a-20a^{-1}-9a^{-3}+a^{-5})Z^4$$
$$+ (6a^2+14+6a^{-2}-2a^{-4})Z^3$$
$$+ (3a+7a^{-1}+4a^{-3})Z^2$$
$$+ (-1-a^{-2})Z^1 + (2a^3+2a+a^{-1})Z^0$$

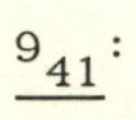

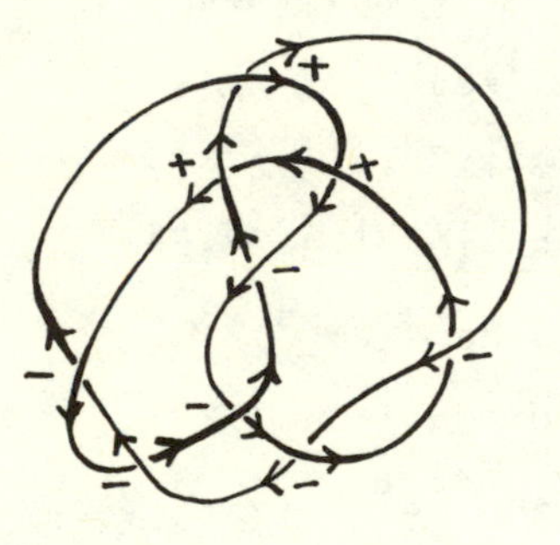

w = -3

$$L = (2a+2a^{-1})Z^8 + (3a^2+9+6a^{-2})Z^7$$
$$+ (a^3-a+5a^{-1}+7a^{-3})Z^6$$
$$+ (-10a^2-26-11a^{-2}+5a^{-4})Z^5$$
$$+ (-3a^3-12a-23a^{-1}-11a^{-3}+3a^{-5})Z^4$$
$$+ (9a^2+19+6a^{-2}-3a^{-4}+a^{-6})Z^3$$
$$+ (3a^3+13a+17a^{-1}+6a^{-3}-a^{-5})Z^2$$
$$+ (-2a^2-4-2a^{-2})Z^1$$
$$+ (-a^3-3a-3a^{-1})Z^0$$

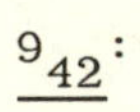

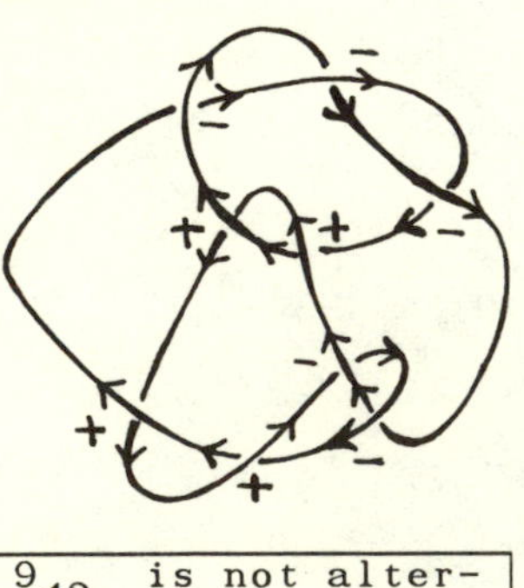

w = -1

$$L = (1+a^{-2})Z^7 + (a+2a^{-1}+a^{-3})Z^6$$
$$+ (-5-5a^{-2})Z^5$$
$$+ (-5a-10a^{-1}-5a^{-3})Z^4$$
$$+ (6+6a^{-2})Z^3$$
$$+ (6a+12a^{-1}+6a^{-3})Z^2$$
$$+ (-2-2a^{-2})Z^1$$
$$+ (-2a-3a^{-1}-2a^{-3})Z^0$$

9_{42} is not alternating
9_{42} _is_ chiral

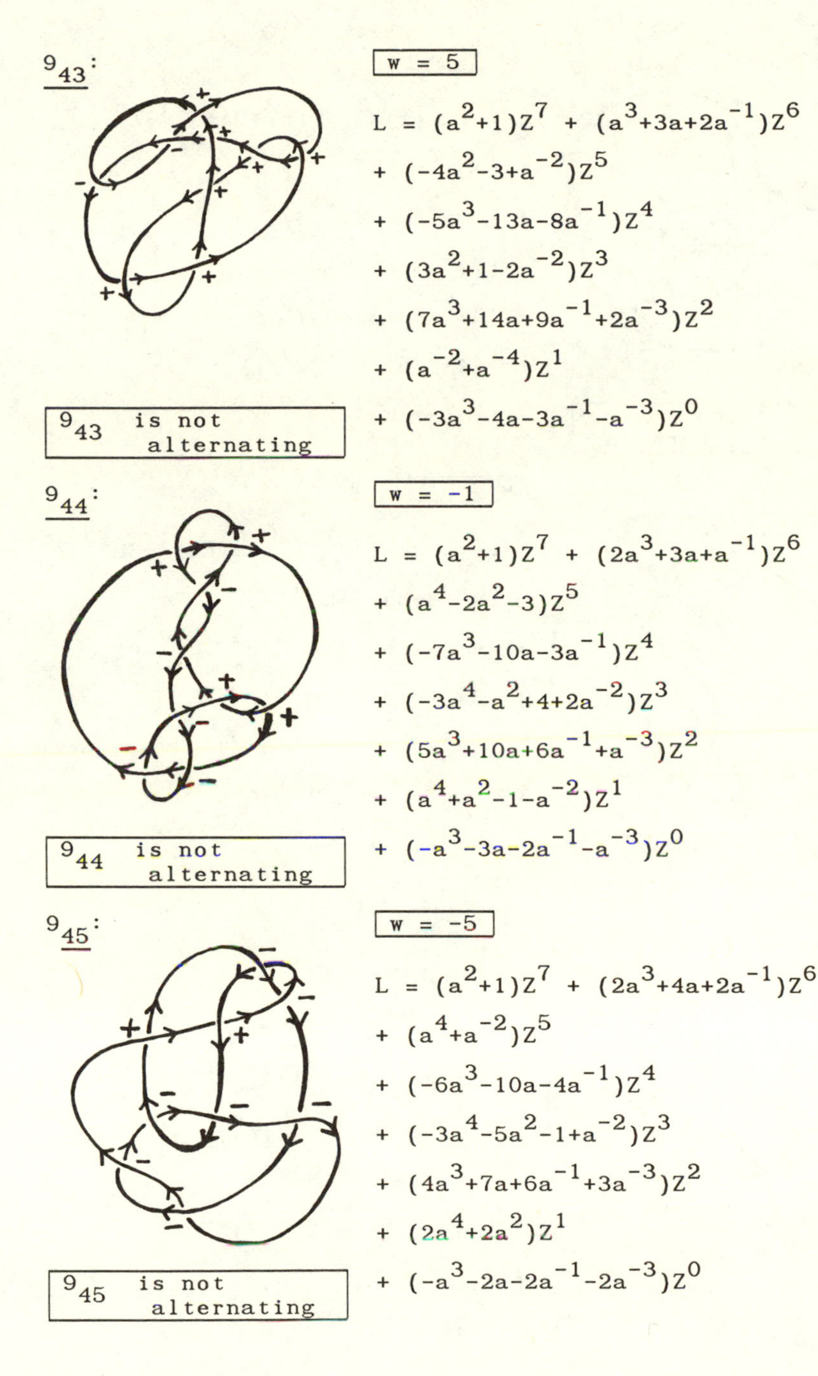

9_{43}:

$w = 5$

$$L = (a^2+1)z^7 + (a^3+3a+2a^{-1})z^6$$
$$+ (-4a^2-3+a^{-2})z^5$$
$$+ (-5a^3-13a-8a^{-1})z^4$$
$$+ (3a^2+1-2a^{-2})z^3$$
$$+ (7a^3+14a+9a^{-1}+2a^{-3})z^2$$
$$+ (a^{-2}+a^{-4})z^1$$
$$+ (-3a^3-4a-3a^{-1}-a^{-3})z^0$$

9_{43} is not alternating

9_{44}:

$w = -1$

$$L = (a^2+1)z^7 + (2a^3+3a+a^{-1})z^6$$
$$+ (a^4-2a^2-3)z^5$$
$$+ (-7a^3-10a-3a^{-1})z^4$$
$$+ (-3a^4-a^2+4+2a^{-2})z^3$$
$$+ (5a^3+10a+6a^{-1}+a^{-3})z^2$$
$$+ (a^4+a^2-1-a^{-2})z^1$$
$$+ (-a^3-3a-2a^{-1}-a^{-3})z^0$$

9_{44} is not alternating

9_{45}:

$w = -5$

$$L = (a^2+1)z^7 + (2a^3+4a+2a^{-1})z^6$$
$$+ (a^4+a^{-2})z^5$$
$$+ (-6a^3-10a-4a^{-1})z^4$$
$$+ (-3a^4-5a^2-1+a^{-2})z^3$$
$$+ (4a^3+7a+6a^{-1}+3a^{-3})z^2$$
$$+ (2a^4+2a^2)z^1$$
$$+ (-a^3-2a-2a^{-1}-2a^{-3})z^0$$

9_{45} is not alternating

$\underline{9_{46}}$:

$\boxed{w = -3}$

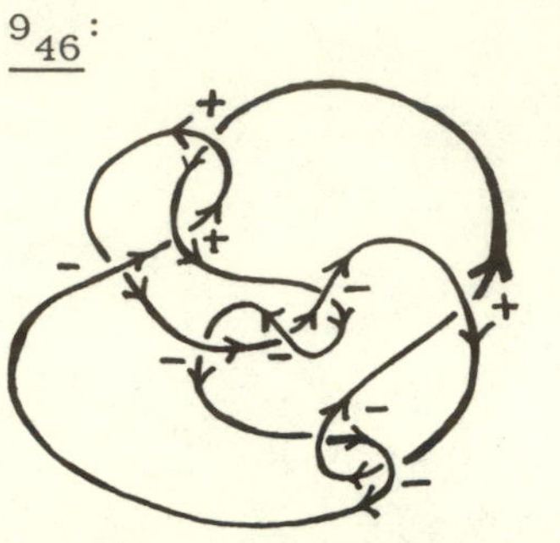

$$L = (a^{2}+1)Z^{7} + (a^{3}+2a+a^{-1})Z^{6}$$
$$+ (-5a^{2}-5)Z^{5}$$
$$+ (-5a^{3}-9a-4a^{-1})Z^{4}$$
$$+ (7a^{2}+8+a^{-2})Z^{3}$$
$$+ (6a^{3}+9a+3a^{-1})Z^{2}$$
$$+ (-4a^{2}-6-2a^{-2})Z^{1}$$
$$+ (-a^{3}-a+a^{-1}+2a^{-3})Z^{0}$$

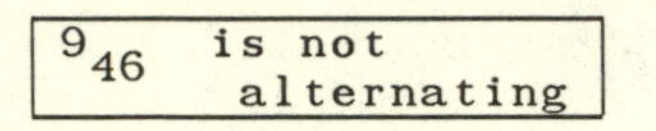

9_{46} is not alternating

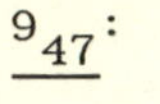

$\underline{9_{47}}$:

$\boxed{w = -3}$

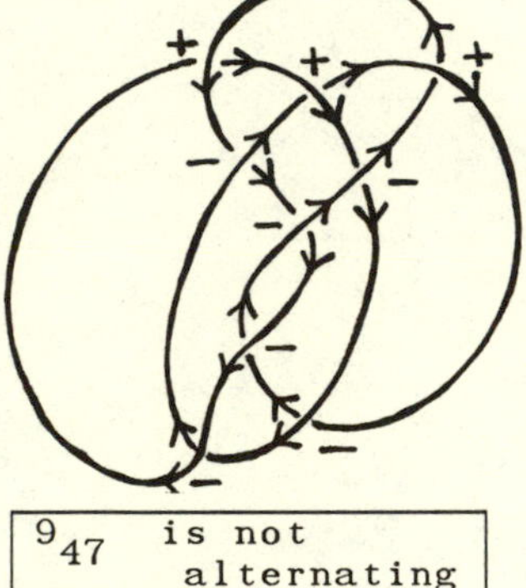

$$L = (2+2a^{-2})Z^{7} + (3a+6a^{-1}+3a^{-3})Z^{6}$$
$$+ (a^{2}-4-4a^{-2}+a^{-4})Z^{5}$$
$$+ (-7a-16a^{-1}-9a^{-3})Z^{4}$$
$$+ (3a^{2}+6+a^{-2}-2a^{-4})Z^{3}$$
$$+ (3a^{3}+9a+11a^{-1}+5a^{-3})Z^{2}$$
$$+ (-3a^{2}-5-2a^{-2})Z^{1}$$
$$+ (-a^{3}-2a-a^{-1}+a^{-3})Z^{0}$$

9_{47} is not alternating

$\underline{9_{48}}$:

$\boxed{w = -5}$

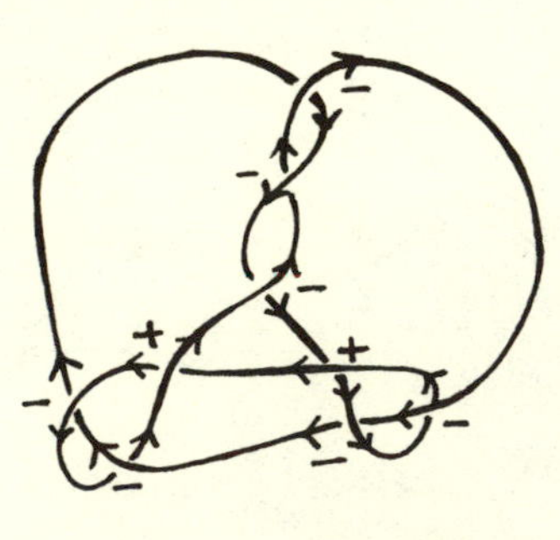

$$L = (1+a^{-2})Z^{7}$$
$$+ (a+4a^{-1}+3a^{-3})Z^{6}$$
$$+ (-1+2a^{-2}+3a^{-4})Z^{5}$$
$$+ (-6a^{-1}-5a^{-3}+a^{-5})Z^{4}$$
$$+ (3a^{2}+5-3a^{-2}-5a^{-4})Z^{3}$$
$$+ (-a+2a^{-1}+2a^{-3}-a^{-5})Z^{2}$$
$$+ (-4a^{2}-5-a^{-2})Z^{1}$$
$$+ (2a+3a^{-1})Z^{0}$$

9_{48} is not alternating

9_{49}:

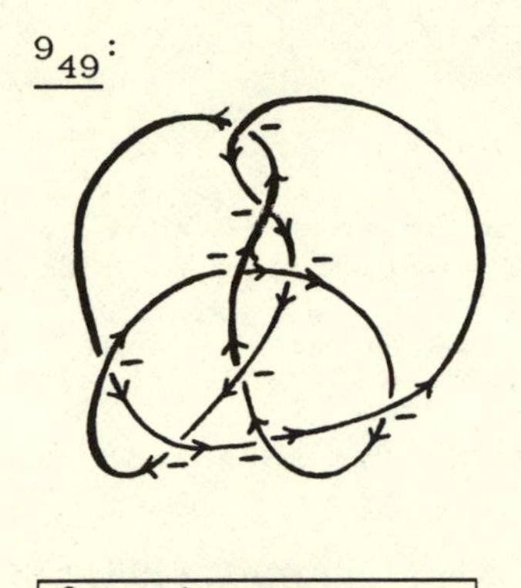

9_{49} is not alternating

w = -9

$$L = (1+a^{-2})Z^7 + (a+4a^{-1}+3a^{-3})Z^6$$
$$+ (-1+a^{-2}+2a^{-4})Z^5$$
$$+ (-9a^{-1}-8a^{-3}+a^{-5})Z^4$$
$$+ (3a^2+3-3a^{-2}-3a^{-4})Z^3$$
$$+ (-a+10a^{-1}+9a^{-3}-2a^{-5})Z^2$$
$$+ (-4a^2-2+2a^{-2})Z^1$$
$$+ (-3a^{-1}-4a^{-3})Z^0$$

REFERENCES

[A1] J.W. Alexander. Topological invariants of knots and links. Trans. Amer. Math. Soc. 30(1923), 275-306.

[AH] K. Appel and W. Haken. Every planar map is four colorable. Bull. Amer. Math. Soc. 82(1976), 711-712.

[A2] C. Ashley. *The Ashley Book of Knots*. Doubleday and Co., New York (1944).

[AS] M.F. Atiyah and I.M. Singer. The index of elliptic operators II. Ann. of Math. 87(1968), 546-604.

[BaM] R. Ball and M.L. Mehta. Sequence of invariants for knots and links. J. Physique 42(1981), 1193-1199.

[BCW] W.R. Bauer, F.H.C. Crick and J.H. White Supercoiled DNA. Sci. Amer. 243(1980), 118-133.

[BI1] J.S. Birman. *Braids, Links and Mapping Class Groups*. Ann. of Math. Studies, No. 82. Princeton University Press, Princeton, N.J. (1976).

[BI2] J.S. Birman and H. Wenzel. Link polynomials and a new algebra (preprint 1986).

[BK] E. Brieskorn. Examples of singular normal complex spaces which are topological manifolds. Proc. Nat. Acad. Sci. U.S.A. 55(1966), 1395-1397.

E. Brieskorn. Beispiele zur Differential topologie von Singularitaten. Invent. Math. 40(1966), 153-160.

[BLM] R. Brandt, W.B.R. Lickorish and K.C. Millett. A polynomial invariant for unoriented knots and links. Invent. Math.84(1986), 563-573.

[BLM] G. Birkhoff and S. Maclane. *Algebra*. MacMillan, New York (1967).

[BR] G. E. Bredon. *Introduction to Compact Transformation Groups*. Academic Press (1972).

[CG] A. Casson and C. McA. Gordon. On slice knots in dimension three. *Geometric Topology*, R.J. Milgram (ed.). Proceedings of Symposia Pure Mathematics XXXII. Amer. Math. Soc., Providence (1978), 39-53.

[C1] J.H. Conway. An enumeration of knots and links and some of their algebraic properties. *Computational Problems in Abstract Algebra*. Pergamon Press, New York (1970), 329-358.

[C] D. Cooper. Signatures of Surfaces in 3-manifolds and Applications to Knot and Link Cobordism. (Thesis, Warwick (1982)).

[CR] R.H. Crowell and R.H. Fox. *Introduction to Knot Theory*. Blaisdell Pub. Co. (1963).

[DK] A. Durfee and L.H. Kauffman. Periodicity of branched cyclic covers. Math. Ann. 218(1975), 157-174.

[DR] G. DeRahm. Introduction aux polynomes d'au noed. Enseignement Math. 13(1967), 187-194.

[DV] P. DuVal. *Quaternions, Homographies and Rotations*. Oxford at the Clarendon Press (1964).

[E] D. Erle. Die quadratische form eines knotens und ein satz uber knotenmannigfaltigkeiten. J. Reine Angew. Math. Band 236(1968), 174-217.

[FO] R.H. Fox. The homology characters of the cyclic coverings of the knots of genus one. Ann. of Math. 71(1960), 187-196.

[F1] R.H. Fox. A quick trip through knot theory. *Topology of Manifolds*. Prentice-Hall (1962), 120-167.

[F2] R.H. Fox and J.W. Milnor. Singularities of 2-spheres in 4-space and cobordism of knots. Osaka J. Math. 3(1966), 257-267.

[F] M.H. Freedman. The topology of four-dimensional manifolds. J. Differential Geometry 17(1982), 357-453.

[FB] F.B. Fuller. Decomposition of the linking number of a closed ribbon: A problem from molecular biology. Proc. Nat. Acad. Sci. U.S.A. 75(1978), 3557-3561.

[FR] G. Francis. Drawing Seifert surfaces that fiber the figure-8 knot complement in S^3 over S^1. Amer. Math. Monthly #9, 90(1983).

[GA] F. Gonzalez-Acuna. Thesis. Princeton Univeristy (1970).

[G1] C. Giller. A family of links and the Conway calculus. Trans. Amer. Math. Soc. 270(1982), 75-109.

[G2] C. McA. Gordon. An elementary proof of the G-signature theorem for 4-manifolds. (To appear.)

[G3] M.J. Greenberg and J.R. Harper. *Algebraic Topology: A First Course*. Benjamin/Cummings (1981).

[GO] R.E. Gompf. An invariant for Casson Handles, disks and knot concordance. Thesis, University of California at Berkeley (1984).

[GL1] P. Gilmer and R. Litherland. The duality conjecture in formal knot theory. Osaka J. Math. 23(1986), 229-247.

[GL2] C. McA. Gordon and R. Litherland. On the signature of a link. Invent. Math. 47(1978), 53-69.

[HOMFLY] P. Freyd, D. Yetter, J. Hoste, W.B.R. Lickorish, K. Millett, and A. Ocneanu. A new polynomial invariant of knots and links. Bull. Amer. Math. Soc. #2, 12(1985), 239-246.

[HNK] F. Hirzebruch, W.D. Neumann and S.S. Koh. *Differentiable Manifolds and Quadratic Forms*. Marcel Dekker, Inc., New York (1971).

[HO] C.F. Ho. A new polynomial invariant for knots and links. Preliminary report. AMS Abstracts, Vol. 6, #4, Issue 39(1985), 300.

[HS] W.C. Hsiang and R.H. Szczarba. On embedding surfaces in four-manifolds. *Proceedings of Symposia in Pure Mathematics*, Vol. XXII, Amer. Math. Soc. (1971), 97-104.

[HZ] F. Hirzebruch and D. Zagier. *The Atiyah-Singer Theorem and Elementary Number Theory*. Mathematics Lecture Series #3. Publish or Perish Press (1974).

[JO1] V.F.R. Jones. A polynomial invariant for knots and links via Von Neumann Algebras, BAMS 12(1985), 103-111.

[JO2] V.F.R. Jones. A new knot polynomial and von Neumann algebras. Notices of AMS(1985).

[JO3] V.F.R. Jones. Hecke algebra representations of braid groups and link polynomials (preprint 1986).

[J] D. Joyce. A classifying invariant of knots, the knot quandle. J. Pure Appl. Algebra 23(1982), 37-65.

[KN] L.H. Kauffman and W.D. Neumann. Products of knots, branched fibrations and sums of singularities. Topology 16(1977), 369-393.

[K1] L.H. Kauffman. *Formal Knot Theory*. Mathematical Notes #30. Princeton University Press (1983).

[K2] L.H. Kauffman. The Conway polynomial. Topology 20(1980), 101-108.

[K3] L.H. Kauffman. Weaving patterns and polynomials. Topology Symposium Proceedings, Seigen (1979). Springer-Verlag Lecture Notes in Mathematics 788, 88-97.

[K4] L.H. Kauffman. Branched coverings, open books and knot periodicity. Topology 13(1974), 143-160.

[K5] L.H. Kauffman. Arf invariant of classical knots. Contemp. Math. (A.M.S.)—*Combinatorial Methods in Topology and Algebraic Geometry*, Vol. 44(1985), 101-116.

[K6] L.H. Kauffman. Products of knots. Bull. Amer. Math. Soc. 80(1974), 1104-1107.

[K7] L.H. Kauffman. Link manifolds and periodicity. Bull. Amer. Math. Soc. 79(1973), 570-573.

L.H. Kauffman. Link manifolds. Michigan Math. J. 21(1974), 33-44.

[K8] L.H. Kauffman. An Invariant of Regular Isotopy. (to appear.)

[K9] L.H. Kauffman. State models and the Jones polynomial. (to appear in Topology).

[K10] L.H. Kauffman. Invariants of graphs in three space. (to appear).

[K11] L.H. Kauffman. New invariants in knot theory. (to appear).

[K12] L.H. Kauffman. Statistical mechanics and the Jones polynomial. (to appear).

[K13] L.H. Kauffman. *Sign and Space, Knots and Physics*. (in preparation).

[K14] L.H. Kauffman. *Map Reformulation*. Princelet Editions #30(1986), 249 pp.

[KI] M. Kidwell. On the degree of the Brandt-Lickorish-Millett polynomial of a link. (preprint 1986).

[KL] F. Klein. *Lectures on the Icosahedron and the Solution of Equations of the Fifth Degree*. Trans-lated by George Gaven Morrice. 2^d ed. London, Paul (1913), 289 pp.

[L] R.D. Laing. *Knots*. Pantheon Books (1970).

[L1] J. Levine. Knot cobordism groups in codimension two. Comm. Math. Helv. 44(1969), 229-244.

[L2] J. Levine. Polynomial invariants of knots of codimension two. Ann. of Math. 84(1966), 537-544.

[LF] S. Lefschetz. *L'analysis Situs et la Geometrie Algebrique*. Paris, Gauthier-Villars (1950). (Collection de Monographies sur la theorie des fonctions.)

[LM] W.B.R. Lickorish and K.C. Millett. A polynomial invariant of oriented links. (to appear in Topology).

[MA] W.S. Massey. *Algebraic Topology: An Introduction*. Harcourt Brace Jovanovich (1967); Springer-Verlag (1977).

[M1] J. Milnor. On the 3-dimensional Brieskorn manifolds M(p,q,r). *Knot Groups and 3-Manifolds - Papers dedicated to the memory of R.H. Fox*, edited by L. Neuwirth. Annals of Mathematics Studies 84. Princeton University Press, Princeton, NJ (1975).

[M2] J.W. Milnor. Infinite cyclic coverings. *Topology of Manifolds* (Michigan State University, 1967). Prindle, Weber and Schmidt, Boston (1968).

[M3] J.W. Milnor. *Singular Points of Complex Hypersurfaces*. Annals of Mathematics Studies. Princeton University Press (1968).

[M] H. Morton. Seifert circles and knot polynomials. Math. Proc. Camb. Phil. Soc. 99(1986), 107-109.

[MU] K. Murasagi. Jones polynomials and classical conjectures in knot theory I and II. (preprints 1986).

[N] W.D. Neumann. Cyclic suspension of knots and periodicity of signature for singularities. Bull. Amer. Math. Soc. 80(1974), 977-981.

[OW] P. Orlik and P. Wagreich. Isolated singularilties of algebraic surfaces with $\mathbb{C}^*$ action. Ann. of Math. 93(1971), 205-228.

[P] A. Plans. "Aportación al estudio de los grupos de homología de los recubrimentos ciclicos ramificados correspondientes a un nudo." Revisita de la Real Academía de Ciencias Exactas, Fisicas y Naturales de Madrid, vol. 47, 161-193.

[PH] F. Pham. Formules de Picard-Lefschetz Généralisées et Ramification des Intégrales. Bull. Soc. Math. France 93(1965), 333-367.

[PR] J. Przytycki and P. Traczyk. Invariants of links of Conway type. (To appear.)

[R1] K. Reidemeister. *Knotentheorie*. Chelsea Publ. Co., New York (1948). Copyright 1932. Julius Springer, Berlin.

[R] R. Riley. Discrete parabolic representations of link groups. Mathematika 22(1975), 141-150.

[R2] D. Rolfsen. *Knots and Links*. Mathematics Lecture Series 7. Publish or Perish Press (1976).

[S] H. Seifert. Uber des geschlecht von knoten. Math. Ann. 110(1934), 571-592.

[ST] H. Seifert and W. Threlfall. Lehrbuch der Topologie. Chelsea, New York (1947).

[SH] I.R. Shafarevich. *Basic Algebraic Geometry*. Springer-Verlag (1974).

[S1] L. Siebenmann. (Unpublished knot notes.)

[S2] L. Siebenmann and B. Morin. (Unpublished notes on the Mobius band and private conversation in an unorientable vein.)

[SI] J. Simon. Topological chirality of certain molecules. Topology, Vol. 25, No. 2 (1986), 229-235.

[SU] D.W. Summers. The role of knot theory in DNA research. *Geometry and Topology*, C. McCrory and T. Schifrin, eds., Marcel Dekker (1986).

[SB] G. Spencer-Brown. *Laws of Form*. The Julian Press. New York (1972).

[TH] M. Thistlethwaite. A spanning tree expension for the Jones polynomial. (to appear in Topology).

[TR] B. Trace. On the Reidemeister moves of a classical knot. Proc. Amer. Math. Soc. Vol. 89, No. 4(1983).

[T] A.G. Tristram. Some cobordism invariants for links. Proc. Cambridge Philos. Soc. 66(1969), 251-264.

[W1] H. Whitney. On regular closed curves in the plane. Compositio Math. 4(1937), 276-284.

[W2] H. Whitney. A logical expansion in mathematics. Bull. Amer. Math. Soc. 38(1932).

[WH] J. White. Self-linking and the Gauss integral in higher dimensions. Amer. J. Math. XCI(1969), 693-728.

[W] S. Winker. Quandles, Knot Invariants, and the n-fold Branched Cover. Thesis. University of Illinois at Chicago (1984).

[Z] H. Zieschang and G. Burde. *Knots*. W. De Gruyter, Berlin and New York (1985), 300 p.

Library of Congress Cataloging-in-Publication Data

Kauffman, Louis H., 1945-
On knots.

(Annals of mathematics studies ; no. 115)
Bibliography: p.
1. Knot theory. I. Title. II. Series.
QA612.2.K38 1987 514′.224 87-3195
ISBN 0-691-08434-3
ISBN 0-691-08435-1 (pbk.)

Louis H. Kauffman is Professor of Mathematics at the University of Illinois at Chicago